D0712624

P. E. Potter · F. J. Pettijohn

PALEOCURRENTS AND BASIN ANALYSIS

Second, Corrected and Updated Edition

With 167 Figures and 30 Plates

Springer-Verlag
Berlin Heidelberg New York 1977

Prof. PAUL EDWIN POTTER, Department of Geology, University of Cincinnati, Old Tech Building, Cincinnati, OH 45221/USA

Prof. FRANCIS J. PETTIJOHN, Department of Earth and Planetary Sciences, The Johns Hopkins University, Baltimore, MD 21218/USA

ISBN 3-540-07952-1 2. Auflage Springer Verlag Berlin Heidelberg New York
ISBN 0-387-07952-1 2nd edition Springer Verlag New York Heidelberg Berlin

ISBN 3-540-03040-9 1. Auflage Springer Verlag Berlin Heidelberg New York
ISBN 0-387-03040-9 1st edition Springer Verlag New York Heidelberg Berlin

Library of Congress Cataloging in Publication Data. POTTER, PAUL EDWIN. Paleocurrents and basin analysis. Includes bibliographies and indexes. 1. Paleocurrents. 2. Sedimentary structures. 3. Sedimentation and deposition. I. PETTIJOHN, FRANCIS JOHN, 1904— joint author. II. Title. QE472.P67 1977 551.4'701 76—30293

Typesetting by Universitätsdruckerei H. Stürtz AG, Würzburg.
2132/3130-543210

To JAMES HALL (1811—1898) whose astute observations on primary sedimentary structures in New York State, and to HENRY CLIFTON SORBY of Great Britain (1826—1906) whose insight into the significance of these structures laid the foundation of paleocurrent analysis.

— Level bedded
  ↘
    ∿ Ripple laminated
    ↘
        ∞ Ripple drifted
      ↘
           ∠ Drift bedded (false bedded)

H. C. Sorby's concept of flow regime and his symbols.

On the Oscillation of the Currents Drifting the Sandstone Beds of the Southeast of Northumberland, and on their General Direction in the Coal Field in the Neighborhood of Edinburgh.

Proc. West Yorkshire Geol. Soc. 3, 232—240 (1851).

Preface to the Second Edition

The study of paleocurrents, since 1963, is now a very routine part of sedimentology, and more and more such studies are finding use in other fields. Thus it seemed appropriate for us to review post-1963 developments and present them in a compact manner for the interested reader. Instead of rewriting a second edition, which thirteen years later we would organize in a completely different way, we have brought each chapter up to date with new material up to 1976. A new update supplement has in this edition been inserted after each one of the original chapters. We have stayed close to the original theme of paleocurrents—how to measure them and how to use them to solve geological problems ranging in scale from the hand specimen to the sedimentary basin and beyond.

We have used many annotated references and tables to help present this information to the reader. The reader will note that we have cited a few 1962 references—publications that appeared too late to be cited in the original 1963 edition. A few times we have also cited a reference which was included in the first edition. These are marked with an asterisk and hence do not appear in the new lists of references.

We have been aided by many. In Cincinnati, WANDA OSBORNE and JEAN CARROL did typing and RICHARD SPOHN, the University's geological librarian, was very helpful in obtaining many references to the literature. In Baltimore, we are indebted to KATHLEEN SHANNON for her typing. A. L. NORRINGTON, JR., of Texas Instruments, Inc., Dallas, Texas kindly provided the computer program of Table 10-3.

We are especially indebted to those who graciously gave us references to the widely scattered literature of paleocurrents. These include BERNARD BEAUDOIN, Ecole des Mines, Paris, France; W. ENGEL, Geologische-Paleontologisches Institut, Göttingen, Germany; R. V. FISHER, University of California-Santa Barbara, Santa Barbara, California; A. DREIMANIS, University of Western Ontario, London, Ontario; T. J. FREEMAN, University of Missouri, Columbia, Missouri; ROBERT N. GINSBURG, Laboratory of Comparative Sedimentology, Miami Beach, Florida; CHARLES HEWITT, Marathon Oil Co., Findlay, Ohio; A. S. HOROWITZ, Indiana University, Bloomington, Indiana; G. RANDY KELLER, University of Texas at

El Paso, El Paso, Texas; ROY C. KEPFERLE,.U.S. Geological Survey, Cincinnati Ohio; SVEN LAUFELD, University of Alaska, Fairbanks, Alaska; DONALD P. McGOOKEY, Texaco, Inc., Houston, Texas; ROGER NEVES, University of Sheffield, Sheffield, Great Britain; WALTER E. PARHAM, U.S. Department of State, Washington D.C.; W. A. PRYOR, University of Cincinnati, Cincinnati, Ohio; ROBERT V. RUHE, Indiana University, Bloomington, Indiana; W. W. SHILTS, Geological Survey of Canada, Ottawa, Ontario; GEORGE SIMPSON, Windsor University, Windsor, Ontario, Canada; CHARLES H. SUMMERSON, Ohio State University, Columbus, Ohio; JAMES T. TELLER, University of Manitoba, Winnepeg, Manitoba; HERBERT VIEBROCK, U.S. Environmental Protection Agency, Research Triangle Park, North Carolina; and WINFRIED ZIMMERLE, Deutsche Texaco Aktiengesellschaft, Hamburg Germany.

Finally, we wish to thank our publishers, Springer-Verlag, for their help and especially Mrs. VON DEM BUSSCHE for her fine copy editing.

Cincinnati, Ohio and Baltimore,   PAUL EDWIN POTTER
Maryland, March 1977   FRANCIS J. PETTIJOHN

Preface to the First Edition

In the past, interest in sedimentary structures has arisen mainly from the expectation that these features might be a guide to the environment of deposition. But many sedimentary structures have also proved useful in determining stratigraphic order in nonfossiliferous, steeply inclined beds especially in Precambrian terranes. As the sequence problem has been reviewed at length by Shrock, it seemed to us, therefore, that the time is now ripe for a new look at sedimentary structures, not with respect to "top and bottom", but with reference to "fore and aft." Much of the present-day interest in these structures stems from their usefulness in mapping of paleocurrents. A stage has been reached where there is need for a work which assembles, digests, and organizes our collective knowledge of the usefulness of directional properties od sediments and their application to basin analysis. This we have attempted to write.

The desirability and need for such a book occurred to both of us independently. Upon discovering our mutual interest, we decided that a better book could be written by collaboration. Fortunately this collaboration became a reality because of support by the Guggenheim Foundation of one of us and the cooperation and support of The Johns Hopkins University of both of us. We acknowledge with thanks this indispensable aid.

We have written this book for both the student new to the subject and for the experienced geologist whose work leads him to the study of sedimentary basins. We believe that the study of paleocurrents provides a new insight into sedimentary geology. Our discussion of sedimentary structures, therefore, is designed with this end in view and much of what has been written about the environmental significance of sedimentary structures is omitted; that which pertains to their relation to transport direction and to basin analysis is emphasized. Additional illustrations of primary sedimentary structures and a glossary of the terms that have been used to describe them are given in our "Atlas and Glossary of Primary Sedimentary Structures."

Whatever success we may have had in our venture would be less were it not for the help we have received from many persons. We are indebted to several members of the Deportments of Geology, Geography, Mechanics, and Physics of The John Hopkins University for their helpful discussion of particular problems. We are grateful for assistance in our search of the literature. In particular we wish to thank COLIN MCANENY and H. E. CLIFTON, graduate students at HOPKINS, and ALMA VEIDENBERGS, MARTHA HUBBARD and ADELAIDE EISENHART, our librarians. We are also grateful to ANDREW VISTELIUS of the Laboratory of Aeromethods, Leningrad, for calling our attention to various Soviet papers and to PETER SARAPUKA and BEVAN FRENCH for translation of this material. G. LUNDQVIST of the Geological Survey of Sweden provided helpful references to the Swedish literature. HANS FÜCHTBAUER of Gewerkschaft Elwerath, Hannover, has also graciously supplied German references. Likewise we are indebted to various authors and journals who gave us permission to reproduce figures and maps from their publications. We would like to thank JOHN W. HUDDLE, U.S. Geological Survey, who kindly assisted with the plane table map of figure 4-4 and L. D. MECKEL, Jr., Shell Development Corp., who computed the data of Table 10-1. We are pleased to acknowledge the many persons and organizations who have donated photographs. Both are fully credited in the proper place. We owe much to WILLIAM HILLER and CHARLES WEBER for their help in preparing photographic copy and for photographs of some specimen material. And we are especially indebted to RANICE W. BIRCH DAVIS for her superb drawings. We express our sincere thanks to MARY GILL for her patience and skill in typing and retyping the manuscript. Others have read parts of the manuscript and made helpful suggestions. These include COLIN BLYTH, University of Illinois; ERNST CLOOS, The Johns Hopkins University; DONALD LINDSLEY, Geophysical Laboratory; RICHARD MAST, Illinois Geological Survey; WAYNE A. PRYOR and LOUIS RIEG, Gulf Development Corporation; DAVID M. RAUP, The Johns Hopkins University; and M. GORDON WOLMAN, The Johns Hopkins University. Lastly we wish to thank our publishers for the large and complicated task of seeing this work through the press.

October 1, 1962   PAUL EDWIN POTTER
Baltimore, Maryland   F. J. PETTIJOHN

Table of Contents

struction; Turbidite Basins; Molasse Basins; Shelf and
Shelf-to-Basin Transition; Cratonic and Other Basins;
References

"... buy yourself stout shoes, get away to the mountains, search the valleys, the shores of the sea, and the deep recesses of the earth ... Lastly, purchase coals, build furnaces, watch and experiment without wearying. In this way, and no other, will you arrive at a knowledge of things and of their properties" — *Petrus Severinus, Idea Medecinae Philisophicae* (1571).

Chapter 1

Introduction

Our objective is to study directional and other primary properties of sedimentary rocks from which we can reconstruct current systems and make a *paleocurrent analysis*. The object of study may be a hand specimen, a sand body, a large basin, or a geologic system over a wide area.

Primary directional properties of sedimentary rocks have been recognized and studied from the beginnings of geology, certainly as early as the first part of the 19th century. The principal objective of this book is to collect, review, and synthesize this knowledge. In contrast, insight about sedimentary basins is an essentially modern 20th century development dependent upon abundant subsurface well data. The secondary objective of this book is to set forth the contribution of primary directional structures and other properties to the analysis and understanding of sedimentary basins. Consideration of the sedimentary basin as a whole provides a truly unified approach to the study of sediments.

Knowledge of paleocurrents can help solve such regional problems as establishing,

1. the direction of initial dip or paleoslope
2. the relations between facies boundaries and paleocurrent direction
3. the direction of sediment supply

and such local problems as,

4. establishing the relations between internal directional structures and the geometry of a lithologic unit such as a sand body or a bioherm
5. evaluating the effect on reservoir performance of inhomogeneities of primary origin that are linked to current direction

and in the hand specimen,

6. specifying the primary depositional fabric that controls the anisotropy of many geophysical properties.

The value of paleocurrent study to basin analysis was early recognized by SORBY when he wrote (1867, p. 285).

"The examination of modern seas, estuaries, and rivers, shows that there is a distinct relation between their physical geography and the currents present in them; currents so impress themselves on the deposits formed under their influence that their characters can be ascertained from those formed in the ancient periods. Therefore their physical geography can be inferred within certain limits."

Paleocurrent analysis has had a long heritage that includes a large portion of geology. It is closely related to and dependent on stratigraphy, sedimentary petrology and structural petrology (fig. 1-1). Stratigraphy has contributed much to basin analysis. Facies analysis began in the early 19th century with the qualitative assessment of the areal variation of outcrop sections. In later years, thickness and facies maps became the basis for more exact paleogeographic interpretations. Today, subsurface facies analysis is highly developed, employing some of the most modern techniques. Sedimentary petrology provides insight to the distribution of detrital mineral provinces across a basin and contributes information about grain size, roundness, and bedding characteristics. Many of the concepts of structural petrology, developed primarily for the study of metamorphic rocks, are applicable to the study of sedimentary fabric; the symmetry concept in sedimentation is an example. Unlike stratigraphy, sedimentary petrology and structural petrology are largely products of the 20th century, especially of the last three decades. Paleocurrent analysis emphasizes the interdependence of these geologic disciplines. Glacial geology was the first field in which primary directional properties such as glacial grooves, striations and till

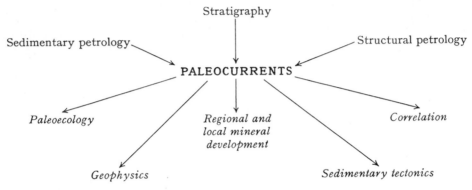

Fig. 1-1. Heritage and some contributions of paleocurrent study

fabric were measured and related to morainal patterns and boulder trains, all of which reflect an integrated dispersal system.

Figure 1-1 shows the fields to which paleocurrent analysis can contribute. These contributions range widely and include topics as varied as regional and local exploration, geophysics, sedimentary tectonics, stratigraphic correlation and paleoecology.

Paleocurrent analysis can aid in prediction of both regional and local sedimentary trends in exploration for petroleum and ground water, can predict the trend of "washouts" or "cut-outs" in coal fields and can aid in mineral exploration where the deposits are related to sedimentary trends as are some gold and uranium deposits. Depositional fabric, a paleocurrent feature, is important to the interpretation of geophysical properties such as dielectric constant, permeability, and paleomagnetism. Paleocurrent analysis can provide useful information about sedimentary tectonics. For example, mapping directional structures may establish the activity of a geologic structure during sedimentation. In stratigraphy, paleocurrent analysis can contribute to improved regional correlation by establishing depositional strike. Knowledge of current direction is also useful in paleoecologic studies since currents supply nutrients, remove waste products, and may disperse and orient organisms.

Paleocurrent analysis depends on evidence obtained from the outcrop, the subsurface and the laboratory. Some studies may rely only on field methods, some may use only laboratory evidence, and others combinations of all three as, for example, do integrated paleocurrent analyses of most sedimentary basins. In any case, however, a distinguishing feature of paleocurrent analysis is its emphasis upon quantitative study. This aspect of paleocurrent analysis is a further development of the accelerating quantification of geologic data.

Paleocurrent analysis depends on the presence in sedimentary rocks of numerous properties from which current direction can be inferred. These properties occur, in varying kinds and abundance, in all sedimentary basins from the present to the far geologic past. These properties are of four types: attributes, scalars, directionals, and tensors.

Attributes are those properties that are specified only by their presence or absence. Neither magnitude nor abundance need be specified. Although compa-

Table 1-1. *Properties used in Paleocurrent Analysis*

Property	Definition	Geologic example	Remarks
Attribute .	Presence or absence	Trace mineral, pebble or boulder in till, gravel or sand. Multicomponent mineral assemblage.	Directional significance only when mapped.
Scalar . . .	Magnitude.	Thickness, lithologic and mineral proportion, grain size, sorting, roundness, etc.	Directional significance only when mapped.
Directional .	Specified by azimuth. May indicate either a line or direction of movement.	All sedimentary structures with directional significance such as sand fabrics, fossil orientation, flute marks, etc.	A vector property, if magnitude could be specified.
Tensor . . .	Directions and lengths of principal axes of ellipsoid	Fluid permeability, dielectric constant, sonic transmissibility, etc.	An anisotropic property that is fabric dependent.

ratively few primary directional properties belong to this class, there are several important ones. One is the diagnostic rock type of critical value in glacial geology. The presence or absence of a distinctive cobble may identify the deposits of a given lobe and a map showing distribution of such cobbles may outline the dispersal pattern (boulder train) and has led to discovery of economically important mineral deposits. The presence or absence of a key mineral plays a similar role in modern or ancient sands. Detrital mineral associations of groups of minerals, either heavy or light, can also be mapped as attributes, if arbitrary limits are placed on the abundance of the different components.

Scalars are those properties that are specified by magnitude alone. Examples are grain size, mineral proportion, formation thickness, clastic ratio, or other combinations of lithologic proportions. Measurement of a scalar quantity at a point has no directional significance. However, when scalars are mapped across a region they define the *directional derivative or gradient* of a scalar function. Mean grain size in gravels, percentage of kaolinite in a mud or shale, or the sand-shale ratio are all examples of scalar properties, which when mapped across a region, may show systematic down-current changes.

Directionals are those properties that indicate either a line of movement or a direction of movement and are specified by azimuth. Ideally, we should prefer to consider such structures as vectors, since each directional structure is indeed the response to a moving fluid which can be represented by a vector or force field. Although fluid force fields are true vector fields, the sedimentary structures they leave behind can usually only be specified by direction, since neither the magnitude of either the applied force or the velocity of the depositing medium commonly cannot be uniquely determined, if at all. Hence the term directional rather than vector is more correct. Cross-bedding, flute marks, sedimentary fabrics, and fossil orientation are all good examples of directional structures. If such structures could be reasonably mapped as vectors, we could then determine the directional derivative or gradient of a vector function. Directional structures may only indicate *a line of movement* (two-dimensional fabrics) or they may indicate *the direction of movement* (cross-bedding).

Sedimentary basin
↑
Depositional facies
↑
Stratigraphic unit
↑
Shapes of sedimentary bodies
↑
Directional structures
↑
Fabric
↑
Single particles

Fig. 1-2. Hierarchy of sedimentation elements in paleocurrent study

Tensors describe fabric-dependent properties such as fluid permeability, dielectric coefficient, magnetic susceptibility and others. The ray velocity surface of crystallography is a familiar example of a physical concept, light transmission in a crystal, that can be described by a tensor, as is the strain ellipsoid of structural geology.

Knowledge of the role of gravity, fluid, and magnetic force fields in deposition of sedimentary rocks provides useful insight to paleocurrent analysis. The fabric of talus is a good example of a property that is the response to only gravity. The vast majority of sediments are the response to fluid plus gravity force fields. Fluid force fields disperse clastics, modify development of sediments of biogenic and chemical origin, and leave behind a recognizable and measurable record of primary directional structures. The earth's magnetic field can also exercise control on the orientation of small, highly magnetic particles under certain conditions.

The deposits of sediment from these force fields can be thought of as forming a hierarchy from small detrital particles to large sedimentary basins (fig. 1-2). The arrangement of the following chapters reflects this natural hierarchy. These chapters set forth the close relations that exist between all the primary directional properties and the analysis of sedimentary basins. Together, the attributes, scalars, directionals, and tensors of primary directional properties form an interrelated set, all of which are the response to regional dispersal patterns. By studying appropriate combinations of such properties, dispersal systems can be identified and defined. Paleocurrent analysis, when combined with the concept of the sedimentary model, thus greatly facilitates basin analysis and prediction of sedimentary trends.

Excluded from consideration are directional structures of glacial ice, an ephemeral metamorphic rock, and of igneous rocks, although the latter contain many directional structures. For example, lava flows contain primary structures which enable one to determine top and bottom (SHROCK, 1948) as well as structures which make possible determination of the direction of flow (HOTCHKISS, 1923; FULLER, 1931, p. 282—287; WATERS, 1960). Graded bedding and cross-bedding have also been reported in large layered basic intrusives (GATES, 1961, figs. 18 and 21) and the relations between magma movement and crystal orientation have been noted repeatedly. Volcanic flows in particular offer opportunities for directional study, for like the currents responsible for the deposition of terrestrial sediments, the flows move down hill and hence direction of lava movement should be consistent with transport direction of interbedded sands (cf. SANDBERG, 1938, p. 818—820).

References

FULLER, R. E., 1931: Aqueous chilling of basaltic lava on the Columbia River Plateau. Am. J. Sci. **221**, 281—300.

GATES, OLCOTT, 1961: Geology of the Cutler and Moose River quadrangles, Washington County, Maine. Maine Geol. Survey, Quad. Mapping Series No. 1, 67 p.

HOTCHKISS, W. O., 1923: The Lake Superior geosyncline. Bull. Geol. Soc. Am. **34**, 669—628.

SANDBERG, A. E., 1938: Section across Keweenawan lavas at Duluth, Minnesota. Bull. Geol. Soc. Am. **49**, 795—830.

SHROCK, R. R., 1948: Sequence in layered rocks. New York: McGraw-Hill Book Co. 507 p.

SORBY, H. C., 1957: On the physical geography of the Tertiary estuary of the Isle of Wight. Edinburgh New Philosophical J., n.s. **5**, 275—298.

WATERS, A. C., 1960: Determining direction of flow in basalts. Am. J. Sci. **258 A**, 350—366.

History of Paleocurrent Investigations up to 1963

Introduction

In order to obtain a proper perspective and thus appreciate better what has been accomplished, the history of the subject is briefly reviewed.

The study of sedimentary deposits is inseparably linked with the study of geology itself. Our way of looking at sediments, however, has changed notably from time to time. In earlier periods, a sedimentary deposit was considered primarily as a stratigraphic unit — a formation — a body characterized by some kind of lithologic unity and having appreciable thickness and lateral extent and a particular position in a rock sequence. This concept is basic to stratigraphy.

Another point of view is that expressed in SORBY's writings, particularly his presidential address before the Geological Society of London in 1879, in which the sedimentary deposit is looked upon as a rock — a significant component of the earth's crust, an entity having certain compositional and textural attributes produced by certain physical and chemical factors in the environment of deposition. This point of view is basic to sedimentary petrology and SORBY is rightly called the "father of petrology."

In more recent years sedimentary deposits have been looked at from other points of view. The concept of a deposit as a "population" — a particulate system — is a relative modern point of view. This approach tends to emphasize operational definitions of rock properties, their measurement and statistical analysis. Sedimentary deposits may also be looked upon as chemical entities — products of the large-scale fractionation related to the external geochemical cycle. This concept is fundamental to the understanding of the chemical evolution of the earth's crust and is a point of view related to the rise of geochemistry as a geological science.

In this book, however, sedimentary deposits are looked at from the point of view of their internal symmetry and organization. The deposit is regarded as an anisotropic body. This concept is concerned with the fabric and directional properties of a sediment which are acquired in response to the earth's gravitational and magnetic fields and to fluid flow systems. This orientation is basic to paleocurrent analysis and paleogeographic considerations as well as to certain geophysical problems.

It is our task not only to trace the development of this point of view but to show also how this approach to the study of sedimentary deposits is related to the more conventional concepts of stratigraphy and petrology and to trace the evolution of the model concept. The latter is an intellectual construct based on the premise that the sedimentary fill of a given basin is organized in such a manner that the sedimentary framework or geometry, the petrology, and the directional properties constitute an integrated whole which, if properly understood, enhances the likelihood of successful prediction of sedimentary trends

It must not be supposed that the various points of view were all developed separately and in an orderly sequence in time. All had early beginnings in the history of geologic thought but the emphasis through time has shifted markedly and some of these concepts have matured only in recent times. This is particularly true of the quantitative study of primary directional properties.

Early Interest in Sedimentary Structures

By the time that HUTTON's "Theory of the Earth" appeared (1788) and certainly with the publication of LYELL's "Principles of Geology," in 1837 (p. 317), many current structures, such as cross-bedding and ripple mark, were recognized as such. Even sole markings, which have been much discussed in recent years, were described, if not understood, by JAMES HALL as early as 1843. The prime interest of the early geologists in such structures arose from the demonstration or proof these features gave of the sedimentary origin of the rocks containing them, despite the present lithified condition, elevation, and tilting of the rock.

Interest was stimulated further by the notion that the several kinds of structures might be indicative of particular agents and/or environments of deposition. A great deal of effort has been expended in attempting to discriminate between aeolian and aqueous cross-bedding and ripple mark, for example, and to recognize the structures characteristic of the strand line and so forth.

The field geologist, too, became aware of the utility of many of these structures in ascertaining stratigraphic order — which enabled him, in the absence of fossils, to determine relative ages of beds, to recognize overturned strata and to unravel complex structures. This interest, most generally cultivated by the students of Precambrian geology, culminated in 1948 with the publication of SHROCK's "Sequence in Layered Rocks" — one of the best summaries in the English language of the various primary structures of sedimentary (and volcanic) rocks.

But it is the use of these structures as a guide to current direction and source of sediments that is of main concern here. The first clearly expressed statement of directional significance of such structures and their potential paleogeographic worth was that of SORBY, who in 1859 (p. 138) wrote:

"... I have shown that many pecularities of physical geography at former epochs may be learned from a knowledge of the directions of the currents in various localities ... I have often felt that scarcely anyone has entered into this field of inquiry, in which the facts are so marked and distinct. If the current structures had been on a small scale requiring the aid of a microscope, there would have been good reason for this; but such is not the case ... Unassisted eyes and a compass are all that are requisite in determining the greater number of the facts ... Moreover, many of the structures have been known long enough, for they are of such a character that no one could overlook them, although sufficient attention may not have been paid to their teachings; and the study of their relations to one another and to other facts in an accurate and business-like manner may have long been neglected."

SORBY (1856, 1857, 1858, 1859) and some of his contemporaries clearly recogniz-ed that the asymmetry of many sedimentary structures with respect to the current system responsible for their formation makes possible the determination of the up-current and down-current direction. JAMIESON (1860, p. 349), for example, observed and fully understood the significance of imbrication of pebbles in a stream bed — an observation evidently made also by the early placer miners (BECKER, 1893, p. 54). Students of the glacial drift not only recognized the meaning of the glacial striations — a directional paleocurrent feature — but also observed the fabric of till itself. HUGH MILLER, son of the HUGH MILLER of Old Red Sandstone fame, for example, noted in 1884, that till stones exhibited a pre-ferred orientation and that the asymmetrical stones were aligned with their blunt ends "up-current" so that they pointed like "index fingers" in the direction of ice flow.

Despite both SORBY's clear insight, over one hundred years ago, into the use-fulness of such studies and his own efforts, which included 20,000 or more obser-vations recorded in his notebooks, it is only in the last decade, or at most two decades, that serious attention has been given to paleocurrent data and analysis.

Mapping of Directional Properties

Despite the many measurements of SORBY, he did not emphasize paleocurrent maps although he published an interpretive one (1857, pl. 4). Nor did any of his contemporaries make such maps. The first map to show sediment dispersal patterns and azimuths of current structures were those made by glacial geologists. The mapping of boulder trains and measurement and plotting of striation azimuths and the like, long antedated the measurement and mapping of either subaqueous or subaerial current structures.

The first paleocurrent map of non-glacial deposits that showed actual measure-ments seems to be that made by RUEDEMANN in 1897. RUEDEMANN plotted the orientation of graptolite rhabdosomes, orthocerid cephalopods, sponge spicules and "mud furrows" in the Utica shales near Dodgeville, New York. He inter-preted the regional alignment as a product of "oceanic currents" and this initiated what KAY (1945, p. 428) later called "paleo-oceanography." HYDE, in 1911, published a map showing the orientation of ripple marks of the Berea and Bedford formations of southern Ohio. The profuse ripple markings in these formations display an astonishing uniformity of direction over a distance in excess of 125 miles.

Most paleocurrent studies have been based on cross-bedding. Although cross-bedding has been seen and its origin as a product of current action has been known for a long time, it has not been mapped in a systematic manner until compara-tively recent years. The earliest such study known to the authors, is that of RUBEY and BASS who mapped cross-bedding in a channel sandstone of Cretace-ous age in Russell County, Kansas in 1925.

BRINKMANN (1933) was, perhaps, the first of the "modern" workers to have a clear concept of the objectives of paleocurrent research and to develop methods to fulfill these aims. He demonstrated (p. 11) how "... sich durch messende Beob-achtung und statistische Auswertung eines einfachen Faziesmerkmals, der

Kreuzschichtung, eine Reihe zuverlässig begründeter paläogeographischer Ergebnisse gewinnen lassen, ..." He plainly saw how the same methodology could be applied to other facies attributes such as grain size, fossil content and orientation, pecularities of bedding, thickness, etc. His own study was based on the cross-bedding of the Triassic Buntsandstein. The studies of SHOTTON (1937) on the Lower Bunter sandstone of England and of REICHE (1938) on the Coconino sandstone of Arizona are two significant areal studies of cross-bedding in sandstones of presumed aeolian origin. REICHE not only mapped the cross-bedding but made important advances in the statistical and graphical treatment of cross-bedding measurements.

Renewed interest in cross-bedding as a tool for paleocurrent analysis followed from the publication of the paper by POTTER and OLSON (1954) on the Caseyville and Mansfield sandstone of early Pennsylvanian age in the Illinois Basin. POTTER and OLSON not only mapped the cross-bedding and drew important paleogeographic conclusions from their data, but made contributions to the sampling and statistical analysis of cross-bedding. POTTER and others (1955, 1956, 1958, 1961, 1962) have continued work in the Illinois Basin area. Although early work on cross-bedding in the Colorado Plateau was done by McKEE (1940), the most extensive work in this area is that undertaken by the U.S. Geological Survey. The work of STOKES (1953), CRAIG et al. (1955) and others on the Saltwash member of the Morrison formation (Jurassic) in Utah and adjacent areas was stimulated by the occurrence of uranium in fluvial channels in this formation. The mapping of cross-bedding has now become commonplace.

Other recent work on cross-bedding in the United States includes that of TANNER (1955) in the Gulf Coast area, BRETT (1955), McDOWELL (1957), and PETTIJOHN (1957) in the Precambrian and HAMBLIN and others (1958, 1961) on the Cambrian and Keweenawan of the Lake Superior region, PRYOR (1960) on the Cretaceous of the Mississippi embayment, and PELLETIER (1958), WHITAKER (1955), and YEAKEL (1962) on the Paleozoic rocks of the mid-Appalachian region. A large part of this work was done on formations presumed to be non-marine, probably fluvial.

European investigators have carried forward the study of cross-bedding so ably begun by BRINKMANN. These include ILLIES (1949), who outlined in some detail methods of investigation, types of cross-bedding, and paleogeographic interpretation with examples from the German Tertiary, SCHWARZACHER (1953), who mapped the cross-bedding of the Lower Greensand in England, VASSOEVICH and GROSSGEYM (1951, cited by STRAKHOV, 1958, p. 401) and GROSSGEYM (1953, cited by STRAKHOV, 1958, p. 401), who mapped the cross-bedding and other directional properties in the Caucasian area, HÜLSEMANN (1955), who studied cross-bedding in North Sea tidal flats and in the Molasse, NIEHOFF (1958), who worked on the Koblenz quartzite (Devonian), and WURSTER (1958), who worked on the Schilfsandstein and other deposits.

There has been renewed interest in the current structures referred to as "sole markings" commonly found on the base of many sandstone beds. These markings, early described by JAMES HALL (1843) and CLARKE (1918), from the Portage beds (Devonian) of New York State, and by FUCHS (1895), VASSOEVICH (1932) and others in flysch sequences elsewhere, were, for a long time *problematica* —

"hieroglyphs" of unknown significance. That some, at least, were of current origin became clear by the time RÜCKLIN published his paper on "Zapfen-wülste" in 1938.

Interest in these structures was stimulated by RICH (1950) who attributed the most common sole markings — groove and flute casts — to the action of turbidity currents. This interpretation has come to be widely accepted, primarily because of the publications of KUENEN (1953a, 1953b, 1956, 1957) and others (KUENEN and TEN HAAF, 1958; KUENEN and SANDERS, 1956; KUENEN et al. 1957).

Of special interest is the systematic measurement of the orientation of sole markings and the preparation of maps to show the movement plan over large areas. As most of these structures are presumed to have been formed by turbidity currents, the mapping of them enables one to trace the turbidity current flow and to delineate also, therefore, the submarine slope. Systematic recording of the orientation of sole markings did not become common until CROWELL's paper on the pre-Alpine Flysch of Switzerland (1955). This paper contained a thorough discussion and classification of the directional current structures, with emphasis on the sole markings, together with a large number of measurements from which a paleocurrent analysis was made. It soon became apparent to many workers that sole markings are the most common and most useful criteria of current direction in flysch facies.

Mapping and paleocurrent analysis based mainly on sole markings was carried out in the Flysch of the Alps by Hsu (1959, 1960a), in the Cambrian of Wales by KOPSTEIN (1954) and BASSETT and WALTON (1960), in the Aberystwyth grits by WOOD and SMITH (1959), in the Oligocene and Miocene of the Apennines by TEN HAAF (1959), in the Maritime Alps of France by KUENEN, et al. (1957) and BOUMA (1959a, 1959b), in the Carpathian flysch by KSIAZKIEWICZ (1957, 1958) and others of the Polish school of investigators (DZULYNSKI and RADOMSKI, 1955; BIRKENMAJER, 1958; DZULYNSKI and SLACZKA, 1958; and DZULYNSKI, et al., 1959) and in the Caucasus by GROSSGEYM (cited by RUKHIN, 1958, p. 392).

Although sole markings have long been known in the United States, little systematic study of them has been made until recent years. REINEMUND and DANILCHIK (1957) seem to have been the first to actually map these structures in the Atoka formation of Pennsylvanian age in the Ouachita Mountains of Arkansas. More recently, maps of sole markings and other structures in upper Devonian strata of the Finger Lakes area of New York, have been made by SUTTON (1959). Extensive mapping of these features in the Paleozoic formations of the mid-Appalachian region has been carried out by McIVER (1961) and McBRIDE (1962).

Directional properties other than cross-bedding and sole markings are little used for paleocurrent mapping. They are accessory criteria. Included here are such features as parting lineation, oriented fossils, and pebble and grain fabric. Some structures may be too uncommon to be used in routine work; others, like grain fabric, are too time-consuming and require too much work for the results they yield. However, if nothing else is available, a fabric analysis may become necessary.

The orientation of pebbles has proved especially useful in the study of the glacial deposits, Pleistocene till in particular. HUGH MILLER (1884) seems to have been the first geologist to report on the orientation of till stones and to comprehend their usefulness in determining the direction of ice-movement. About 50 years later RICHTER (1932, 1933, 1936a, 1936b) made extensive use of this observation in studying ice-movement. RICHTER systematically measured and plotted the long-axis orientation. The ice-movement pattern inferred from his maps was wholly consistent with known bedrock striations and morainic and other ice-constructed land forms. Since RICHTER'S work was published, various papers have appeared which have applied his techniques to problems of ice-movement in the United States (KRUMBEIN, 1939; HOLMES, 1941), in Finland (OKKO, 1949; VIRKKALA, 1951, 1960; KAURANNE, 1960), and more recently in Great Britain (WEST and DONNER, 1956). There is a considerable literature on the techniques of investigation and the causes of tillstone orientation. Long antedating the studies of till fabric, however, the direction of ice-movement, itself a paleocurrent problem, had been inferred from the bedrock striations, morainic and other land forms, and from dispersal patterns of drift boulders. HITCHCOCK (1843) made an early regional map in eastern United States and Canada. MILLER (1884) made a detailed local map in Scotland.

Work on the fabric of nonglacial sediments has been much less complete and only rarely has the fabric of such deposits been used to ascertain current direction. That the sedimentary processes should impart a depositional fabric to the deposit was recognized by BRUNO SANDER (1930).

The fabric pattern of river gravels has been given attention, perhaps because of the ease of study of these unconsolidated materials (JOHNSTON, 1922; WADELL, 1936; CAILLEUX, 1938; KRUMBEIN, 1939; KALTERHERBERG, 1956; SCHLEE, 1957a; UNRUG, 1956, 1957). To a limited degree knowledge of gravel fabric has been applied to the determination of the transport direction in ancient gravels, even Precambrian gravels (WHITE, 1952), and to mapping of the paleocurrent systems in upland gravels (SCHLEE, 1957b). Although some consideration has been given to the paleogeographic significance of gravel fabrics (SARKISIAN and KLIMOVA, 1955; RUKHIN, 1958, p. 413—421), few, if any, regional studies of the fabric of ancient conglomerates have been made.

Owing, perhaps, to lack of suitable methods of study, investigation of sand fabrics is a comparatively recent development. DAPPLES and ROMINGER (1945) studied and experimentally produced dimensional orientation of sand grains. SCHWARZACHER (1951) and RUSNAK (1957b) have continued the experimental approach. ROWLAND (1946) and GRYSANOVA (1947, 1949, and 1953) were among the first to measure systematically and to evaluate the fabric of actual sandstones. The discovery that the permeability of a natural sand deposit was a vector property (JOHNSON and HUGHES, 1948) led to the conclusion that the pore pattern was in some way controlled by a grain orientation. This notion led to further studies of sand fabrics (GRIFFITHS, 1949, 1952; GRIFFITHS and ROSENFELD, 1950). Subsequently there has been a greatly expanded interest in the orientation of sand grains both in modern sands (NANZ, 1955; CURRAY, 1956) and in ancient sands and the relation of such orientation to the shapes of sand bodies (NANZ, 1960). The relation between fabric and various vector properties other than permeability has led to new methods of fabric analysis. MARTINEZ (1958) investigated the relations between optical orientation and photometric anisotropy.

ARBOGAST *et al.* (1960) investigated the relations between dimensional fabrics and dielectric anisotropy. POTTER and MAST (1963) related fabric to anisotropic wetting pattern of sands.

The study of sand fabrics has been largely confined to single samples or small groups of samples from a restricted area, although KOPSTEIN (1954) relied largely on grain orientation in his study of the Cambrian of the Harlech dome of Wales. RUSNAK (1957a) applied fabric techniques to the study of a channel sandstone in the Illinois basin.

Although the fabric of sandstones is tedious to work out, unless the newer optical or geophysical methods are used, many sandstones display a "parting lineation" (CROWELL, 1955, p. 1361) which is probably related to and dependent on the grain orientation. This structure, first described SORBY (1857, p. 279), was subsequently recognized by HANS CLOOS (1938, fig. 2), STOKES (1947) and others, is readily seen in the field on the bedding planes of certain sandstone facies — those which make good flagstones. Like many other features it is used to supplement the more common indicators of current flow.

Shells and other skeletal structures respond to current flow and come to assume a preferred orientation. Arrangement of organic structures with respect to current flow was utilized by RUEDEMANN in 1897 in his studies of the Ordovician of the Albany area in New York State. Likewise MATTHEW (1903, p. 52) observed the orientation of brachiopod shells in the Cambrian rocks of the Cape Breton area. He noted (p. 54) that some elongate forms, *Acrothyra* in particular, showed a strong preferred orientation and that this genus tended to become oriented "... presenting its smallest and heaviest end to the current ..." because "... the shell would swing on this as a pivot." In his study of the Maikop beds in the northeast part of the Caucasus, GROSSGEYM (cited by STRAKHOV, 1958, figs. 105 and 106) plotted the orientation of elongated skeletal structures. Their orientation was consistent with the current system deduced from a study of cross-bedding. The orientation of fusulinids was mapped by KING (1948, p. 84) in the Guadaloupe area in New Mexico. SEILACHER (1959, 1960) has summarized much of what is known about the orientation of skeletal structures in response to currents and developed some principles of interpretation.

One of the most commonly seen orientations is that of plant fragments — the so-called "charcoal fragment lineation" (CROWELL, 1955, p. 1361). It has been utilized in some paleocurrent studies.

The investigation of ripple mark orientation has been less thorough than that of other structures. There seem to be few systematic measurements of this feature in either modern or ancient deposits. Mention has been made of the early work of HYDE (1911) on the Berea sandstone in Ohio. KING (1948, p. 84), HUNTER (1960, fig. 67), MCIVER (1961, fig. 43), and PELLETIER (1961) have made good use of ripple orientation. The measurement of ripple orientation, however, has generally been subordinated to measurement of other current structures.

A review of the literature shows that cross-bedding has proved to be the most useful structure for mapping the paleocurrent systems of shallow water and fluvial sands be they Precambrian or Recent. The sole markings, so characteristic of the graded flysch sandstones, are most useful in indurated but not metamorphosed turbidites. Poorly-consolidated beds fail to separate along the bedding planes;

metamorphic strata part along cleavage. Hence there are no systematic studies of the paleocurrents of the Precambrian flysch; nor are there any of the unconsolidated younger counterparts.

It is noteworthy that studies of primary directional structures of sediments deal largely with those of sandstones. The directional structures of limestones are less well known even though cross-bedding and ripple marks are not uncommon in limestones. Information about directional structures in shale is virtually nil. It is also worthy of note that, excepting some presumed aeolian sandstones, the fluvial sandstones display the best cross-bedding — the most investigated primary directional structure.

Integrated Paleocurrent Analysis

The concept of a completely integrated paleocurrent analysis, utilizing all criteria of current flow, although expressed by BRINKMANN (1933), received a more complete formulation and realization by HANS CLOOS (1938) and BAUSCH VAN BERTSBERGH (1940). Just as the lineations and related structures in igneous and metamorphic rocks depict a movement pattern, so also do the directional structures of sediments. It is not surprising, therefore, that the student of granite tectonics, HANS CLOOS, should turn to the systematic measurement and mapping of the primary directional properties of sediments.

The map on the Devonian of the Rhineland by BAUSCH VAN BERTSBERGH seems to be the first study in which different directional structures were used and mapped to infer paleocurrents. CLOOS wrote (1953, p. 296): "Finally we had a wind- and wave-map of the Devonian Rhineland though there was no Rhine and no land on our chart. The land was to the north, in the present lowland. And where the mountains are today, there was nothing but water. On this map of the shallow Devonian sea, there are many arrows, numbers and symbols. We had to use a little imagination to see, from a dry map, in black and white, water and wind, the play of waves on the sand, the swirling and slow settling of the mud. But vision is essential to all scientific research."

Thought has been given to the signs and symbols that could be used on such maps as described by CLOOS. SORBY (1857, p. 281) first suggested the use of symbols. SANDERS (1946) and BOUMA and NOTA (1960), in particular, have devised and suggested symbolism.

Mapping of Attribute and Scalar Properties

The direction of current flow can be deduced from other characteristics of sedimentary rocks other than those usually characterized as directional. Some of the earlier efforts at paleogeographic analysis have utilized such properties. Among these are pebbles or mineral grains in the sediment known to be derived from a unique source rock. In general, tracing of minerals to their source has not been particularly successful except in the glacial drift. In Finland and Sweden, in particular, geologists have been unusually adept in locating source ledges and hidden ore bodies by mapping dispersal patterns ("boulder trains") of distinctive rocks or ores.

As a result of selective abrasion, dilution, or selective sorting, the mineral composition of sands shows a down-current change. The composition of gravels is likewise systematically modified during transport (PLUMLEY, 1948, p. 552). Such changes should be amenable to cartographic representation and the direction of flow deduced from such maps. Extended and important work on mineral composition of ancient sands in relation to source and to transport direction has been done by FÜCHTBAUER (1954 and 1958) on the sandstones of the Molasse basin of Germany. Here specific mineral assemblages characterize specific Alpine source regions and areas over which these assemblages were spread are well-defined dispersal "fans". They are mappable units. For examples of modern studies illustrating the relations between sand mineralogy and sand movement, see HSU (1960b) and VAN ANDEL (1960).

Theoretically any property which shows a distinct down-current change, such as roundness or size of the constituent cobbles, can be used to determine direction of current flow. The down-current increase in roundness of river gravels could be used to determine current direction in ancient fluvial deposits. But as field and laboratory studies show, this increase is most rapid in the first few miles of transport and the roundness rapidly becomes asymptotic to some particular value. Thereafter it shows little or no change. Consequently roundness has not yet proved a useful property in paleocurrent analysis.

On the other hand, the down-current decline in size seems to be very regular in some deposits. SCHLEE (1957b) mapped the downcurrent decline in both maximum and mean pebble sizes in the upland gravels ("Brandywine") in the vicinity of Washington, D.C. The current direction inferred was substantiated by mapping the cross-bedding. Because of the success achieved by SCHLEE, PELLETIER (1958, fig. 14) mapped the maximum pebble size in the Pocono (Mississippian) conglomerates in the mid-Appalachian area. A significant size decline, closely correlated with the paleocurrent systems inferred from cross-bedding directions, was found. Similar relations were demonstrated by McDOWELL (1957) for the Precambrian Mississagi quartzite of the north shore of Lake Huron and by YEAKEL (1962) for late Ordovician and early Silurian conglomeratic quartzites in the central Appalachians.

The observation that the size decline of fluvial gravels of alluviating streams seems to follow a regular law, led to application of this principle to determination of the *distance* of transport as well as its *direction*. SCHLEE, PELLETIER, YEAKEL and McDOWELL all applied this concept to the estimation of the distance to the margin of the basin of sedimentation.

The Facies Model

The success in making paleocurrent maps based on both directional and scalar properties, has added a new dimension to the study of sedimentary basins. It has given us a new insight into basin filling. It has led to the concept of a facies model based on a synthesis of stratigraphic geometry, sedimentary lithology, and paleocurrent data. The data on thickness (isopach maps), lithology (lithofacies maps), and directional and scalar properties (paleocurrent maps) are all integrated into a facies model. The basic premise is that *a clastic dispersal system produces attribute, scalar and directional properties forming an interrelated set* that can be used to reconstruct the original conditions of sedimentation. With adequate data and an understanding of the principles involved, one should be able to make more successful predictions of sedimentary trends.

The model concept has gradually evolved. It was expressed in a partial way by BRINKMANN in 1933, by CLOOS in 1949, by KUENEN and CAROZZI in 1953, and by others. The studies of BAUMBERGER (1934) of molasse basins in Switzerland illustrate early documentation of the basin geometry and fill of a recurring type of sedimentation without, however, any study of directional properties. This and other similar studies illustrate the qualitative and "static" stage of basin analysis that preceded development of most of the main elements of the model concept. The work of PELLETIER (1957) on the Pocono (Mississippian) and YEAKEL (1962) on the Tuscarora (Silurian) in the central Appalachian basin embodies a synthesis of data which are the main elements of the model concept. In effect these investigators formulated models without explicitly saying so. The first formal presentation of the concept, however, was made by PRYOR (1960, 1961) in the study of the Cretaceous sedimentation in the upper Mississippi embayment. POTTER (1962) also used the model approach in his work on sedimentation in the Illinois basin. The model concept has also been presented by SLOSS (1962).

Thus we see that the interrelated nature of sedimentary petrology, stratigraphic geometry and directional properties has led to a return to the field on the part of sedimentary petrologists and the development of paleocurrent analysis. The return to the field carried with it the quantitative methodology which has characterized sedimentary petrology during the last several decades.

The emphasis on mapping is a further earmark of paleocurrent research. This approach was stimulated by the interest in sedimentary basins as a whole, in part because they are potential producers of oil and gas and, in part because of the practice of depicting thickness and structures on maps — a development made possible only by the vast quantity of subsurface data spawned by the proliferation of oil wells and other deep borings.

General Observations and Summary

From this review it is clear that the current origin of many sedimentary structures was known to LYELL and his contemporaries and that the *paleogeographic* value of a systematic study of these structures was understood by SORBY in the 1850's but that systematic study did not really get under way until the thirties with the work of BRINKMANN, CLOOS, FORSCHE, REICHE and others. Only in the period following World War II did such studies become commonplace. The number of published papers dealing with this topic has shown a phenomenal increase in the last decade (fig. 2-1).

This great expansion is, in one respect, unlike that in many other branches of geology which followed from the introduction of a new concept from physics or chemistry, the isotope concept for example, or the invention of a new tool such as the polarizing microscope. The tools for study of directional structures are simple — a compass and clinometer — and the essential idea was clearly expressed by SORBY over one hundred years ago. The significance and value of mapping directional structures was never properly appreciated until recently. Although nearly all geologists had seen cross-bedding, ripple-mark and the like, and some had studied them in serious fashion, almost none had mapped them. Failure to heed LAPWORTH's dictum: "Map it, and it will all come out right" arrested our understanding and greatly handicapped our progress. The systematic mapping of these features has fully vindicated SORBY's prophecy that such procedure would shed much light on the geography of past epochs.

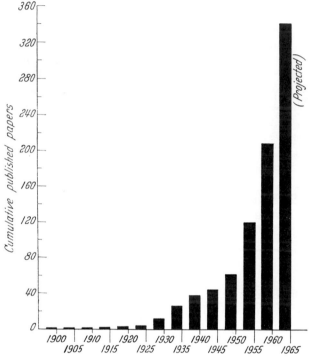

The concept of an integrated analysis of sedimentary basin, combining the data of stratigraphy, sedimentary petrology, and directional properties is a more recent development — one associated with a return to the field and the growth of sedimentology as a discipline independent of stratigraphy. Such studies, and the concept of a sedimentary model, are now just taking shape and their value in making predictions concerning sedimentary trends is still being tested.

It is noteworthy that most of what we now know about the directional structures of sediments is the result of field study of ancient sedimentary rocks — not the product of the study of modern sediments. Surprisingly little of what is known about the orientation of such structures and the currents responsible for them is a product of study of present-day sediments. Even the origin of these structures, rather incompletely understood, has been little advanced by the study of existing environments of sedimentation. Moreover, it is unlikely that study of modern sediments can elucidate the origin of certain structures, notably the sole-markings of many sandstone beds — structures which are formed and simultaneously covered in a single short-lived episode of sedimentation. It is only in the lithified rocks that separation of strata along bedding planes is possible so that these directed structures can be seen and studied. Here, indeed is a case where the past is the key to the present.

Another approach to the study of primary current structures has been the experimental one. Mention has been made of the work of DAPPLES and ROMINGER

Fig. 2-1. Cumulative bar graph, by five-year intervals, of published papers which reported measurements of directional structures. Total for 1960—1965 projected from papers published through 1962

(1945), SCHWARZACHER (1951), and RUSNAK (1957b) on experimentally produced sand fabrics. To these can be added the work of RÜCKLIN (1938) on flute casts, McKEE (1957) and BRUSH (1958) on cross-bedding, KUENEN (1958) on "pseudo-nodules" and load-casting, and many others. Experimentation in this field, unlike in most of geology, is not hampered by factors of scale, time, temperature or pressure.

Future developments are difficult to forecast. Experimental studies may enable us to relate current structures to the dynamics of fluid flow and perhaps even make possible mapping of current *vectors* rather than just current *direction*. Possibly also the principles of soil mechanics can be applied to the problems of penecontemporaneous deformation of sediments and shed light on such structures as convolute bedding, slump, load casts, and the pillow-and-ball structure of sandstones.

In conclusion, it is worth noting that although many have contributed to the study of directional properties and paleocurrents, there are two who stand out as pioneers. JAMES HALL in America saw, and perhaps correctly interpreted, more primary directional structures in sediments than anyone else. He described and figured flute casts, groove casts, current crescents, sand shadows, oriented fossils, cross-bedding and ripple mark. HENRY CLIFTON SORBY in England most clearly saw the usefulness of the study of directional properties and measured and used them in a paleogeographic synthesis. A large portion of the writings of these two early workers are very modern in outlook. They anticipated developments which came a hundred years later.

References

ANDEL, T. H. VAN, 1960: Sources of Recent sediments in the northern Gulf of Mexico. J. Sediment. Petrol. **30**, 91—122.

ARBOGAST, J. L., C. H. FAY and S. KAUFMAN, 1960: Method and apparatus for determining directional dielectric anisotropy in solids. U.S. Patent No. 2,963,641.

BARRELL, J., 1925: Marine and terrestrial conglomerates. Bull. Geol. Soc. Am. **36**, 279—342.

BASSETT, D. A., and E. K. WALTON, 1960: The Hell's Mouth Grits; Cambrian grey-wackes in St. Tidwall's Peninsula, North Wales. Quart. J. Geol. Soc. London **116**, 85—110.

BAUMBERGER, ERNST, 1934: Die Molasse der Schweizerischen Mittellands-Jura-gebiete, *in*: Guide Geol. Suisse. Soc. Geol. Suisse Fasc. I, 57—75.

BAUSCH VAN BERTSBERGH, J. W. B., 1940: Richtungen der Sedimentation in der rheinischen Geosynkline. Geol. Rundschau **31**, 328—364.

BECKER, G. F., 1893: Finite homogeneous strain, flow and rupture of rocks. Bull. Geol. Soc. Am. **4**, 13—90.

BIRKENMAJER, K., 1958: Oriented flowage casts and marks in the Carpathian flysch and their relation to flute and groove casts. Acta Geol. Polon. **8**, 139—146 [Polish, English summary].

BOUMA, A. H., 1959a: Flysch Oligocène de Peira-Cava (Alpes-Maritimes, France). Eclogae Geol. Helv. **51**, 893—900.

— 1959b: Some data on turbidites from the Alpes Maritimes (France). Geol. en Mijnbouw **21**, 223—227.

—, and D. J. G., NOTA, 1960: Detailed graphic logs of sedimentary formations. 21st Internat. Geol. Congr. Session Norden, pt. 23, 52—74.

BRETT, G. W., 1955: Cross-bedding in the Baraboo quartzite of Wisconsin. J. Geol. 63, 143—148.

BRINKMANN, R., 1933: Über Kreuzschichtung im deutschen Buntsandsteinbecken. Nachr. Ges. Wiss. Göttingen, Math.-physik. Kl. Fachgruppe IV, Nr. 32, 1—12.

BRUSH jr., L. N., 1958: Study of stratification in a large laboratory flume (abstract). Bull. Geol. Soc. Am. 69, 1542.

CAILLEUX, A., 1938: La disposition individuelle des galets dans les formations détritiques. Rev. Geogr. Phys. et Géol. Dynam. 11, 171—198.

CLARKE, J. M., 1918: Strand and undertow markings of Upper Devonian time. N.Y. State Museum, Bull. 196, 199—238.

CLOOS, H., 1938: Primäre Richtungen in Sedimenten der rheinischen Geosynkline. Geol. Rundschau 29, 357—367.

— 1953: Conversation with the earth. New York: A. A. Knopf. 413 p. (translated from the German).

CRAIG, L. C. and others, 1955: Stratigraphy of Morrison and related formations, Colorado Plateau region. A Preliminary report. U.S. Geol. Survey, Bull. 1009 E, 125—168.

CROWELL, J. C., 1955: Directional-current structures from the Prealpine Flysch, Switzerland. Bull. Geol. Soc. Am. 66, 1351—1384.

CURRAY, J. R., 1956: Dimensional grain orientation studies of Recent coastal sands. Bull. Am. Assoc. Petrol. Geologists 40, 2440—2456.

DAPPLES, E. C., and F. J. ROMINGER, 1945: Orientation analysis of fine-grained clastic sediments. J. Geol. 53, 246—261.

DZULYNSKI, S., M. KSIAZKIEWICZ and PH. H. KUENEN, 1959: Turbidites in flysch of the Polish Carpathian mountains. Bull. Geol. Soc. Am. 70, 1089—1118.

—, and A. RADOMSKI, 1955: Origin of groove casts in the light of the turbidity current hypothesis. Acta Geol. Polon. 5, 47—66.

—, and A. SLACZKA, 1958: Directional structures and sedimentation of the Krosno beds. Ann. soc. géol. Pologne 28, 205—260.

FORSCHE, F., 1935: Stratigraphie und Paläogeographie des Buntsandsteins im Umkreis der Vogesen. Mitt. geol. Staatsinst. Hamburg 15, 15—55.

FUCHS, T., 1895: Studien über Fucoiden und Hieroglyphen. Kon. Akad. Wiss. Wien 62, 371—374.

FÜCHTBAUER, H., 1954: Transport und Sedimentation der westlichen Alpenvorlandsmolasse. Heidelberger Beitr. Mineral. u. Petrog. 4, 26—53.

— 1958: Die Schüttungen im Chatt und Aquitan der deutschen Alpenvorlandsmolasse. Eclogae Geol. Helv. 51, 928—941.

GRIFFITHS, J. C., 1949: Directional permeability and dimensional orientation in Bradford sand. Penn. State Coll. Mineral Inds. Exp. Sta., Bull. 54, 138—143.

— 1952: A review of dimensional orientation of quartz grains in sediments. 16th Tech. Conf. Petroleum Prod., Penn. State Univ., Bull. 60, 47—55.

—, and M. ROSENFELD, 1950: Progress in measurement of grain orientation in Bradford sand. Penn. State Coll. Mineral Inds. Exp. Sta., Bull. 56, 202—236.

GRYSANOVA, T. E., 1947: Methods for studying the orientation of grains in sandy deposits. Doklady Acad. Sci. U.S.S.R. 58, 647—650 [Russian].

— 1949: Orientation of sand grains, methods of studying it and utilizing it in geology. Leningrad Univ. Herald, Students Scientific Paper, No. 2, 97—105 [Russian].

— 1953: Oriented textures of the sandstones in a productive series of the Apsheron Peninsula. Geol. Collection of the All-Union Petrol. Geol. Res. Inst. No. 2, 224—240 [Russian].

HAAF, E. TEN, 1959: Graded beds of the northern Apennines. Ph.D. thesis, Rijks University, Groningen, 102 p.

HALL, J., 1843: Remarks upon casts of mud furrows, wave lines, and other markings upon rocks of the New York System. Assoc. Am. Geol. Rept. 422—432.

HAMBLIN, W. K., 1958: The Cambrian sandstones of northern Michigan. Geol. Survey Michigan, Pub. 51, 149 p.

— 1961: Paleogeographic evolution of the Lake Superior region from Late Keweenawan to Late Cambrian time. Bull. Geol. Soc. Am. 72, 1—18.

HITCHCOCK, E., 1843: The phenomena of drift, or glacio-aqueous action in North America between the Tertiary and alluvial periods. Assoc. Am. Geologists, Repts., 164—221.

HOLMES, C. D., 1941: Till fabric. Bull. Geol. Soc. Amer. 51, 1299—1354.

HSU, K. J., 1959: Flute- and groove-casts in the Prealpine Flysch; Switzerland. Am. J. Sci. 257, 529—536.

— 1960: Paleocurrent structures and paleogeography of the Ultra-helvetic Flysch Basins, Switzerland. Bull. Geol. Soc. Am. 71, 577—610.

— 1960: Texture and mineralogy of the Recent sands of the Gulf Coast. J. Sediment. Petrol. 30, 380—403.

HÜLSEMANN, J., 1955: Großrippeln und Schrägschichtungs-Gefüge im Nordsee-Watt und in der Molasse. Senckenbergiana 36, 359—388.

HUNTER, R. E., 1960: Iron sedimentation in the Clinton group of the central Appalachian basin. Unpublished Ph.D. thesis, The Johns Hopkins University, 416 p.

HUTTON, J., 1788: Theory of the earth. Trans. Roy. Soc. Edinburgh 1, 209—304.

HYDE, J. E., 1911: The ripples of the Bedford and Berea formations of central Ohio, with notes on the paleogeography of that epoch. J. Geol. 19, 257—269.

ILLIES, H., 1949: Die Schrägschichtung in fluviatilen und litoralen Sedimenten, ihre Ursachen, Messung und Auswertung. Mitt. Geol. Staatsinst. Hamburg 19, 89—109.

JAMIESON, T. F., 1860: On the drift and rolled gravel of the north of Scotland. Quart. J. Geol. Soc. London 16, 347—371.

JOHNSON, W. E., and R. V. HUGHES, 1948: Directional permeability measurements and their significance. Penn. State Coll. Mineral. Inds. Exp. Sta., Bull. 52, 180—205.

JOHNSTON, W. A., 1922: Imbricated structure in river gravel. Am. J. Sci., ser. V, 4, 387—390.

KALTERHERBERG, J., 1956: Über Anlagerungsgefüge in grobklastischen Sedimenten. Neues Jahrb. Geol. Paläont., Abhandl. 104, 30—57.

KAURANNE, L. K., 1960: A statistical study of stone orientation in glacial till. Bull. comm. géol. Finlande No. 188, 87—97.

KAY, M., 1945: Paleogeographic and palinspastic maps. Bull. Am. Assoc. Petrol. Geologists 29, 426—450.

KING, P. B., 1948: Geology of the southern Guadaloupe Mountains, Texas. U.S. Geol. Survey Prof. Paper 215, 183 p.

KOPSTEIN, F. P. H. W., 1954: Graded bedding of the Harlech dome. Ph.D. thesis, Rijks University, Groningen. 97 p.

KRUMBEIN, W. C., 1939: Preferred orientation of pebbles in sedimentary deposits: J. Geol. 47, 673—706.

KSIAZKIEWICZ, M., 1957: Tectonics and sedimentation in the Northern Carpathians. 20th Intern. Geol. Congr., sec. V, 1, 227—252.

— 1958: Sedimentation in the Carpathian Flysch sea. Geol. Rundschau 47, 418—424.

KUENEN, PH. H., 1953a: Graded bedding with observations on Lower Paleozoic rocks of Britain. Verhandel. Konikl. Ned. Akad. Wetenschap., Afdeel. Natuurk. 20, 1—47.

— 1953b: Significant features of graded bedding: Bull. Am. Assoc. Petrol. Geologists 37, 1044—1066.

— 1956: Problematic origin of the Naples rocks around Ithaca, New York. Geol. en Mijnbouw 18, 277—283.

— 1957: Sole markings of graded graywacke beds. J. Geol. 65, 231—258.

— 1958: Experiments in geology. Trans. Geol. Soc. Glasgow 23, 1—28.

—, and A. CAROZZI, 1953: Turbidity currents and sliding in geosynclinal basins in the Alps. J. Geol. 61, 314—317.

— A. FAURE-MURET, M. LANTEAUME and P. FALLOT, 1957: Observations sur les Flyschs des Alpes Maritimes Françaises et Italiennes. Bull. soc. géol. France, ser. VI, 7, 11—26.

—, and E. TEN HAAF, 1958: Sole markings of graded beds: A reply. J. Geol. 66, 335—337.

—, and J. E. SANDERS, 1956: Sedimentation phenomena in Kulm and Flözleeres graywackes, Sauerland and Oberharz, Germany. Am. J. Sci. 254, 649—671.

LYELL, CH., 1837: Principles of geology (1st Amer. ed., vol. 2). Philadelphia: Jas. Kay, Jr. and Bro. 553 p.

MARTINEZ, J. D., 1958: Photometer method for studying quartz grain orientation. Bull. Am. Assoc. Petrol. Geologists 42, 588—608.

MATTHEW, G., 1903: Cambrian rocks of Cape Breton. Geol. Survey Canada, Rept. No 797, 246 p.

McBRIDE, E. F., 1962: Flysch and associated beds of the Martinsburg formation (Ordovician), central Appalachians. J. Sediment. Petrol. 32, 39—91.

McDOWELL, J. P., 1957: The sedimentary petrology of the Mississagi quartzite in the Blind River area. Ontario Dept. Mines, Cir. No. 6, 31 p.

McIVER, N. L., 1961: Sedimentation of the Upper Devonian marine sediments of the central Appalachians. Unpublished Ph.D. thesis, The Johns Hopkins University, 347 p.

McKEE, E. D., 1940: Three types of cross-lamination in Paleozoic rocks of northern Arizona. Am. J. Sci. 238, 811—824.

— 1957: Primary structures in some Recent sediments. Bull. Am. Assoc. Petrol. Geologists 41, 1704—1747.

MILLER, HUGH, 1884: On boulder glaciation. Proc. Roy. Phys. Soc. Edinburgh 8, 156—189.

NANZ, R. H., 1955: Grain orientation in beach sands: a possible means for predicting reservoir trend (abstract). J. Sediment. Petrol. 25, 130.

— 1960: Exploration of earth formations associated with petroleum deposits. U.S. Patent No. 2,963,641.

NIEHOFF, W., 1958: Die primär gerichteten Sedimentstrukturen, insbesondere die Schrägschichtung im Koblenzquartzit am Mittelrhein. Geol. Rundschau 47, 252—321.

OKKO, V., 1949: Explanation to the map of surficial deposits. Sheet B4, Kokkola, Geol. Survey of Finland.

PELLETIER, B. R., 1958: Pocono paleocurrents in Pennsylvania and Maryland. Bull. Geol. Soc. Am. 69, 1033—1064.

— 1961: Triassic stratigraphy of the Rocky Mountains and Foothills northeastern British Columbia. Geol. Survey Canada, Paper 61-2, 32 p.

PETTIJOHN, F. J., 1957: Paleocurrents of Lake Superior Precambrian quartzites. Bull. Geol. Soc. Amer. 68, 469—480.

PLUMLEY, W. J., 1948: Black Hills terrace gravels: a study in sediment transport. J. Geol. 56, 526—577.

POTTER, P. E., 1955: Petrology and origin of the Lafayette gravel. Part I. Mineralogy and Petrology. J. Geol. 63, 1—38.

— 1962: Regional distribution pattern of Pennsylvanian sandstones. Bull. Am. Assoc. Petrol. Geologists 46, 1890—1911.

—, and R. F. MAST, 1963: Sedimentary structures, sand shape fabrics and permeability. J. Geol. 71, 441—471.

—, E. NOSOW, N. M. SMITH, D. H. SWANN and F. H. WALKER, 1958: Chester cross-bedding and sandstone trends in Illinois basin. Bull. Am. Assoc. Petrol. Geologists 42, 1013—1046.

—, and J. S. OLSON, 1954: Variance components of cross-bedding direction in some basal Pennsylvanian sandstones of the Eastern Interior Basin: Geological applications. J. Geol. 62, 50—73.

—, and W. A. PRYOR, 1961: Dispersal centers of the Paleozoic and later clastics of the Upper Mississippi Valley and adjacent areas. Bull. Geol. Soc. Am. 72, 1195—1250.

—, and R. SIEVER, 1956: Sources of basal Pennsylvanian sediments in the Eastern Interior Basin: Cross-bedding. J. Geol. 64, 225—244.

PRYOR, W. A., 1960: Cretaceous sedimentation in Upper Mississippi embayment. Bull. Am. Assoc. Petrol. Geologists 44, 1473—1504.

— 1961: Sand trends and paleoslope in Illinois basin and Mississippi embayment: Geometry of Sandstone Bodies; Tulsa. Am. Assoc. Petrol. Geologists, p. 119—133.

REICHE, P., 1938: An analysis of cross-lamination: the Coconino sandstone. J. Geol. **46**, 905—932.

REINEMUND, J. A., and W. DANILCHIK, 1957: Preliminary geologic map of the Waldron Quadrangle and adjacent areas, Scott County, Arkansas. U.S. Geol. Survey, OM 192.

RICH, J. L., 1950: Flow markings, groovings, and intrastratal crumplings as criteria for recognition of slope deposits, with illustrations from Silurian rocks of Wales. Bull. Am. Assoc. Petrol. Geologists **34**, 717—741.

RICHTER, K., 1932: Die Bewegungsrichtung des Inlandeises, rekonstruiert aus den Kritzen und Längsachsen der Geschiebe. Z. Geschiebeforschung **8**, 62—66.

— 1933: Gefüge und Zusammensetzung des norddeutschen Jungmoranengebietes. Abh. Geol.-Paläont. Inst. Univ. Greifswald **11**, 1—63.

— 1936 a: Gefügestudien im Engebrae, Fondalsbrae und ihren Vorlandsedimenten. Z. Gletscherk. **24**, 22—30.

— 1936 b: Ergebnisse und Aussichten der Gefügeforschung pommerschen Diluvium. Geol. Rundschau **27**, 197—206.

ROWLAND, R. A., 1946: Grain shape fabrics of clastic quartz. Geol. Bull. Soc. Am. **59**, 547—564.

RUBEY, W. W., and N. W. BASS, 1925: The geology of Russell County, Kansas. Kansas Geol. Survey, Bull. **10**, 1—86.

RÜCKLIN, H., 1938: Strömungsmarken im unteren Muschelkalk des Saarlandes. Senckenbergiana **20**, 94—114.

RUEDEMANN, R., 1897: Evidence of current action in the Ordovician of New York. Amer. Geologist **19**, 367—391.

RUKHIN, L. B., 1958: Grundzüge der Lithologie. Berlin: Akademie-Verlag. 806 p. (translated from Russian).

RUSNAK, G. A., 1957 a: A fabric and petrologic study of the Pleasantview sandstone. J. Sediment. Petrol. **27**, 41—55.

— 1957 b: The orientation of sand grains under conditions of "unidirectional" fluid flow. 1. Theory and experiment. J. Geol. **65**, 384—409.

SANDER, B., 1930: Gefügekunde der Gesteine. Vienna: Springer. 325 p.

SANDERS, J. E., 1956: Oriented phenomena produced by sedimentation from turbidity currents and in subaqueous slope deposits. Program 30th Ann. Meeting Soc. Econ. Paleon. Mineralogists, Chicago, p. 46—47.

SARKISIAN, S. G., and L. T. KLIMOVA, 1955: The orientation of pebbles and methods of studying it for paleogeographical reconstructions. Petroleum Institute, Publishing House of the Academy of Sciences, U.S.S.R., Moscow, 165 p. [Russian].

SCHLEE, J., 1957 a: Fluvial gravel fabric. J. Sediment. Petrol. **27**, 162—176.

— 1957 b: Upland gravels of southern Maryland. Bull. Geol. Soc. Am. **68**, 1371—1410.

SCHWARZACHER, W., 1951: Grain orientation in sands and sandstones. J. Sediment. Petrol. **21**, 162—172.

— 1953: Cross-bedding and grain size in the lower Cretaceous sands of East Anglia. Geol. Mag. **90**, 322—330.

SEILACHER, A., 1960: Strömungsanzeichen im Hunsrückschiefer. Notizbl. hess. Landesamtes Bodenforsch. Wiesbaden **88**, 88—106.

— 1959: Fossilien als Strömungs-Anzeiger. Aus der Heimat **67**, 170—177.

SHOTTON, F. W., 1937: Lower Bunter sandstones of north Worcestershire and east Shropshire. Geol. Mag. **74**, 534—553.

SHROCK, R. R., 1948: Sequence in layered rocks. New York: McGraw-Hill Book Co. 507 p.

SORBY, H. C., 1856: On the physical geography of the Old Red Sandstone sea of the Central District of Scotland. Edinburgh New Philosophical J. n.s. **3**, 112—122.

— 1857: On the physical geography of the Tertiary estuary of the Isle of Wight. Edinburgh New Philosophical J., n.s. **5**, 275—298.

— 1858: On the ancient physical geography of the southeast of England. Edinburgh New Philosophical J., n.s. **7**, 226—237.

Sorby, H. C., 1859: On the structures produced by the current present during the deposition of stratified rocks. The Geologist 2, 137—147.
— 1879: Presidential Address. Quart. J. Geol. Soc. London 35, 56—77.
Stokes, W. L., 1947: Primary lineation in fluvial sandstones: A criterion of current direction. J. Geol. 55, 52—54.
— 1953: Primary sedimentary trend indicators applied to ore-finding in the Carrizo Mountains, Arizona and New Mexico. U.S. Atomic Energy Comm. RME-3043, pt. 1, 48 p.
Strakhov, N. M., ed., 1958: Méthodes d'étude des roches sédimentaires, vol. 2: Bur. Rech. Géol., Geophys. et Minières. Ann. Ser. d'Information Géol., No. 35, 535 p. (translated from the Russian).
Sutton, R. G., 1955: Use of flute casts in stratigraphic correlation. Bull. Am. Assoc. Petrol. Geologists 43, 230—237.
Tanner, W. F., 1955: Paleogeographic reconstructions from cross-bedding studies. Bull. Amer. Assoc. Petrol. Geologists 39, 2471—2483.
Unrug, R., 1956: Preferred orientation in Recent gravels, etc. Bull. Acad. Sci. Polonaise 4, 469—473.
— 1957: Recent transport and sedimentation of gravels in the Dunajec Valley (western Carpathians). Acta Geol. Polon. 7, 217—257 [Polish, English summary].
Vassoevich, N. B., 1932: Some data allowing us to distinguish the overturned position of flysch sedimentary formations from the normal ones. Academy Science, U.S.S.R., Trudy Geol. Inst. 2, 47—63 [Russian].
Virkkala, K., 1951: Glacial geology of the Suomussalmi area, East Finland. Bull. comm. géol. Finlande No. 155, 66 p.
— 1960: On the striations and glacier movements in the Tampere region, southern Finland. Bull. comm. géol. Finlande No. 188, 161—176.
Wadell, H., 1936: Shape and shape position of rock fragments. Geog. Annaler 18, 74—92.
West, R. G., and J. J. Donner, 1956: The glaciation of East Anglia and the East Midlands, A differentiation based on stone orientation measurements of the tills. Quart. J. Geol. Soc. London 112, 69—81.
White, W. S., 1952: Imbrication and initial dip in a Keweenawan conglomerate bed. J. Sediment. Petrol. 22, 189—194.
Whitaker, J. C., 1955: Direction of current flow in some lower Cambrian clastics in Maryland. Bull. Geol. Soc. Am. 66, 763—766.
Wood, A., and A. J. Smith, 1959: The sedimentation and sedimentary history of the Aberystwyth Grits (Upper Llandoverian). Quart. J. Geol. Soc. London 115, 163—195.
Wurster, P., 1958: Geometrie und Geologie von Kreuzschichtungskörpern. Geol. Rundschau 47, 322—358.
Yeakel, L. S. jr., 1962: Tuscarora, Juniata and Bald Eagle paleocurrents and paleogeography in the central Appalachians. Bull. Geol. Soc. Am. 73, 1515—1540.

History of Paleocurrent Investigations (1963—1976)

Widely diversified post-1963 paleocurrent studies range in scale from microdepositional environments to dispersal in the world ocean as well as experiments

The years since 1963 have seen rapid progress in all aspects of paleocurrents — primary sedimentary structures, the study of dispersal patterns, and in basin analysis. Today paleocurrent studies are made not only in terrigenous sandstones, but in carbonate rocks, volcaniclastic deposits, and even in argillaceous sediments and evaporites. And, unlike 1963, there are today many more studies of paleocurrents in Recent and Pleistocene sediments.

Since 1963, there has been an upsurge of interest in primary sedimentary structures. An expression of this interest is the publication of a number of lavishly illustrated, largely descriptive, monographic works. They are annotated at the end of this section.

Although our concern here is those sedimentary structures (and other properties) that yield evidence on the direction of paleocurrent flow, such structures are but a subset of the larger class of primary sedimentary structures. Efforts to organize our knowledge of these structures have led to various classifications of them. Although all major works include some sort of classification, several authors have addressed themselves particularly to the classification problem.

Two approaches to the problem have been used. One is morphologic, the other genetic. The first involves setting up classes based on form or geometry and on the place of occurrence — base of bed, etc. In the genetic approach, structures are grouped according to process of formation. A genetic classification presupposes that we know the origin — an assumption not always warranted. A purely morphological classification, however, leads to classes in which structures unrelated to one another and of greatly different origin are placed together. Although perhaps providing a suitable key for identification, such classification does not contribute to our understanding.

A good example of a purely morphologic classification is that of CONYBEARE and CROOK (1968). Classifications more largely genetic are those of NAGTEGAAL (1965), ELLIOTT (1965), and ALLEN (1968). Elliott's classification is based on rheological and kinematic parameters; that of Allen is based on bed form and fluid flow (fig. 2-2). Inasmuch as the primary sedimentary structures, most of which are indicators of paleocurrent flow are related to the bed forms developed at the sediment-fluid interface, ALLEN's 1968 paper is a major contribution. His classification is both descriptive and genetic. The forms are classified according to their orientation (longitudinal or transverse), polarity (unipolar versus bipolar), character of the interface (cohesive or cohesionless), and scale dependence. Allen's classification can be directly related to the dynamics of formation of the bed form and its related structure.

By now all the common sedimentary structures have been described and their nomenclature is more or less stabilized. Only a few new and somewhat obscure or uncommon structures have been reported since 1963. References to some of these are made in the appropriate parts of this supplement.

Interest in sedimentary structures has moved into new areas. Indicative of this new interest is the symposium on the hydrodynamic interpretation of such

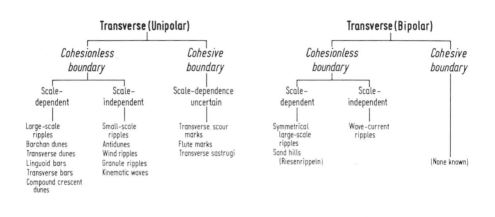

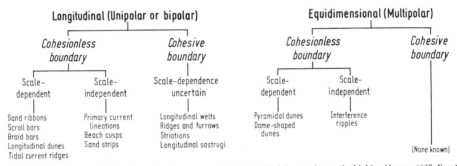

Fig. 2-2. Comprehensive classification of bedforms, a classification embodying much genetic thinking (ALLEN, 1968, figs. 3 and 4). Sastrugi are erosion forms produced in hardened snow by the action of wind driven ice particles. For an example of a purely morphologic classification of a sedimentary structure see fig. 4-2

structures (MIDDLETON, 1965), the basic goal being a more complete understanding of the *intensity and type of flow* as well as its direction. The recent suggestions of weighting paleocurrent measurements by their volume is one example of this effort (IRIONDO, 1973). A further expression of this interest is seen in the various experimental studies of ALLEN (1970a, 1971a, 1971b, 1971c, 1971d), DZULYNSKI (1963), HARMS (1969), SOUTHARD and DINGLER (1971) and SOUTHARD and BOGUCHWAL, 1973). Of special interest are the techniques for producing sole marks and for rendering them visible for study (especially ALLEN, 1966, 1971 a).

The mapping of directional structures is now commonplace and to compile a list of such studies would yield little value. The most useful structure in the coarse sediments is cross-bedding, both large- and small-scale, and in all sediments, fabric, particularly imbrication. In turbidites virtually all paleocurrent maps continue to be based on the measurement of flutes and grooves. The developments in this area are covered in the appropriate parts of this supplement.

Use of directional structures in paleogeographic reconstruction and basin analysis, as noted above, is widespread. Paleocurrent patterns, used in conjunction with other data, have proved not only helpful in looking at sedimentary basins as a whole, but also in looking at lesser areas and in environmental interpretation. The radial paleocurrent patterns on alluvial fans, for example, differ markedly from those of shallow marine shelves under the influence of reversing tidal currents. The paleocurrent patterns characteristic of the basic environments have been identified by SELLEY (1968). He describes one fanglomerate model, two fluvial, one eolian, six shoreline, and two turbidite models. The discovery of bipolar cross-bedding and ripple orientation in carbonates in the middle 1960s — and its interpretation as forming from reversing tidal currents — is particularly noteworthy.

Paleocurrent studies of modern environments have been another area of significant post-1963 activity. Good examples of such study for alluvial sediments are those of WILLIAMS (1971), PICARD and HIGH (1973) and SMITH (1972), McKEE (1966) and BIGARELLA (1970/71) for eolian dunes and KLEIN (1970) for inter-tidal sand bars.

As the collection of paleocurrent data has become more commonplace, it has become possible to study the history of sedimentary basins and to reconstruct the sediment transport patterns for successive stages in the evolution of the basin. One example of such studies is that of CONTESCU (1974) for the Carpathians.

Inevitably the study of basin evolution and the paleocurrent patterns related to it leads to the concept of plate tectonics. The revision of our thinking about geosynclines, growing out of the plate tectonic theory has, for the most part, yet to take account of sediment provenance and transport. An exception is the paper by GRAHAM et al. (1975) on the history of the several basins formed by the closure of the protoAtlantic in eastern North American during the Paleozoic. These authors attempt, among other things, to integrate the various paleocurrent studies of the Appalachian Paleozoic and the tectonic evolution of the eastern margin of North America. Paleoglacial flow and continental fragmentation of Gondwanaland has long been the prime example of the relation of paleocurrents to plate movement and separation (see, for example, BIGARELLA, 1971).

In short, we now have paleocurrent studies at varying scales, from the smallest environmental unit such as a point bar, through larger basin-wide analysis, to even full continental-scale paleocurrent maps. The latter, for the most part, are those showing the paleoglacial flow, which as in the case of Gondwanaland, extend over continental areas that have since broken up and drifted apart. Some paleowind analyses also are on a similar grand scale. Dispersal studies of scaler properties have also attained a comparable and even larger size, the most notable being the distribution of clay minerals in the world ocean (RĀTEEV et al., 1969). Can you think of a larger basin to study?

The economic value of paleocurrent studies is largely indirect. Better understanding of the paleocurrents in a basin cannot help but enable us to predict sand trends and the like better. Of more direct economic interest is the dispersion of distinctive rock types and ores. For the most part boulder trains and glacial flow patterns, deduced from striations and till fabric, have been helpful in prospecting in areas of poor bedrock exposure. So keen is the interest in this subject that cash prizes have been offered in Sweden for ore-bearing boulders. Trained dogs have been employed to detect such boulders! In Finland 10 mines have been discovered by follow-up work on boulders found by the public (ALLCOCK, 1974). Paleocurrent data has proved helpful in interpretation of both modern and ancient placers (SESTINI, 1973).

Of interest is the growing recognition of primary sedimentary structures in relatively high-grade metamorphic rocks — even in rocks of the sillimanite grade. Not only are such structures as graded bedding recognizable and useful, but such directional structures as cross-bedding are preserved and may make possible sedimentological analyses considered heretofore impossible.

And what of technological developments? These have been very diverse and include space photography that has provided the first complete maps of the dune systems of continental deserts, side scan sonar for the mapping of dune systems on the sea bottom, bottom photography to help define the orientation of current structures on modern shelves and in the deep sea, high resolution marine seismic profiling that has added immeasurably to our knowledge of offshore sedimentary basins, the dipmeter log for possible identification of primary depositional dip and cross-bedding, devices to measure either directly or indirectly clast orientation, and the now widespread use of the computer to help in the analysis of both scalar and vectoral data for dispersal studies. To all of this we would like to add the electronic hand calculator, which now permits one to calculate almost effortlessly the vector mean in the field.

In appraising post-1963 paleocurrent studies we note that they are most prolific in North America and western Europe. However, ardent converts exist almost everywhere, so that there is a paleocurrent bibliography for every continent and even in Antarctica where among other paleocurrent studies a vertical profile — Devonian to Triassic — has been made (BARRETT and KOHN, 1975). In Japan, for example, there is even a paper on the history of paleocurrent research (NAGAHAMA et al., 1968). Thus the study of paleocurrents in all their aspects is today an integral, well-accepted part of sedimentology, one that shows every sign of continuing to be active for many, many years to come.

Annotated References

ANGELUCCI, A., E. DE ROSA, G. FIERRO, M. GNACCOLINI, G. B. LaMONICA, B. MARTINIS, G. C. PAREA, T. PESCATORE, A. RIZZINI and F. C. WEZEL, 1967: Sedimentological characteristics of some Italian turbidites. Geol. Romana VI, 345—420.

One of the most comprehensive documentations of a turbidite — or almost any other — sequence ever made. Ninety-seven described sections from all over Italy and Sicily plus 65 figures most of which are sedimentary structures. A splendid reference for a library field trip.

CONYBEARE, C. E. B., and K. A. W. CROOK, 1968: Manual of sedimentation structures. Australian Dept. Nat. Devel., Bur. Mineral. Res. Geol. Geophys. **102**, 327 p. Contains a discussion of classification of sedimentary structures. Structures are classed on a purely morphological basis. Such a classification makes possible a dichotomous key for identification.

DIMITRIJEVIĆ, M. N., M. D. DIMITRIJEVIĆ and B. RADOSĚVIĆ, 1967: Sedimente teksture u turbiditima. Zavod Geološka Geofizička Istraživanja. **16**, 70 p.
This small volume contains a brief description and 52 beautiful line drawings of the inorganic and organic structures of turbidites. Serbian, but key ideas also in English, French, and German.

DZULYNSKI, S., 1963: Directional structures in flysch. Polska Akad. Sci., Studia Geol. Pol., **12**, 61 plates, 136 pages. Polish and English.
A beautifully illustrated treatment of sole marks based on the pioneer work of the Polish school of which DZULYNSKI is a leader.

—, and J. E. SANDERS, 1962: Current marks on firm mud bottoms. Connecticut Acad. Arts and Sci. **42**, 75—96.
Thirty seven pages of text and 22 excellent plates dealing mainly with sole marks that occur on the undersides of sandstones.

—, and E. K. WALTON, 1965: Sedimentary features of flysch and greywackes. Developments in Sedimentology 7, Amsterdam: Elsevier Publ. Co., 274 p.
Devoted mainly (188 pages) to sedimentary structures, most of which are directional. A well-illustrated summary by two active workes in the field.

GUBLER, Y., D. BUGNICOURT, J. FABER, B. KUBLER and R. NYSSEN, 1966: Essai de nomenclature et caractérisation des principales structures sédimentaires. Paris: Éditions Technip, 291 p.
Contains a short review on stratification and stratigraphic terminology followed by a long section (186 pages) on sedimentary structures. Very systematic and complete. Given for cach structure are definition, description, measurement, frequency of occurrence, origin, and utility; well illustrated.

KHABAKOV, A. V., ed., 1962: Atlas tekstur i struktur osadochnykh gornykh porod (An atlas of textures and structures of sedimentary rocks. Pt. 1, Clastic and argillaceous rocks), Moscow: VSEGEI, 578 p. (Russian with French translation of plate captions).
Perhaps the first comprehensive picture book of sedimentary structures and textures.

LANTEUME, M., B. BEAUDOIN and R. CAMPREDON, 1967: Figures sédimentaires du flysch. Grés d'Annot: du synclinal de Peira-Cava. Paris, Éditions Centre National de la Recherche Scientifique, 97 p.
Sixty-one fine plates, mostly of sole marks, with full captions in French, English, German, Italian, and Spanish. Full cross index and all the previous relevant literature. Essential companion for the field study of turbidity deposits.

PETTIJOHN, F. J., and P. E. POTTER, 1964: Atlas and glossary of primary sedimentary structures. New York: Springer, 117 plates, 370 p. (English, French, German, Spanish).
Primarily a picture book prefaced by a short essay on classification and followed by a four-language glossary of 360 entries.

RICCI LUCCHI, F., 1970: Sedimentografia. Bologna: Zonichelli, 288 p.
A beautifully illustrated picture book of primary sedimentary structures, mainly sole marks of flysch sandstones. 170 plates with marginal text. Italian with English-Italian lexicon.

References

ALLCOCK, J., 1974: Prospecting in areas of glacial terrain (book review). Econ. Geol. **69**, 277.

ALLEN, J. R. L., 1966: Flow visualization using Plaster of Paris. J. Sediment. Petrol. **36**, 806—811.

—, 1968: On the character and classification of bed forms. Geol. Mijnbouw, **47**, 173—185.

—, 1970a: A quantitative model of climbing ripples and their cross-laminated deposits. Sedimentology **4**, 5—26.

—, 1970b: Physical processes of sedimentation. New York: American Elsevier Publ. Co., 248 p.

—, 1971a: Some techniques in experimental geology. J. Sediment. Petrol. **41**, 695—702.

—, 1971b: Transverse erosional marks of mud and rock: their physical basis and geological significance. Sediment. Geol. **5**, 165—385.

—, 1971c: A theoretical and experimental study of climbing-ripple cross-lamination, with a field application to the Uppsula esker. Geog. Annaler, Ser. A, **53**, 157—187.

—, 1971d: Instantaneous sedimentation deposition rates deduced from climbing-ripple cross-lamination. J. Geol. Soc. London **127**, 553—561.

BARRETT, P. J., and B. P. KOHN, 1975: Changing sediment transport directions from Devonian to Triassic in the Beacon Super-Group of South Victoria Land, Antarctica *in* K. S. W. CAMPBELL, ed., Gondwana geology. Canberra, Australian Nat. Univ. Press, 15—35.

BIGARELLA, J. J., 1970/1971: Wind pattern deduced from dune morphology and internal structures. Bol. Paranaense de Geociencias **28/29**, 73—113.

—, 1971: Continental drift and paleocurrent analysis (A comparison between Africa and South America). Int. Union Geol. Sci., Comm. Stratigr. Subcomm. Gondwana Stratigr. Palaeontol. Gondwana Symp. Proc. Paper **2**, 73—97.

CONTESCU, L. R., 1974: Geologic history and paleogeography of eastern Carpathians, example of Alpine geosyncline evolution. Bull. Am. Assoc. Petrol. Geologists **58**, 2436—2476.

CONYBEARE, C. E. B., and K. A. W. CROOK, 1968: Manual of sedimentary structures. Australian Dept. Nat. Devel., Bur. Mineral Res. Geol. Geophys. **102**, 327 p.

DZULYNSKI, S., 1963: Directional structures in flysch. Pol. Akad. Sci., Studia Geol. Pol. **12**, 136 p.

—, and E. K. WALTON, 1965: Sedimentary features of flysch and greywacke. Devel. Sedimentol. **7**, Amsterdam: Elsevier Publ. Co., 274 p.

ELLIOTT, R. E., 1965: A classification of subaqueous sedimentary structures based on rheological and kinematical parameters. Sedimentology **5**, 193—209.

GRAHAM, S. A., W. R. DICKINSON and R. V. INGERSOLL, 1975: Himalayan-Bengal model for flysch dispersal in the Appalachian-Ouachita system. Bull. Geol. Soc. Am. **86**, 273—286.

HARMS, J. C., 1969: Hydraulic significance of some sand ripples. Bull. Geol. Soc. Am. **80**, 363—396.

IRIONDO, M. H., 1973: Volume factor in paleocurrent analysis. Bull. Am. Assoc. Petrol. Geologists **57**, 1341—1342.

KLEIN, G. DeVRIES, 1970: Depositional and dispersal dynamics of intertidal sand bars. J. Sediment. Petrol. **40**, 1095—1127.

McKEE, E. D., 1966: Structures of dunes at White Sands National Monument, New Mexico. Sedimentology **7** Sp. Issue, 69 p.

MIDDLETON, G. V., ed., 1965: Primary sedimentary structures and their hydrodynamic interpretation. Soc. Econ. Paleontol. Mineral., Spec. Pub. **12**, 265 p.

NAGAHAMA, H., O. HIROKAWA and T. ENDA, 1968: History of researches on paleocurrents in reference to sedimentary structures — with paleocurrent maps and photographs of sedimentary structures. Geol. Survey Japan, Bull. **19**, 1—17.

NAGTEGAAL, P. J. C., 1965: An approximation to the genetic classification of non-organic sedimentary structures. Geol. MIJNBOUW, **44**, 347—352.

PICARD, M. D., and L. R. HIGH, JR., 1973: Sedimentary structures of ephemeral streams. Devel. Sedimentol., **17**, Amsterdam: Elsevier Publ. Co., 223 p.

RATEEV, M. A., Z. N. GORBUNOVA, A. P. LISITZYN and G. L. NOSOV, 1969: The distribution of clay minerals in the oceans. Sedimentology **13**, 21—43.

SELLEY, R. C., 1968: A classification of paleocurrent models. J. Geol. **76**, 99—110.

SESTINI, G., 1973: Sedimentology of a paleoplacer: The gold-bearing Tarkwaian of Ghana *in* G. C. AMSTUTZ and A. J. BERNARD, eds. Ores in sediments. Berlin-Heidelberg-New York: Springer, 275—305.

SMITH, N. D., 1972: Some sedimentological aspects of planar cross-stratification in a sandy braided river. J. Sediment. Petrol. **42**, 624—634.

SOUTHARD, J. B., and L. A. BOGUCHWAL, 1973: Flume experiments on the transition from ripples to lower flat bed with increasing sand size. J. Sediment. Petrol. **43**, 1114—1121.

—, and J. R. DINGLER, 1971: Flume study of ripple propagation behind mounds on flat sand beds. Sedimentology **16**, 257—263.

WILLIAMS, G. E., 1971: Flood deposits of the sand-bed ephemeral streams of central Australia. Sedimentology **17**, 1—40.

Fabrics and Geophysical Properties up to 1963

Introduction

Observations of the fabrics of clastic sediments were made as early as the middle of the 19th century. In 1860 JAMIESON (p. 349) commented on imbrication of stones in creek beds in Scotland. Systematic study of sedimentary fabrics did not begin, however, until after the appearance of " Gefügekunde der Gesteine" by BRUNO SANDER in 1930. His book, primarily concerned with the fabric of tectonites, provided a systematic methodology and developed principles of interpretation which are readily applied to sedimentary rocks.

The principal object of the study of sedimentary, primary fabrics has been the reconstruction of current direction. Unlike the study of tectonites, fabric studies of sedimentary rocks have been little used to understand the transport process itself. Fabric studies have chiefly been used to infer current direction in sands and gravels and to reconstruct direction of glacial motion in tills, both Pleistocene and pre-Pleistocene. In addition, fabrics have an important bearing on physical properties of rocks such as thermal-, electrical-, fluid-, and sonic conductivity.

Fabric, as used by sedimentary petrologists, refers to the spatial arrangement and orientation of fabric elements. This definition is more restrictive than SANDER'S (1948, p. 4—5) use of the term *Gefüge*, which included properties such as grain size, sorting, porosity, etc. or what are usually regarded as textural properties by sedimentary petrologists. A *fabric element* of a sedimentary rock may be a single crystal, a detrital fragment, a fossil, or any component that behaves as a single unit with respect to the applied force (FAIRBAIRN, 1949, p. 3).

Fabrics may have no orientation and be *isotropic* or they may have preferred orientation and be *anisotropic*. An anisotropic fabric is produced by alignment of fabric elements in a force field. The earth's gravitational field and force fields related to current flow, either water, air, or ice, are the principal causes of such alignment in sedimentary rocks. Such force fields have both magnitude and direction and constitute *vector fields*. Primary depositional fabrics may be partially or completely modified by soft-sediment and tectonic deformation. Examples of soft-sediment deformation include flowage, sliding, and disturbance by benthonic organisms. Primary sedimentary fabric, rather weak at best, is usually destroyed by such deformation. In a few cases, however, a new fabric may be imposed, such as that observed in some clastic dikes. The fabric found along the borders of injected clastic dikes appears to be analogous to fabrics formed at the boundaries of many deep seated plutons.

Reference Systems

Primary directional structures are described in relation to transport direction and bedding. A standard reference system, not only for fabrics but for all the directional properties, is essential (fig. 3-1). The system used is a modification of SANDER'S (1930, p. 57) coordinates. Following SANDER, the a axis is defined as parallel to the direction of transport, b as the depositional strike and c as perpendicular to the ab plane. This choice of axes brings out as simply as possible the symmetry of sedimentary fabrics and structures. The ac plane of figure 3-1

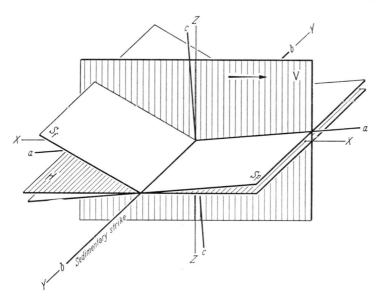

Fig. 3-1. Reference system for directional structures. The line of movement a is parallel to the average transport vector, b is the sedimentary strike perpendicular to a and c is perpendicular to the ab plane. H is horizontal, S_p lies in the principal surface of accumulation, and S_f is the foreset surface of cross-bedding

is the vertical plane that contains the average current vector of the transporting medium. This is the plane of symmetry in most sedimentary fabrics and is also the plane of symmetry of many directional structures.

The ab plane does not always correspond to the horizontal plane, H, which is usually assumed to be the principal surface of sedimentary accumulation. Although this assumption is substantially correct, several complicating factors should be recognized. Both are illustrated by cross-bedding. The bounding surfaces of tabular cross-bedded layers may make angles of a few degrees with the horizontal and actually define the plane S_p of figure 3-1 rather than H. Subsequent erosion or original depositional thinning can also cause S_p to dip down-current or even in other directions at similar low angles.

Foreset bedding dips in the down-current direction at relatively high angles, usually between 10 and 30°. Obviously, foreset bedding, plane S_f of figure 3-1, is not to be confused with the principal surface of accumulation, S_p. Because some massive appearing rocks are in fact cross-bedded (HAMBLIN, 1962a, 1962b), unrecognized foresets can introduce complications, if a specimen is referenced to plane S_p rather than S_f. Failure to recognize S_f in massive sandstones has been a source of confusion in same fabric and geophysical studies.

Symmetry Concepts

Symmetry concepts play an important role in the study of sedimentary fabrics and are useful with the other primary directional structures.

Fabric elements may be considered as spheres, disks, or rods, even though the actual shapes of most elements only approximate these three forms.

Spheres cannot have a preferred orientation and thus are not useful fabric elements for reconstructing current direction. They can have, however, different types of packing arrangements. They may be completely disordered and non-repetitive or may be at least partially ordered and repetitive and packed in any one of six ways (SLICHTER, 1899, p. 350), of which rhombohedral packing has the least pore volume. In natural sands, combinations of mixtures of these different packing types are probably common. Because fabric elements are commonly nonspherical and of unequal size, the packing arrangement is very difficult to specify in natural gravel and sand. It does not hold much promise as a directional structure.

KAHN (1959, Table 5) found *packing density* (KAHN, 1956, p. 390—392), the ratio of length of grains intercepted to length of traverse, to vary in different directions on the different faces of a cube-shaped sample. But because the ratio of length of grains intercepted to length of traverse is the compliment of porosity, it is a scalar and not a directional property. R. F. MAST has suggested that number of grains per unit length of traverse, measured in $30°$ intervals in the thin section, may have potential directional significance. This suggestion assumes the thin section to have a relatively homogeneous size distribution of particles in the section.

Disks and disk-like particles, such as mica flakes, pebbles of slate and flagstones, commonly display markedly preferred orientation. Orientation of a disk is specified by the dip and strike of its planar surface. This plane may be represented on a Schmidt net by plotting the projection of the pole perpendicular to it. The lower hemisphere is always used.

Disk-like particles may either receive their primary orientation by gravity or by a combination of gravity and current action. Figure 3-2 shows idealized relations of disks oriented by gravity alone and by disks oriented by a combination of gravity and current forces. In each diagram, the surface of accumulation is parallel to the plane of projection.

Gravity alone produces axial or orthorhombic symmetry, with three mutually perpendicular symmetry planes and three twofold axes. The more variable the orientation of the disks, the less they approach being coplanar, and therefore the less pronounced the maximum on the fabric diagram. Because gravity is the only force orienting the disks, only S_p is defined and a unique vertical plane of symmetry does not exist. One can identify only the ab plane. Orientation of disk-like particles by gravity alone corresponds to sedimentation in a stagnant medium as may occur in sedimentary basins with negligible current action on the bottom. Many muds and shales, deposited largely from suspension are good examples of such deposition. The arrangement of detrital micaceous minerals of such muds should possess orthorhombic symmetry.

The combination of gravity plus current action results in *imbrication* and shift of the maximum concentration in the down-current direction, away from the center of the diagram (fig. 3-2). This reduces the symmetry to a single symmetry

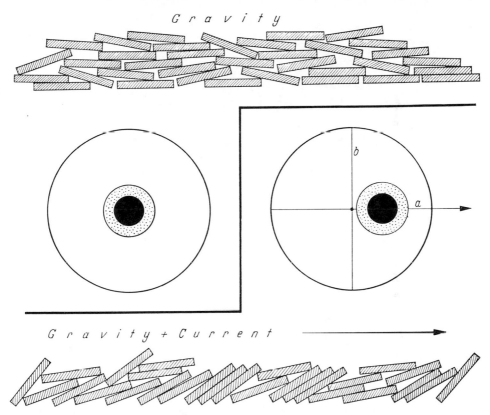

Fig. 3-2. Orientation of disks by gravity (orthorhombic symmetry) and by gravity plus current action (monoclinic symmetry)

plane *ac* and one two-fold symmetry axis, *b*. The monoclinic symmetry of figure 3-2 is characteristic of current deposited fabrics of sediments whether the medium is air, water or ice and is an example of what SANDER (1930, p. 57) has called tangential transport — transport that has appreciable horizontal as well as vertical components.

Comparison with structural petrology is instructive. Structural petrologists generally consider undeformed sediments as having orthorhombic symmetry which, with deformation or tectonic transport, is transformed into monoclinic symmetry. Although both the scale and transport mechanism are quite different, there is an exact analogy with the symmetry resulting from gravity sedimentation alone and the symmetry resulting from gravity plus currents.

An example of orientation of disks in a gravity plus current field is the strong imbrication of slabs of Ordovician limestone in a creek bed (pl. 1 A). Its corresponding fabric diagram with monoclinic symmetry, obtained by plotting the poles of the slabs, is shown in Plate 1 B. Many other primary directional structures, such as cross-bedding and flute marks, also have monoclinic symmetry.

If the plane of the Schmidt net and the surface of accumulation are not parallel, the symmetry plane ac may not pass through the center of the net and the monoclinic symmetry of the diagram may not be so readily seen.

Gravity

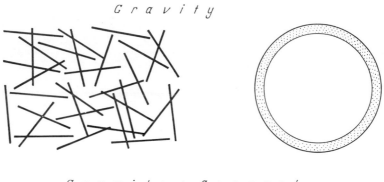

Gravity + Current

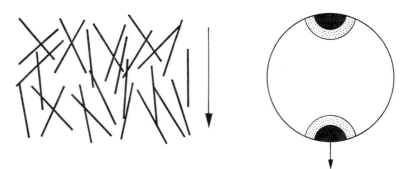

Fig. 3-3. Orientation of rods by gravity and by gravity plus current action (without imbrication)

Fabrics with triclinic symmetry, with no symmetry planes, are produced by unidirectional flow with asymmetrical boundary conditions. Such flow may occur around or behind an obstacle at the sediment-water interface. Disturbance by organisms can also produce such symmetry. Convolute bedding (Chapter 6) also provides an example.

Rod and rod-like particles also show contrasts in orientation when they accumulate in a gravity field or in a gravity and current force field. The orientation of a rod is represented on a Schmidt net by the projection of the intersection of its long axis with the lower hemisphere.

Figure 3-3 shows examples of rods oriented in both a gravity field and in a combined gravity and current force field. A gravity field produces a girdle with orthorhombic symmetry. Gravity plus currents produces orthorhombic symmetry, if the rods are not imbricated, and monoclinic symmetry, if they are. Asymmetrical boundary conditions will also lead to triclinic symmetry.

Symmetry of the Deposit and the Transporting Medium

SANDER, in his interpretation of the fabric of tectonites, drew heavily on analogies with flow in metals and in water. He developed the idea that the symmetry of the deposit reflects the symmetry of the depositing medium, or, as he stated it (SANDER, 1930, p. 280),

"... daß ein monoklines Vektorensystem eines Vorgangs, ..., nur ein monoklines Gebilde schaffen kann, sofern seine Vektoren das Gefüge überhaupt beeinflussen."

This idea has proved useful in the interpretation of tectonites where symmetry has emerged as the basic means of correlating fabric with movement (TURNER, 1957, p. 2; PATTERSON and WEISS, 1961, p. 844—845).

SANDER's idea has direct application to the interpretation of water-deposited primary directional structures because they, unlike tectonites, can be directly related to the flow of the depositing current. In unidirectional flow, both linear (sand grains, plant-fragments, etc.) and planar (disk-like pebbles, etc.) fabric elements and cross-bedding have monoclinic symmetry. If flow in a stream is represented by a vector field and plotted on a Schmidt net, the plot would have a monoclinic symmetry. Different forms of bed roughness and their associated turbulent flow and secondary currents would increase the overall dispersion of the diagram but would not basically alter its monoclinic symmetry.

In actual practice, symmetry concepts have been chiefly used in the study of primary sedimentary fabrics to infer the ac symmetry plane of the depositing current. They have also been applied to sedimentary structures such as cross-bedding and ripple mark for the same purpose. Beyond this, however, the application of symmetry concepts to sedimentary rocks has yielded comparatively little insight to either the nature or the mechanics of the depositing currents.

Till and Gravel Fabric

Method of Study

The fabric of tills and gravels can be readily measured principally because of the coarseness of these materials. Following the terminology of KALTERHERBERG (1956, p. 32), the longest axis of a pebble is designated A, the intermediate one B, and the shortest C. The maximum projection plane of the particle is the AB plane. Capital letters distinguish particle axes from the coordinate axes of SANDER designated by lower case italic letters.

BURACHEK (1933), using very simple equipment, was apparently the first to publish a method of specifying pebble orientation. KRUMBEIN and PETTIJOHN (1938, p. 268—274), modifying a procedure of WADELL (1936, p. 76—80), used a frame to mark pebbles in the field so that its orientation could be reconstructed and precisely determined in the laboratory. A modification, by KARLSTROM (1952, fig. 1), of this frame is shown in figure 3-4.

A vertical undisturbed face is selected and its azimuth is recorded. An adequate number of pebbles is systematically selected from a portion of the exposed face. Usually 100—200 are sufficient. The frame is held parallel to the face and perpendicular lines, forming an upright L, are drawn on the pebbles and extended around their corners and curved surfaces. In the laboratory the pebbles are placed on a two circle contact goniometer. The pebble has its original orientation when it lies in the azimuth of the face and when the two perpendicular lines again form an upright L. Orientation of an axis of the AB plane is determined by direct goniometer measurement (KRUMBEIN and PETTIJOHN, 1938, p. 268—274, and KARLSTROM, 1952, p. 491—492). SARKISIAN and KLIMOVA (1955, p. 45—47) and RUKHIN (1958, p. 418) describe a more complex goniometer. SCHMALL and BENNETT (1961) describe an axiometer—a mechanical device for locating and measuring pebble and cobble axes for studies of macrofabric. This device is

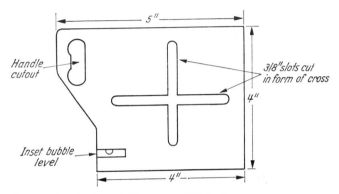

Fig. 3-4. Orientation template (KARLSTROM, 1952, fig. 1)

intended to supplement the contact goniometer. UNRUG (1957, pl. 42) shows a variant of KARLSTROM's method. Frame-goniometer methods have been used for both tills and gravel deposits. A photoelectric method for determining maximum projection plane of a pebble was described by KRUMBEIN (1939).

Because the orientation of each fragment must be reconstructed in the laboratory, the above methods are slow. SCHLEE (1957a, p. 163—166) devised a modification, using a reorienting board, that eliminated the need for laboratory study. He found his method saved one-half to one-third of the time required for the frame-goniometer method. Rod-like and disk-like pebbles on the face of the outcrop were embedded in plasticene, placed on the reorienting board, and both azimuth and inclination measured with a Brunton compass. KARLSTROM (1952, p. 492—493) also describes a procedure, requiring only simple equipment, that permits measuring both inclination and orientation in the field.

In tills, both A-axis orientation and inclination of elongate pebbles can usually be measured in the field. HOLMES (1941, p. 1307—1308) drove a horizontal pole, with north-south orientation, into a vertical face and used a protractor to measure azimuths and inclinations of elongate fragments relative to it.

KALTERHERBERG (1956, p. 33) describes a special "pebble compass" that permits the orientation of either the AB plane or any axis to be measured in the field, if the fragment is of sufficient size. Such a device is very time-saving and should be used whenever possible.

Working in clay-rich tills, HARRISON (1957a, p. 98—100) cut a small oriented block, usually less than 4 inches or 100 mm on an edge, from undisturbed till and reoriented the block in the laboratory on a two-circle contact goniometer. Particles were measured directly, after progressively removing the matrix of the till with a stylus. Similar techniques can also be used with glacial ice (HARRISON, 1957a, p. 98—100).

If only the azimuth of an elongate pebble is desired, a compass can be used or a device such as a till-fabric rack (MACCLINTOCK, 1959) can be employed. DREIMANIS (1959) has described a simple transparent plaque that speeds determination of A-axis orientation. The plaque is laid over a horizontal till or tillite surface and orientation is measured with a compass. Most till fabric investigators have studied only long-axis orientation. Normally this is all that is necessary.

Microfabrics of till, the fabrics of the sand- and silt-sized particles in the till matrix, have also been studied (SEIFERT, 1954; SITLER and CHAPMAN, 1955; HARRISON, 1957b, fig. 8; and GRANEVOR and MENELEY, 1958, p. 722—726). Microfabric analysis requires oriented thin sections cut from impregnated till samples. The method of fabric measurement is essentially similar to that used in the study of sand-sized particles.

The methods so far described are applicable primarily to unconsolidated gravels and tills. In conglomerates and tillites, fragments generally cannot be removed and orientation of either the *apparent* longest axis, A', or the apparent maximum projection plane $A'B'$ only can be determined. Unless unusual exposures exist, however, only the general direction of imbrication or the orientation of apparent long axes, A', can be determined and only a compass is needed. The pebble compass of KALTERHERBERG (1956, fig. 1) may be useful for some conglomerates. Measurement is usually more difficult on a vertical face than on a bedding plane, S_p, the principal surface of accumulation. Care should be taken to observe and record S_p when studying imbrication of conglomerates. Failure to recognize the principal surface of accumulation may lead to misinterpretation of the true direction of imbrication.

In all fabric studies, the effect of later rotation of bedding, if present, should be considered (Chapter 10).

The number of pebbles measured per sample has varied widely. Many investigators have used 100 pebbles per sample, others as many as 400 or more. RICHTER (1936a, p. 204) used at least 100 pebbles per sample. HARRISON (1957b, p. 279) has suggested that as few as 75 pebbles at till outcrops might be sufficient to define a preferred direction. Even a few measurements at an outcrop can be useful, especially, if no other directional property can be measured.

Size of fragments utilized has also varied widely. Without a binocular microscope, the orientation of particles as small as 3 mm has been determined in the laboratory. The only upper limit to size is the requirement that all or a large part of the boulder can be observed. Using a microscope, the lower size limit has commonly been fine sand or coarse silt.

An irregularly shaped pebble has a more complex relation to the transporting medium than either a rod or disk-like fragment. In addition, the orientation of irregularly shaped particles is difficult to specify. Hence some investigators utilize only rod-like and disk-like particles. Use of all particles tends to increase variability and, in extreme cases, to obscure the current direction. A fabric of selected shapes, good as it may be for inferring current direction, is only a portion of the fabric of the entire coarse fraction, however.

RICHTER (1936a, fig. 3) appears to have been the first to plot gravel and till fabrics on a Schmidt net. Although the Schmidt net is now almost universally used, some early workers plotted their results on polar coordinate paper.

If desired, the orientation of any two axes, or an axis and pole to the $A B$ plane, may be plotted on the same net (fig. 3-5-1). Although the Schmidt net is best for gaining an insight into symmetry relations and for exploring certain processes, calculation of a vector mean from a frequency distribution of azimuths is more rapid and more suitable for systematic mapping. Current-rose diagrams (Chapter 10) may also be appropriate for some studies.

That till and gravel fabrics can be objectively determined is demonstrated by their good correlation with other directional properties and by some reproducibility experiments. HYYPPÄ (1948, p. 110—111) and KAURANNE (1960, fig. 2) show good reproducibility for tills. SCHLEE (1957a, p. 163—166) discusses the comparison of two methods of fabric analysis for gravels.

Outcrops for till-fabric studies should, of course, be sampled well below the frost line. Moreover, MOSS and RITTER (1962, p. 102) found that, in areas of glaciated, mature topography, down-slope creep may affect the fabric of hillside exposures. They also noted that valley bottom exposures may have a different movement pattern than hilltop exposures.

Fabric of Tills

Hugh MILLER (1884, p. 175), the younger, appears to have been the first to observe the preferred orientation of pebbles and cobbles in till. Shortly thereafter, BELL (1888, p. 341) noted the preferred orientation of fragments in glacial ice. It was not until almost half a century later, however, that RICHTER (1932, p. 62—65) used till fabrics to infer direction of ice movement in former ice sheets in Germany.

RICHTER'S (1932, 1933 and 1936a, 1936b) studies were the first quantitative ones, and appear to have been inspired by the work of SANDER. Subsequently, many studies of till fabric have been published. In North America KRUMBEIN (1939, p. 682—685), HOLMES (1941), SITLER and CHAPMAN (1955), WRIGHT (1957; 1962, p. 89—93), HARRISON (1957b), GRAVENOR and MENELEY (1958, p. 722—726), MACCLINTOCK (1958, fig. 2, 3 and 4), MACCLINTOCK and TERASAMAE (1960), FLINT (1961, figs. 3 and 5), and MOSS and RITTER (1962, p. 101—103) have studied till fabrics; PETTIJOHN (1962) recorded the fabric of a Precambrian tillite. The studies of HOLMES and HARRISON are the most comprehensive and are entirely devoted to till fabrics. In Europe till-fabric studies have been somewhat more numerous. Following RICHTER'S studies, DEWARD (1945), KIVEKÄS (1946, cited in HYYPPÄ, 1948, p. 110), MOLDER (1948, fig. 1), LUNDQVIST (1948; 1949, p. 335—347; 1951, p. 77—81), HYYPPÄ (1948, p. 109—111), VIRKKALA (1951, p. 22—23, 1960, p. 170—172; 1961, fig. 2), HOPPE (1952, p. 19—24), JÄRNEFORS (1952, p. 192—194), SEIFERT (1954), DONNER and WEST (1955), WEST and DONNER (1956), GLEN, et al. (1957), KAURANNE (1960), LEE (1959),

DUMANOWSKI (1960, p. 320—324), GILLBERG (1961, fig. 6), GROTH (1961, figs. 5, 6, and 10), HANSEN, *et al.* (1961, p. 1417) and SCHULZ (1961, fig. 4) have published data on till fabrics. Most of these studies have specified the *A*-axis orientation of elongate particles.

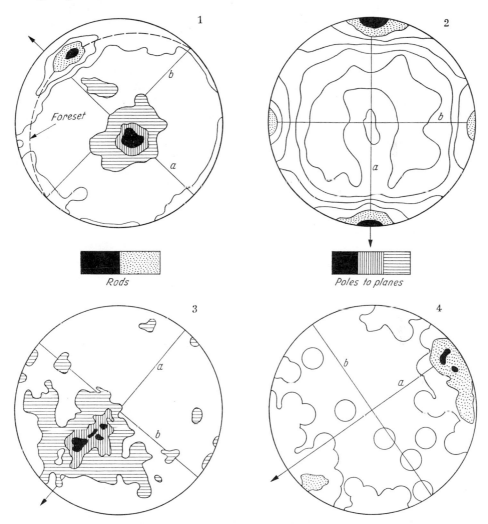

Fig. 3-5. Diagram of gravel and till fabrics: 1) Long axes of pebbles on a foreset bed and poles to their maximum projection plane (modified from KALTERHERBERG, 1956, fig. 3a), 2) Long axes of till stones (modified from HOLMES, 1941, fig. 5), 3) Poles to platy particles in till (modified from HARRISON, 1957b, pl. 2, 9a, and, 4) Long axes of pebbles in fluvial gravel (modified from SCHLEE, 1957a, fig. 3)

Rod-like particles in tills of ground moraine have a relatively simple relation to direction of glacial flow. This is demonstrated by comparison of fabric pattern with moraine configuration, glacial striations, roches moutonnées, chatter marks and other directional features of glacial origin. Rod-like particles lie chiefly subhorizontally in till and have their principal long-axis concentration parallel to *a* and a lesser, secondary concentration parallel to *b*. GILLBERG (1961, fig. 6) shows a well-defined secondary concentration parallel to b in Swedish moraines. The secondary mode develops only under certain conditions according to GLEN, el al. (1957, p. 202). RICHTER (1932, p. 64—65) observed striations on elongate tillstones parallel to their long axes and concluded that such elongate stones lie parallel to the direction of ice movement. Earlier HUGH MILLER (1850, p. 93), the elder, had also commented on these lengthwise striations*. Subsequently, RICHTER (1933, fig. 4; 1936a, fig. 1), HOLMES (1941, fig. 5b), LUNDQVIST (1948, fig. 7), and VIRKKALA (1960, fig. 2; 1961, fig. 2) working with tills, confirmed this conclusion (figs. 3-5, 2 and 3). Modern glaciers also show the A-axes of elongate particles to be dominantly parallel to ice movement (BELL, 1888, p. 341; RICHTER, 1936b, p. 25, OKKO, 1955, p. 44; HOPPE, 1953).

HARRISON (1957b, p. 279—288) has shown that both rod- and blade-like particles in clay-rich tills, lie parallel to flow direction and are imbricated 20—25° up-current (fig. 3-11). WRIGHT (1962, Table 3) also found comparable angles of inclination for rod-like particles in till. Similar relations of sand- and silt-sized particles in the microfabrics show that both coarse and fine till fabrics are produced by a common orienting force (HARRISON, 1957b, p. 285).

The diagrams of figure 3-5 have monoclinic symmetry (except figure 3-5-2).

The fabrics of end-moraines may have more complicated relations to direction of ice movement because of more complex processes of mudflow, sliding, and pushing that occur in end-moraine construction (cf. HARRISON, 1957b, figs. 3 and 5), although consistent relations have been reported from lodgement moraines (GILLBERG, 1961, fig. 6). LUNDEGÅRDH and LUNDQVIST (1959, fig. 35) also show that relations similar to those of ground moraine can prevail.

Mudflow and solifluction deposits have fabrics similar to those of ground moraine (LUNDQVIST, 1949, 1951, p. 77—81; HARRISON, 1957b, p. 284).

Some joints in till appear to be related to direction of ice movement (RICHTER, 1933, p. 38—42). But because till jointing unrelated to ice flow is also known, the recognition of jointing due to movement is difficult.

Till fabric has been mapped to infer regional pattern of ice movement. RICHTER (1936a, fig. 1) mapped pebble orientation in the calcareous ground moraine over a large area of northern Germany. He had many sample points. His map shows a rather large variability of orientation comparable to that shown by cross-bedding. Other good examples of fabric mapping come from Finland and England. Figure 3-6 is a map of till fabric in the Tampere region of southern Finland. This map shows the similarity of ice movement indicated by both till fabrics and by bedrock striations and position of the stoss sides of bedrock hillocks. It also demonstrates the low variability or variance in direction of ice movements. Moss and RITTER (1962, fig. 7) also show good agreement between bedrock striations and till fabric. In southeastern England, DONNER and WEST (1956, figs. 6 and 7) studied fabrics in three ice advances over an area of 120 by 100 miles (192 by 160 km). Their maps display a low variance and correspond, in both number of samples and size of area, to many regional studies of cross-bedding.

---

* Presumably the relative proportions of elongate particles parallel to a and b could easily be estimated by observing the proportions of particles that have striations parallel or perpendicular to A.

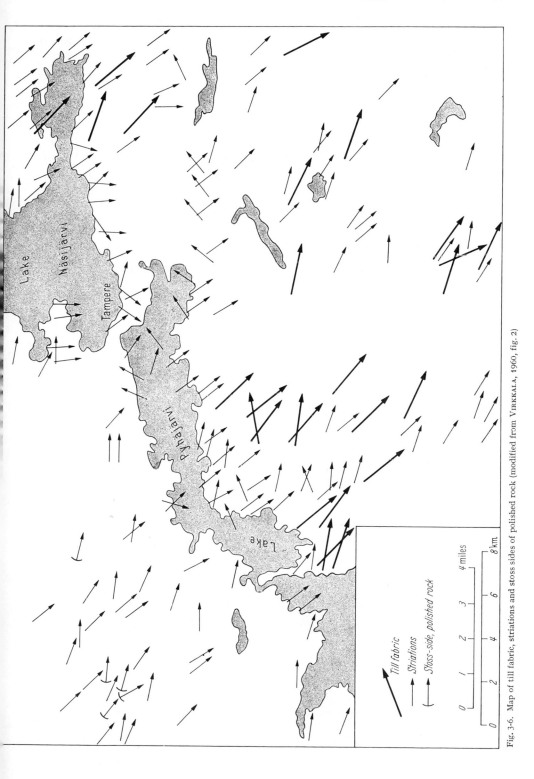

Fig. 3-6. Map of till fabric, striations and stoss sides of polished rock (modified from Virkkala, 1960, fig. 2)

SIEFERT'S (1954, figs. 5, 6, 7 and 8) maps of the microscopic orientation of sand-sized till particles show very consistent directions in a 25 by 35 km area in northern Germany.

Till fabrics have played a supplementary but useful role in prospecting for possible bedrock mineral deposits in some glaciated areas (Chapter 8).

Fabric of Gravels

Although earlier literature contains some references to imbrication, it was not until 1933 that RICHTER *measured* the fabric of glacial gravels. There have been fewer studies of gravel fabric than of till fabric, probably because other primary directional properties, such as cross-bedding, are usually present. Relations of particle orientation to flow direction also appear to be more complicated in gravels than in tills.

In western Europe FORCHE (1935, p. 51—52), RICHTER (1936a, p. 202—206; 1936b, p. 25—26), CAILLEUX (1938 and 1945), PICARD (1950, fig. 5), WIESER (1954, p. 354—359), KALTERHERBERG (1956), UNRUG (1956 and 1957), GRADZINSKI, (1957, fig. 8), GRADZINSKI and UNRUG (1959, p. 185—190), WRIGHT (1959, p. 611—613), SCHIEMENZ (1960, p. 13—15), KOSZARSKI (1956, p. 316—397), and KÜRSTEN (1960) have published studies of gravel orientation. The Russian literature is summarized by SARKISIAN and KLIMOVA (1955), PIDOPLICHKO (1956), STRAKHOV, *et al.* (1957, p. 85—88) and RUKHIN (1958, p. 413—421). In North America, WADELL (1936), KRUMBEIN (1939, p. 698—705; 1940, p. 668—669; 1942, p. 1386—1391), WHITE (1952), LANE and CARLSON (1954, p. 459) and SCHLEE (1957a, 1957b, p. 1377) have studied gravel fabrics.

Observation of modern streams indicates that the overwhelming majority of disk-like and ellipsoidal particles have their maximum projection plane dipping up-current (FRASER, 1935, p. 985; KRUMBEIN, 1940, fig. 8; LANE and CARLSON, 1954, p. 458—459). Plate 1A shows an unusually good example. The average angle of inclination of the limestone slabs shown in this example is 25°. Imbrication angles generally vary between 10 and 30°. UNRUG (1957, fig. 21) found that they decreased downstream over a distance of 200 km.

Imbrication in the down-current direction of modern streams has been reported (KRUMBEIN, 1940, p. 653), but it is exceedingly rare. KALTERHERBERG (1956, figs. 8b and 8c) and RUKHIN (1958, fig. 160) report anomalous imbrication in some terrace and ancient gravels. KALTERHERBERG (1956, p. 53—55) thought such imbrication depends, in part, upon packing density. High packing density has upstream imbrication and low packing density downstream imbrication. However, apparent downstream imbrication may also result from failure to recognize cross-bedding. If inclination of the AB plane is measured from the horizontal rather than a foreset plane, S_f, apparent down-current imbrication might be found. Fragments are always imbricated with respect to a surface of accumulation, which may not be the principal surface of accumulation, S_p. In coarse clastics, cross-bedding is sometimes difficult to recognize and may have very large dimensions. Figure 3-5-1 shows an example of pebbles apparently imbricated down-current because they lie on a foreset.

In beach gravels the available information indicates dominant inclination away from the land (RUKHIN, 1958, p. 416) and a tendency for elongate pebbles to be parallel to the strand line (FRASER, 1935, Table 8; RUKHIN, 1958, fig. 161).

If care is taken to distinguish between S_p and S_f, imbrication is an excellent guide to direction of current flow in the vast majority of gravel deposits. Hence, finding the face of an exposure that exhibits the maximum angle of imbrication is a quick and easy way to estimate qualitatively direction of current flow in gravels and conglomerates devoid of cross-beds.

The long axes of rod-like and ellipsoidal particles in beach gravel are believed to generally be parallel to the strand line b (FRASER, 1935, p. 978; RUKHIN, 1958, p. 416). In streams, relations between long axis and direction of flow are variable, as RICHTER (1936b, p. 25—26) noted in the ice-contact and outwash gravel of modern glaciers in Norway.

Review of the literature confirms RICHTER's early observations: long axes may either lie parallel or perpendicular to stream direction. FORCHE (1935, p. 50—51) found long axes of pebbles approximately perpendicular to direction of current as defined by cross-bedding in the Triassic Buntsandstein. KRUMBEIN (1940, p. 653) found A-axes parallel to current direction in flood gravels. PICARD (1950, fig. 5), LANE and CARLSON (1954, p. 459), UNRUG (1957, p. 255—256), KÜRSTEN (1960) and UNGER and ZIEGENHARDT (1961, p. 470—471) report A-axes perpendicular to a. KALTERHERBERG (1956), RUKHIN (1958, p. 414—415) and SCHIEMENZ (1960, fig. 1a) show A-axes both perpendicular and parallel to a. WRIGHT (1959, p. 611—613) found, on foreset beds in gravel, that long axes of pebbles had variable relations to maximum dip direction. UNRUG (1957, p. 255—256) suggested that size is an important factor because the largest ellipsoidal pebbles roll along the bottom with A perpendicular to a. He thought that many of the smaller pebbles show less distinct orientation, inasmuch as they fill the void space between the larger ones. KALTERHERBERG (1956, p. 51—53) has emphasized shape control: the more rod-like a particle the more likely will A be parallel to b except on foresets where it will tend to be parallel to a. Perhaps packing density, sorting, and stream gradient, in addition to shape and size, all affect long axis orientation. More study of the contrasts in orientation of modern stream and beach gravels, as was done by CAILLEUX (1938), would also be useful. The fabric of gravels deserves more study than it has received.

The fabric of talus and avalanche boulder tongues has received some study, principally by RAPP (1959, fig. 9; 1960a, fig. 63; 1960b, pl. 9), who found A axes of oblong boulders to be parallel to direction of slope in talus and to direction of mass movement in avalanche tongues.

There have been only a few areal studies of gravel fabrics as it is generally easier to measure cross-bedding in gravels and conglomerates. Some, however, have been made. REINCKE (1928, pl. 4) early made a map of pebble orientation in ore bodies of the Witwatersrand conglomerates of South Africa and APRODOV (1949, cited by ZHEMCHUZNIKOV, 1954, fig. 5) shows the relations between pebble orientation and channels in a Permian delta. UNRUG and ZIEGENHARDT (1961, fig. 1) related gravel orientation to channel pattern of glacial outwash. Imbrication has been used as a guide to current direction in some conglomerates of economic interest (Chapter 7).

Fossil Orientation

In 1843 JAMES HALL (p. 52—54 and pl. 2), carefully described orientation of single brachiopod valves in the Medina formation of New York and related their orientation to current structures. Over half a century later, RUEDEMANN (1897) published the first map of fossil orientation. Since that time, the study of primary fossil orientation has been largely a novelty that only occasionally has attracted the attention of paleontologists, stratigraphers and sedimentologists, perhaps because fossils exhibiting preferred orientation are neither as common nor as simple to interpret as are many directional sedimentary structures. Actually, the study of deformed fossils in metamorphic rocks has probably received more study than has primary fossil orientation.

Methods of measurement of fossil orientation differ little, if at all, from those used in other directional studies. Field observation with a compass is usually adequate.

Specification of fossil orientation presents a few problems in elongate forms. If no distinction is made between the ends of the elongate form, it specifies the line of movement only. Such is the case with wood charcoal fragments, crinoid stems, graptolites, spindle-shaped fusulinids and others. Elongate forms that have clearly differentiated ends, such as orthoceracone cephalopods, *Tentaculites* and *Turritella*, probably indicate a direction of movement, the pointed end usually thought to point up-current into the direction of flow.

In more equant and complex forms, specification of orientation is more difficult. For example, the orientation of molluscs and brachiopods may be specified by the position of an appropriate plane of symmetry of the fossil. Figure 3-7 shows examples of fossil orientation on a slab from the Röt formation (Triassic) of Germany. The more complex the form, the more difficult it usually is to specify an orientation and in many cases none at all can be designated. In some cases, the average direction of many fossil fragments rather than the orientation of individual fossils is stated.

Most of the studies of invertebrate fossil orientation have been restricted to very local areas — usually only one or two outcrops and some cases even to single slabs! Some of this literature is cited.

HALL (1843, p. 52—54) commented on and illustrated fossil orientation. MATTHEW (1903, p. 52—63) studied orientation of brachiopods. KLÄHN (1929) estimated current direction and strength from position of sea lilies and sea stars. KING (1948, figs. 6, 8, and 10) studied fusulinid orientation in the Guadaloupe (Permian) deposits of west Texas. Among others are BULACH (1951) who studied belemnite orientation, KAY (1945, fig. 1), PETRANEK and KOMARKOVA (1953) and KRINSLEY (1960), who recorded orientation of orthoceracone cephalopods, CHENOWETH (1952, p. 556—559) who measured orientation of both gastropods and cephalopods, and SCHWARZACHER (1961, fig. 1) who measured crinoid stems. SEILACHER (1959 and 1960) has described the orientation in a variety of fossil assemblages and has advanced some principles of interpretation.

The interpretation of fossil orientation is generally difficult. Is the fossil in a growth position or was it moved as a detrital particle? Living organisms can be responsive to current direction (rheotaxis) because currents bring nourishment, remove waste products and in some cases may be strong enough to produce marked symmetry in some sessile forms. On the other hand, fossil debris that has been transported by currents presumably attains a hydrodynamically stable rest position.

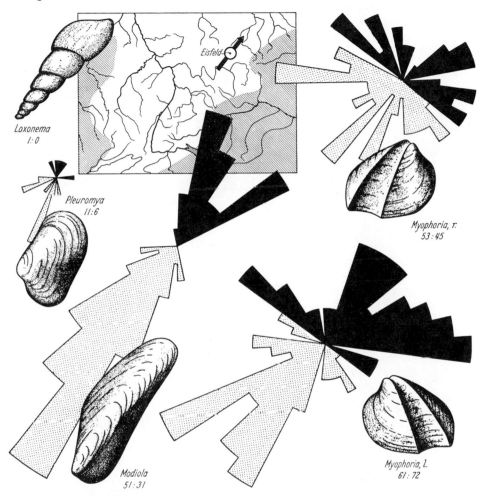

Fig. 3-7. Fossil orientation in the Röt formation (Triassic) near Eisfeld, Germany. Numbers refer to proportions of shells that point into and away from the current (SEILACHER, 1960, fig. 8)

Bivalve shells provide a useful criterion to distinguish between traction and suspension transport. In traction transport, bivalve shells dominantly lie with their convex side up. In suspension transport, as occurs in turbidity currents, bivalve shells tend to be deposited with their convex side down. In the absence of currents, such shells lie in either position and no preferential arrangement is present.

Relation of particle orientation to current direction principally depends upon shape and uniformity of density of the particle. According to SEILACHER (1960, p. 59), many oriented fossils tend to be either parallel or perpendicular to current direction. He also suggested that the equally developed opposing modes of a current rose represent orientation of detrital shells perpendicular to current direction whereas unequally developed, opposing modes, have their larger mode pointing up-current. Even the orientation of simple forms such as orthoceracone cephalopods or *Turritella* is of uncertain significance. The less symmetrical the shape, the more difficult the interpretation. Displaced centers of gravity due to uneven filling prior to burial can also affect orientation.

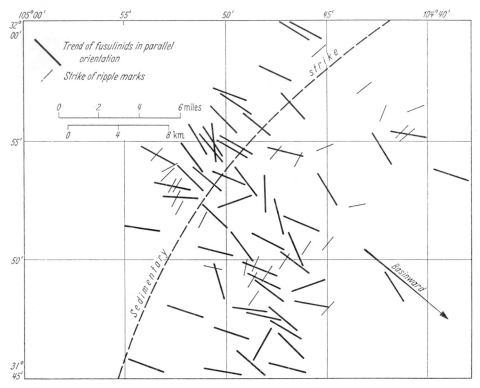

Fig. 3-8. Orientation of fusilinids, ripple mark and sedimentary strike in the Guadaloupe series (Permian) of west Texas, U.S.A. (modified from King, 1948, figs. 6, 8 and 10)

Even as homogeneous and as simply shaped particles as wood and charcoal fragments present interpretative problems. Orientation both perpendicular (Pelletier, 1957, p. 1051—1052; Cloos, 1938, fig. 1) and parallel (Crowell, 1955, p. 1361; Colton and DeWitt, 1959, p. 1759; McIver, 1961, p. 149—155) to current direction, as inferred from other sedimentary structures, has been observed. Nor did Sullwold (1960, p. 447) find consistent relationships. Gradazinski et al. (1959, fig. 6) also show poor correlation between plant debris and current direction. Size may well play an important role. Fine debris can accumulate in troughs of ripple marks where small turbulent eddies may be the controlling factor in the transverse orientation of the wood fragments.

An interesting but rather special case is that of Pleistocene forest beds that have been overridden by advancing glacial ice. A well-known example, in which the trees are oriented and record the direction of ice movement, occurs in the Two

Creeks forest bed in Wisconsin (WILSON, 1932, p. 36). The orientation of the smaller englacial wood debris associated with a forest bed should provide a sensitive indicator of glacial movement.

Obviously, inferences about currents from oriented fossils are of variable reliability. Associated directional sedimentary structures, if present, should be used to help interpret fossil orientation.

A good example of the contribution of oriented fossils to basin analysis, however, is provided by KING (1948, p. 50—52 and p. 83—84). In the marine Permian Guadaloupe series of west Texas, on the east flank of the Capitan reef, KING used both ripple mark and fusulinid orientation to supplement stratigraphic evidence in interpreting the sedimentation. He found (fig. 3-8) that ripple mark is parallel to sedimentary strike and the fusulinids are perpendicular to it. He attributed these relations to currents moving generally perpendicular to a strand line that lay to the northwest.

Oriented fossils can thus provide helpful supplementary information in a regional study of directional structures and in basin analysis as well. Fossil orientation should be sought for and recorded. The maps of RUEDEMANN (1897) and KING (1948) clearly demonstrate the utility of the systematic mapping of fossil orientation. The distribution and orientation of individual organisms and even communities may be closely related to current system and hence knowledge of these systems has potential ecologic value.

Fabric of Sand and Sandstone

Methods of Study

The fabric displayed by sand-sized particles has received less study than that of the larger fragments of tills and gravels, largely because the orientation of the smaller size is more difficult to measure. Although many investigators have noted quartz orientation in sandstone, only recently have satisfactory methods of measurement been developed. However, the effort expended on developing methods of measurement greatly exceeds the geologic results so far published.

Sand fabrics can be analyzed either by *particulate* or by *aggregate* methods. Particulate methods attempt to determine the fabric by measurement of the orientation of individual grains. Aggregate methods involve measurement of a bulk or aggregate property that is controlled by and integrates over the orientation of many grains. Most methods of sand fabric measurement have been particulate ones but in recent years some aggregate methods have been developed.

The earliest studies, following the methods developed by SANDER, specified orientation of the crystallographic c axes of calcite and dolomite (SANDER, 1936) and quartz (INGERSON, 1940). Fabrics were measured on a three-axis universal stage and plotted on a Schmidt equal area net. Others who studied c-axis orientation include ROWLAND (1946), INGERSON and RAMISCH (1954), HUCKENHOLZ (1959) and IVANOV (1959). Although the method is well established and straightforward, it is not entirely satisfactory for quartz grains because the crystallographic c axis need not coincide with A, the longest shape axis. Figure 3-9 shows the fabric diagrams of the same sample — one of A-axes and the other of the crystallographic c axes. The A-axes have the best correlation with current direction. The discrepancy between the two diagrams stems from the fact that quartz is principally elongate both parallel to its c-axis as well as to the rhombohedral $\{10\bar{1}1\}$ zone faces (INGERSON and RAMISCH, 1942, fig. 6-1; BLOSS, 1957, p. 221). Others who have investigated relations between dimensional and crystallographic orientation include WAYLAND (1939), SCHUMANN (1942), ROWLAND (1946), VOLLBRECHT (1953a) and BONHAM (1957, p. 258). Because orientation of detrital quartz is dependent on shape rather than crystallography, study of crystallographic orientation is less valuable, even though shape orientation is usually more difficult to determine. Hence most sand fabric analyses have been based on shape rather than on crystallographic directions.

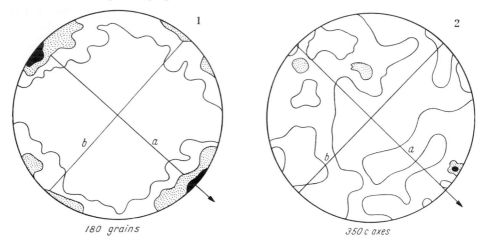

Fig. 3-9. Dimensional (1) and crystallographic orientation (2) in the same sample of a beach sand from the Baltic Sea (modified from WENDLER, 1956, figs. 4 and 5)

A few investigators have used a binocular microscope and a modified (SCHWARZ-ACHER, 1951, p. 162—164) or unmodified (RUSNAK, 1957b, p. 400) multiaxis stage to determine the three-dimensional orientation of long axes. Although only a three-dimensional investigation can display the symmetry of sand fabrics on a Schmidt net, most investigators have chosen to make a two-dimensional fabric analysis. Such analysis generally require either a current rose to display the fabric or the calculation of a vector mean. Properly designed, two-dimensional orientation studies can be very rewarding.

Many fabric studies have been based on direct measurement of the apparent long-axes seen in thin sections. Most investigators have made their measurements along traverses under the microscope but some have projected the thin section on a screen and measured grain orientation from a reference line (GRIFFITHS and ROSENFELD, 1951 and 1953; CURRAY, 1956, p. 2441; RUSNAK, 1957a, p. 48—49; MCBRIDE, 1960, p. 164; BOUMA, 1962, p. 84—85; GANGULY, 1960). Others (DAPPLES and ROMINGER, 1945; GRYAZNOVA, 1947, 1949, and 1953; NAIRN, 1958; POTTER and MAST, 1963) have photographed thin sections and measured grain orientation from a reference line on the photograph. Photographs have several advantages. They provide a permanent record and they are usually easier to work with than are thin sections themselves. POTTER and MAST (1963) obtained photographs by using a photographic enlarger to project the thin section on to 8 by 10 inch (20 by 25 cm) photographic paper.

If thin sections are used, one can measure all the available grains in randomly selected fields of view, sample by linear traverse or sample by point count. CURRAY (1956, p. 2441—2442) preferred either fields of view or point counts to linear traverses. An alternative procedure was followed by POTTER and MAST (1963, pl. 1), who placed a 1 inch (25 mm) grid over an 8 by 10 inch (20 by 25 cm) photograph and selected grains from each square.

Specification of the longest apparent axis, A', of a grain requires some care as quartz grains have neither perfect ellipsoidal nor rectangular shapes. DAPPLES and ROMINGER (1945, p. 250—251) recognized three ways of measuring apparent long dimensions of quartz grains. One could measure the azimuth of the apparent longest dimension. An alternative procedure was to utilize the least projection elongation. Least projection elongation is the direction of two parallel lines drawn tangent to the grain image and separated by the least possible distance. Still another way is center of area elongation. Center of area elongation is the azimuth of the longest line that can be drawn through the center of area, the equivalent of the two-dimensional center of mass, of the grain. Although these three directions will in general not coincide, they will usually be very close in most grains. However, a unique direction cannot be specified in some irregular grains and in those that are essentially circular.

In practice these difficulties are only of minor importance, especially, if only the more elongate grains are utilized and if azimuths are grouped into class intervals. Variations of 10° and even 15° in selecting the "long" axis of single grains have comparatively little effect on the mean azimuth when 100 or more grains per sample are measured and grouped into classes. As shown by operator experiments (POTTER and MAST, 1963, Table 3), use of elongate (axial ratio of 1.3 or larger) grains yields a readily reproducible mean azimuth. Another factor favoring elongate grains is their greater sensitivity to current orientation (WALKER, 1955, fig. 9; RUSNAK, 1957b, p. 407). However, just as with coarser particles, the fabric displayed by elongate grains is only a subfabric of the sandstone.

DAPPLES and ROMINGER (1945, p. 251—253) specified the blunter of the two ends of grains. The blunt end of the grain was believed to lie up-current. MILLER (1884, p. 178) had earlier remarked that the blunt end of elongate cobbles and boulders in glacial till pointed "down-current," although he added that this was by no means always so. Asymmetry has also been used in the study of fossil orientation. Quartz grains, however, are less asymmetrical than many fossils. Although GRYAZNOVA (1947, 1949 and 1953), HELMBOLD (1952, p. 283), WALKER (1955, p. 125—126), and NAIRN (1957, p. 62) identified the blunt end of the grain, few others have done so. Neither SCHWARZACHER (1951, p. 164) nor RUSNAK (1957b, Table 3), who studied fabrics deposited in a flume, found little evidence that the blunt end pointed up-current. Moreover inspection of any thin section of sandstone will show the difficulty of defining the blunt end of a quartz grain.

Number of grains counted per sample has varied widely, from as few as 60 to as many as 500 grains. Commonly counts of between 100 to 200 grains have been used and found to give satisfactory results.

Aggregate methods of fabric analysis depend on the anisotropy of a bulk or volume measurement that, in some way, correlates with the grain fabric. Aggregate methods have great appeal because they integrate over a volume of sandstone that may contain 10^6 to 10^8 more grains than a thin section and because they are usually less time-consuming. These methods usually only determine the direction of the ac symmetry plane. Before an aggregate method can be used with confidence, however, it must be shown to correlate with the grain fabric.

MARTINEZ (1958) devised an optical method of measuring the varying intensity of monochromatic light passing through a thin section. The method depends on the statistical correlation between crystallographic c axes and longest shape axes of quartz grains. His data (p. 605) generally showed fair to good agreement with particulate fabrics, although variations in excess of 60° were obtained between the two methods on some samples. A modification of MARTINEZ's photometer method was made by PIERSON (1959).

ARBOGAST et al. (1960) describe a method for determining dielectric anisotropy in sediments such as sandstones and limestones which are relatively poor conductors or *dielectrics*. Dielectric coefficient of a heterogeneous solid depends not only on the dielectric coefficients of the different mineral components but also on their arrangement and shape. Thus, in a sandstone or limestone, the dielectric coefficient should be principally dependent on the fabric of either the grains and/or the pore geometry because both mineralogy and grain shape are essentially constant in any given sample. Hence an anisotropic fabric should give rise to an anisotropic dielectric coefficient.

ARBOGAST et al. (1960) claim that sandstones and limestones have dielectric anisotropy that correlates with fabric anisotropy. Their data show excellent agreement with direction of inhomogeneity in simple solids and good to fair agreement with the particulate shape fabrics of sandstones, although some substantial deviations do occur (ARBOGAST et al. 1960, 7 and 8).

NANZ (1960, 2) suggests the use of a device to measure acoustic anisotropy that is implied to be principally dependent on fabric anisotropy. Such devices facilitate rapid identification of maximum or minimum direction of anisotropy of the sample.

HIGGS et al. (1960, p. 284—287) found good agreement between crystallographic orientation of quartz and x-ray diffraction pattern in the Poughquag quartzite.

ZIMMERLE and BONHAM (1962) describe an electronic spot scanner which in effect counts the number of grain-matrix contacts. Number of grain-matrix contacts varies in different directions and is related to the particulate, dimensional fabric of the sandstone.

MAST and POTTER (1963, pl. 1) found that the ability of a sandstone to imbibe water is an anisotropic property dependent, in part, on grain fabric. Maximum imbibition of water tends to be parallel to the average grain fabric and offers some promise as a simple means of determining fabric anisotropy in the plane of the bedding. The method is more effective with relatively permeable sandstones.

Internal cross-bedding, large or small scale, can cause difficulties, especially if not recognized, because it introduces internal discontinuities. Grains may either be imbricated with respect to S_f or flow may be refracted by internal laminations that are not parallel to the plane of measurement. Thus care should be taken to specify fully the bedding features of the samples used for study of aggregate as well as particulate fabrics.

Another aggregate measure that may be of use is number of grains per unit length of traverse. Measurement can be made in thin sections, oriented in either vertical or horizontal directions, at 30 degree intervals. The data will plot, perhaps as an ellipse whose axes should correspond to maximum and minimum directions of fabric anisotropy. Although this method requires further investigation, it is attractive because measurement is relatively simple. In addition, because it is an aggregate property, estimates based on it will be much more reliable and stable than estimates based on a particulate property such as apparent long axes.

CLEARY (1959, p. 30) suggests that compressibility measurements in the ab plane might be used as a measure of grain orientation.

Unconsolidated sands, as well as sandstones, can be studied by using an impregnating compound. BROWN and PATNODE (1953) describe how, with a thermosetting plastic, unconsolidated samples may be obtained for laboratory study.

Fabrics and Sands and Sandstones

BRUNO SANDER (1936), in his study of the Triassic carbonates in Austria, was apparently the first to publish on the fabric of a sand-sized sedimentary rock. From these beginnings, late in comparison to the study of other directional structures, a widely scattered, but not too voluminous, literature has developed. The greatest portion of this literature is concerned with methods and techniques of measurement. Another portion reports on fabrics, incidental to other sedimentation study. Only a comparatively few papers are primarily devoted to the analysis and understanding of the fabric of sands and sandstones. A modest number of applications have been made. Sands deposited artificially in flumes as well as modern sands and ancient sandstones demonstrate that:

1. the long axes of sand grains, in response to unidirectional flow, tend to parallel the ac transport plane and are imbricated up-current defining a monoclinic symmetry;

2. variability in orientation is greater in sections parallel to the ab plane than in sections perpendicular, or nearly so, to the ab plane;

3. sand shape fabrics appear to have a rational relationship to primary constructional directional sedimentary structures such as cross-bedding and current parting.

SCHWARZACHER (1951), VOLLBRECHT (1953b), and RUSNAK (1957b) found by flume study that long axes of quartz grains are principally aligned parallel to current direction (fig. 3-10). SCHWARZACHER (1951, pl. 1) and VOLLBRECHT (1953b, pls. 2 and 3) also show some diagrams with minor modes of long axes perpendicular to a. Apparent long axes in sandstones, graywackes to clean quartz sands, are parallel to current direction as defined by sedimentary structures such as flute marks, sole marks, cross-bedding, and parting lineation (TEN HAAF, 1959, p. 25; PRYOR, 1960, fig. 19; McBRIDE, 1960, p. 164—180; SMOOR, 1960, fig. 25; McIVER, 1961, p. 140—148; POTTER and MAST, 1963, fig. 8). FRAZIER and OSANIK (1961, p. 135), using the photometer developed by MARTINEZ (1958), also found good agreement between fabric and cross-bedding in a Mississippi River point bar. Beach deposits also demonstrate that the long dimension of a quartz grain tends to be parallel to the direction of the oncoming surf (CURRAY, 1956). An exception is reported by BOUMA (1962, p. 84—85), who found that long axes of quartz grains tend to be at right angles to current direction as defined by sole marks in turbidites.

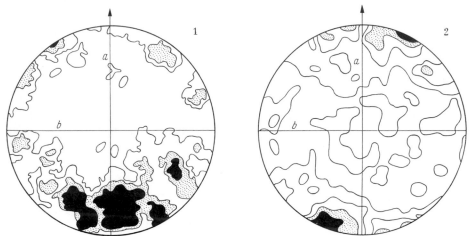

Fig. 3-10. Dimensional fabric of sands deposited in flumes: (1) 200 grains (modified from RUSNAK, 1957b, fig. 9d) and (2) 400 grains (modified from SCHWARZACHER, 1951, pl. 1, D 2)

SCHWARZACHER (1951, p. 165) reports up-current imbrication of grains in the principal modal class, generally at angles of 25 to 30 degrees. RUSNAK (1957b, Table 3) found average grain imbrication to be between 10 and 20 degrees with

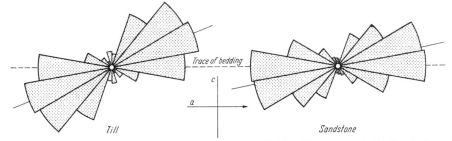

Fig. 3-11. Imbrication of 104 sand sized particles in glacial till (modified from HARRISON, 1957b, fig. 8) and 540 quartz grains in sandstone showing parting lineation (POTTER and MAST, 1963, fig. 7). Solid line is average angle of imbrication

the horizontal. Similar up-current imbrication has been observed in many sandstones and even in the sand-sized components of till (fig. 3-11). POTTER and MAST (1963, Table 5) report that average imbrication values generally lie between 10 and 25 degrees. Schmidt net diagrams also show minor modes imbricated down-current. CURRAY (1956, p. 2447) reports that on beaches grains are imbricated toward the shore, because the ebb flow apparently has more affect on dimensional fabric than the oncoming surf.

Both McBRIDE (1960, fig. 74) and POTTER and MAST (1963, fig. 5) found that grains are also imbricated up-current on foresets. SCHWARZACHER (1951, p. 168) had earlier observed similar up-slope imbrication in experiments by heaping sand underwater. These observations suggest that grains are always imbricated with respect to a surface of accumulation but not necessarily, S_p, the principal surface of accumulation. This may explain some anomalous down-current imbrication that has been reported.

Long-axis orientation and imbrication of sand-sized particles display a monoclinic symmetry, as is well illustrated by the fabric diagrams of figure 3-12. Both diagrams were obtained from thin sections cut parallel to the ab plane. The fabric plotted on the Schmidt net, of thin sections cut parallel to ac, bc and ab, is shown in figure 3-12. This diagram displays the general monoclinic symmetry of fabrics of clastic sedimentary rocks. NAIRN (1958, fig. 5) shows a somewhat comparable diagram. The fabric diagrams would be more complex if the plane of section were cut at some other angle.

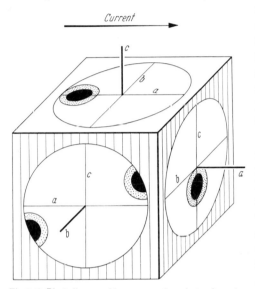

Fig. 3-12. Block diagram of homogeneously, anisotropic sandstone with monoclinic symmetry. (A axes of quartz grains.) The ac plane is the plane of symmetry and b is the two-fold axis (modified from POTTER and MAST, 1963, fig. 15)

The three perpendicular planes of figure 3-12 are planes of maximum fabric contrast. Variance of apparent long axes of all the grains was estimated by POTTER and MAST (1963) to be approximately 1.4 times greater in the ab plane than the ac or bc planes. In other words, shape fabrics specified by apparent long axes are weaker in the ab plane than in either the ac or bc planes. This reflects the fact that constraints on grain orientation are greater in the vertical than in the horizontal. If only elongate grains from foreset beds, rather than the entire fabric, were considered, a typical standard deviation for an ab section was 45° and only 32° for an ac section (POTTER and MAST, 1963, Table 5).

Both SCHWARZACHER (1951, p. 168) and VOLLBRECHT (1953b, p. 271) thought that more rapid deposition produces weaker and more variable sand fabrics. Rapidly flocculated aggregates of clay minerals likewise have less well defined symmetry than better laminated marine clays of presumed slower deposition (VAN STRAATEN, 1954, p. 60).

By relating the grain fabric to maximum dip direction of foresets and the trend of parting lineation (fig. 3-13), it was possible to show that for each structure the average fabric direction is closely parallel to the current direction. MCBRIDE (1960, fig. 77) and MCIVER (1961, p. 140—148) also found good correlation between fabric anisotropy in the ab plane and orientation of directional sole marks of the same bed. BOUMA also (1962, figs. 20 and 21) found relatively consistent relationships between grain orientation and sedimentary structures. Perhaps many primary sedimentary structures have a relatively simple internal, anisotropic fabric, which is an essential, integral part of the structure.

The tendency of small groups of quartz grains to exhibit apparent similarity of orientation has been termed "nesting" or "clustering." SCHWARZACHER (1951, p. 165—167) thought such lodgement orientation was the result, in part at least, of crystallographic controls on face development. On the other hand, POTTER and MAST (1963, Table 6) did not find much evidence for existence of clusters in sandstones of the Illinois basin.

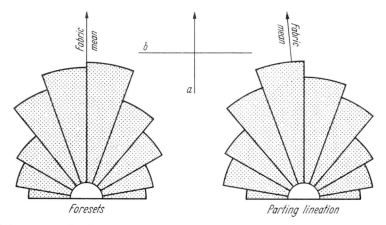

Fig. 3-13. Relations between current direction and apparent long axes of 3,493 quartz grains from foreset beds and 1,102 quartz grains from parting lineation. Average direction of long axis is closely parallel to current direction, a, defined by directional structures. (POTTER and MAST, 1963, fig. 5)

Applications and areal fabric studies have been comparatively few. INGERSON (1940) effectively used crystallographic fabrics to distinguish ripple mark from pseudo-ripple mark. Almost all subsequent workers have studied only dimensional orientation. GRIFFITHS (1949 and 1950) and GRIFFITHS and ROSENFELD (1953) investigated the relation between fabric and vectorial permeability. Others who have explored the relation between fabric and permeability include HUTTA (1956) and MAST and POTTER (1963).

Fabrics have been studied in primary deformational structures by VAN STRAATEN (1949). He found the fabric to be parallel to the direction of movement. VITANAGE (1954) examined the fabric of a sandstone dike.

The fabrics of modern sediments have also been studied. CURRAY (1956) examined beach and littoral sands along the Gulf Coast. His study clearly indicated that long axes of quartz grains lie perpendicular to the trend of the beach. WENDLER (1956, fig. 2) measured fabrics from the Bock River in East Germany and found good parallelism between stream direction and long axes. He also (fig. 3) presented some data on beach sands. YOUNG and MANKIN (1961) examined crystallographic fabrics in sands of the Canadian River of Oklahoma and found parallelism to stream direction. MAXWELL et al. (1961, p. 227—228) studied fabrics in modern carbonate sediments around an island of the Great Barrier Reef. They found a complicated pattern of orientation whose complexity arises from interference between normal tidal currents and reef flat currents.

Studies that have used fabric primarily to determine current direction include those by GRYAZNOVA (1953), KOPSTEIN (1954, p. 68—75), RUSNAK (1957a), SRIRAMADAS (1957), GANGULY (1960). HENNINGSEN (1961, p. 616—618), and STAUFFER (1962, figs. 19 and 23). The presence of other directional structures in sandstones has tended to minimize the role of dimensional fabrics in most regional and even local studies. Fabric studies do have, however, great potential use in subsurface exploration especially, if measurement can be made in the well wall (NANZ, 1960) rather than in oriented cores.

Care should be used to distinguish primary depositional fabrics from those made by either soft-sediment or tectonic deformation. Depending on the intensity of deformation and competency of the sediment, tectonic deformation can either partially or completely alter and erase the original depositional fabric and impose a grain orientation that is the product of deformation (BASSETT and WALTON, 1960, p. 99). The less competent the sediment the more likely the transition from a primary to a tectonic fabric. CLOOS' (1947) study of deformation of an oolitic limestone in an Appalachian fold provides a good example of how a sedimentary fabric can be modified. Moreover, quartz, the almost universal fabric element of sandstone, is very sensitive to deformation. Knowledge of both structural field relations and primary directional structures is essential in attempting to interpret sedimentary fabrics in areas of deformation. In contrast, depth of burial appears to have little effect on fabric anisotropy in the ab plane.

Post-depositional or diagenetic fabrics of carbonate have received some study, especially their qualitative aspects (cf. SANDER, 1936; BATHURST, 1958; SCHWARZ-ACHER, 1961, p. 1492—1501).

Fabric of Argillaceous Sediments

Some study has been given to the fabric of argillaceous sediments.

Because of fineness of grain, optical methods of fabric determination of argillaceous sediments are not widely used, although some observations have been recorded (VAN STRAATEN, 1954, p. 60). X-ray techniques that have been proposed include those of FAIRBAIRN (1943), HO (1947), KAARSBERG (1959), and SILVERMAN and BATES (1960). KAARSBERG (1959) investigated the response of argillaceous samples to sonic transmissibility.

LAMBE (1958) and ROSENQUIST (1959) review fabric concepts as applied to soils and clays. Primary effort was directed at classifying fabric types rather than determining anisotropy in the plane of the bedding.

Geophysical Properties

In a homogeneous anisotropic porous medium, the magnitude of geophysical properties, such as dielectric constant, electrical resistivity, fluid permeability, magnetic susceptibility, sonic transmissibility and thermal conductivity depends on the direction in which they are measured. These properties are represented by second rank tensor quantities as has been recognized by SCHEIDEGGER

(1960, p. 76—79) and COLLINS (1961, p. 62—65) for fluid permeability and ISING (1942, p. 6—7) for magnetic susceptibility*.

Insight to the relations between the anisotropic fabric of a sedimentary rock and anisotropy of a tensor geophysical property is obtained by considering a homogeneous, non-laminated sandstone with monoclinic fabric symmetry. The symmetry of the fabric controls the symmetry of the geophysical property and hence its tensor representation.

The second rank tensor associated with a quadratic form, an ellipsoid, has three principal axes, longest, intermediate, and shortest. These three axes, in the case of fluid permeability, correspond to the greatest, intermediate and least directions of flow. These axes are called the principal axes of the medium with respect to permeability. This ellipsoid, assuming the property is a fabric dependent one, bears a simple relation to the reference system of figure 3-14. Its intermediate axis, β, is colinear with b and its long axis, α, lies in the ac symmetry plane parallel to the three dimensional fabric mean. Hence the ellipsoid is inclined to the ab plane at an angle Θ corresponding

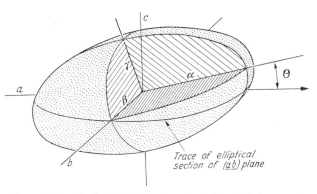

Fig. 3-14. Triaxial ellipsoid of dielectric constant with principal axes α, β and γ and standard reference system a, b, and c assuming complete fabric dependence. The angle Θ is the mean angle of inclination of the dimensional fabric

to the average angle of inclination of the fabric. If the average angle of inclination is zero, the principal axes of ellipsoid coincides with the a, b, c axes of the reference system.

Commonly in paleocurrent studies, only anisotropy in the ab plane is of interest. For an ellipsoid of given eccentricity, the anisotropy in the ab plane is given by the eccentricity of the elliptical section in that plane.

If an arbitrary reference system is used, x_1, x_2 and x_3 the edges of a cube, the tensor for dielectric coefficient, is

$$K_{ij} = \begin{bmatrix} K_{11} & K_{12} & K_{13} \\ K_{21} & K_{22} & K_{23} \\ K_{31} & K_{32} & K_{33} \end{bmatrix}$$

where K_{11} is the dielectric coefficient measured in the direction normal to the face cutting x_1 when an electric field is applied parallel to x_1, K_{23} is the dielectric coefficient measured in the direction normal to the face cutting x_2 when a field is applied parallel to x_3, and so on.

---

* NYE (1957) shows the use of tensors in crystal physics. TEMPLE (1960) provides a more mathematical treatment.

If, however, the a, b and c axes of the standard reference system of figure 3-14 are used as the reference system, the tensor has the form

$$K_{ij} = \begin{bmatrix} K_{11} & 0 & K_{13} \\ 0 & K_{22} & 0 \\ K_{31} & 0 & K_{33} \end{bmatrix} \qquad K_{13} = K_{31}$$

Similar relations prevail for other fabric dependent geophysical properties that are described by symmetrical second rank tensors.

The presence of unrecognized internal laminations can introduce complications. These are well illustrated by fluid permeability (MAST and POTTER, 1963).

If the laminations are essentially parallel to the ab plane, they, of course, markedly reduce flow in the vertical but have little effect on flow in the ab plane, provided each lamination has an essentially similar fabric. If, however, the laminations are inclined to the ab plane, measurements in the ab plane will no longer coincide with the plane of flow, which is now chiefly controlled by the inclined laminations rather than their internal fabric. Because grains are imbricated with respect to a surface of accumulation, but not necessarily the principal one, inclined laminations can cause the ellipsoid of figure 3-14 to be rotated about the b axis so that it may dip down rather than up-current.

To what extent have aggregate geophysical properties actually been correlated with anisotropic fabrics?

Directional permeability has received most attention. Experiments by SULLI-VAN and HERTEL (1940, p. 761—765) on cotton and WYLLIE and SPANGLER (1952, p. 391—395) on sandstones have linked tortuosity variations, which can depend on fabric, to directional permeability.

NEWELL (cited in KING, 1899, p. 126), reported that permeability in sandstone was greater parallel to bedding than perpendicular to it. FETTKE (1938) also noted the anisotropic character of permeability in sandstones, when he reported that permeability varied in different directions in the plane of the bedding and was least perpendicular to the bedding, as was confirmed by PRESSLER (1947, p. 1859—1862). JOHNSON and HUGHES (1948) and RUHL and SCHMIDT (1957) found that the ratio of maximum to minimum permeability in the plane of the bedding of sandstones varied between 1.0 and 1.5. Hence the anisotropic character of permeability in the ab plane of sandstones is well established.

In sandstones GRIFFITHS (1949, 1950) and GRIFFITHS and ROSENFELD (1953) found the strongest grain shape orientation in a plane parallel to direction of maximum permeability and perpendicular to bedding. HUTTA (1956) made a more comprehensive study and found no strong correlation between dimensional orientation, studied in vertical section, and direction of maximum permeability. MAST and POTTER (1963) also failed to find good correlation between fabrics and permeability, both measured in the plane of the bedding. They attributed poor correlation to (1) weak anisotropic permeability contrasts in the ab plane of the sandstones they studied and (2) to their use of a subfabric of elongate grains rather than an aggregate fabric. Better correlation was obtained between fabric and a permeability dependent property, imbibition or the ability of a rock to imbibe water, in vertical sections cut parallel to the ac plane. Long axes of elliptical imbibition pattern showed good agreement with average angle of imbrication. Better correlation in the ac plane was attributed to greater fabric anisotropy.

Judged by the published literature, directional permeability in the ab plane does not appear to offer much promise as an aggregate measure of fabric anisotropy. JOHNSON and BRESTON (1951), however, do show a uniform pattern of permeability obtained from oriented cores of the Third Bradford Sand in Pennsylvania. Unfortunately, no other directional evidence was presented.

ARBOGAST et al. (1960) report good correlation between dielectric coefficient and directional inhomogeneities in artificial media and show fair to good correlation with particulate fabrics of sandstones. McIVER (1961, p. 141—149) shows good correlation between dielectric anisotropy and several types of oriented sole marks. NANZ (1960) implies good agreement between sonic anisotropy and particulate fabrics. Both anisotropy of dielectric constant and sonic transmissibility can be affected by presence of unrecognized cross-laminations in a sample.

There appear to be few published studies of directional variation of either thermal conductivity or electrical resistivity in the ab plane of sedimentary rocks, although as early as 1884, JANNETTAZ (p. 500—505) showed the dependence of thermal conductivity upon direction of schistosity in metamorphic rocks. BRINKMANN et al. (1961, p. 29) found thermal conductivity to be parallel to modal grain shape orientation in three samples from a folded sandstone bed. Electrical resistivity should correlate closely with permeability variations. However, both elasticity and consolidation in a porous anisotropic media have been considered. BIOT (1955) and CLEARY (1959, p. 30) have recognized compressibility in sandstone as a fabric dependent and hence anisotropic property.

Magnetic properties of sedimentary rocks have received considerable attention.

If a test sample is introduced to a magnetic field, it is characterized by its magnetic moment M which consists of two parts,

$$M = M^i + M^r$$

where

$M^i =$ is the induced moment, a vector quantity which depends on the magnetic suspectibility of the material

and

$M^r =$ is the remanent moment, a vector quantity.

The induced moment is that magnetism acquired by a particle in an external magnetic field. The components of the induced moment are obtained from

$$(1) \qquad M^i = \sum_{j=1}^{3} \sum_{k=1}^{3} S_{kj} H_j i_k,$$

where H_j is the jth component of the applied magnetic field and S_{kj} are the Cartesian components of the magnetic susceptibility tensor. Magnetic susceptibility of elongate ferrimagnetic grains in a weak magnetic field is greatest parallel to long dimension (ISING, 1942, p. 5; RUNCORN, 1956, p. 479—480). Magnetite is usually the most important mineral in determining the susceptibility of a sedimentary rock, although in some sediments hematite is dominant. The susceptibility of sediments is less than that of basic and acidic igneous rocks (RUNCORN, 1956, p. 496).

Remanent magnetism exists independently of weak applied magnetic fields and depends only on the inherent magnetization of a particle. Remanent magnetism is specified by two angles, declination and inclination.

Induced and remanent magnetism are independent of one another and are measured by different means. In nature, however, both affect the magnetic moment of a particle because sedimentation occurs in the magnetic field of the earth.

ISING (1942, p. 10) found that in varved glacial clays magnetic susceptibility is considerably higher in the bedding than perpendicular to it. In the plane of the bedding, there was a slight but measurable anisotropy. GRAHAM (1954) also commented on the tensor character of magnetic susceptibility in bedded rocks and noted that it was least perpendicular to bedding. HOWELL et al. (1958, p. 297) also found magnetic susceptibility greatest approximately parallel to bedding. This anisotropy has been attributed to the preferred arrangement of magnetic particles in sediments and is believed to be the response of elongate magnetic particles to alignment by the earth's magnetic field. Two other factors that may influence the susceptibility of ferrimagnetic minerals are mechanical orientation related to current at the time of deposition and post-depositional growth orientation.

Relationships between particulate fabrics of quartz grains, current direction and magnetic susceptibility have been investigated by WENDLER (1961). He found (fig. 7) that maximum and minimum direction of magnetic susceptibility, in the plane of the bedding, usually makes an angle of 45° with transport direction in beach sands. In one sample, he found poor correlation between magnetic susceptibility and crystallographic fabric of quartz.

Investigators of natural remanent magnetism in sedimentary rocks have obtained best results with fine grained sandstones and silts and noted that coarse grained sandstones have often yielded less satisfactory results (GRAHAM, 1949, p. 146—147; CLEGG et al., 1954, p. 594). This suggests that both declination and inclination can be affected at the time of deposition by hydraulic as well as magnetic forces. As particle size decreases, the orienting effect of the magnetic moment will become large in comparison to the moment of inertia (RUNCORN, 1956, p. 479). Thus the smaller the particle the more sensitive it is to magnetic forces. Magnetic particles of detrital origin range in size from 0.10 mm in sandstones to as small as 30 to 0.1 μ in glacial varves (GRAHAM, 1949, p. 143; GRIFFITHS et al., 1960, p. 374).

When glacial varves were redispersed in quiet water in a tank, their declination was essential parallel to the present field but their inclination was usually less (KING, 1955, p. 120; GRIFFITHS et al., 1960). This appears to be related to the fact that the remanent moment is parallel to long dimension and, upon deposition, long dimension rotates into the plane of the bedding. Declination was also affected by the attitude or dip of the surface of accumulation. The greater the initial dip the greater the deviation of the declination. CLEGG et al. (1954) also ground and redispersed in quiet water thinly laminated sandstones of the Keuper Marl series of England. They found that polarization of the remanent magnetism was also essentially parallel to the present magnetic field. Using redispersed glacial varves, GRIFFITHS *et al.* (1960, p. 364—367) found that deposition in running water produced deviations in declination up to 20°, although earlier KING (1955, fig. 15) had found some deflections in excess of 40°. Thus experimental evidence does indicate that currents can affect magnetic declination and that larger magnetic grains are more susceptible to hydrodynamic forces than small ones. Sedimentary fabric, bedding or directional structures have received little attention, however, in most studies of remanent magnetism. Instead, most investigators have largely relied on indirect evidence such as the similarity of remanent polarization of interbedded flows and sediments (OPDYKE, 1961). Direct evidence could be obtained by sampling a formation over a small area and determining remanent magnetism in directional structures with marked contrasts in orientation.

Remanent magnetism can be either altered or acquired after deposition by deformation, heating, and weathering. GRAHAM (1949, p. 132—134) and COX and DOELL (1960, p. 655—661) provide guides for the recognition of these possibilities. GRAHAM (1954) also comments on the effect of deformation on magnetic susceptibility.

Summary

The study of sedimentary fabrics developed from structural petrology and the study of metamorphic rocks.

The primary depositional fabric of sedimentary rocks is the response of clastic particles to orientation by gravity, fluid, and magnetic force fields. Although gravity has the strongest affect, primary anisotropic fabrics of current origin are present in the vast majority of clastic rocks ranging in size from the coarsest conglomerates to fine silts. Such primary anisotropic fabrics almost always have monoclinic symmetry regardless of size of grain or agent of deposition. Triclinic symmetry can occur, however, in clastics deposited behind or around obstacles. Orthorhombic symmetry is characteristic of the fabric deposits such as some muds that accumulated in response to weak, variable currents. Magnetic force fields principally affect small ferrimagnetic particles usually ranging from 30 to 0.1 μ in size.

Methods of fabric study vary widely. Particulate methods of fabric study, those that specify orientation of individual particles, are the most widely used. Orientation studies of conglomerates, gravels, tills and fossils are usually made in the field by simple direct methods. Sand-sized particles of quartz and carbonate sands are usually studied by particulate methods in the laboratory using either a microscope or photographs. Only relatively recently, however, have satisfactory methods for particulate, dimensional fabric study of sandstones been developed. Some aggregate methods, methods that measure a bulk property, such as dielectric constant or sonic transmissibility, are available. Their success depends upon the extent to which such properties are clearly fabric dependent. Relations between fabric anisotropy and anisotropy of geophysical properties provide a fruitful field of research. Whether fabrics or geophysical properties are studied, however, care should always be taken to recognize and specify the bedding, for unrecognized bedding laminations can cause interpretative difficulties.

The relations between direction of sediment transport and particle shape have been chiefly established by the study of ancient sediments. However, both modern sediments and experimental flume studies have provided useful supplementary knowledge. The fabric of till is best understood, followed by that of sands and gravels. Fossil orientation is not yet well understood.

The maximum projection plane of particles dips into the oncoming current in the vast majority of sediments. Long dimension usually lies parallel or sub-parallel to the current. Although rolling transverse to current direction does occur in practically all clastics, it is usually unimportant, except in some gravel deposits and fossil assemblages.

Currents can affect the orientation of ferrimagnetic particles at the time of deposition, especially the larger ones. As particle size decreases, however, the orienting effect of the magnetic moment becomes large in comparison to the moment of inertia. Fabric, bedding, and directional structures have received comparatively little attention in studies of remanent magnetism in ancient sediments. Magnetic susceptibility has received little attention.

The available evidence suggests that relatively simple, consistent relations exist between many of the principal primary sedimentary structures and their constituent sand grains. This implies that many primary sedimentary structures can be thought of as having a relatively uniform, collective fabric.

Care should be taken to distinguish primary depositional fabrics from later ones of either soft sediment or tectonic deformation. Either can partially or totally alter and erase original depositional fabric. Depth of burial alone, however, appears to have little affect on fabric anisotropy in the plane of the bedding.

References

ARBOGAST, J. L., C. H. FAY and S. KAUFMAN, 1960: Method and apparatus for determining directional dielectric anisotropy in solids. U.S. Patent 2,963,642.

BASSETT, D. A., and E. K. WALTON, 1960: The Hell's Mouth grits: Cambrian grey-wackes in St. Tudwal's Peninsula, North Wales. Quart. J. Geol. Soc. London 116, 85—110.

BATHURST, R. G. C., 1958: Diagenetic fabrics in some British Dinantian limestones. Liverpool Manchester Geol. J. 2, 11—36.

BELL, DUGALD, 1888: Additional notes to Mr. Bell's papers (p. 237—261). Trans. Geol. Soc. Glasgow 8, 341.

BIOT, M. A., 1955: Theory of elasticity and consolidation for a porous anisotropic solid. J. Appl. Phys. 26, 182—185.

BLOSS, F. D., 1957: Anisotropy of fracture in quartz. Am. J. Sci. 225, 214—225.

BONHAM, L. C., 1957: Structural petrology of the Pico anticline, Los Angeles County, California. J. Sediment. Petrol. 27, 251—264.

BOUMA, A. H., 1962: Sedimentology of some flysch deposits. Amsterdam: Elsevier Publishing Co. 168 p.

BRINKMANN, R., W. GIESEL and R. HOEPPENER, 1961: Über Versuche zur Bestimmung der Gesteinsanisotropie. Neues Jahrb. Geol. Paläontol. Monatsh. 22—33.

BROWN, W. E., and H. W. PATNODE, 1953: Plastic lithification of sands *in situ*. Bull. Am. Assoc. Petrol. Geologists 37, 152—157.

BULACH, M. KH., 1951: Orientation of belemnites in Jurassic deposits of South Daghestan. All Union Petroleum Geological Prospecting Scientific Research Institute, Geological Collection, No. 1 (4), 144—145 [Russian].

BURACHEK, A. G., 1933: On the method of measurement of orientation of pebbles and cross-bedding. Repts. All-Russian Mineral. Soc. 62, 432—434 [Russian].

CAILLEUX, A., 1938: La disposition individuelle des galets dans les formations detritiques. Rev. Geogr. Phys. et Géol. Dynam. **11**, 171—198.

— 1945: Distinction des galets marins et fluviatiles. Bull. soc. géol. France [5] **15**, 375—404.

CHENOWETH, P. A., 1952: Statistical methods applied to Trentonian stratigraphy in New York. Bull. Geol. Soc. Am. **63**, 521—560.

CLEARY, J. M., 1959: Hydraulic fracture theory, Part III, Elastic properties of sandstone. Illinois Geol. Survey, Cir. 281, 44 p.

CLEGG, J. A., M. ALMOND and P. H. S. SUBBS, 1954: The remanent magnetism of some sedimentary rocks in Britain. Phil. Mag. **45**, 583—598.

CLOOS, E., 1947: Oölite deformation in the South Mountain fold, Maryland. Bull. Geol. Soc. Am. **58**, 843—918.

CLOOS, H., 1938: Primäre Richtungen in Sedimenten der rheinischen Geosynkline. Geol. Rundschau **29**, 357—367.

COLLINS, R. E., 1961: Flow of fluids through porous materials. New York: Reinhold Publishing Corp. 270 p.

COLTON, G. W., and W. DE WITT jr., 1959: Current oriented structures in some Upper Devonian rocks in western New York (abstract). Bull. Geol. Soc. Am. **70**. 1759—1760.

COX, A., and R. R. DOELL, 1960: Review of paleomagnetism. Bull. Geol. Soc. Am. **71**, 645—768.

CROWELL, J. C., 1955: Directional-current structures from the Prealpine flysch, Switzerland. Bull. Geol. Soc. Am. **66**, 1351—1384.

CURRAY, J. H., 1956: Dimensional orientation studies of Recent coastal sands. Bull. Am. Assoc. Petrol. Geologists **40**, 2440—2456.

DAPPLES, E. C., and J. F. ROMINGER, 1945: Orientation analysis of fine-grained clastic sediments. J. Geol. **53**, 246—261.

DONNER, J. J., and R. G. WEST, 1956: The Quaternary geology of Brogeneset, Nordanstlandet, Spitsbergen. Norsk Polarinstitutt Skrifter No. 109, 29 p.

DREIMANIS, A., 1959: Rapid macroscopic fabric studies in drillcores and hand specimens of till and tillite. J. Sediment. Petrol. **29**, 459—463.

DUMANOWSKI, B., 1961: Zagadnienie dwudzielnosci moren w Sudetach. Ann. soc. géol. Pologne **31**, 317—333.

EDELMAN, N., 1951: Glacial abrasion and the movement in the area of Rosala-Nötö, S.W. Finland. Bull. comm. géol. Finlande No. 154, 157—169.

FAIRBAIRN, H. W., 1943: X-ray petrology of some fine-grained foliated rocks. Am. Mineralogist **28**, 246—256.

— 1949: Structural petrology of deformed rocks. Cambridge: Addison-Wesley Press, Inc. 344 p.

FETTKE, C. R., 1938: The Bradford oil field. Penn. Geol. Survey, Bull., M-21, 211—228.

FLINT, R. F., 1961: Two tills in southern Connecticut. Bull. Geol. Soc. Am. **72**, 1687—1691.

FORCHE, F., 1935: Stratigraphie und Paläogeographie des Buntsandsteins im Umkreis der Vogesen. Mitt. geol. Staatsinst. Hamburg **15**, 16—55.

FRASER, H. J., 1935: Experimental study of the porosity and permeability of clastic sediments. J. Geol. **43**, 910—1010.

FRAZIER, D. E., and A. OSANIK, 1961: Point-bar deposits, Old River Locksite, Louisiana. Trans. Gulf. Coast Ass. Geol. Soc. **11**, 121—137.

GANGULY, S., 1960: Dimensional fabric of Barakar and Barren Measure sandstone in eastern part of Ramgarh Coal-field, Hazaribagh. Quart. J. Geol. Mining Met. Soc. India **32**, 39—47.

GILLBERG, G., 1961: The Middle-Swedish moraines in the province of Dalsland, W. Sweden. Geol. Föreningens I. Stockholm Förhandlingar **83**, 335—369.

GLEN, J. W., J. J. DONNER, and R. G. WEST, 1957: On the mechanism by which stones in till become oriented. Am. J. Sci. **255**, 194—205.

GRADZINSKI, R., 1957: Uwagi o sedymentacji Miocenu w okolicy Proszowiz. Rocznik Polsk. Towarz. Geol. **25**, 1—27 [Polish, English summary].

GRADZINSKI, R., A. RADOMSKI and R. UNRUG, 1959: Preliminary results of sedimento-logical investigations in Silesian coal basin. Bull. Acad. Sci. Polonaise, Series chim. géol., et geogr. **7**, 433—440.

—, and R. UNRUG, 1959: Origin and age of the "Witow series" near Cracow. Ann. soc. géol. Pologne **29**, 181—195 [Polish, English summary].

GRAHAM, J. W., 1949: The stability and significance of magnetism in sedimentary rocks. J. Geophys. Research **54**, 131—167.

— 1954: Magnetic susceptibility anisotropy, an unexploited petrofabric element (abstract). Bull. Geol. Soc. Am. **65**, 1257—1258.

GRAVERNOR, C. P., and W. A. MENELEY, 1958: Glacial flutings in central and northern Alberta. Am. J. Sci. **256**, 715—728.

GRIFFITHS, D. H., R. F. KING, A. I. REES and A. E. WRIGHT, 1960: The remanent magne-tism of some recent varved sediments. Phil. Trans. Roy. Soc. (London) A **256**, 359—383.

GRIFFITHS, J. C., 1949: Directional permeability and dimensional orientation in the Bradford sand. Penn. State Coll. Mineral. Inds. Exp. Sta., Bull. **54**, 138—163.

— 1950: Directional permeability and dimensional orientation in Bradford sand. Producers Monthly **14**, 26—32.

—, and M. A. ROSENFELD, 1951: Progress in measurement of grain orientation in Bradford sand. Producers Monthly **15**, 24—26.

— — 1953: A further test of dimensional orientation of quartz grains in Bradford sand. Am. J. Sci. **251**, 192—214.

GROTH, K., 1961: Beiträge zur Gliederung des Saaleglazials bei Halle (Saale) und im Mansfelder Seekreis. Geologie **10**, 169—184.

GRYAZNOVA, T. E., 1947: Methods of studying the orientation of grains in sandy deposits. Doklady Acad. Sci. U.S.S.R. **58**, 647—650 [Russian].

— 1949: Orientation of sand grains, methods of studying it and utilizing it in geology. Leningrad University Herald, Student's Scientific Papers No. 2, 97—105 [Russian].

— 1953: Oriented textures of the sandstones in a productive series of the Apsheron Peninsula. Geological Collection of the All-Union Petroleum Geology Research Inst. No. 2 (5), 224—210 [Russian].

HAAF, E., TEN, 1959: Graded beds of the Northern Apennines. Ph.D. thesis, Rijks University, Groningen, 102 p.

HALL, J., 1843: Geology of New York, Part IV. Albany: Carroll and Cook. 683 p.

HAMBLIN, W. K., 1962a: X-ray radiography in the study of structures in homogeneous sediments. J. Sediment. Petrol. **32**, 201—210.

— 1962b: Staining and etching techniques for studying obscure structures in clastic rocks. J. Sediment. Petrol. **32**, 530—533.

HANSEN, E. H., S. C. PORTER, B. A. HALL and A. HILLS, 1961: Décollement structures in glacial-lake sediments. Bull. Geol. Soc. Am. **72**, 1415—1418.

HARRISON, P. W., 1957a: New technique for three dimensional fabric analysis of till and englacial debris containing particles from 3 to 40 mm in size. J. Geol. **65**, 98—105.

— 1957b: A clay-till fabric: its character and origin. J. Geol. **65**, 275—303.

HELMBOLD, REINHARD, 1952: Beitrag zur Petrographie der Tanner Grauwacken. Heidelberger Beitr. Mineral. u. Petrogr. **3**, 253—288.

HENNINGSEN, D., 1961: Untersuchungen über Stoffbestand und Paläogeographie der Gießener Grauwacke. Geol. Rundschau **51**, 600—626.

HIGGS, D. V., M. FRIEDMAN and J. E. GEBHART, 1960: Petrofabric analysis by means of the X-ray diffractometer. Geol. Soc. Am., Mem. **79**, 275—292.

HO, T. L., 1947: Petrofabric analysis by means of X-rays. Bull. Geol. Soc. China **27**, 389—398.

HOLMES, C. D., 1941: Till fabric. Bull. Geol. Soc. Am. **52**, 1299—1354.

HOPPE, G., 1952: Hummocky moraine regions. Geog. Annaler **34**, 1—72.

— 1953: Några iakttagelser vid isländska jöklar sommaren 1952. Ymer. **4**, 241—265.

HOWELL, L. G., J. D. MARTINEZ and E. H. STATHAM, 1958: Some observations on rock magnetism. Geophysics **23**, 285—298.

HUTTA, J. J., 1956: Relation of dimensional orientation of quartz grains to directional permeability in sandstones. Unpublished Master's thesis, Pennsylvania State University, 97 p.

HUCKENHOLZ, H. G., 1959: Sedimentpetrographische Untersuchungen an Gesteinen der Tanner Grauwacke. Beitr. Mineral. u. Petrogr. 6, 261—298.

HYYPPÄ, E., 1948: Tracing the source of the pyrite stones from Vihanti on the basis of glacial geology. Bull. comm. géol. Finlande No. 142, 97—122.

INGERSON, EARL, 1940: Fabric criteria for distinguishing pseudoripple marks from ripple marks. Bull. Geol. Soc. Am. 51, 557—574.

—, and J. L. RAMISCH, 1942: Origin of shapes of quartz sand grains. Am. Mineralogist 27, 595—606.

— — 1954: Studying of unconsolidated sediments: I. Quartz fabric of current and wind ripple marks. Tschermak's mineral. u. petrog. Mitt. 4, 117—124.

ISING, G., 1942: On the magnetic properties of varved clay. Arkiv Mat. Astron. Fysik A 29, 1—37.

IVANOV, D. N., 1959: Orientation of the optic axes of quartz grains in the Red Sandstone of the Chelehen Peninsula. Repts. Acad. Sci. Ukr.S.S.R. 128, 604—606 [Russian]. English translation by Am. Geol. Inst. 1960, 986—988.

JÄRNEFORS, B., 1952: A sediment petrographic study of glacial till from the Pajala District, Northern Sweden. Geol. Fören. in Stockholm Förh. 74, 185—211.

JAMIESON, T. F., 1860: On drift and rolled gravel of the north of Scotland. Quart. J. Geol. Soc. London 16, 347—371.

JANNETTAZ, ED., 1884: Les Roches. Paris: Rothschild. 486 p.

JOHNSON, W. E., and J. W. BRESTON, 1951: Directional permeability of sandstones from various states. Producers Monthly 15, 10—19.

—, and R. V. HUGHES, 1948: Directional permeability measurements and their significance. Penn. State Coll. Mineral. Inds. Exp. Sta., Bull. 52, 180—205.

KAARSBERG, E. A., 1959: Introductory studies of natural and artificial aggregates by sound propagation and x-ray diffraction methods. J. Geol. 67, 447—472.

KAHN, J. S., 1956: Analysis and distribution of the properties of packing in sand-size sediments. 1. On the measurement of packing in sandstones. J. Geol. 64, 385—395.

— 1959: Anisotropic sedimentary parameters. Trans. N.Y. Acad. Sci. [2] 21, 376—386.

KALTERHERBERG, J., 1956: Über Anlagerungsgefüge in grobklastischen Sedimenten. Neues Jahrb. Geol. Paläontol. 104, 30—57.

KARLSTROM, T. N. V., 1952: Improved equipment and techniques for orientation studies of large particles in sediments. J. Geol. 60, 489—493.

KAURANNE, L. K., 1960: A statistical study of stone orientation in glacial till. Bull. comm. géol. Finlande, No. 188, 87—97.

KAY, M., 1945: Paleogeographic and palinspastic maps. Bull. Am. Assoc. Petrol. Geologists 29, 426—450.

KING, F. H., 1899: Principles and conditions of the movements of ground water. U.S. Geol. Survey, 19th Annual Report, 1898—1899, p. 59—294.

KING, P. B., 1948: Geology of the southern Guadaloupe Mountains, Texas. U.S. Geol. Survey Prof. Paper 215, 183 p.

KING, R. F., 1955: The remanent magnetism of artificially deposited sediments. Monthly Notices Roy. Astron. Soc., Geophys. Suppl. 7, 115—134.

KLÄHN, H., 1929: Die Bedeutung der Seelilien und Seesterne für die Erkennung von Wasserbewegung nach Richtung und Stärke. Paleobiologica 2, 28—73.

KOPSTEIN, F. P. H. W., 1954: Graded bedding of the Harlech Dome. Ph.D. thesis, Rijks University of Groningen, 97 p.

KOSZARSKI, L., 1956: Observations on the sedimentation of the Ciezkowice Sandstone near Ciezkowice (Carpathian Flysch). Acad. Bull. Sci. Polonaise, Cl. III 4, 393—398.

KRINSLEY, D., 1960: Orientation of orthoceracone cephalopods at Lemont, Illinois. J. Sediment. Petrol 30, 321—323.

KRUMBEIN, W. C., 1939: Application of photo-electric cell to the measurement of pebble axes for orientation analysis. J. Sediment. Petrol. 9, 122—130.

— 1939: Preferred orientation of pebbles in sedimentary deposits. J. Geol. 47, 673—706.

— 1940: Flood gravel of San Gabriel Canyon. Bull. Geol. Soc. Am. 51, 639—676.

KRUMBEIN, W. C., 1942: Flood gravel of Arroyo Seco, Los Angeles County, California. Bull. Geol. Soc. Am. **53**, 1355—1402.

—, and F. J. PETTIJOHN, 1938: Manual of sedimentary petrography. New York: Appleton-Century Co. 549 p.

KÜRSTEN, M., 1960: Zur Frage der Geröllorientierung in Flußläufen. Geol. Rundschau **49**, 498—501.

LAMBE, T. W., 1958: The structure of compacted clay. J. Soil Mech. and Foundation, Div., Proc. Am. Soc. Civil Engrs. **84**, Paper 1654, 34 p.

LANE, E. W., and E. J. CARLSON, 1954: Some observations of the effect of particle shape on movement of coarse sediments. Trans. Amer. Geophys. Union **35**, 453—462.

LEE, H. A., 1959: Surficial Geology New Brunswick sheet 21 J/13 Aroostook, New Victoria County, New Brunswick (Preliminary series). Geol. Survey Canada.

LUNDEGÅRDH, P H., and G. LUNDQVIST, 1959: Beskrivning till Kartbladet Eskilstuna. Sveriges Geol. Undersökn., Ser. Aa No. 200, 125 p.

LUNDQVIST, G., 1935: Blockundersökningar: Historik och metodik. Sveriges Geol. Undersökn., Ser. C No. 390, 45 p.

— 1948: Blockens orientering i olika jordarter. Sveriges Geol. Undersökn., Ser. C No. 497, 1—29.

— 1949, The orientation of block material in certain species of flow earth *in* Glaciers and Climate. Geog. Annaler HI-2, 335—347.

— 1951: Beskrivning till jordartskarta över Kopparbergs, Län. Sveriges Geol. Undersökn., Ser. Ca No. 21, 213 p.

MacCLINTOCK, P., 1958: Glacial geology of the St. Lawrence Seaway and power projects. New York State Museum and Science Service, Univ. State New York, State Education Department, Albany, 26 p.

— 1959: A till-fabric rack. J. Geol. **67**, 709—710.

—, and J. TERASAMAE, 1960: Glacial history of Covey Hill. J. Geol. **68**, 232—241.

MARTINEZ, J. D., 1958: Photometer method for studying quartz grain orientation. Bull. Am. Assoc. Petrol. Geologists **42**, 588—608.

MAST, R. F., and P. E. POTTER, 1963: Sedimentary structures, sand shape fabrics, and permeability, Part II. J. Geol. **71**, 548—565.

MATTHEW, G., 1903: Report on the Cambrian rocks of Cape Breton. Geol. Survey Canada, Rept. 797, 246 p.

MAXWELL, W. G. H., R. W. DAY and P. J. G. FLEMING, 1961: Carbonate sedimentation on the Heron Island Reef, Great Barrier Reef. J. Sediment. Petrol. **31**, 215—230.

McBRIDE, E. F., 1960: Martinsburg flysch of the central Appalachians. Unpublished Ph.D. thesis, The Johns Hopkins University, 375 p.

McIVER, N. L., 1961: Upper Devonian marine sedimentation in the central Appalachians. Unpublished Ph.D. thesis, The Johns Hopkins University, 346 p.

MILLER, H., 1850: On peculiar scratched pebbles and fossil specimens from boulder clay, and on chalk flints and oolitic fossils from the boulder clay in Caithness. Rept. British Assoc. Advance. Science (Edinburgh), Trans. 93—96.

— 1884: On boulder-glaciation. Proc. Roy. Phys. Soc. Edinburgh **8**, 156—189.

MOLDER, K., 1948: Die Verbreitung der Dacitblöcke in der Moräne in der Umgebung des Sees Lappajärvi. Bull. comm. géol. Finlande No. 142, 45—52.

MOSS, J. H., and D. F. RITTER, 1962: New evidence regarding the Binghamton substage in the region between the Finger Lakes and the Catskills, New York. Am. J. Sci. **260**, 81—106.

NANZ, R. H., 1960: Exploration of earth formations associated with petroleum deposits. U.S. Patent, 2,963,641.

NAIRN, A. E. M., 1958: Petrology of the Whita Sandstone, Southern Scotland. J. Sediment. Petrol. **28**, 57—64.

NYE, J. F., 1957: Physical properties of crystals, their representation by tensors and matrices. Oxford: Clarendon Press. 322 p.

OKKO, V., 1955: Glacial drift in Iceland: its origin and morphology. Bull. comm. géol. Finlande No. 170, 133 p.

OPDYKE, N. D., 1961: The paleomagnetism of the New Jersey Triassic: a field study of the inclination error in red sediments. J. Geophys. Research 66, 1941—1950.

PATTERSON, M. S., and L. E. WEISS, 1961: Symmetry concepts in the structural analysis of deformed rocks. Bull. Geol. Soc. Am. 72, 841—882.

PELLETIER, B. R., 1958: Pocono paleocurrents in Pennsylvania and Maryland. Bull. Geol. Soc. Am. 69, 1033—1064.

PETRAMEK, YA., and YE. KOMARKOVA, 1953: The orientation of cephalopod shells in the limestones of Barandien and its paleogeographical significance. Sborník ustřed ústavu geol. 20, 129—148.

PETTIJOHN, F. J., 1962: Dimensional fabric and ice flow, Precambrian (Huronian) glaciation. Science 135, 442.

PICARD, K., 1950: Beobachtungen im Diluvium des Stadtgebietes Essen. Geol. Jahrb. 65, 573—588.

PIDOPLICHKO, I. G., 1956: The glacial period, Part 4. The origin of conglomeritic deposits. Ukrainian Acad. Sci. U.S.S.R., Inst. Zoology, Kiev, 336 p. [Russian].

PIERSON, III, A. L., 1959: A photomultiplier photometer for study of quartz grain orientation. J. Sediment. Petrol. 29, 98—103.

POTTER, P. E., and R. F. MAST, 1963: Sedimentary structures, sand shape fabrics and permeability, Part I. J. Geol. 71.

PRESSLER, E. D., 1947: Geology and occurrence of oil in Florida. Bull. Am. Assoc. Petrol. Geologists 31, 1851—1862.

PRYOR, W. A., 1960: Cretaceous sedimentation in Upper Mississippi Embayment. Bull. Am. Assoc. Petrol. Geologists 44, 1473—1504.

RAPP, A., 1959: Avalanche boulder tongues in Lappland. Geog. Annaler 41, 34—48.

— 1960a: Recent development of mountain slopes in Kärkevagge and surroundings, northern Scandinavia. Geog. Annaler 42, 71—200.

— 1960b: Talus slopes and mountain walls at Tempelfjorden, Spitsbergen. Norsk Polarinst., Skrifter No. 119, 96 p.

REINECKE, L., 1928: The location of payable ore-bodies in the gold-bearing reefs of the Witwatersrand. Trans. Geol. Soc. S. Africa 30, 89—119.

RICHTER, K., 1932: Die Bewegungsrichtung des Inlandeises, rekonstruiert aus den Kritzen und Längsachsen der Geschiebe. Z. Geschiebeforsch 8, 62—66.

— 1933: Gefüge und Zusammensetzung des norddeutschen Jungmoränen-Gebietes. Abh. Geol.-Paläont. Inst. Univ. Greifswald 11, 1—63.

— 1936a: Ergebnisse und Aussichten der Gefügeforschung im pommerschen Diluvium. Geol. Rundschau 27, 196—206.

— 1936b: Gefügestudien in Engebrae, Fondalsbrae und ihren Vorlandsedimenten. Z. Gletscherkunde 24, 22—30.

ROSENQVIST, I. TH., 1959: Physio-chemical properties of soils: Soil-water systems. J. Soil Mech. and Found. Div., Proc. Am. Soc. Civil Engrs. 85, Paper 2000, 31—52.

ROWLAND, R. A., 1946: Grain shape fabrics of clastic quartz. Bull. Geol. Soc. Am. 59, 547—564.

RUEDEMANN, R., 1897: Evidence of current action in the Ordovician of New York. Am. Geologist 19, 367—391.

RÜHL, W., and CH. SCHMID, 1957: Über das Verhältnis der vertikalen zur horizontalen absoluten Permeabilität von Sandsteinen (mit Lagerstätten-Beispielen aus dem Gifhorner Trog). Geol. Jahrb. 74, 447—462.

RUKHIN, L. B., 1958: Grundzüge der Lithologie. Berlin: Akademie-Verlag. 806 p. [Translated from the Russian].

RUNCORN, S. K., 1956: Magnetization of rocks: FLÜGGES Handbuch der Physik, vol. 47, p. 470—497. Berlin-Göttingen-Heidelberg: Springer.

RUSNAK, G. A., 1957a: A fabric and petrologic study of the Pleasantview sandstone. J. Sediment. Petrol. 27, 41—55.

— 1957b: Orientation of sand grains under conditions of "unidirectional" fluid flow. 1. Theory and experiment. J. Geol. 65, 384—409.

SANDER, B., 1930: Gefügekunde der Gesteine. Wien: Springer. 352 p.

SANDER, B., 1936: Beiträge zur Kenntnis der Anlagerungsgefüge (Rhythmische Kalke und Dolomite aus der Trias). Mineral. u. petrog. Mitt. **48**, 27—139 (Translated by ELEANORA KNOPF, 1951, Tulsa, Am. Assoc. Petrol. Geologists, 160 p.).

— 1948: Einführung in die Gefügekunde der Geologischen Körper, Erster Teil, Allgemeine Gefügekunde und Arbeiten im Bereich Handstück bis Profile. Vienna and Innsbruck: Springer. 215 p.

SARKISIAN, S. G., and L. T. KLIMOVA, 1955: The orientation of pebbles and methods of studying it for paleogeographical reconstructions. Petroleum Institute, Publishing House of the Academy of Sciences, U.S.S.R., Moscow, 165 p. [Russian].

SCHEIDEGGER, A. E., 1960: The physics of flow through porous media. New York: The Macmillan Co. 313 p.

SCHIEMENZ, S., 1960: Fazies und Paläogeographie der Subalpen-Molasse zwischen Bodensee und Isar. Beih. Geol. Jahrb. **38**, 114 p.

SCHLEE, J., 1957a: Fluvial gravel fabric. J. Sediment. Petrol. **27**, 162—176.

— 1957b: Upland gravels of southern Maryland. Bull. Geol. Soc. Am. **68**, 1371—1410.

SCHMALL, H. R., and R. H. BENNETT, 1961: Axiometer — mechanical device for locating and measuring pebble and cobble axes for macrofabric studies. J. Sediment. Petrol. **31**, 617—622.

SCHULZ, W., 1961: Sedimentpetrographische Untersuchungen im Pleistozän westlich von Halle (Saale). Geologie **10**, 30—49.

SCHUMANN, H., 1942: Zur Korngestalt der Quarze in Sanden. Chem. Erde **14**, 131—151.

SCHWARZACHER, W., 1951: Grain orientation in sands and sandstones. J. Sediment. Petrol. **21**, 162—172.

— 1961: Petrology and structure of some Lower Carboniferous reefs in northwestern Ireland. Bull. Am. Assoc. Petrol. Geologists **45**, 1481—1503.

SEIFERT, G. V., 1954: Das mikroskopische Korngefüge des Geschiebemergels als Abbild der Eisbewegungen, zugleich Geschichte des Eisabbaues in Fehmarn, Ost-Wagrien und dem dänischen Wohld. Meyniana **2**, 126—189.

SEILACHER, A., 1959: Fossilien als Strömungs-Anzeiger. Aus der Heimat **67**, 170—177.

— 1960: Strömungsanzeichen im Hunsrückschiefer. Notizbl. hess. Landesamtes Bodenforsch. Wiesbaden **88**, 88—106.

SILVERMAN, E. N., and T. F. BATES, 1960: X-ray diffraction study of orientation in the Chattanooga shale. Am. Mineralogist **45**, 60—68.

SITLER, R. F., and C. A. CHAPMAN, 1955: Microfabrics of till in Ohio and Pennsylvania. J. Sediment. Petrol **25**, 262—269.

SLICHTER, C. S., 1899: Theoretical investigation of the motion of ground water. U.S. Geol. Survey, 19th Ann. Report, Pt. II, p. 295—384.

SMOOR, P. B., 1960: Dimensional grain orientation studies in turbidite graywackes. Unpublished M.Sc. thesis, McMaster University, 97 p.

SRIRAMADAS, A., 1957: Appositional fabric study of the coastal sedimentaries, East Godavari District, Andhra, India. J. Sediment. Petrol. **27**, 447—452.

STAUFFER, K. W., 1962: Quantitative petrographic study of Paleozoic carbonate rocks, Caballo Mountains, New Mexico. J. Sediment. Petrol. **32**, 357—398.

STRAATEN, L. M. J. K. VAN, 1949: Occurrence in Finland of structures due to subaqueous sliding of sediments. Bull. comm. géol. Finlande No. 144, 9—18.

— 1954: Composition and structure of recent marine sediments in the Netherlands. Leidse Geol. Mededel. **29**, 108 p.

STRAKHOV, N. M., G. I. BUSHINSKII, L. V. PUSTOVALOV, A. V. KHABAKOV and I. V. KHVOROVA, 1957: Methods of studying sedimentary rocks. Moscow, Geological Institute, Acad. Sci., U.S.S.R., vol. 1, 611 p. [Russian].

SULLIVAN, R. R., and K. L. HERTEL, 1940: Flow of air through porous media. J. Appl. Phys. **11**, 761—765.

SULLWOLD jr., W. H., 1960: Tarzana fan, deep submarine fan of late Miocene age of Los Angeles County, California. Bull. Am. Assoc. Petrol. Geologists **44**, 443—457.

TEMPLE, G., 1960: Cartesian tensors, an introduction. London: Methuen & Co. 92 p.

TURNER, F. J., 1957: Lineation, symmetry, and internal movement in monoclinic tectonite fabrics. Bull. Geol. Soc. Am. **68**, 1—18.

Unger, K. P., and W. Ziegenhardt, 1961: Periglaziale Schotterzüge und glazigene Bildungen der Mindel-(Elster-)Eiszeit im zentralen Thüringer Becken. Geologie 10, 469—479.

Unrug, R., 1956: Preferred orientation of pebbles in Recent gravels of the Dunajec River in the western Carpathians. Bull. Polish Acad. Sci. Cl. III 4, 469—473.

— 1957: Recent transport and sedimentation of gravels in the Dunajec Valley (Western Carpathians). Acta Geol. Polon. 7, 217—257 [Polish, Russian and English summary].

Virkkala, K., 1951: Glacial geology of the Suomussalmi area, east Finland. Bull. comm. géol. Finlande No. 155, 1—66.

— 1960: On the striations and glacier movements in the Tampere region, southern Finland. Bull. comm. géol. Finlande No. 188, 161—176.

— 1961: On the glacial geology of the Hämeenlinna region, southern Finland. Bull. comm. géol. Finlande No. 196, 214—241.

Vitanage, P. W., 1954: Sandstone dikes in the South Platte area, Colorado. J. Geol. 62, 493—500.

Vollbrecht, K., 1953: Zur Quarzachsen-Regelung sandiger Sedimente. Acta Hydrophys. 1, 1—87.

— 1953: Zur Untersuchung von Sinkstoff und Geschiebebarren. Geologie 2, 268—276.

Wadell, H., 1936: Shape and shape position of rock fragments. Geog. Annaler 18, 74—92.

Walker, C. T., 1955: Current-bedding directions in sandstones of Lower *Reticuloceras* age in the Millstone Grit of Wharfedale, Yorkshire. Proc. Yorkshire Geol. Soc. 30, 115—132.

Ward, D. de, 1945: Bidraje tot de kennis van het glaciale Diluvium in het Gooi. Verhandel. Geol. Mijnbouwk Genootschap Ned. en Kolonien (Geol. Series) 14, 551—555.

Wayland, R. G., 1939: Optical orientation in elongate quartz. Am. J. Sci. 237, 99—109.

Wendler, R., 1956: Zur Frage der Quarz-Kornregelung von Psammiten. Wiss. Z. Karl-Marx-Univ. Leipzig 5, 421—426.

— 1961: Beziehungen zwischen Fluidaltexturen und der magnetischen Anisotropie der Gesteine. Notizbl. hess. Landesamtes Bodenforsch. 89, 420—437.

West, R. C., and J. J. Donner, 1956: The glaciation of East Anglia and the East Midlands: A differentiation based on stone orientation measurement of tills. Quart. J. Geol. Soc. London 112, 69—91.

White, W. S., 1952: Imbrication and initial dip in a Keweenawan conglomerate bed. J. Sediment. Petrol. 22, 189—199.

Wiesser, T., 1954: Spostrzezenia nad sedymentacja zlepiencow fliszu kapackiego. Acta Geol. Polon. 4, 341—360.

Wilson, L. R., 1932: The Two Creeks Forest Bed, Manitowoc County, Wisconsin. Trans. Wisconsin Acad. Sci. 27, 31—46.

Wright jr., H. E., 1957: Stone orientation in Wadena drumlin field, Minnesota. Geog. Annaler 39, 19—31.

— 1962: Role of the Wadena lobe in the Wisconsin glaciation of Minnesota. Bull. Geol. Soc. Am. 73, 73—100.

Wright, M. D., 1959: The formation of cross-bedding by a meandering or braided stream. J. Sediment. Petrol. 29, 610—615.

Wyllie, M. R. J., and M. B. Spangler, 1952: Application of electrical resistivity measurements to the problem of fluid flow in porous media. Bull. Am. Assoc. Petrol. Geologists 36, 359—403.

Young, L. M., and C. J. Mankin, 1961: Dimensional grain-orientation studies of recent Canadian River sands. Oklahoma Geol. Survey, Okla. Geol. Notes 21, 99—107.

Zhemchuzhnikov, Yu. A., 1954: The possibility and conditions of burial of alluvial sediments in fossil strata *in* Alluvial deposits in coal measures in the Middle Carboniferous of the Donets Basin. Trans. Inst. Geol. Sci. U.S.S.R. No. 151, 9—29 [Russian].

Zimmerle, W., and L. C. Bonham, 1962: Rapid methods for dimensional grain orientation measurements. J. Sediment. Petrol. 32, 751—763.

Fabrics and Geophysical Properties (1963—1976)

Persistent, very diversified activity with much interest in fossil orientation

The study of particle orientation to determine flow direction has been actively pursued since 1963, even though fabrics are generally used only when nothing else suffices. This activity includes some new techniques and some limited experimental studies. Much more significant, however, is the wide range of sediments in which dimensional fabrics have been studied. These include carbonate rocks, volcaniclastic deposits, and loess. Thus, after seeing the results from glacial tills and sandstones, sedimentologists and paleoecologists have applied these methods to other lithologies. On the other hand, there has been little or no progress in relating the orientation of framework grains to directional permeability and we know of only one earth science paper since 1963 on the theoretical aspects of particle orientation. This paper, by BHATTACHURYYA (1966), is an amplification of an earlier study by RUSNAK* (1957b), who in turn followed JEFFERY (1922). Many theoretical papers, of potential value to those sedimentologists who understand mathematics and theoretical hydrodynamics, are found in the chemical engineering literature (CONDIFF and BRENNER, 1969; HENLINE and CONDIFF, 1970; BRENNER and CONDIFF, 1972, 1974). These papers, mostly concerned with orientation of molecular particles in gases, are very well referenced and probably contain the basis for some significant contributions to sedimentology by those who are willing to take the trouble to master them.

The guiding principles for the orientation of particles have perhaps been best stated by JOHANSSON (1965) and elaborated by KELLING and WILLIAMS (1967), RUST (1972) and others. Their results can be summarized as follows.

1. The preferred orientation of disk-like particles transported in contact with a frictional substrate has monoclinic symmetry with an up-dip inclination of 10° to 30°; the long axes of such particles tend to be subhorizontal and perpendicular to flow direction.

2. Under the same conditions scattered, isolated, elongate particles however, tend to have their long axes parallel to flow.

3. Particles immersed in the transporting media tend to align themselves parallel to and dipping into the flow, because of the media's shearing stress.

4. Particle shape and size as well as local geometry of the substrate can introduce some variations in the above generalizations; commonly large particles are better oriented than smaller ones, simple shapes better than complicated ones, and clasts better than fossils.

5. Orientation of the particles should always be related to the depositional surface — which may not be identical with the local structural dip.

---

* Cited in 1963 edition.

6. Bioturbation can be a major factor affecting grain orientation in sands, silts, and clays and even of some pebble-sized shells.

Terrigenous Sandstones

Most studies involve correlation of grain fabric with sedimentary structures and only a few regional paleocurrent fabric maps have been published.

SPOTTS (1964) made a very careful study of the orientation of framework grains (mostly glaucophane) and their relation to sole mark orientation in a Miocene turbidite and found that there could be differences up to 40 to 60° between the

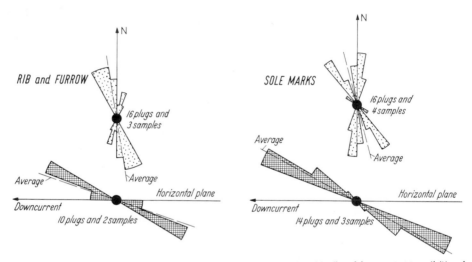

Fig. 3-15. Current rose diagrams of dielectric grain orientation from samples with rib-and-furrow structure *(left)* and sole marks *(right)* show excellent correlation between structures and grains. *Light stippled current rose:* plan view; *ruled current rose:* vertical section of transport plane. (Redrawn from SHELTON and MACK, 1970, figs. 2 and 5; figs. 5—7, J. Sediment. Petrol., Soc. Econ. Paleontol. Mineral.)

two. He attributed this to the difference between the erosional and depositional currents that produced the same bed. Rather similar studies were also published by SESTINI and PRANZINI (1965). ONIONS and MIDDLETON (1968) also noted the variability of grain fabric in relation to the sole marks of turbidites.

The most important paper is that of SHELTON and MACK (1970) who studied grain orientation, using the dielectric anisotropy technique of NANZ* (1960) on oriented plugs, in a variety of environments (fluvial and barrier) and with sedimentary structures as well. The correlation of dielectric anisotropy with sedimentrary structures by SHELTON and MACK is the most comprehensive ever published (fig. 3-15). Plugs have the great advantage of using more grains. This is clearly seen in a second study by SHELTON et al. (1974), who used 190 plug samples to evaluate the relative variability of grain fabric, parting lineation and cross-bedding in a field study of the sands of a modern braided stream in Oklahoma (fig. 3-16). It is estimated that over 5×10^{10} single grains were involved. For this study they used a device which measures the anisotropic electrical conductivity of the pore system, a property dependent on the pore system of a sand — which in turn depends on orientation of the framework.

One of the few regional fabric studies of sandstones is that of SESTINI (1964), who studied carbonate turbidites of Eocene age in Italy and found good uniformity of flow direction. In another fabric study, MARTINI (1971a), found two paleo-current systems — one deltaic and the other longshore. MARTINI (1971b) also suggested that 50 to 100 grains are sufficient to determine the direction of maximum imbrication whereas 150 to 250 are needed to estimate grain orientation in the plane of the bedding.

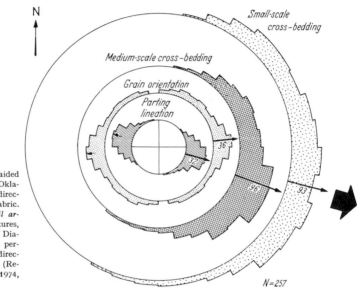

Fig. 3-16. Sand bar of braided Cimarron River at Perkins, Oklahoma and orientation of its directional structures and grain fabric. *Numbers:* measurements, *small arrows:* averages of different features, *large arrow:* grand average. Diagrams such as this show the pervasiveness and uniformity of directional fabrics and structures. (Redrawn from SHELTON *et al.*, 1974, fig. 2)

As is also true with other sediments, especially tills, efforts have been made to automate fabric analysis of sandstones with the X-ray (BAKER *et al.*, 1969; STARKEY, 1974). Commonly the resulting data are fed into a computer to yield a compuzerized orientation diagram. Magnetic suspectibility of arenaceous sediments has also been investigated as a means to determine primary depositional fabric both experimentally by REES (1971) and in the field (REES *et al.*, 1968). His laboratory experiments show good correlation with orientation of framework grains; unfortunately, however, success in the field has been less striking, probably because of a deficiency of magnetic minerals. PICARD and BECKMAN (1966), in one of the few studies of its kind, compared fabrics of opaque and nonopaque silt grains, both in the plane of the bedding and perpendicular to it, with orientation of ripple marks and the predominant (normal and reverse) magnetic directions. In the plane of the bedding both fabrics were essentially perpendicular to the strike of ripple orientation and the opaque fabrics closely parallel magnetic directions. Additional technique studies include those of SIPPEL (1971), who presented a mathematical explanation of the optical method of grain orientation proposed by MARTINEZ* (1958). WINKELMOLEN *et al.* (1968) also used an optical technique that gave good agreement with grain orientation. Automation of fabric analysis is described by DELFINER et al. (1972) and by RÜHL (1974) and MÜLLER (1974), all three of whom use an optical-electronic textural analyser.

Gravels, Conglomerates, Tills and Diamictites

SEDIMENTATION SEMINAR (1965) made a very thorough study of imbrication of large slabs of limestone in an ephemeral creek in southern Indiana. The larger slabs show a very strong monoclinic symmetry; the variation in size and flatness of particles larger than 32 mm had virtually no effect on orientation. In ancient conglomerates, the direction of dip of the AB plane of flat pebbles is the best guide to paleocurrent direction. JOHANSSON (1965) is the best single source of additional papers; his 1963 study is one of the most complete experimental ones ever made on pebble orientation. A later general reference is by KATZUNG (1971), who discusses types of gravel fabric and their relation to stream gradients. More recently, TEISSEYRE's (1975) study of pebble fabric in braided stream deposits, modern and ancient, seems particularly noteworthy.

Fabric studies of ancient water-laid conglomerates are not too common: CANUTI et al. (1966) related cobble shape and size to direction of imbrication in a Pliocene conglomerate in Italy, and BLUCK (1965) correlated cobble fabric with paleocurrents and bed thickness in Triassic conglomerates in Wales. SCHLAGER and SCHLAGER (1973) used clast orientation in Jurassic breccias to help infer paleoslope in a deep-water carbonate sequence. ENGEL (1974) studied pebble fabrics in carbonate turbidites.

Fabrics of tills and other diamictites have continued to receive attention. A substantial study of till is that of KRÜGER (1970). DRAKE (1974) confirmed that rods are parallel to flow, whereas blades and disks are less so, although even their long axes tend to be parallel to flow direction. Technique papers are provided by ANDREWS (1974), ANDREWS and SMITH (1969) and McGOWEN and DERBYSHIRE (1974), the last noting that a scanning electron microscope can be used to examine the orientation of the clay matrix of diamictites to supplement field study of pebble and cobble fabrics. LIBORIUSSEN (1973) suggested the use of two radiographs, one horizontal and one perpendicular to bedding to give the flow direction of embedded grains; and he also found good correlation between the X-ray and macrofabric. GRAVENOR and STUPAVSKY (1975) note that magnetic anisotropy may obviate particulate study of till fabric. Some examples of paleocurrent study in ancient diamictites include those of LINDSEY (1966) and CASSHYAP and QIDWAI (1974). Normally, it is good practice to measure other paleocurrent structures when they are available, in addition to clast orientation.

A significant paper is LINDSAY's (1968) computer simulation study of clast orientation in mud flows (fig. 3-17). He found a strong long axis orientation parallel to flow. This orientation probably formed just before cessation of movement in response to a strong vertical velocity gradient.

CAINE (1967) found a weak to isotropic fabric in talus (cf. RAPP*, 1959, 1960a and 1960b); later CAINE (1972) published a very interesting paper wherein he showed that variance of long axes of talus blocks becomes smaller downslope with increasing distance. In the same paper he also comments on sample size.

Earlier block orientation has also been measured from airphotographs by WASTEN-
SON (1969). McSAVENEY (1971), studying the surficial fabric of rockfall talus,
found fabric strength to be a function of surface roughness and to decrease with
increasing particle size downslope.

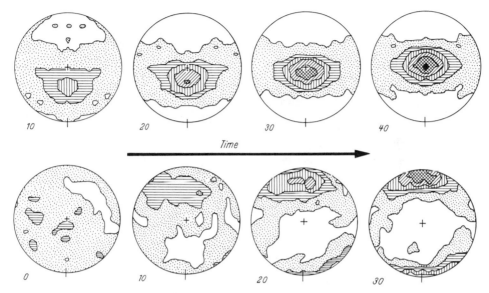

Fig. 3-17. Computer-generated equal-area fabric diagram of clast orientation in a mud flow deposited from laminar flow
(LINDSAY, 1968, figs. 3 and 4, J. Sediment. Petrol. Soc. Econ. Paleontol. Mineral.). *Above:* short axis fabric; *below:* long
axis fabric. *Arrow:* flow direction; *numbers:* progressive sequential development; *white areas:* lowest concentration; *tick
mark:* flow direction. Computer simulation of fabrics is one of paleocurrent's significant post-1963 developments. But
see also computer modeling of oceanic circulation patterns in Basin Analysis and Beyond (Part 2.9)

Volcaniclastics

Orientation studies of clasts and phenocrysts in ash and lava flows by sedimentologists are not abundant, but definitely do represent a new post-1963
development. SCHMINCKE and SWANSON (1967) describe oriented flattened and
stretched pumic fragments, due to laminar, viscous flow in welded tuffs from the
Canary Islands. The orientation of the stretched pumice is correlated with lineation and tension cracks. In lava flows, SMITH and RHODES (1972) made paleocurrent maps from over 20,000 measurements of microscopic and megascopic
phenocrysts. Their study is also a good source for the earlier literature. GREEN-
WOOD et al. (1972) even have reported on the phenocryst fabric of a lunar basalt!
CUMMINGS (1964) noted that eddies on the downcurrent side of inclusions in lava
give a unique sense to the local flow direction and are, indeed, the exact analog
to eddies behind obstacles in water and air. Paleocurrents in late Precambrian
flows in Minnesota are reported by GREEN (1972, Fig. V–12).

Fossil Orientation

Interest in fossil orientation continues and some very innovative applications
have emerged. Fossils having cylindrical, conical, and oblate shapes, have proved

Table 3-1. *Studies of Fossil Orientation*

Author and year	Remarks
ZANGERL and RICHARDSON, 1963	Oriented driftwood and fossils, level by level, in a thin Pennsylvanian black shale
SEILACHER and MEISCHNER, 1964	Much quantitative and qualitative data on fossil orientation beautifully integrated into the sedimentology of a lower Paleozoic geosyncline in Norway. Their Table 1 is a superb model to follow for the paleoecologist
COLTON, 1967a	Integrated paleocurrent study of turbidites: flute casts, ripples, plant fragments and a problematica trace fossil, *Fucoides graphica*, the latter possibly related to transport of waterlogged fecal pellets of fish. Short and well done.
KISSLING and LINEBACK, 1967	One of the most unusual paleocurrent studies ever made — orientation of 2483 fragments of *Favostinae* was measured and indicates an east-west orientation, believed to be the response to gentle tidal currents. Other interesting results are dispersion maps and good discussion of environmental dynamics
WOBBER, 1967	The bivalve *Donax* orients nearly perpendicular to the present shoreline with its blunt, anterior margin pointing shoreward
ERNST and WACHENDORF, 1968	Oriented gastropods are part of a comprehensive paleocurrent-sedimentologic study of 12 quarries of a Middle Triassic limestone in Germany
REYMENT, 1968	Orthoconic nautiloids, when stranded in soft carbonate sediment, orient in the swash perpendicular to the shoreline
SCHLEIGER, 1968	Field studies supplemented by theoretical justification for longitudinal, transverse, and sub-parallel modes of graptolite orientation using harmonic analysis. Correlates graptolite orientation with orientation of other paleocurrent structures
MOORS, 1969	Graptolites in the pelitic shale above the sandy part of a turbidite are oriented parallel to emplacement of underlying turbidite. Good source of earlier references
TROMPETTE, 1969	Many photographs and measurements of the long axes of elongate, domal stromatolites. Associated ripple and sole marks indicate similar orientation of currents. Well referenced
EDER, 1970	Brachiopod orientation used to assess orientation of currents that deposited the calcareous detritus of turbidites
HUBBARD, 1970	Careful analysis of coral orientation and its relation to paleoecology and sedimentology. Percent of corals lying in plane of bedding, inclined to it, in growth position as well as overall orientation. Two interesting tables and one very informative flow diagram. Outstanding

Table 3-1 (Cont.)

Author and year	Remarks
JONES and DENNISON, 1970	Over 13,000 measurements of fossil orientation (graptolites, brachiopods, cricconarids, gastropods, ostracods, and *Tasmanites*) in Athens (Ordovician) and Chattanooga (Devonian) shales. Authors suggest that while hundreds of fossils must be measured at a locality to yield significant results this is generally not a serious problem in shales, because the fossils are small and usually a single bedding plane contains sufficient numbers. Five useful suggestions for additional research. Short but significant paper
REYMENT, 1970	Orientation studied in a single cutcrop — primarily a methodology paper with well-worked examples
ERICKSON, 1971	A line constructed across the aperatural opening of *Helminthoglypta arrosa* (and tangent to its last whorl) makes an angle of about 95° with the wind
LAWRENCE, 1971	Scattered individual oysters have no preferred orientation of their planes of commissure, but become oriented as packing density increases with orientation responding to tidal flow
DEWINDT, 1972	Preferentially oriented ostracoderm shields in fine grained redbeds — the only study of oriented vertebrate debris known to us.
BAILEY and ERICKSON, 1973	*Tancredia americana*, a bivalve, orients with long axis parallel to strandline of low energy beach and also with long axis parallel to unidirectional flow (as judged by flume studies)
WENDT, 1973	Orientation and imbrication of cephalopod shells by strong currents during periods of minimum sedimentation
CONRAD and BEAUDOIN, 1974	Thin section study of sponge spicules reveals good preferred orientation. One plate with four figures plus current roses. One of the few orientation studies using sponge spicules as oriented particles
LAUFELD, 1974	Fossil orientation integrated with local paleogeography, which suggests long axes of orthoconic nautiloids are perpendicular to shore line
YOUNG and LONG 1976	Almost perfect orientation of elongate stromatolitic columns, ridges and grooves, and intermound fillings.

exceptionally useful (Table 3-1). Certainly oriented fossils deserve more attention by sedimentologists than they seem to have received, especially because small fossils such as graptolites have been used to infer paleocurrents in shales, an observation first made by RUEDEMANN* (1897). Small fossils and carbonaceous debris of light density appear to be very sensitive to even the weakest currents. Unfortunately, bioturbation commonly destroys the orientation of small as well as larger body fossils. In a number of studies, fossil orientation was believed to be the response to tidal currents. Particularly unusual is the use of oriented gastropod shells on beaches as paleowind indicators (ERICKSON, 1971). CONRAD and BEAUDOIN's (1974) study of orientation of siliceous sponge spicules in Creta-

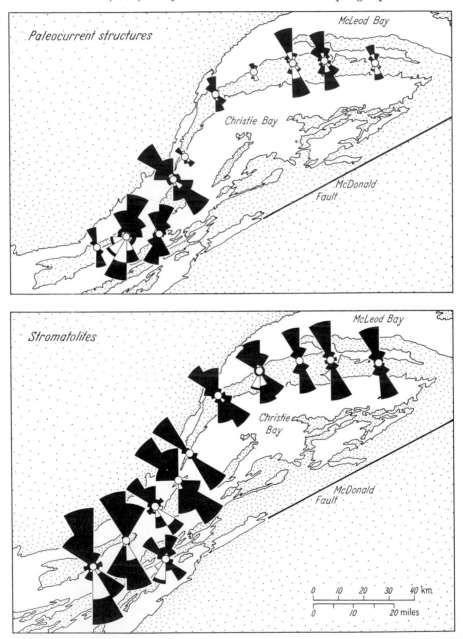

Fig. 3-18. Similarity between paleocurrent orientation (679 structures at top) in sediments of Aphebian (Precambrian) age and long axes of 1 377 interbedded stromatolites (below) along East Arm, Great Slave Lake, Northwest Territories, Canada. (Redrawn from HOFFMAN, 1967, fig. 1, Science, **157** Published by permission). In addition to responding to the local current direction, stromatolite morphology has also been used as a means of correlation in the Precambrian

ceous cherts in France is unusual as is IMOTO and FUKUTOMI's (1975) study of sponge spicules in Permian-Triassic cherts in Japan. Another aspect of fossil orientation that is little studied is the orientation of algal stromatolites as was done by HOFFMAN (1967) in the Precambrian of Canada (fig. 3-18).

Flume studies of fossil orientation include those by KELLING and WILLIAMS (1967), NAGLE (1967), BRENCHLEY and NEWELL (1970) and CLIFTON (1971). Some of these investigators considered the relative stability of concave and convex shells under different conditions of flow (concave down in traction currents: concave up in turbidity currents) whereas others noted that cylindrical fossils, lying on the bottom, are oriented parallel to wave crests (and thus parallel to shoreline) and rodlike fossils are oriented parallel to the flow of undirectional and bidirectional currents. If the fossil is elongate, but with an apex, the apex will point into or away from a unidirectional current, depending on its geometry and on its internal distribution of density.

Argillaceous Sediments and Loess

In 1964 BREWER provided, in his Chapter 8, a good introduction to the fabrics of soils and mudstones, one that is worth reading for background even though it does not contain too much on the orientation of clay size particles. Since then more studies of clay mineral orientation by the X-ray and scanning electron microscope have been made. Using the latter, GIPSON (1965) showed that the bedding planes of shales contain small oriented particles. Using optical, SEM, and X-ray methods, KRIZEK et al. (1975) studied the fabric of experimentally deformed clays. Somewhat earlier OERTAL and CURTIS (1972) related the clay mineral fabric of a clay ironstone concretion to its cementation and compactional history — a study somewhat peripheral perhaps to paleocurrents, but an interesting technique paper that might be of value. For those wishing to explore the use of the polarizing microscope for determining clay fabrics, the paper by MORGENSTERN and TCHALENKO (1967) is fundamental.

MATALUCCI et al. (1969) determined paleowind direction in Wisconsin loess using oriented plugs measuring their dielectric anisotropy. This is the only study of its kind known to us.

Miscellaneous Fabric Structures

COLTON (1967b) measured the long axes of carbonate concretions in the Upper Devonian of New York along a 110 mile (176 km) outcrop (fig. 3-19); he believed this orientation, which is very well defined, was a response to the anisotropic fabric of the host, which in turn was the response to the original marine currents. The paleocurrents of these sediments do in fact indicate a general westward flow. Independently, SPERLING (1967, p. 342) noted a similar correlation between currents and elongation of concretions in the Carboniferous of Germany. Certainly, we as sedimentologists, have not extracted all the information from concretions! HENNINGSMOEN (1974) provides a good insight into some of the factors controlling the shape of concretions, including permeability.

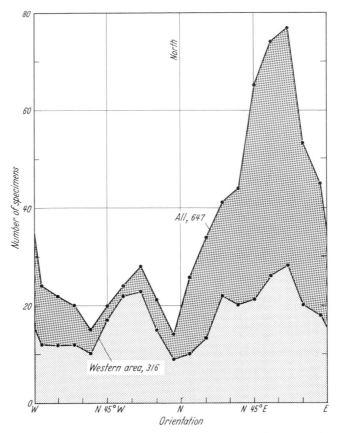

Fig. 3-19. Frequency diagram of elongate concretions in shale and mudrock of Devonian age in western New York. (Redrawn from Colton 1967b, fig. 3.) Vector mean of N 55° E closely parallels that of plant debris and associated sole marks. How often have we measured concretions? Would not such information be most helpful for better inferring their origin and perhaps even contribute significantly to our insight into the depositional environment of the host sediment?

Another way to determine the initial dip of a bed is to collect oriented specimens showing geopetal structures and measure the angle between its cement-mud contact and the bedding. This is done fairly commonly by carbonate petrologists, one of but many examples being the study of Broadhurst and Simpson (1973), who also used crinoid orientation to help establish the dip direction of the sloping shelf on which the bed was deposited.

Future Problems

The foremost is to obtain more paleocurrent information from mudstones and shales, which form about 60 percent of all sediments. Oriented scanning electron photomicrographs of bedding planes and orientation studies of small and micro-fossils or carbonaceous debris offer promise, as possibly do X-ray and some microscope methods. It is often said that the paleocurrent system of muds — be they of carbonate or argillaceous composition — *could be* significantly different from that of the interbedded sandstones or conglomerates, be they terrigenous or carbonate. So why not find out? And when doing so, why not examine the orientation of their associated elongate concretions?

Another opportunity is reconstruction of the Pleistocene paleowinds from the fabric of loess as was done by MATALUCCI *et al.* (1969). In this time of a possible new cycle of colder climates and of Pleistocene paleoclimate data acquired from deep sea cores, studies such as this seem especially timely.

More theoretical and/or computer simulation papers such as the work of Lindsey's on the fabric of debris flows seem to be worthwhile, particularly if the basic theory can be borrowed from elsewhere. For example, what explanations, other than Bagnold's suggestion of internal friction (1956, p. 293) can be offered for the commonly observed imbrication angle of 10° to 30° of flat pebbles, which tend to occur as oriented chains or packets of 3 to 5?

Finally, because an anisotropic mass property measured on an oriented plug is nearly always quicker and better than direct measurement of fine silt or carbonaceous debris, devices such as those used by SHELTON *et al.* (1974) should be used much more widely. However, as Table 1 shows, conventional studies of fossil orientation can yield high rewards, so that very much can still be done by simple means. And in defence of classical methods what is more rewarding, on a pleasant day in the spring or fall, than leisurely measuring fossil orientation in the field musing all the while on its paleoecologic significance?

References

ANDREWS, J. T., 1974: Techniques of till fabric analysis. Brit. Geomorph. Res. Group, Tech. Bull. **6**, 43 p.
—, and D. I. SMITH, 1969: Statistical analysis of till fabric: methodology, local and regional variability. Quart. J. Geol. Soc. London **125**, 503—542.
BAGNOLD, R. A., 1956: Flow of cohesionless grains in fluids. Proc. Roy. Philos. Soc. Trans. **249**, 235—297.
BAILEY, L. T., and J. J. ERICKSON, 1973: Preferred orientation of bivalve shells in the Upper Timber Lake Member, Fox Hills Formation in North Dakota — Preliminary interpretations. Compass **50**, 23—37.
BAKER, D. W., H. R. WENK and J. M. CHRISTIE, 1969: X-ray analysis of preferred orientation in fine-grained quartz aggregates. J. Geol. **77**, 144—172.
BHATTACHARYYA, D. S., 1966: Orientation of mineral lineation along the flow direction in rocks. Tectonophysics **3**, 29—33.
BLUCK, B. K., 1965: The sedimentary history of some Triassic conglomerates in the Vale of Glamorgan, South Wales. Sedimentology **4**, 225—245.
BRENCHLEY, P. J., and G. NEWELL, 1970: Flume experiments on orientation and transport of models and shell valves. Paleogeog., Palaeoclimatol., Palaeoecol. **7**, 185—220.
BRENNER, H., and D. W. CONDIFF, 1972: Transport mechanics in systems of orientable particles III. Arbitrary particles. J. Colloidal Interface Sci. **41**, 228—274.
—, 1974: Transport mechanics in systems of orientable particles IV. Convective transport. J. Colloidal Interface Sci. **47**, 199—264.
BREWER, R., 1964: Fabric and mineral analysis of soils. New York: John Wiley and Sons, 470 p.
BROADHURST, F. M., and I. M. SIMPSON, 1973: Bathymetry on a Carboniferous reef. Lethaia **6**, 367—372.
CAINE, N., 1967: The texture of talus. J. Sediment. Petrol. **37**, 796—803.

—, 1972: Air photo analysis of blockfield fabrics in Talus Valley, Tasmania. J. Sediment. Petrol., **42**, 33—48.

CANUTI, P., G. PRANZINI and G. SESTINI, 1966: Provenienza ed ambiente di sedimentazione dei ciottolami del Pliocene di S. Casciano (Firenze). Mem. Soc. Geol. Ital. **5**, 340—364.

CASSHYAP, S. M., and H. A. QIDWAI, 1974: Glacial sedimentation of the Late Paleozoic Talchir diamictite, Pench Valley Coalfield, Central India. Bull. Geol. Soc. Am. **85**, 749—760.

CLIFTON, H. E., 1971: Orientation of pelecypod shells and shell fragments in quiet waters. J. Sediment. Petrol. **41**, 671—682.

COLTON, G. W., 1967a: Late Devonian current directions in western New York with special reference to *Fucoides graphica*. J. Geol. **75**, 11—22.

—, 1967b: Orientation of carbonate concretions in the Upper Devonian of New York. U.S. Geol. Survey Prof. Paper **575**, B 57—B 59.

CONDIFF, D. W., and H. BRENNER, 1969: Transport mechanics in systems of orientable particles. Physics Fluids **12**, 539—551.

CONRAD, M., and M. B. BEAUDOIN, 1974: Orientation de spicules de spongiaires dans le Crétacé Supérieur du Bassin de l'Esterou (Alpes-Maritimes); Application à la réconstitution de la dynamique du dépôt et de la configuration du bassin. C.R. Acad. Sci. Paris **278** D, 3047—3050.

CUMMINGS, D., 1964: Eddies as indicators of local flow direction in rhyolite. U.S. Geol. Survey Prof. Paper **475-D**, 70—72.

DELFINER, P., J. ETIENNE and J. M. FONCK, 1972: Application de l'analyseur de textures à l'étude morphologique des réseaux poreux en lames minces. Rev. Inst. Français Petrol. **27**, 535—556.

DeWINDT, J. T., 1972: Vertebrate fossils as paleocurrent indicators in the Upper Silurian of the central Appalachians. Compass **49**, 125—138.

DRAKE, L. D., 1974: Till fabric control by clast shape. Bull. Geol. Soc. Am. **85**, 247—250.

EDER, F. W., 1970: Genese Riff-naher Detritus-Kalke bei Balve im Rheinischen Schiefergebirge (Garbecker Kalk). Verh. Geol. B.-A. Jg. 1970, 551—569.

ENGEL, W., 1974: Sedimentologische Untersuchungen im Flysch des Beckens von Ajdovščina (Slowenien). Göttinger Arb. Geol. Paläontol. **16**, 65 p.

ERICKSON, J. M., 1971: Wind-oriented gastropod shells as indicators of paleowind direction. J. Sediment. Petrol. **41**, 589—593.

ERNST, G., and H. WACHENDORF, 1968: Feinstratigraphisch-fazielle Analyse der „Schaumkalk-Serie" des Unteren Muschelkalkes im Elm (Ost-Niedersachsen). Beih. Ber. Naturh. Ges. **5**, 165—205.

GIPSON, M., JR., 1965: Application of the electron microscope to the study of particle orientation and fissility in shale. J. Sediment. Petrol. **35**, 408—414.

GRAVENOR, C. P., and M. STUPAVSKY, 1975: Convention for reporting magnetic anisotropy of till. Canadian J. Earth Sci. **12**, 1063—1069.

GREEN, J. C., 1972: North Shore Volcanic Group *in* P. K. SIMS and G. B. MOREY, eds., Geology of Minnesota: Centennial Volume Minn. Geol. Sur., 294—332.

GREENWOOD, W. R., D. A. MORRISON and A. L. CLARK, 1972: Phenocryst fabric in lunar basalt sample 12052 from the Ocean of Storms. Bull. Geol. Soc. Am. **83**, 2809—2816.

HENLINE, W. D., and D. W. CONDIFF, 1970: Transport mechanics in systems of orientable particles. II. Kinetic theory of orientation specific transport for hardcore models. J. Chemical Physics **52**, 5027—5043.

HENNINGSMOEN, G., 1974: A comment. Origin of limestone nodules in the Lower Paleozoic of the Oslo region. Norsk Geol. Tidsskrift **54**, 401—412.

HOFFMAN, P., 1967: Algal stromatolites: use in stratigraphic correlation and paleocurrent direction. Science **157**, 1043—1045.

HUBBARD, J. A. E. B., 1970: Sedimentological factors affecting the distribution and growth of Visean Caninioid corals in North-West Ireland. Paleontology **13**, 191—209.

IMOTO, N., and M. FUKUTOMI, 1975: Genesis of bedded cherts in the Tamba Belt, southwest Japan *in* Editorial Committee, 1975, Problems on geosynclines in Japan, Mon. **19**, p. 35—42, (Japanese with Engl. summ.)

JEFFERY, G. B., 1922: The motion of ellipsoidal particles immersed in a viscous fluid. Proc. Roy. Soc. London **102**A, 6—79.

JOHANSSON, C. E., 1963: Orientation of pebbles in running water. A laboratory study. Geog. Annaler **45**, 85—112.

—, 1965: Structural studies of sedimentary deposits. Lund Studies in Geography, Ser. A., Phys. Geog. **32**, 61 p.

JONES, M. L., and J. M. DENNISON, 1970: Oriented fossils as paleocurrent indicators in Paleozoic lutites of southern Appalachians. J. Sediment. Petrol. **40**, 642—649.

KATZUNG, G., 1971: Zur fluviatilen Gerölleinregelung. Z. angew. Geol. **17**, 39—42.

KELLING, G., and P. F. WILLIAMS, 1967: Flume studies on the reorientation of pebbles and shells. J. Geol. **75**, 243—267.

KISSLING, D. L., and J. A. LINEBACK, 1967: Paleoecological analysis of corals and stromatoporoids in a Devonian biostrome, Falls of the Ohio, Kentucky, Indiana. Bull. Geol. Soc. Am. **78**, 157—174.

KRIZEK, R. J., T. B. EDIL and I. K. OZAYDIN, 1975: Preparation and identification of clay samples with controlled fabric. Eng. Geol. **9**, 13—38.

KRÜGER, J., 1970: Till fabric in relation to direction of ice movement. A study from the Fakse Banke, Denmark. Geog. Tidsskr. **69**, 133—170.

LAUFELD, S., 1974: Preferred orientation of orthoconic nautiloids in the Ludlovian Hemse Beds of Gotland. Geol. Fören. Stockholm Förh. **96**, 157—162.

LAWRENCE, D. R., 1971: Shell orientation in recent and fossil oyster communities from the Carolinas. J. Paleontol. **45**, 347—349.

LIBORIUSSEN, J., 1973: Till fabric analysis based on X-ray radiography. Sediment. Geol. **10**, 249—260.

LINDSAY, J. F., 1968: The development of clast fabric in mudflows. J. Sediment. Petrol. **38**, 1242—1253.

LINDSEY, D. A., 1966: Sediment transport in a Precambrian ice age: the Huronian Gowganda Formation. Science **154**, 1442—1443.

MARTINI, I. P., 1971a: Grain orientation and paleocurrent systems in the Thorold and Grimsby Sandstones (Silurian), Ontario and New York. J. Sediment. Petrol. **41**, 425—434.

—, 1971b: A test of validity of quartz grain orientation as a paleocurrent and paleoenvironmental indicator. J. Sediment. Petrol. **41**, 60—68.

MATALUCCI, R. V., J. W. SHELTON and M. ABDEL-HADY, 1969: Grain orientation in Vicksburg loess. J. Sediment. Petrol. **39**, 969—979.

McGOWEN, A., and E. DERBYSHIRE, 1974: Technical developments in the study of particulate matter in glacial tills. J. Geol. **82**, 225—235.

McSAVENEY, EILEEN R., 1971: The surficial fabric of rockfall talus *in* M. MORISAWA, ed., Quantitative geomorphology: some aspects and applications. Binghamton State Univ. New York: Publ. in Geomorphol., 181—197.

MOORS, H. T., 1969: The position of graptolites in turbidites. Sediment. Geol. **3**, 241—261.

MORGENSTERN, N. R., and J. S. TCHALENKO, 1967: The optical determination of preferred orientation in clays and its application to the study of microstructure in consolidated kaolin. I and II. Proc. Roy. Soc. London **300**A, 218—250.

MÜLLER, W., 1974: The Leitz texture-analysing system. Leitz Sci. Tech. Information, Suppl. **1**, 4, 101—136.

NAGLE, J. S., 1967: Wave and current orientation of shells. J. Sediment. Petrol. **37**, 1124—1138.

OERTAL, G., and C. D. CURTIS, 1972: Clay-ironstone concretions preserving fabrics due to progressive compaction. Bull. Geol. Soc. Am. **83**, 2597—2606.

ONIONS, D., and G. V. MIDDLETON, 1968: Dimensional grain orientation of Ordovician turbidite greywackes. J. Sediment. Petrol. **38**, 164—174.

PICARD, M. D., and D. D. BECKMAN, 1966: Non-opaque and opaque grain fabrics of siltstones in Red Peak Member (Triassic), Central Wyoming. J. Sediment. Petrol. 36, 506—521.

REES, A. I., 1971: The magnetic fabric of a sedimentary rock deposited on a slope. J. Sediment. Petrol. 41, 307—327.

—, U. VON RAD and F. P. SHEPARD, 1968: Magnetic fabric of sediments from the La Jolla Submarine Canyon and Fan, California. Marine Geol. 6, 145—178.

REYMENT, R. A., 1968: Orthoconic nautiloids as indicators of shoreline surface currents. J. Sediment. Petrol. 38, 1387—1389.

—, 1970: Quantitative paleoecology of some Ordovician orthoconic nautiloids. Palaeogeog., Palaeoclimatol., Palaeoecol. 7, 41—49.

RÜHL, H., 1974: Das Leitz Texture-Analyse-System (T.A.S.). Leitz-Mitt. Wiss. Techn. 6, 113—116.

RUST, B. R., 1972: Pebble orientation in fluvial sediments. J. Sediment. Petrol. 42, 384—388.

SCHLAGER, W., and M. SCHLAGER, 1973: Clastic sediments associated with radiolarites (Tauglboden-Schichten, Upper Jurassic, Eastern Alps). Sedimentology, 20, 65—89.

SCHLEIGER, N. W., 1968: Orientation distribution patterns of graptolite rhabdosomes from Ordovician sediments in central Victoria, Australia. J. Sediment. Petrol. 38, 462—472.

SCHMINCKE, H.-U., and D. A. SWANSON, D. A., 1967: Laminar viscous flowage structures in welded ash flow tuffs from Gran Canaria, Canary Islands. J. Geol. 75, 641—664.

SEDIMENTATION SEMINAR, 1965: Gravel fabric in Wolf Run. Sedimentology 4, 273—238.

SEILACHER, A., and D. MEISCHNER, 1964: Faziesanalyse im Paläozoikum des Oslo-Gebietes. Geol. Rundschau 54, 596—619.

SESTINI, G., 1964: Paleocorrenti eoceniche nell'area tosco umbra. Bol. Soc. Geol. Ital. 83, 1—54.

—, and G. PRANZINI, 1965: Correlation of sedimentary fabric and sole marks in turbidites. J. Sediment. Petrol. 35, 100—108.

SHELTON, J. W., and D. E. MACK, 1970: Grain orientation in determination of paleocurrents and sandstone trends. Bull. Am. Assoc. Petrol. Geologists 54, 1108—1119.

—, H. R. BURMAN and R. L. NOBLE, 1974: Directional features in braided-meandering-stream deposits, Cimarron River, north-central Oklahoma. J. Sediment. Petrol. 44, 1114—1117.

SIPPEL, R. F., 1971: Quartz grain orientation — 1 (The photometric method). J. Sediment. Petrol. 41, 38—59.

SMITH, E. I., and R. C. RHODES, 1972: Flow direction determination of lava flows. Bull. Geol. Soc. Am. 83, 1869—1874.

SPERLING, H., 1967: Sedimentstrukturen und Strömungsmarken im höheren Kulm III β Geologische Beobachtungen im Erz-Bergwerk Grund/Westharz. Neues Jahrb. Geol. Paläontol. Abh. 127, 337—349.

SPOTTS, J. H., 1964: Grain orientation and imbrication in Miocene turbidity current sandstones, California. J. Sediment. Petrol. 34, 229—253.

STARKEY, J., 1974: The quantitative analysis of orientation data obtained by the Starkey method of X-ray fabric analysis. Canadian J. Earth Sci. 11, 1507—1516.

TEISSEYRE, A. K., 1975: Pebble fabric in braided stream deposits from Recent and "frozen" Carboniferous channels (Intrasudetic Basin, Central Sudetes). Geol. Sudetica 10, 1—46 (Polish and English).

TROMPETTE, P., 1969: Les Stromatolites du "Précambien Supérieur" de L'Adrar de Mauritianie (Sahare Occidental). Sedimentology 13, 123—154.

WASTENSON, L., 1969: Blockstudier i flygbilder. Sveriges Geol. Unders. Ser. CNR 638, 63, 95 p.

WENDT, J., 1973: Cephalopod accumulations in the Middle Triassic Hallstatt-Limestone in Jugoslavia and Greece. Neues Jahrb. Geol. Paläontal. Mh., 189—206.

WINKELMOLEN, A. M., W. VAN DER KNAAP and R. EIJPE, 1968: An optical method of measuring grain orientation in sediments. Sedimentology **11**, 183—196.

WOBBER, F. J., 1967: The orientation of *Donax* on an Atlantic coast beach. J. Sediment. Petrol. **37**, 1233—1235.

YOUNG, G. M. and D. G. F. LONG, 1976: Stromatolites and basin analysis: an example. from the Upper Proterozoic of northwestern Canada. Palaeogeogr. Palaeoclimatol. Palaeoecol. **19**, 303—318.

ZANGERL, R., and E. S. RICHARDSON JR., 1963: The paleoecological history of two Pennsylvanian black shales. Fieldiana, Geol. Mem. **4**, 352 p.

Cross-Bedding and Ripple Marks up to 1963

Introduction

Cross-bedding and ripple mark have received more attention from geologists than all other directional structures combined.

Cross-bedding is shown rather commonly on geologic sections and drawings of the early 19th century indicating that it was recognized almost from the beginnings of modern observational geology. Subsequently, it has been mentioned and described in many studies of ancient sediments. Ripple marks may have been recognized as a current structure as early as the Renaissance, if not in antiquity. Descriptions of ripple marks, in both modern and ancient sediments, far exceed those of cross-bedding and are truly voluminous. Beginning in the later part of the 19th century, much effort has been devoted to qualitative appraisal of the environmental significance of both cross-bedding and ripple marks. In contrast, maps of cross-bedding and ripple marks were not made until the 20th century.

Although cross-bedding is defined by its internal character and ripple marks by external form, both are considered in the same chapter because of their relation to sand waves at the sediment-water interface.

Cross-Bedding

Introduction

SORBY (1853) appears to have been the first to systematically measure cross-bedding, although he apparently neither published a current rose nor plotted his measurements on a map. However, he made several thousand observations on various formations and in his lifetime accumulated many measurements in his notebooks. RUBEY and BASS (1925, pl. 3) were apparently the first to actually plot cross-bedding measurements on a map.

From these beginnings cross-bedding has been measured and mapped by many. Table 4-1 lists all those who have published maps of cross-bedding. Others who have measured cross-bedding, usually presenting their data in a frequency distribution or current rose, include JÜNGST (1928, figs, 7, 8, 9 and 10), KNIGHT (1929, figs. 30 and 31), KIDERLEN (1931, p. 298), SCHMITT (1935, fig. 2), JÜNGST (1938), McKEE et al. (1953, figs. 3, 4 and 5), AGATSON (1954, p. 522), EDWARDS (1955, fig. 26), DOTT (1955, fig. 10), TANNER (1955, figs. 3 and 7), WHITAKER (1955, fig. 2), MELLEN (1956, fig. 30), CROOK (1957, Table 1), PETTIJOHN (1957), RADOMSKI (1958, fig. 15), GRADZINSKI et al. (1959, fig. 5), TANNER (1959), THOME (1959, fig. 4), BASSETT and WALTON (1960, fig. 9a), MARTIN et al. (1960, fig. 2),

MIKKOLA (1960, fig. 9), BALL and DINELEY (1961, p. 185), HAMBLIN (1961b, fig. 9), McIVER (1961, fig. 122), POTTER and PRYOR (1961, Table 7), POWER (1961, fig. 3), RADOMSKI (1961, figs. 1 and 3), KANEKO (1958, figs. 1 and 2), ALLEN (1962, fig. 18) and STEWART (1962, figs. 2 and 10).

These studies demonstrate that cross-bedding has an ubiquitous occurrence. It is found in every major environment: fluvial, littoral, marine, and aeolian. It occurs in clastic deposits varying from fine silt to coarse conglomerate. It is found in mechanically deposited carbonates such as oolites and in sandstones ranging from graywackes of turbidite origin to the cleanest quartz sand. Cross-bedding may also be present in massive silts and sands and difficult to see except by staining of by X-ray techniques (HAMBLIN, 1962). Cross-bedding occurs in sediments that range in age from Precambrian to Recent. With a few exceptions, such as the studies by JÜNGST (1938, p. 267—270), SZÁDECZKY-KARDOSS (1938, figs. 7 and 8), HÜLSEMANN (1955), RUKHIN (1958, fig. 290), FRAZIER and OSANIK (1961) and several others, most investigations are of cross-bedding in ancient sediments, principally of Paleozoic and Mesozoic age, rather than modern sediments. Most of the cross-bedding mapped occurs in fluvial and aeolian sands, probably because it is more abundant in these deposits than it is marine and littoral sediments.

The above cited references probably represent well over 150,000 measurements. These studies demonstrate that, with few exceptions, cross-bedding in a sand body, a formation, or even an entire basin, nearly always has well defined preferred orientation. Practically every clastic formation, whatever its age, may be safely assumed to have a preferred orientation of cross-bedding. This direction is estimated by systematic mapping.

Table 4-1. *Published Maps of Cross-bedding*

Author and year	Formation and age	Measurements	Environment	Other properties studied
RUBEY and BASS, 1925	Dakota (Cretaceous)	At least 11	Fluvial	Subsurface Mapping
JÜNGST, 1928 . .	Tertiary	Unspecified	Marine	
JÜNGST, 1929 . .	Tertiary	Unspecified	Marine	
KNIGHT, 1929 . .	Casper (Pennsylvanian)	Approx. 125	Marine	
BRINKMANN, 1933	Bunter (Triassic)	4,000	Fluvial	
FORCHE, 1935 . .	Bunter (Triassic)	3,235	Fluvial	Maximum pebble size, petrology and isopach
SHOTTON, 1937 .	Bunter (Triassic)	1,142	Aeolian	Grain size and sorting
JÜNGST, 1938 . .	Modern dunes (Recent)	Unspecified	Aeolian	Wind velocities
REICHE, 1938 . .	Coconino and DeChelly (Permian)	909	Aeolian	
SZÁDECZKY-KARDOSS, 1938 . .	Pliocene, Pleistocene and Recent	Unspecified	Fluvial	

Table 4-1 (Continued)

Author and year	Formation and age	Measurements	Environment	Other properties studied
BAUSCH VAN BERTSBERGH, 1940	Siegner Beds (Devonian)	Unspecified	Marine	Ripple and sole mark, plant remains and heavy mineral provinces
McKEE, 1940 . .	Supai (Cambrian)	510	Deltaic	
	Tapeats (Cambrian)	300	Marine	
SEIFERT, 1942 .	Bunter (Triassic)	1,090	Fluvial	
ALLEN, 1949 . .	Top Ashdown Pebble Bed (Cretaceous)	Unspecified	Fluvial	Grain size, heavy minerals
ILLIES, 1949 . .	Kaolin sand of Sylt (Pleistocene)	Approx. 200	Fluvial	
KIERSCH, 1950 .	Navajo (Jurassic)	180	Aeolian	
PICARD, 1950 . .	Bunter (Triassic)	1,652	Fluvial	
VASSOEVICH and GROSSGEYM, 1951	Chokraksk (Tertiary)	2,896	Marine	Grain size
	Karagansk (Tertiary)	Unspecified	Lacustrine	
STOKES, 1952 . .	Dakota (Jurassic)	Unspecified	Fluvial	Ore deposits in channel
ILLIES, 1952a . .	Outwash (Pleistocene)	Unspecified	Fluvial	Terraces and moraines
ILLIES, 1952b . .	Outwash (Pleistocene)	Unspecified	Fluvial	Terraces and moraines
BIEBER, 1953 . .	Mansfield (Pennsylvanian)	76	Fluvial	Channel system at unconformity
LEMCKE, et al., 1953	Upper Molasse (Tertiary)	4,250	Fluvial	Heavy mineral provinces and isopach
PICARD, 1953 . .	Upper Fresh-Water Molasse (Tertiary)	Unspecified	Fluvial	
SCHWARZACHER, 1953	Woburn (Cretaceous)	Approx. 600	Marine, delta platform	
STOKES, 1953 . .	Salt Wash (Jurassic)	Unspecified	Fluvial	
WILSON et al., 1953	Moine Series (Precambrian)	Unspecified	Unspecified	
JONES, 1954 . .	Salt Wash (Jurassic)	Unspecified	Fluvial	Sand-shale ratio map
OSMOND, 1954 . .	Sevy (Silurian)	Unspecified	Marine	Grain size, sorting and isopach
POTTER and OLSON, 1954 . .	Caseyville, Mansfield (Pennsylvanian)	542	Fluvial	Channel system at unconformity
STOKES, 1954a .	Morrison (Jurassic)	Unspecified	Fluvial	

Table 4-1 (Continued)

Author and year	Formation and age	Measurements	Environment	Other properties studied
BRETT, 1955 . .	Baraboo (Precambrian)	283	Unspecified	
CRAIG et al., 1955	Salt Wash (Jurassic)	Unspecified	Fluvial	Isopach and facies map
HÜLSEMANN, 1955	Modern beach (Recent)	Unspecified	Littoral	
	Upper Marine Molasse (Tertiary)	Unspecified	Littoral	
LOWELL, 1955 . .	Morrison (Jurassic)	Unspecified	Fluvial	Ore deposits in channel
MIKKOLA, 1955 .	Karelian (Precambrian)	110	Fluvial	
POTTER, 1955 . .	Lafayette (Pliocene)	158	Fluvial	Heavy mineral provinces, grain size
REINEMUND, 1955	Triassic	Unspecified	Fluvial	
TANNER, 1955 . .	Tuscaloosa (Cretaceous	378	Marine and non-marine	
	Providence (Cretaceous)	154	Fluvial	
WALKER, 1955 .	Millstone Grit (Lower Carboniferous)	114	Deltaic	Fabric
POTTER and SIEVER, 1956 . .	Basal (Pennsylvanian)	950	Fluvial	Mineral provinces and channel system at unconformity
ROBSON, 1956 . .	Fell (Lower Carboniferous)	Unspecified	Unspecified	
SHOTTON, 1956 .	New Red (Permian)	Unspecified	Aeolian	
McDOWELL, 1957	Mississagi (Precambrian)	1,230	Fluvial	Ripple mark and pebble size
REINEMUND and DANILCHIK, 1957	Atoka and post-Atoka (Pennsylvanian)	Unspecified	Unspecified	Sole and ripple mark and slumping
RUSNAK, 1957 . .	Pleasantview (Pennsylvanian)	22	Marine	Sand fabrics and channel system
SCHLEE, 1957 . .	Brandywine (Pleistocene)	86	Fluvial	Fabric, texture and petrology
STEWART et al., 1957	Wingate (Jurassic)	Unspecified	Aeolian ?	
	Navajo (Jurassic)	Unspecified	Aeolian ?	
HAMBLIN, 1958 .	Jacobsville (Cambrian)	476	Fluvial	Ripple mark and heavy mineral provinces
	Chapel Rock (Cambrian)	96	Marine	Ripple mark and heavy mineral provinces
	Miner's Castle (Cambrian)	334	Marine	Ripple mark and heavy mineral provinces

Table 4-1 (Continued)

Author and year	Formation and age	Measurements	Environment	Other properties studied
HOPKINS, 1958 .	Anvil Rock (Pennsylvanian)	Unspecified	Fluvial	Sand thickness map and petrology
NICKELSEN, 1958	Pennsylvanian	876	Unspecified	
NIEHOFF, 1958 .	Coblenz (Devonian)	Unspecified	Marine	Ripple and current mark, plant debris and pebble orientation
PELLETIER, 1958	Pocono (Mississippian)	4,971	Deltaic	Plant debris, pebble size and isopach
POTTER and GLASS, 1958 .	Pennsylvanian	172	Fluvial	
POTTER et al., 1958	Chester (Mississippian)	1,341	Fluvial and deltaic	Local and regional sand thickness maps
RUKHIN, 1958 . .	Devonian	Unspecified	Unspecified	
WURSTER, 1958 .	Schilfsandstein (Triassic)	Unspecified	Unspecified	
FRANKS et al., 1959	Dakota (Cretaceous)	1,733	Unspecified	
KNILL, 1959 . .	Dalradian (Precambrian)	Unspecified	Marine	
TANNER, 1959 . .	Miami oolite (Pleistocene)	48	Marine	
TOWE, 1959 . . .	Pennsylvanian	138	Fluvial	Plant debris
WILLIAMS, 1959 .	Pottsville (Pennsylvanian)	Unspecified	Fluvial	
	Kittanning (Pennsylvanian)	Unspecified	Fluvial	
CAZEAU, 1960 . .	Dockum (Triassic)	278	Unspecified	
DZULYNSKI and GRADZINSKI, 1960	Triassic	500	Fluvial	
EMRICH, 1960 . .	Hawthorne (Tertiary)	284	Unspecified	Texture
FARKAS, 1960 . .	Franconia (Cambrian)	1,023	Marine	Regional stratigraphic relations
GANGULY, 1960 .	Barakar and Barren Measures (Gondwana)	Unspecified	Unspecified	
OPDYKE and RUNCORN, 1960 . .	Tensleep (Pennsylvanian)	239	Aeolian	
	Casper (Permian)	87	Aeolian	
	Weber (Pennsylvanian)	161	Aeolian	
MOBERLY, 1960 .	Cloverly (Cretaceous)	Unspecified	Fluvial	Light, heavy and clay minerals
PELLETIER, 1960.	Llard, Toad, and Grayling (Triassic)	Unspecified	Marine	Ripple mark

Table 4-1 (Continued)

Author and year	Formation and age	Measurements	Environment	Other properties studied
Pryor, 1960	McNairy (Cretaceous)	920	Deltaic	Heavy mineral provinces, texture and facies maps
Sullwold, 1960	Modelo (Miocene)	Unspecified	Marine	
Sutton and Watson, 1960	Epidotic Grits (Precambrian)	Unspecified	Unspecified	
Wurster, 1960	Bunter (Triassic)	191	Unspecified	
Baars, 1961	Cedar Mesa (Permian)	Unspecified	Littoral and marine	Isopach
Bigarella and Salamuni, 1961	Botucatú (Mesozoic)	2,892	Aeolian	
Fahrig, 1961	Athabaska (Precambrian)	1,244	Fluvial	Ripple marks and pebble fabrics
Frazier and Osanik, 1961	Point bar (Recent)	137	Fluvial	Sand fabric and modern Mississippi River
Greensmith, 1961a	Calciferous (Carboniferous)	134	Unspecified	
Greensmith, 1961b	Oil-shale Group (Carboniferous)	105	Deltaic	
Hamblin, 1961a	Freda (Precambrian)	294		
	Jacobsville and Bayfield (Cambrian)	554	Fluvial	
	Dresbach (Cambrian)	425		Isopach and bedding thickness
	Franconia (Cambrian)	925	Marine	Isopach
Hamblin and Horner, 1961	Keweenawan (Precambrian)	241	Fluvial	
Hannemann, 1961	Glacial outwash (Pleistocene)	Unspecified	Fluvial	
Lerbekmo, 1961	Blue (Tertiary)	Unspecified	Unspecified	
Naha, 1961	Precambrian	Unspecified	Marine	
Poole, 1961	Moenkopi (Triassic)	Unspecified	Fluvial	
	Chinle (Triassic)	Unspecified	Fluvial	
	Kayenta (Triassic)	Unspecified	Fluvial	
Schlee and Moench, 1961	"Jackpile" (Jurassic)	Unspecified	Fluvial	Channel system
Mackenzie and Ryan, 1962	Cloverly-Lakota (Cretaceous)	Unspecified	Fluvial-deltaic	Mineral provinces
	Fall River (Cretaceous)	Unspecified	Fluvial-deltaic	Mineral provinces

Table 4-1 (Continued)

Author and year	Formation and age	Measurements	Environment	Other properties studied
MAPEL and PILL- MORE, 1962 . .	Lakota (Cretaceous)	1,184	Fluvial	
McBRIDE and HAYES, 1962 .	Modern dunes (Recent)	130	Aeolian	
POOLE, 1962 . .	Weber, Cutler, De Chelly and Coconino (Pennsylvanian and Permian)	Unspecified	Aeolian	
	Chinle, Moenave (Triassic)	Unspecified	Aeolian	
	Nugget, Navajo and Aztec (Jurassic and Triassic)	Unspecified	Aeolian	
	Carmel, Entrada Bluff, Junction Creek, and Cow Springs (Jurassic)	Unspecified	Aeolian	
POTTER, 1962a .	Palzo (Pennsylvanian)	Unspecified	Fluvial	
POTTER, 1962b .	Aux Vases (Mississippian)	Unspecified	Fluvial	Isopach
POTTER, 1962c .	Pennsylvanian	1,887	Fluvial	Isopach and facies maps
ROSS, 1962 . . .	Yellowknife (Precambrian)	64	Marine	
SHACKLETON, 1962	Rough Rock (Carboniferous)	1,738	Deltaic	
VAN HOUTEN	Frontier (Cretaceous)	Unspecified	Marine-littoral	Pebble size and parting lineation
YEAKEL, 1962 . .	Tuscarora (Silurian)	6,753	Fluvial	Pebble size and facies maps
	Juniata (Ordovician)	2,254	Fluvial	Pebble size
	Bald Eagle (Ordovician)	1,792	Fluvial	Pebble size

Descriptive Aspects

Definition. Most sedimentary structures, with the exception of sedimentary fabric and graded bedding, cannot be effectively defined in abstract, analytic terms. Most sedimentary structures, like organisms, land forms, and shapes of lithologic units, have a morphology that requires *illustration* for effective definition. Cross-bedding is no exception. Our inability to formulate an exact definition in no way, however, affects specification and measurement of many of the particular properties as, for example, direction of maximum dip of a foreset bed. We can commonly specify quantitatively many of the properties of a structure but only define the structure itself in a qualitative way or, in some cases, only by an illustration. Definitions of inclined bedding and cross-bedding show some of these difficulties.

Bedding of primary origin inclined to the principal surface of accumulation of a formation (McKee and Weir, 1953, p. 382), forms a continuous series from low to high dip and is termed inclined bedding, a completely descriptive term. Cross-bedding is a particular type of inclined bedding. Cross-bedding is a structure confined to a *single sedimentation unit* (Otto, 1938, p. 575) consisting of *internal bedding, called foreset bedding, inclined to the principal surface of accumulation.* Figure 4-1 shows several cross-bedded units and their systematic, internally arranged foreset beds. This definition is independent of scale. Thus a cross-bedded unit may vary in thickness from 3 mm (Hamblin, 1961, p. 390) to over

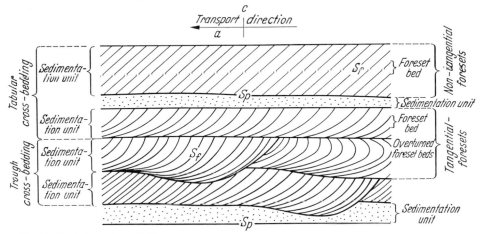

Fig. 4-1. Terminology of cross-bedding

33 meters (Knight, 1929, p. 59) in ancient sandstones, a variation of over 11,000 times. Plate 2 illustrates large-scale cross-bedding and plates 3 and 4A small-scale cross-bedding. Single aeolian dunes that produce cross-bedding have been reported as high as 120 feet (40 m) in the Sahara (Beadnell, 1910, p. 386) and 200 feet (66 m) by Norris and Norris (1961, p. 608) in southern California. Cloos (1953, Pl. 12) illustrates a dune 800 feet (267 m) high in the Namib desert of Southwest Africa. The definition of cross-bedding is also independent of any particular angle of dip of a foreset bed, although most foresets are in excess of 10°.

By restricting cross-bedding to a single sedimentation unit, inclined bedding of other origins, in beaches, in spits or bars, in lateral accretion deposits, in deltas and in talus, is not considered as cross-bedding because such inclined bedding usually involves many sedimentation units, some of which may themselves be cross-bedded. Cross-bedding, as defined above, may include the inclined bedding formed by lee side accumulation in some sand or snow drifts, in micro-deltas of sedimentation unit size, in some reef flank deposits, in the back fill of an erosional scour pit and in sand waves. The last-named is the common cross-bedding of sedimentary rocks.

The definition of cross-bedding given above is widely used but is by no means universal. For example, McKee and Weir (1953, p. 382) define the foreset bed as a "cross-stratum" and the cross-bedded unit "a set of cross-strata." They distinguish between cross-bedding, having foreset beds greater than 1 cm in thickness, and cross-lamination as having foresets less than 1 cm in thickness.

Other terms that have been used for cross-bedding are cross-lamination, cross-stratification, current bedding, diagonal bedding, false stratification, lee side concentration, foreset bedding, oblique bedding, oblique stratification and many

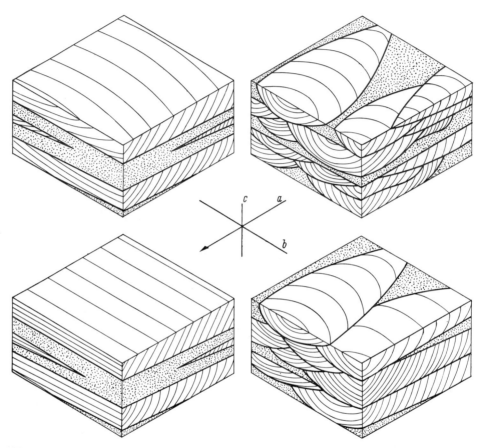

Fig. 4-2. Block diagrams of the two end members of cross-bedding, tabular (left) and trough (right) and standard reference system

terms with self-evident modifiers. Both imprecise original definition and subsequent usage make the exact equivalence of many of these terms uncertain. All, however, imply the essential idea of deposition at an angle to the principal surface of accumulation. Some terms have been defined on the appearance of cross-bedding as seen in single section, usually vertical, rather than on a three dimensional analysis of the structure. Some investigators (ANDRÉE, 1915, p. 388—395; ANDERSEN, 1931, fig. 38; BIRKENMAJER, 1959, p. 86—88 and others) have distinguished diagonal bedding from cross-bedding, simply on the *apparent* uniformity of foreset inclination. "Diagonal bedding" was applied to cross-bedding in longitudinal section whereas "cross-bedding" was applied to the appearance of the structure in transverse section (fig. 4-2).

The terms topset and bottomset have frequently been applied to cross-bedding. The topset bed is the layer immediately overlying the cross-bedded unit and the bottomset bed is the layer immediately underlying the same. These terms probably arose from the study of deltas and were applied to cross-bedding because of a presumed delta-like origin of this structure. In practice, however, they are of little value. The term "bedding" or "true" bedding is adequate to designate the principal surface of accumulation. Moreover, if an even-bedded unit lies between two cross-bedded units, the use of the term "topset" or "bottomset" is ambiguous. The terms suppose that a given topset layer can be traced through a foreset into a bottomset layer in a manner analogous to the presumed delta structure. This analogy is almost certainly incorrect for most cross-bedding. A migrating sand wave such as a dune can leave behind a record of neither bottomsets or topsets.

Classification and Shape. Cross-bedding presents challenging problems of classification.

ZHEMCHUZHNIKOV (1926, p. 60—63), ILLIES (1949, p. 94—98) and JOPLING (1960, p. 220—241) provide genetic classifications of cross-bedding whereas ANDRÉE (1915, p. 382—393), ANDERSEN (1931, fig. 38), LAHEE (1952, p. 88—96), MCKEE and WEIR (1953), RUKHIN (1958, Table 2), BIRKENMAJER (1959), p. 86—88, BOTVINKINA (1959) and PETTIJOHN (1962, p. 1471) are descriptive to varying degrees.

Cross-bedding has been classified by such internal properties as shape of its foreset beds, whether they are concave, convex or straight in vertical section, as well as angle of inclination. The character of the basal contact, whether it is erosional or non-erosional, planar, or curved, has also been used for classification. And, of course, cross-bedding has been classified by presumed hydrodynamic origin and by environment of deposition, aeolian, deltaic, or neritic.

The applicability of the classifications based on geometrical characteristics, in both ancient and modern sediments is, to varying degrees, unfortunately difficult to apply because of vagaries of exposure. For example, if cores are studied, only thickness and angle of inclination can be measured. In outcrops even such a simple property as foreset curvature may have a misleading appearance. Figure 4-2 shows how cross-bedded units will appear in different sections.

The cross-bedding of figure 4-2 consists of units that have sensibly *planar contacts* and are essentially *tabular* bodies and those that have *curved basal contacts* and are *trough*-shaped. These are the two principal types of cross-bedding. Curved basal contacts are generally concave and have been called festoon cross-bedding (KNIGHT, 1929, p. 56) but convex basal contacts have also been figured by ANDERSEN (1931, fig. 38) and MCKEE and WEIR (1953, fig. 2). Gradations between tabular and trough cross-bedding probably exist and hence these two types can best be thought of as end members. Both can occur in the same outcrop and even in the same sedimentation unit. If additional properties are described, say tangency of the foresets (pl. 2), they become simply modifiers as, for example, tabular, tangential, cross-bedding.

In outcrop, distinction between tabular and trough type of cross-bedding depends on the extent to which one can see the structure in three dimensions and on the size of the cross-bedded layers. If the beds are small and can be viewed in three dimensions, comparatively little difficulty may be encountered. Very large units, however, may be incompletely exposed. Difficulty occurs in many outcrops, however, because only two-dimensional sections, commonly neither parallel nor perpendicular to bedding are observed. The geometry of cross-bedding is best understood by considering the appearance of cross-bedding on the different sections of figure 4-2.

If longitudinal *ac* sections show cross-beds with troughs (DAVIS, 1890, fig. 4; KNIGHT, 1929, fig. 21), the cross-beds belong to the trough type because these

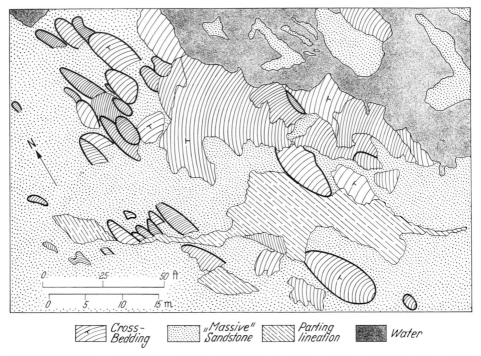

Cross- Bedding   "Massive" Sandstone   Parting lineation   Water

Fig. 4-3-1. Plane table map (above) of cross-bedding and parting lineation in Pennsylvanian sandstone in spillway of Crab Orchard Lake, Williamson Co., Illinois, U.S.A. Heavy black lines represent original boundaries of cross-bedded units; light lines eroded or overlapped boundaries (Illinois Geological Survey)

beds will show trough form in transverse *bc* sections. If *bc* sections show tabular cross-bedding with planar basal contacts, the cross-beds are planar. In *ac* sections, however, apparently tabular cross-bedding such as shown in plates 4B and 5A may have concave basal contacts. Hence the *ac* section is not definitive for identification of tabular cross-bedding. Tabular cross-beds may or may not have erosional basal contacts. Trough cross-beds usually do.

Exposures parallel to *ab* may be helpful in distinguishing between tabular and trough cross-bedding. The vast majority of the foresets of both tabular and trough cross-beds are concave downdip. Foresets with sensibly straight traces may be a diagnostic feature of planar cross-beds; foresets with markedly curved traces indicate trough cross-bedding. On the other hand, gently curved foreset traces can occur in tabular units. Foresets that are, in plan, convex in the down-current direction appear to be rare.

Mapping of cross-bedding in outcrop provides useful insight to the shapes and internal structure of cross-bedded units.

Plate 5 B shows a view, approximately parallel to maximum foreset dip direction, of cross-bedding in limestone. It illustrates the change in apparent dip direction of foresets with different planes of section. The principal surface of the outcrop is approximately parallel to the bc plane and on this surface the foresets traces are approximately horizontal. Apparent dip angle increases in sections more nearly parallel to ac.

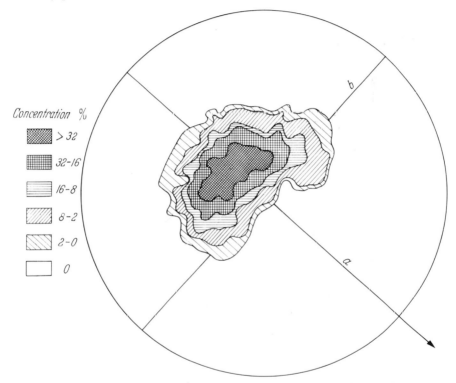

Concentration %

> 32

32-16

16-8

8-2

2-0

0

Fig. 4-3-2 SCHMIDT net shows 200 poles to the foreset beds shown in fig. 4-3-1. Note monoclinic symmetry

Two large-scale planetable maps of outcrops exposed in river beds illustrate the geometry of cross-bedded units as seen in plan view. Figure 4-3-1 shows the appearance of an outcrop in Pennsylvanian sandstone in Illinois. With the possible exception of the widest cross-bed, most of the cross-bedding is of the trough type, is less than 2 feet (0.6 m) thick and displays the following properties:

1. the original boundary of many units has an elliptical outline;

2. the foresets are concave in the direction of dip and are symmetrical with respect to the ac symmetry plane of the trough, wherever the trough is fully exposed;

3. fully exposed cross-beds vary in width from 1 foot (0.3 m) to about 15 feet (5 m).

4. axes of cross-bedded troughs have subparallel orientation and are parallel to parting lineation.

Plate 6 A shows the appearance of the trough cross-bedding of figure 4-3-1. In plan view, many of the foresets have an angular discordance with the edges of the trough, although they may also be tangential. KNIGHT (1929, fig. 25), STOKES (1953, fig. 13; 1954b, fig. 2B), McDOWELL (1957, p. 18), FAHRIG (1961, pl. 9) (BEUF *et al.*, 1962, fig. 2), and HAMBLIN (1961b, fig. 3) also illustrate troughs filled with foreset bedding.

Figure 4-4 shows a map of cross-bedding in a Pennsylvanian sandstone in Kentucky. The cross-bedded layers are all less than 3 feet (1 m) thick. Although

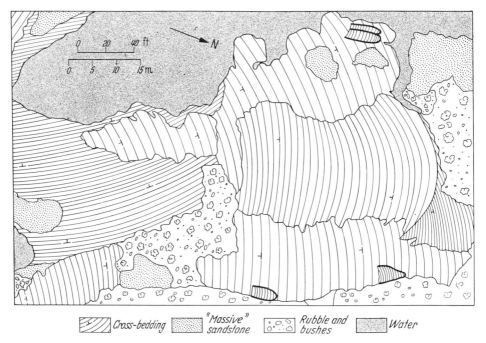

Fig. 4-4. Plane table map of cross-bedding in "falls maker" sandstone of the Lee formation (Pennsylvanian) at Cumberland Falls, Whitley County, Kentucky, U.S.A. Current from right to left. Note that foresets are concave in down-current direction. Wide cross-bedded units are essentially tabular

several small troughs with elliptical outline are present, some foresets are traceable for at least 130 feet (43 m). Basal contacts of these units are sensibly planar. All foresets are concave and all but one have comparable direction of dip. The extent of these large units underscores the difficulties of determining, in small outcrops, whether the foresets in plan are sensibly straight or curved.

A distinctive type of small-scale, trough cross-bedding is micro-cross-lamination (pl. 3) a name suggested by HAMBLIN (1961b) for what GÜRICH (1933, figs. 1, 2, 4 and 6) called, "Schrägschichtungsbögen" and later was designated "rib-and-furrow" by STOKES (1953, p. 17—21). Micro-cross-lamination, as originally drawn by GÜRICH (fig. 4-5), has many similarities to figures 4-3-1 and 4-4 even though its size is but a small fraction of most cross-bedded units.

Although dimensions of cross-bedded units and radius of curvature of their foresets do vary widely, figures 4-3-1 and 4-4 are typical of the appearance of the vast majority of cross-beds in the ab plane. The foreset beds in plan are generally concave in direction of maximum dip, although both straight and convex ones, especially the former, are known. Foresets concave in plan in the down-dip direction are commonly symmetrical with respect to the ac plane of the trough, but asymmetry is also reported (KNIGHT, 1929, fig. 29; McKEE and WEIR, 1953, Table 4). As seen in plan view, foresets are either tangential or angularly discordant to the edge of a trough. KNIGHT (1929, figs. 24, 28 and 29), however, illustrates foresets as being subparallel to trough boundaries.

In vertical longitudinal sections, foreset beds may be straight (pl. 4B), concave (pl. 4A), or very rarely convex. Within the same unit, they may pass from con-

Fig. 4-5. Micro-cross-lamination as originally figured by GÜRICH (1933, fig. 8). Current from top to bottom. Cross-bedded units are trough-shaped. Note that foresets are concave in the down-current direction

cave to straight and less commonly to convex. Foresets may be tangential or non-tangential (angular or discordant) with the underlying bed; they are almost never tangential with overlying bed. Foresets will usually appear to be quite different in the transverse bc than in the longitudinal ac plane.

Foreset beds are usually thought of as single, homogeneous layers within a cross-bedded sedimentation unit. Commonly, however, such layers may consist of a series of fine laminations. If coarse-grained, foresets may be graded upward, from coarse to fine, perpendicular to their thickness and, in some cases, *along* the foreset, being coarsest at the toe and finest at the upper end (REINECK, 1961, fig. 2). Foreset beds that contain internal inclined bedding occur but rarely. Unusual occurrences of cross-bedding, such as that shown in figure 4-6, raise the question as to what is to be considered the sedimentation unit in defining cross-bedding.

Measurement. Measurement of angle of inclination, maximum dip, of azimuth or direction of maximum dip of the foresets, and thickness of the cross-bedded layer is straightforward. The accuracy with which these properties can be measured contrasts sharply with the difficulty of defining cross-bedding.

In undisturbed beds the angle between the foreset bed and the horizontal is called the inclination or dip. It is usually measured with a clinometer. In tilted cross-beds one measures the dihedral angle between the foreset plane and the true bedding. If the true bedding originally had not been deposited horizontally, this dihedral angle may be slightly larger than the angle of inclination (fig. 4-7). The bearing of the maximum dip direction is the azimuth of the foreset bed.

Orientation of cross-bedding is specified by dip direction of its foreset bedding (tilt-corrected, if necessary). Figure 4-3-2 shows a plot of poles of 200 systematic

Fig. 4-6. An unusual type of cross-bedding, called "furious" or doubly cross-laminated by REICHE (1938, p. 926) in sands of McNairy formation (Cretaceous). Life quadrangle (15,500 F.W.L.: 20,500 F.S.L.), Henderson County, Tennessee, U.S.A. (drawn from a photograph by R. W. B. DAVIS)

measurements of foresets of cross-beds in the outcrop mapped in figure 4-3-1. The diagram shows strong monoclinic symmetry and reflects the pronounced directional homogeneity of the outcrop. WURSTER (1958, figs. 16 to 25) shows similar diagrams also obtained from single exposures.

Rarely, however, will the geologist wish to obtain as many as 200 measurements at a single outcrop. To be effective, the study of cross-bedding requires a simple method of sampling an outcrop that yields reproducible results. Cross-bedding direction should always be specified by measurements made according to some sampling rule, and never by a qualitative estimate of the entire outcrop. Cross-beds should never be omitted from a sample because they appear to have an abnormal direction.

Because variability at an outcrop is usually greater between than within cross-bedded units (POTTER and OLSON, 1954, Table 3), the best practice is to measure one foreset in each cross-bedded layer. This practice yields the best estimates of foreset dip direction for a given number of measurements. An exception to this rule may occur in sections parallel to ab, in stream beds, where it may be possible to specify the orientation of the ac symmetry plane of a cross-bedded unit (fig. 4-3-1). The orientation of both large as well as small cross-bedded units may be determined in this way. If the entire unit cannot be seen in plan, the bisectrix direction of the exposed concave foresets should be measured.

In vertical exposures, the sampling procedure is somewhat different, although the sedimentation unit remains the basic sampling element as before. In exposures parallel to either ac or bc, sedimentation units may be sampled by either of two rules. An arbitrary location may be chosen and the desired number of sedimentation units measured in vertical section starting at either the base or top of the outcrop. This procedure is especially good for sampling relatively thin units in unconsolidated sands and for bc exposures in general. Another simple procedure is to select the first, best defined, accessible units encountered at the outcrop. In a carefully controlled experiment of operator variation (POTTER *et al.*, 1958,

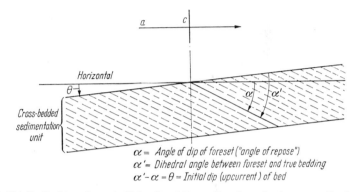

$\alpha =$ *Angle of dip of foreset ("angle of repose")*
$\alpha' =$ *Dihedral angle between foreset and true bedding*
$\alpha' - \alpha = \theta =$ *Initial dip (upcurrent) of bed*

Fig. 4-7. Relations between initial dip Θ of top of cross-bedded sedimentation unit, α the angle of repose or dip of the foreset and α' the dihedral angle between the foreset and the principal bedding. Arrow shows current

p. 1042—1043), this rule was shown to yield reproducible cross-bedding means. Either sampling procedure may, of course, be applied to randomly oriented exposures.

The simple rules for the objective systematic collection of cross-bedding dip measurements given above have maximum applicability in large outcrops with moderate sized sedimentation units where the observer is faced with a choice of which beds to measure. In many outcrops, however, the conditions of exposure effectively dictate the measurement to be made, as is also true of both cores and outcrops with very large, thick cross-bedded units.

In consolidated rocks the strike and dip, or azimuth of cross-bedded foresets, is made in the usual manner with compass and clinometer. In unconsolidated beds determining direction of maximum dip of inclined bedding is facilitated by use of a simple level called a dip-direction indicator (PRYOR, 1958). Figure 4-8-1 shows the instrument used by PRYOR for unconsolidated sands and a modification for consolidated sandstones. A section is cut into the sand face and the circular, dip-direction indicator is placed coplanar with a foreset bed, rotated until its level bubble defines the strike and maximum dip direction read by a compass. In sandstones, the modification (fig. 4-8-2) of PRYOR's of dip-direction indicator is laid directly on the foreset bed. Such devices greatly improve determination of maximum dip direction. GLASHOFF (1935) illustrates another simple device for measuring cross-bedding. BERRY (1960) also shows an instrument for determining planar orientation. If a dip-direction indicator is not available, the loose sand can be cut with a knife so that the foreset beds trace horizontal lines on a vertical face, the azimuth of this face being the strike of the foreset bed (BURACHEK, 1933, p. 432).

If the sandstones are well indurated, dip direction of a foreset bed may have to be determined by measuring its apparent dip in two intersecting faces. The azimuth of each face is recorded along with the apparent dip of the trace of the

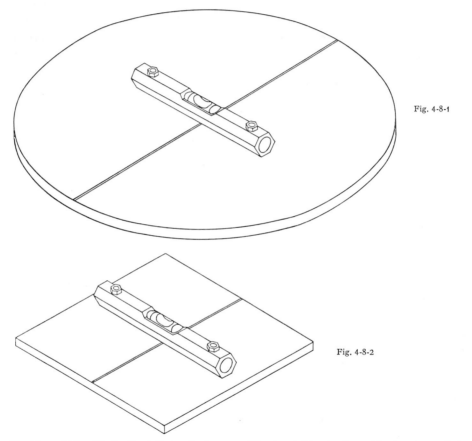

Fig. 4-8-1

Fig. 4-8-2

Fig. 4-8-1 and 4-8-2. Dip direction indicators for the unconsolidated sands (above) and consolidated sandstones (below)

foreset in each face, and maximum dip direction is determined with a stereonet (BUCHER, 1944, p. 195—196). STEINLEIN (1950) constructed a nomograph for determining maximum dip direction from two apparent dips. Use of a stereonet can be avoided by holding a field notebook parallel with the two apparent dips of a foreset and determining the direction of maximum dip of the notebook.

Azimuths may be recorded directly or grouped into an appropriate class interval, usually 30° or 40°. An arithmetic mean is adequate for many outcrops. However, if many measurements are in three or more quadrants, a vector mean should be calculated (see Chapter 10/I). In any case, recording the data initially into class intervals expedites computation. Some investigators have also calculated a measure of variability for each outcrop mean. Others have plotted current roses (Chapter 10) and based their conclusions on modal class intervals (BAUSCH VAN BERTSBERGH, 1940, p. 342—346; NIEHOFF, 1958, pl. 9; TANNER, 1959).

Some investigators have plotted the poles to the cross-bedding foresets on a stereographic polar projection (REICHE, 1938, p. 907) or on a Schmidt net (WURSTER, 1958, figs. 16—25).

If there is appreciable structural dip, the dip and strike of the bedding, S_f, should be recorded so the dip direction of the foreset bedding can be rotated to original position (Chapter 10/I). Any structural dip will, of course, affect the angle of inclination of the foreset.

The number of cross-bedded units measured at an outcrop has usually varied from 1 to as many as 20 in most modern regional studies, although many more have been taken. No outcrop should ever be ignored because it only contains one or two cross-bedded units.

Measurement of dip angle of foreset beds has been made by a number of investigators, although it has not been universally recorded. Factors that affect angle of dip of foreset beds include: granular texture (size, roundness, shape, and moisture content) of the material (VAN BURKALOW, 1945), conditions of flow (JOPLING, 1960, p. 158—160), possibly the agent or environment of deposition (McKEE, 1957a, p. 131) and post-depositional deformation. Hence interpretation of dip angle depends on the degree to which these factors can be segregated. It is probably best to record *maximum* angle of inclination of *undeformed* foresets in a cross-bedded unit, for it doubtless has some interpretative significance. Moreover, recording the maximum, undeformed angle of inclination would standardize field data. Maximum angle of inclination probably occurs in the *ac* symmetry plane of curved foresets.

KNIGHT (1929, p. 66), McKEE (1940, p. 822), POTTER and OLSON (1954, fig. 5), POTTER (1955, fig. 8), WALKER (1955), WRIGHT (1956, fig. 2), CROOK (1957, Table 1), SCHLEE (1957, p. 1377), YEAKEL (1959, p. 85—91), CAZEAU (1960, fig. 5), FARKAS (1960, fig. 3), OPDYKE and RUNCORN (1960, fig. 3), BIGARELLA and SALAMUNI (1961, Table 1-7), FRAZIER and OSANIK (1961, fig. 11) and SHACKLETON (1962, fig. 2) have published distributions of foreset inclination angles which indicate that the average angle of inclination (but not the average maximum angle) in undeformed rocks is usually in the range 18 to 25°. YEAKEL (1959, p. 85—91), after reviewing the literature, concluded that angle of inclination was not a reliable means to distinguish aeolian from subaqueous cross-bedding. Angles of sands in excess of 34°—36° commonly appear to represent either soft sediment or tectonic deformation. However, McBRIDE and HAYES (1962, fig. 7) found a substantial number of angles greater than 35° in modern dune sands.

Thickness of cross-bedded units, like foreset dip angle, is another property of cross-bedding that may be helpful for its interpretation and probably should be consistently noted in either outcrops or cores. Although some workers have emphasized length of the foreset as a measure of scale, most investigators have specified scale by thickness of the cross-bedded unit.

Distributions of cross-bedding thickness have been tabulated by BAUSCH VAN BERTSBERGH (1940, fig. 2), SCHWARZACHER (1953, fig. 3), CROOK (1957, fig. 4), PELLETIER (1958, fig. 7), YEAKEL (1959, p. 91—93), PRYOR (1960, fig. 15),

McIver (1961, fig. 111), and Shackleton (1962, fig. 2c) and have been judged by some to have a log normal distribution.

Most investigators have not specified the sampling rule they followed. Measurement of thickness of cross-bedded units in outcrops, requires a definite sampling procedure. Thickness of cross-bedded units could be obtained by measurement, in vertical section, at a randomly selected site in the outcrop, perhaps its mid-point, or it could be obtained where the foreset dip azimuth was measured. Maximum exposed thickness of each cross-bedded unit is simpler, however, and probably more meaningful. Brush (1958) found experimentally that the maximum height of subaqueous sand waves was dependent on flow conditions. In ancient sediments the observed thickness of the cross-beds may represent the erosional remanents rather than the original height of the sand waves. Unlike thickness of foresets, thickness of cross-bedded units is very well defined, except perhaps in some coarse gravels and conglomerates.

Deformation. Cross-bedding may be deformed both prior to and after consolidation.

Plates 4 A and 6 B show good examples of soft-sediment deformation of cross-bedding. The cross-bed above the hammer of Plate 6 B is overturned in the down-dip direction as are those in plate 4 A. The absence of reverse drag at the base of the overlying cross-bedded layer proves that this deformation took place before the overlying bed was deposited. This kind of deformation is not rare; it has been noted by many (Geikie, 1882, fig. 189; Miller, 1922, fig. 8; Knight, 1929, figs. 36 and 40; McKee, 1938, fig. 4e; Kiersch, 1950, figs. 10 and 11; Konyukhov, 1951, fig. 5; Botvinkina et al., 1954, figs. 76 and 77; Fuller, 1955, Plate 2, fig. 1; Robson, 1957, fig. 6; Valdyanakhan, 1957, p. 195; Potter and Glass, 1958, fig. 9; Wood and Smith, 1959, fig. 5; Fahrig, 1961, pl. 11; Frazier and Osanik, 1961, fig. 11; Beuf et al., 1962, pl. 136; Jones, 1962) and has been produced experimentally (McKee et al., 1962). Overturning may be restricted to a limited portion of a cross-bed or may extend throughout its entire exposed length. In addition to being overturned, the foresets may be buckled and crumpled in the down-dip direction, especially if they are large. Knight (1929, fig. 40), shows overturning and crumpling in cross-bedded units over 300 feet (100 m) wide. He attributed it to mechanical instability resulting from asymmetrical filling of a trough. However, overturned and crumpled cross-bedding may be related more generally, to other processes that produce soft-sediment deformation (Chapter 6/I). Faults appear never to have been reported in the soft-sediment deformation of cross-bedding.

Folding of consolidated sandstones affects the inclination of foreset bedding causing it to be oversteepened in some situations and reduced in others (Brett, 1955, p. 146—147; Pettijohn, 1957, fig. 2; and Pelletier, 1958, Table 3). Apparently foreset dip angle tends to be reduced on the limbs of either synclines or anticlines, if foreset dip direction opposes the structural dip of the limb of the fold, and tends to be oversteepened, if its dip is in the same direction as structural dip. Change in inclination of the foreset is believed to be caused by shearing within the bed. Appreciable change of foreset inclination of structural deformation has not been found, however, in all cases (Yeakel, 1959, p. 85—91).

Principles of Interpretation

The interpretation of cross-bedding can be made at several levels: at an outcrop in a single sand body, in a formation, or across an entire basin. In the development of principles of interpretation each evel has contributed to the other.

Interpretation at the Outcrop. In the early 19th Century, false or inclined bedding was recognized as caused by currents, generally assumed to have acted parallel to maximum dip direction of the inclined beds. This assumption is usually but not universally correct. For example, the maximum dip direction of the inclined bedding of a spit or bar prograding into deeper water, of a beach, or of a lateral accretion deposit (*cf.* REINECK, 1958, fig. 1) can develop at a large angle and even at right angles to current direction. Such possibilities are minimized by focusing attention on inclined bedding within a sedimentation unit, cross-bedding, which constitutes the vast bulk of inclined bedding.

Cross-bedding is produced by the migration of sand waves, filling of scour pits, lee-side accumulations, and possibly in some reef-flank deposits. Observation of cross-bedding in both modern and ancient sediments, shows that the maximum dip of a foreset bed, S_f of figure 4-1, is parallel or subparallel to the direction of the average local current vector recognizing, of course, that minor secondary or eddy currents may display marked deviation from this trend. Figure 4-9 shows the relations between orientation of cross-bedding and current direction under dominantly unidirectional conditions of current flow in streams. There is little evidence to support the idea of WRIGHT (1959, fig. 1) that cross-bedding in streams usually forms by lateral accretion at right angles to stream direction (*cf.* REINECK, 1958, fig. 1).

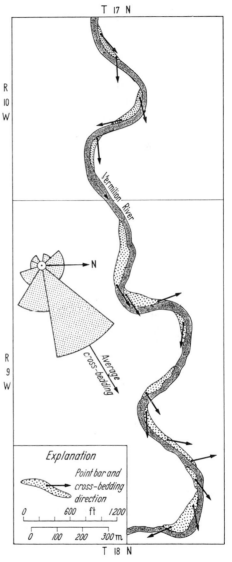

Fig. 4-9. Cross-bedding and point bars along the Vermillion River near Eugene, Vermillion County, Indiana, U.S.A. Note both local and overall general correspondence between cross-bedding direction and direction of river flow. Plate 8 B illustrates one of the cross-beds

Backset bedding refers to inclined surfaces, which dip into the direction of flow. A good example is the deposition on the inclined windward side of a dune. In the absence of independent evidence of current flow, recognition of backset beds is uncertain. However, they have been reported by DAVIS (1890, p. 197—199) and POWER (1961). BRETZ *et al.* (1956, p. 985) also report a possible occurrence. SPURR (1894) indicates how apparent backset beds may form.

Interpretation Across the Basin. The relation between cross-bedding or any other directional structure and the initial sedimentary dip or slope at the time of deposition is critical to regional interpretation. This slope is called the *paleoslope*. Paleoslope is at right angles to depositional strike which has usually been considered to lie subparallel to the strand line. The term paleoslope implies a subaqueous origin, either marine littoral or fluvial, and should not be applied to sandstones of presumed aeolian origin. The existence of preferred direction of cross-bedding in a deposit is independent of whether or not cross-direction reflects paleoslope.

The interpretation of maps of cross-bedding direction, as well as other directional structures, in ancient sandstones has generally followed one of two paths.

One approach has been to interpret the directional structure by independently determining the environment of deposition. Once the environment has been established, then the structure is interpreted as it is either known or assumed to behave in the corresponding modern environment. This is an interpretation by *direct analogy* with modern sedimentation. The other approach has been to examine the relations between direction of sediment transport implied by the structure and other features of the deposit such as lateral variations in grain size, formation thickness, proportion of marine sediment, erosional channel systems developed at unconformities and so on. This approach primarily relies on the *internal consistency* of related variables. Some studies have relied almost completely on the modern sediment analogy, some on internal consistency and others on combinations of the two. Reliance on internal consistency is usually the best procedure.

FORCHE (1935), utilizing some of BRINKMANN'S (1933) cross-bedding data, made the first integrated study of cross-bedding in the Triassic Buntsandstein in France and Germany (pl. 5 A). He showed (fig. 4-10) that formation thickness increased northward in the direction of transport and that pebble size decreased in the same direction.

Subsequently, other examples of close relations between cross-bedding and other sedimentary properties have been found in fluvial and deltaic sandstones. Two fluvial gravels (POTTER, 1955, fig. 12; SCHLEE, 1957, figs. 11 and 12) and three sandstones of fluvial-deltaic origin (McDOWELL, 1957, fig. 12; PELLETIER, 1958, figs. 10 and 14; YEAKEL, 1959, figs. 64—66 and 68) show a decrease in pebble size in the transport direction deduced from cross-bedding. A decrease in thickness of individual cross-bedded units in the same direction has also been demonstrated by PELLETIER (1958, fig. 16) and McIVER (1961, fig. 113). In some cases the formation thickness decreases in down-current direction as inferred from cross-bedding (CRAIG *et al.*, 1955, figs. 21 and 26; and PELLETIER, 1958, figs. 3 and 9). In other cases, it increases (FORCHE, 1935, pl. 5 A; HAMBLIN, 1958, figs. 2 and 29). Good agreement also exists between cross-bedding direction and ancient stream patterns mapped in both outcrop and subsurface (HOPKINS, 1958, fig. 6; ANDRESEN, 1961, fig. 7; SCHLEE and MOENCH, 1961, fig. 4). PRYOR

(1960, fig. 18) showed that cross-bedding in the sands of the deltaic McNairy formation of Cretaceous age dips in the direction of the marine facies. Mineral provinces also have a useful role in documenting the directional significance of cross-bedding in fluvial sands. POTTER (1955, fig. 17) and SIEVER and POTTER (1956, fig. 3) found that areas with contrasting directions of cross-bedding had

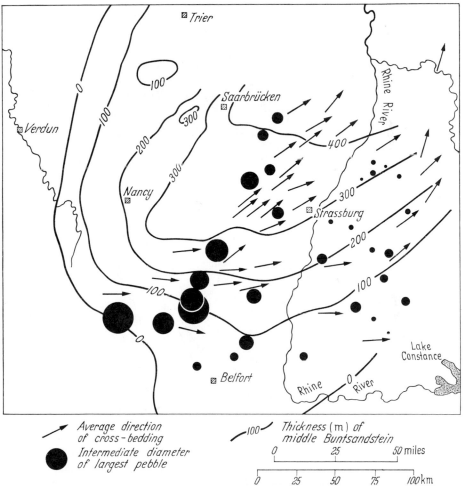

Fig. 4-10. Cross-bedding, pebble size and thickness of Triassic Buntsandstein in adjacent parts of France and Germany. Note decrease in pebble size and increase in thickness in downdip direction of cross-bedding (modified from FORCHE, 1935, figs. 4 and 8)

contrasts in mineral associations. Integrated studies such as these thus provide important supporting evidence for the *a priori* assumption that cross-bedding in dominantly fluvial and deltaic sands, if mapped over a sufficiently large area, accurately reflects the regional paleoslope.

Problems of paleoslope interpretation vary markedly with the different types of marine sands.

SULLWOLD (1960) interpreted small scale cross-bedding, generally less than 4 inches (10 cm) thick, in immature, graded sandstones of turbidite origin, as a product of currents issuing from a point source and moving down a subaqueous slope into deeper water. He found (fig. 8) cross-bedding to be essentially perpendicular to generalized isopach lines, as it is in alluvial fans.

Cross-bedding in marine shelf sands has been studied more frequently, the principal studies being those of KNIGHT, BAUSCH VAN BERTSBERGH, SCHWARZACHER, NIEHOFF and HAMBLIN.

KNIGHT (1929, p. 79—80) related cross-bedding direction to paleoslope by paleogeography. He noted that cross-bedding in the Pennsylvanian Fountain

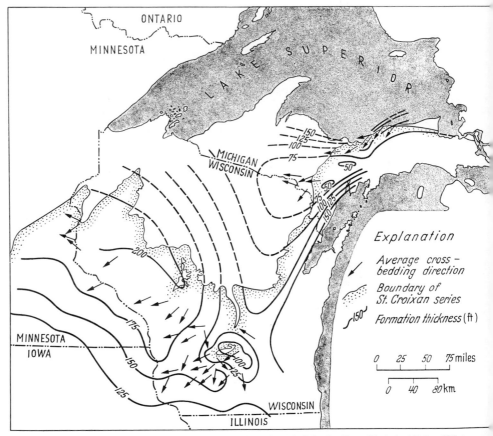

Fig. 4-11. Cross-bedding and thickness of Franconia formation (Cambrian) in the Upper Mississippi Valley, U.S.A. (modified from HAMBLIN, 1961, fig. 7)

and Casper formations dipped northwestward, in the direction of increasing limestone thickness, away from a coastal plain to the southwest (fig. 2). SCHWARZACHER (1953, p. 329) believed the cross-bedding of the Cretaceous Woburn sand to be a product of downslope transport and noted the decrease in size of its cross-bedded units in the same direction, as has been observed in fluvial-deltaic deposits by PELLETIER (1958, fig. 16). KNILL (1959, p. 280—281) interpreted the cross-bedding in the presumed marine Precambrian metasediments of the Dalradian series as reflecting paleoslope. HAMBLIN (1958, figs. 30 and 54; 1961a, figs. 5 and 7) related cross-bedding direction in the Cambrian Munising, Dresbach and Franconia formations to regional thickness variations and found that in all three cross-bedding direction was essentially perpendicular to isopach lines as is shown by figure 4-11 for the abundantly glauconitic Franconia formation. In the southern half of the map area of figure 4-11, the total Cambrian section expands southward, in the down-dip direction of cross-bedding, as the section becomes progressively more marine. FARKAS (1960, fig. 8) found a similar direction of cross-bedding in the Franconia formation in southern Wisconsin. HAMBLIN (1961a, p. 11) also noted that cross-bedded units in the Dresbach formation become thinner in their down-dip direction. A paleoslope interpretation of these Cambrian sands is consistent with other evidence in the upper Mississippi valley area (POTTER and PRYOR, 1961, p. 1219—1221).

In contrast, two studies of essentially similar design, those of BAUSCH VAN BERTSBERGH and NIEHOFF, did not find strong evidence for a paleoslope interpretation of cross-bedding in the marine environment. BAUSCH VAN BERTSBERGH (1940, p. 343—346) presented two alternative interpretations for cross-bedding direction in the Devonian Siegener beds; downslope deposition on the distal marine flanks of an infilling delta or migrating sand accumulation, essentially unrelated to paleoslope, as in the present North Sea. NIEHOFF (1958, p. 314—316) believed that cross-bedding orientation in the Devonian Coblenz quartzite was more strongly controlled by tidal currents and orientation of active tectonic structures in the basin than by paleoslope.

Cross-bedding in carbonate sediments has been observed by many. Unfortunately, few have mapped it, even though SORBY (p. 227) made 2,150 measurements of the Permian Magnesian limestone in England as long ago as 1858. OSMOND (1954, fig. 9) made a few measurements in the Silurian Sevy dolomite as did TANNER (1959, fig. 1) in the Pleistocene Miami oolite. He found mean inclination to dip seaward. Could it be that cross-bedding orientation in the Miami oolite and similar formations is more dependent on ebb than on flood tide currents?

Meager evidence indicates a dominant paleoslope control of cross-bedding in turbidites. Most of the available evidence indicates a downslope transport for some marine sands, although not all have been so interpreted. With little sand input, orientation may be variable and not closely related to paleoslope. Little is known of the relationship of cross-bedding in carbonate sediments to paleoslope.

In any case, the foregoing studies emphasize the importance of integrating a cross-bedding study of a marine sand with other geologic evidence. Without it, interpretation is uncertain. Even integration with other directional structures may prove helpful. Cross-bedding orientation in modern marine shelf sands and in ancient carbonates deserves more study than it has received.

HÜLSEMANN (1955) has made the only extended, published study of cross-bedding in a modern littoral deposit. He found (fig. 7) a relatively complicated transport pattern largely parallel to the coast in the North German Watt. He considered the cross-bedding in the upper marine Molasse in Schwaben as also predominantly reflecting longshore transport. TANNER (1955, figs. 3e and 3f) found landward dipping cross-bedding in a modern lagoon and beach sands. He also recognized (p. 2478) the possibility that longshore currents may have formed the cross-bedding in portions of the Tuscaloosa formation in Georgia and in some Pennsylvanian and Permian sandstones in Oklahoma (p. 2482).

Cross-bedding in sandstones of presumed aeolian origin has been studied by SHOTTON (1937), REICHE (1938), KIERSCH (1950), WRIGHT (1956), SHOTTON (1956), OPDYKE and RUNCORN (1960) and BIGARELLA and SALAMUNI (1961), the last covering the largest area and having the greatest number of measurements.

Cross-bedding of aeolian origin has no relation to paleoslope but is instead dependent on some measure of wind direction. BAGNOLD (1941, p. 69) developed a formula relating threshold wind velocity for sand movement to wind direction. LANDSBERG (1956, p. 181—189), using a somewhat similar formula, found good correlation between weighted wind direction and orientation of modern coastal parabolic dunes.

JÜNGST (1938) compared cross-bedding in modern coastal dunes, in a small area in the Kurische Nehrung of East Germany, with wind direction. He found (fig. 22) good correlation as did FINKEL (1959, p. 642) for a barchan field in Peru, and McBRIDE and HAYES (1962, fig. 1) on Mustang Island along the Gulf of Mexico. No regional studies relating prevailing wind direction and dune orientation in modern deserts are known to the authors.

Both tabular and trough cross-bedding identical to that of subaqueous sands have been reported from ancient sandstones that have been considered aeolian. Although not all investigators have been explicit, foreset beds in ancient aeolianites appear to be generally concave in their down-dip direction, as was noted by SHOTTON (1956, p. 461) in the New Red Sandstone. These aeolianites seem not to have inclined bedding of the type as is presumed to occur in longitudinal dunes (BAGNOLD, 1941, fig. 83). Thickness of cross-bedded units ranges from approximately a meter to over 33 meters in some deposits (SHOTTON, 1956, p. 461). The dimensions of presumed aeolian cross-bedding thus apparently differ little from some subaqueous cross-bedding. HAMBLIN (1958, p. 84), for example, reported troughs as wide as 1,000 feet (330 m) in the marine Chapel Rock member of the Munising formation and noted (1961, p. 10) troughs as wide as 600 feet (200 m) in the fluvial and conglomeratic Jacobsville sandstone.

The published studies of ancient cross-bedding of presumed aeolian origin have usually not been closely integrated with other features of the deposits such as has been done for many fluvial and some marine sands. Principal reliance has been placed on grain size, angle of inclination and absence of both fossils and inter-bedded marine sediments. The difficulties of environmental recognition of ancient aeolianites, combined with the lack of knowledge between prevailing winds and regional dune orientation, suggest caution in evaluating paleowind belts and the position of the earth's poles presumed to be related to them.

Variability of Cross-bedding

The dispersion or variability of cross-bedding has received minor but persistent attention almost from the beginnings of modern study of cross-bedding. The principal papers are those by BRINKMANN (1933, p. 5—11); JÜNGST (1938), HAMBLIN (1958, p. 51—54), PELLETIER (1958, p. 1044—1046) and PRYOR (1960, Table 4). Others who have presented data on the subject include McDOWELL

(1957, fig. 15), PETTIJOHN (1957, Table 4) and FARKAS (1960, fig. 9). These investigators have examined variability qualitatively by visual comparison of current roses and quantitatively by using either the standard deviation or its square, the variance. Interest has been chiefly directed at two questions:

1. variability of cross-bedding direction and stream gradient;

2. possible contrasts in variability of cross-bedding in the fluvial, marine and aeolian environments.

BRINKMANN (1933, p. 5—9) first compared the standard deviation (Chapter 10) of cross-bedding in the Buntsandstein with standard deviations of streams with different meander patterns. The more complex the meander pattern the greater the standard deviation of tangents to stream direction. JÜNGST (1938, p. 245), using the data of BRINKMANN and FORCHE (1935, p. 48), found that the standard deviation of the Buntsandstein increased in the direction of dip of its cross-bedding. HAMBLIN (1958, fig. 28) found standard deviations of several modern streams to vary from 20° to as high as 83°, depending upon the degree of meandering and thus gradient. Because the Jacobsville sandstone had a relatively low standard deviation, he concluded that it was deposited by streams with fairly high gradients.

JÜNGST (1938) appears to have been the first to evaluate the possible environmental significance of cross-bedding current roses. He emphasized the range of cross-bedding orientation in several major environments. He concluded that in most stream deposits most cross-bedding lies within a 90° to 120° sector (p. 242) and in delta and littoral deposits it is usually more variable commonly lying within a 180° to 220° sector (p. 256). Later, PELLETIER (1958, p. 1045) concluded that the standard deviation of aeolian cross-bedding differed little from those of fluvial deposits. PRYOR (1960, Table 4) found that the regional variances of cross-bedding in three marine deposits varied between 7,000 to 8,000 (standard deviation of 83° to 89°), whereas those of four fluvial-deltaic deposits were 5,000 to 6,000 (standard deviation of 71° to 77°).

The published data on the variability of cross-bedding in individual outcrops of aeolian and fluvial sandstones does not show any pronounced differences between the two (cf. REICHE, 1938, Table 1; PELLETIER, 1958, Table 1; FRANKS et al., 1959, Table 1).

Table 4-2 lists the variances computed from the regional distributions of cross-bedding in sandstones that have been considered to be of fluvial-deltaic, marine and aeolian origin. The variance was computed using the vector mean of the regional distribution. For comparison, the variance of a uniform distribution is 10,800, the square of 103.92 (Chapter 10/I).

Evaluation of these results is beset with a number of difficulties. Estimates of cross-bedding direction are clearly dependent on the sampling procedure at the outcrop. Thus measurements of trough axes of cross-bedding (HAMBLIN, 1958, p. 51 and 91; 1961, p. 11) should not be compared with measurements in random vertical outcrop sections. This may in part explain the unusually low variances of the fluvial Jacobsville and the marine Chapel Rock and Miners Castle formations. Another factor is the size of the area studied. The greater the area, the greater the likelihood of its having a more variable, inhomogeneous transport direction. This is illustrated by the variances of direction of river flow in Table 4-2, 1830,

Table 4-2. *Variance of Cross-bedding and Environment*

Unit	Source	Variance
A. Fluvial and deltaic		
Direction of river flow, Georgia	TANNER (1955, fig. 3a)	1830
Direction of river flow, Texas .	TANNER (1955, fig. 3b)	2305
Direction of river flow, Upper Mississippi Valley	POTTER and PRYOR (1961, Table 7)	8049
Post-Pleistocene channels, Mississippi delta.	PRYOR (1960, Table 4)	3418
Brandywine	SCHLEE (1957, fig. 3)	2889
Lafayette	POTTER (1955, fig. 7)	4166
Providence	TANNER (1955, fig. 4)	6000
McNairy	PRYOR (1960, Table 4)	4768
Dakota	FRANKS et al. (1959, Table 1) and unpublished data	5438
Supai	McKEE (1940, unpublished data)	2083
Hardinsburg	POTTER et al. (1958, Table 2)	7778
Palestine	POTTER et al. (1958, Table 2)	4795
Pocono	PELLETIER (1958, pl. 1)	5267
Caseyville and Mansfield . . .	POTTER and SIEVER (1956, fig. 4)	4972
Lee and Sharon	POTTER and SIEVER (1956, fig. 4)	5241
Jacobsville	POTTER and PRYOR (1961, Table 7)	1148
Bayfield	POTTER and PRYOR (1961, Table 7)	1456
Freda.	POTTER and PRYOR (1961, Table 7)	3249
Copper Harbor.	POTTER and PRYOR (1961, Table 7)	7026
Mississagi	McDOWELL (1957, fig. 15)	4741
B. Marine		
Hawthorne	EMRICH (1960, fig. 4)	7391
Tapeats.	McKEE (1940, unpublished)	5027
Glenwood-St. Peter.	POTTER and PRYOR (1961, Table 7)	7666
Casper	KNIGHT (1929, figs. 30 and 31)	6724
Cincinnatian	This book, fig. 4-15	7287
Chapel Rock.	HAMBLIN (1958 and unpublished data)	2482
Franconia	FARKAS (1960, Table 1) and HAMBLIN (1961, fig. 7)	7120
Ironton	EMRICH (1962)	10,404
Miner's Castle	HAMBLIN (1958, fig. 68)	4149
C. Aeolian		
Recent	McBRIDE and HAYES (1962, Table 1)	5854
Chuska	WRIGHT (1956 and unpublished data)	4503
Navajo	POOLE (personal communication)	2589
Wingate	POOLE (personal communication)	2850
Botucatú	BIGARELLA and SALAMUNI (1961, Tables 2, 3 and 5)	5271
Weber	OPDYKE and RUNCORN (1961, Table 1)	1560
Tensleep	OPDYKE and RUNCORN (1961, Table 1)	2694

2305 and 8049 the last being obtained from the greater upper Mississippi valley region (POTTER and PRYOR, 1961. fig. 1), which includes several major water sheds. Moreover, variances of direction of stream or channel flow cannot be directly compared with those obtained from cross-bedding distributions. The former do not contain any estimates of variability within the stream or channel (PRYOR, 1960, p. 1492).

The above factors suggest that a qualitative appraisal of the data of Table 4-2 would be more appropriate than a more precise statistical evaluation. As judged by Table 4-2, the variances of all three major environments overlap to at least some degree. However, some differences are suggested. The most common variance of fluvial-deltaic deposits is in the range 4,000 to 6,000 (standard deviations between 63° and 78°) whereas the smaller sample of marine deposits is in the range 6,000 to 8,000 (standard deviations between 78° and 89°). The variances of the deposits reported as aeolian are comparable to those of many fluvial-deltaic deposits. More study of cross-bedding in marine and aeolian sands is necessary, however, before these apparent differences should be accepted as final.

Thickness of the stratigraphic unit does not appear to be a major factor affecting regional variance.

Ripple Marks

Introduction

Papers describing, illustrating or discussing the origin of ripple marks are more numerous than those on any other sedimentary structure. KINDLE (1917), BUCHER (1919), KINDLE and EDWARDS (1924), KINDLE and BUCHER (1932, p. 632—668), SHROCK (1948, p. 92—127), KUENEN (1950, p. 288—297) and MII (1955) summarize much of this literature.

Comparatively few investigators, however, have systematically measured and mapped ripple marks. HYDE (1911) appears to have been the first to do so. Others who have published maps of ripple mark orientation are listed in Table 4-3. Those who have measured it, usually presenting their data in a frequency distribution or current rose, include BUCHER (1919, fig. 14), BELL (1929, figs. 2 and 3), CULEY (1932, figs. 4 and 5), SCHMITT (1935, figs. 10 and 11), CHENOWETH (1952, p. 550—556), HAMBLIN (1958, figs. 54 and 68), KELLING (1958, figs. 2 and 5), LIPPITT (1959, fig. 4), WOOD and SMITH (1959, fig. 9), HUNTER (1960, fig. 72) and McIVER (1961, fig. 43).

These studies demonstrate that ripple marks have been studied chiefly in the marine environment. The aforementioned studies also show that, where ripplemarks have been mapped, they commonly have a preferred rather than random orientation.

Descriptive Aspects

Definition and Occurrence. Ripple marks, like cross-bedding, are difficult to define without illustration. As commonly used by geologists, however, ripple marks refer to rhythmic or periodic undulations that occur on bedding planes. Generally these undulations are less than 1 meter long. VAN STRAATEN (1953, p. 1—2) used the term megaripple for ripple marks with wave lengths (fig. 4-13) greater than 1 meter.

Ripple marks form chiefly in granular materials at the sedimentary interface in response to a moving fluid, either air or water. They are found in quartz and carbonate sands and in fluvial, littoral, marine and aeolian deposits of all ages.

Table 4-3. *Published Maps of Ripple Marks*

Author and year	Formation and age	Measurements	Environment	Other properties studied
HYDE, 1911 . .	Berea (Mississippian)	Unspecified	Marine	
HÄNTZSCHEL and SEIFERT, 1932 .	(Cretaceous)	Unspecified	Marine	
SEIFERT, 1935 . .	Turon (Cretaceous)	Unspecified	Marine	
HÄNTZSCHEL, 1938	North Sea Watt (Recent)	Unspecified	Littoral	Relationship to tides
BARTENSTEIN and BRAND, 1938 .	Muschelkalk (Triassic)	Unspecified	Marine	
BAUSCH VAN BERTSBERGH, 1940	Siegener Beds (Devonian)	Unspecified	Marine	Cross-bedding, sole mark, plant orientation and heavy mineral associations
KING, 1948 . . .	Guadaloupe Series (Permian)	Unspecified	Marine	Fusilinid orientation and facies distribution
MCKEE, 1949 . .	Moenkopi (Triassic)	Unspecified	Unspecified	
VAN STRAATEN, 1953	Tidal flat (Recent)	Unspecified	Littoral	
MII, 1955	Tidal flat (Recent)	Unspecified	Littoral	
MCKEE, 1957b .	Tidal flat (Recent)	23	Littoral	
REINEMUND and DANILCHIK, 1957	Atoka and post-Atoka (Pennsylvanian)	Unspecified	Unspecified	Cross-bedding, sole mark and slumping
NIEHOFF, 1958 .	Coblenz (Devonian)	Unspecified	Marine	Cross-bedding, plant debris and pebble orientation
VAUSE, 1959 . .	Shelf sands (Recent)	Unspecified	Marine	Texture and water depth
ALLEN, 1960 . .	Weald (Cretaceous)	Unspecified	Deltaic	Flute marks, heavy mineral associations, and channels
PELLETIER, 1960.	Llard, Toad and Grayling (Triassic)	Unspecified	Marine	Cross-bedding
FAHRIG, 1961 . .	Athabaska (Precambrian)	62	Fluvial	Cross-bedding and pebble orientation
GREINER, 1962 .	Albert (Mississippian)	Unspecified	Lacustrine	

Ripple marks occur primarily on the principal surface of accumulation but can also occur on the inclined surfaces of small channels or on foresets of either aeolian or subaqueous cross-beds, where they generally strike parallel to foreset dip (*cf.* MCKEE, 1945, p. 318). Ripple marks are probably most abundant in relatively shallow water deposits but have been reported at depth in the ocean (ERICSON *et al.*, 1955, pl. 1). Ripple marks have also been reported in muds (VAN STRAATEN, 1951) and shales (FREBOLD, 1928) but such occurrences are unusual.

Classification. Ripple marks may be classified in a number of different ways: by appearance in plan view, whether the crests are straight or curved or continuous or discontinous; by the symmetry of the crests, whether they are symmetrical or asymmetrical; by relation of internal grain fabric to external form; by either wave length, amplitude or by ripple index; by orientation with respect to current direction (VAN STRAATEN, 1953, p. 1—2); and by presumed hydrodynamic conditions of origin. Although there are only a relatively small number of recurring types of ripple marks in the geologic column, their variation and the variety

Table 4-4. *Crest Patterns of Ripple Marks, Relationships to Current and Measurement*

Crest Pattern	Relations of Crest Pattern to Current	Measurement	Principal Types
Relatively straight and continuous	Transverse Parallel	Strike of crests, if symmetric; direction of current, if asymmetric	Transverse, current and oscillation Longitudinal
Markedly curved and discontinuous	*ac* symmetry plane parallel to current	azimuth of *ac* symmetry plane	Cuspate, linguoid
Complex	—	—	Rhombohedral, Interference

of many of the minor types have made a comprehensive descriptive classification difficult. Thus BUCHER (1919, p. 208), KINDLE and BUCHER (1932, Table 90), KUENEN (1950, p. 292—293), MII (1955, fig. 1), VAN STRAATEN (1953, p. 1—2), and RUKHIN (1958, Table 31) have listed genetic characteristics of ripple marks whereas SHROCK (1948, p. 92—127) has described some of the principal types, emphasizing their value for determining top and bottom of vertical or overturned beds.

If only orientation is of interest, classification is simplified, for only relatively simple ripple mark patterns can be related to the *ac* symmetry plane (fig. 3-1) of the depositing current. Table 4-4 summarizes the relations between pattern of ripple crests and the current.

Ripples with relatively straight, subparallel, crests, either continuous or discontinuous, are commonly transverse to current direction but have also been reported parallel to current direction. The latter are called longitudinal ripples (VAN STRAATEN, 1953, p. 2). Distinction between longitudinal and transverse ripples depends on independent knowledge of current direction which, in ancient sediments, can only be obtained from other directional structures. The vast majority of ripple marks with relatively straight crests appear to be transverse to current direction, however. In addition to the principal crests, minor secondary ones may also be present (fig. 4-13). The cross-sections of the crests of transverse ripple marks may be symmetrical, or asymmetrical, if one side, the lee side, is steeper than the other. If asymmetrical, the term current ripple mark has been commonly applied. Current ripples, regardless of size, have a cross-bedded internal structure, of which micro-cross-bedding is an example (HAMBLIN, 1961 b, p. 399—400). Asymmetrical climbing ripples have also been reported (McIVER, 1961, fig. 31). Transverse ripples with symmetrical cross-section have been called

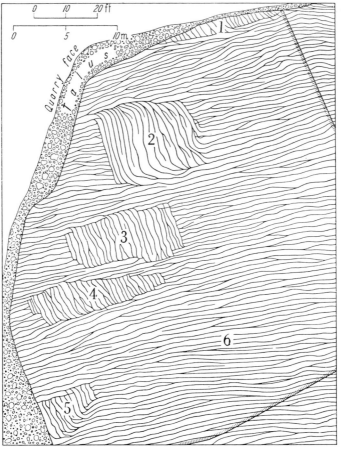

Fig. 4-12. Sketch of ripple marks in quarry of Muschelkalk (Triassic) limestone at Gersheim, Saarland, Germany. Numbers refer to different sets of ripples, 6 in all (modified from BARTENSTEIN and BRAND, 1938, fig. 1)

oscillation ripples (BUCHER, 1919, p. 182—190) or wave ripples (pls. 7A and B). Symmetrical transverse ripples define a line of transport or movement, perpendicular to the strike of their crests. MANOBAR (1955, fig. 8) has called attention to the fact that oscillation, wave-formed ripples can also have, under some circumstances, asymmetrical cross-sections, but this occurrence appears to be exceptional.

Ripple marks with markedly curved crests, linguoid (BUCHER, 1919, fig. 1) or cuspate ripples (pl. 8A and 10A) have a well defined ac symmetry plane and have asymmetrical crests with the steeper side indicating the down-current direction of transport.

More complex ripple mark patterns, such as rhomboid (BUCHER, 1919, p. 153) or interference (BUCHER, 1919, p. 190—195) ripple marks, consist of two or more sets of intersecting crests and usually have a more complex relationship to current direction.

Measurement. Plate 9B shows a plot on a Schmidt net of 196 poles to the flanks of ripples shown in plate 9A. The resultant pattern suggests an overall orthorhombic symmetry, although ac is not quite a full symmetry plane. Typically, oscillation ripples will have orthorhombic symmetry and markedly asymmetric ripples a monoclinic symmetry.

Ripples with relatively straight crests are specified by the azimuth of each crest (Table 4-4). Adjacent ripples that have similar orientation, on the same bedding plane, are termed a *set*. Figure 4-12 shows a sketch of ripple crests on a limestone bed that has six sets. One to six measurements of crests per set are

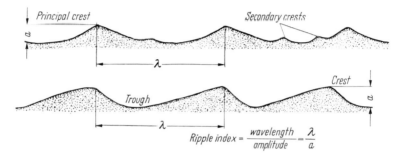

Fig. 4-13. Wave length, amplitude and ripple index

usually adequate to specify orientation. Most outcrops have one or two sets exposed per bed. If present, ripples should be measured on more than one bed at an outcrop. If asymmetry is present, the direction of movement rather than the strike of the crest should be recorded.

Orientation of cuspate and linguoid ripple marks is specified by the azimuth of the ac symmetry plane. Like cross-bedding, this azimuth indicates the direction of current flow, the steeper side of the crest indicating the direction of movement. Cuspate ripples also occur in sets but these are not as well defined as those of ripples with relatively straight crests. Usually one to six measurements per set are sufficient, although some investigators have only recorded the general ripple trend on a bed, as is commonly done for glacial striations.

In cores, crest pattern usually cannot be seen but ripple profile can be determined.

The average strike of the crests of relatively straight ripples or the average direction of the ac symmetry plane of cuspate ripples should be computed and plotted at each outcrop.

The ripple index (BUCHER, 1919, p. 154), the ratio of wave length to amplitude (fig. 4-13), has been used to describe quantitatively ripple mark with relatively straight crests. CORNISH (1902, p. 105), KINDLE and BUCHER (1932, p. 635—636) and others have considered that the ripple index is larger for subaqueous than for aeolian current ripples. However, there appears to be no recent critical evaluation of ripple index as a guide to enviroment of deposition.

Deformation. Tectonic deformation appears to have little affect on ripple marks, as seen on bedding planes. Tectonic deformation, however, can produce small regularly spaced crenulations, called pseudoripple-mark (INGERSON, 1940, p. 558) which can be confused with ripple marks. Another type of pseudoripple-mark, apparently related to closely spaced jointing in limestone, has been called solution ripple mark (SCHMITT, 1935, p. 338—348).

Principles of Interpretation

Interpretation of ripple mark orientation in ancient sediments has developed by relating its orientation to other directional structures, to facies distribution,

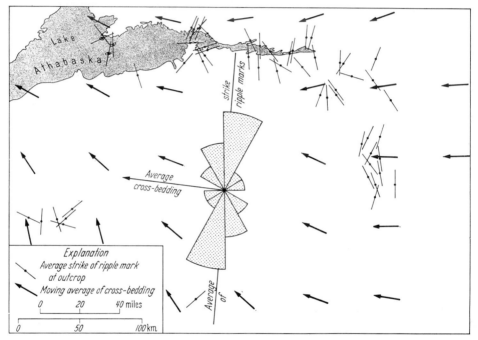

Fig. 4-14. Ripple marks and cross-bedding in fluvial sandstone of Athabaska formation (Precambrian) in portions of Alberta and Saskatchewan, Canada. Note that average cross-bedding direction is approximately at right angles to average strike of ripple marks. Compare with figure 4-15 (modified from FAHRIG, 1961, fig. 2)

to regional isopach, as well as by direct analogy with ripple mark orientation mapped in modern environments.

Where it has been studied in fluvial sediments, transverse ripple mark appears to be commonly oriented at right angles to cross-bedding direction and thus parallel to despositional strike. FAHRIG (1961, fig. 2) found such relations in a regional study of the sandstones of the Precambrian Athabaska formation (fig. 4-14) as did BRETT (1955, p. 147—148) in the Precambrian Baraboo quartzite. Similar relations prevail on many modern point bars. Thus the available evidence suggests that systematic regional mapping of transverse ripple mark in fluvial deposits will delineate depositional strike and hence paleoslope.

Figure 4-15 shows a map of transverse ripple mark and cross-bedding orientation in a marine sequence of interbedded carbonate and shale, approximately 500 feet (167 m) thick, in the Ordovician Cincinnatian series of portions of Ohio

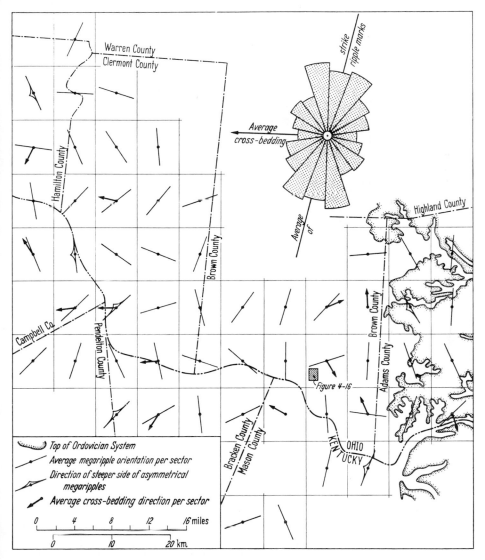

Fig. 4-15. Orientation of ripple marks and cross-bedding in marine limestones of Cincinnatian series (Ordovician) in portions of Ohio and Kentucky, U.S.A. Note that average direction of cross-bedding (59 measurements) is approximately perpendicular to average strike of ripple marks (576 measurements)

and Kentucky. Figure 4-16 shows a detailed local map. The ripples measured are similar to those illustrated in plate 9A. The ripple marks of the Cincinnatian series have an average wave length of approximately 75 cm and an average amplitude of approximately 6 cm.

These two maps illustrate several important points. First, throughout this marine sequence there is a preferred orientation of ripple marks. A measure of the homogeneity of the sequence is the similarity of the average strike of the ripples, 26° in the small area of figure 4-16, and 13° in the larger area of figure 4-15.

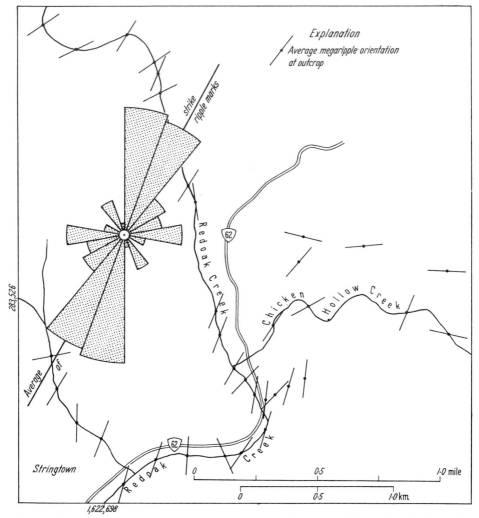

Fig. 4-16. Detailed map of ripple mark orientation (120 measurements) in limestone of Cincinnatian series (Ordovician) in part of Union Township, Brown County, Ohio, U.S.A. Note similarity of average strike to that shown in figure 4-15

Secondly, the vector mean of 56 cross-bedding measurements is approximately at right angles to the average strike of the ripple marks. The asymmetrical ripples present generally indicate a current to the west. Orthogonal relationships of cross-bedding and ripple mark have been found in some other marine sandstones (HAMBLIN, 1958, figs. 54 and 63; PELLETIER, 1960, fig. 1) and, as shown by FAHRIG's study of the Athabaska formation, in fluvial deposits. Thus, if one recalls that cross-bedding in a number of marine shelf sands points down the paleoslope, ripple mark should define depositional strike. Other studies of ripple mark in both modern and ancient marine sands support this conclusion.

VAUSE (1959, figs. 8 and 10) measured ripple mark, chiefly the oscillation type, underwater in modern marine shelf sands of the Gulf of Mexico and found that it tends to be parallel to water depth and hence to strand line (fig. 4-17).

In ancient marine sediments, SCOTT (1930, p. 53) thought the symmetrical and asymmetrical ripple marks of the Cretaceous Comanchean limestones of Texas were oriented parallel to strand line, although he made only a few measurements. In other marine shelf sandstones, KING (1948, figs. 6, 8 and 10) found

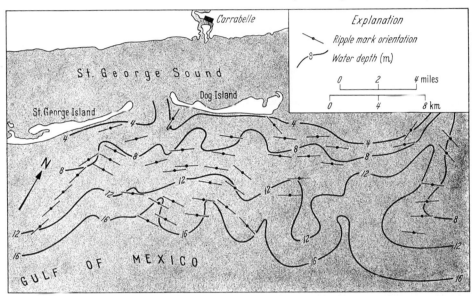

Fig. 4-17. Orientation of ripple marks and water depth in modern marine sands of Gulf of Mexico. Note tendency of ripple to strike parallel to contours and hence strand line (modified from VAUSE, 1959, figs. 8 and 10

oscillation ripple marks to closely parallel depositional strike as inferred from facies distribution. GREINER (1962, p. 232—233) reached a similar conclusion for oscillation ripples in the Albert shale of Mississippian age, a deposit he considered to be lacustrine. Parallelism of ripple marks to strand line of shelf deposits appears to reflect the fact that in shoal waters the wave normal tends to be perpendicular to strand line. HYDE (1911), the first to systematically map and measure ripple marks, reached a similar conclusion (p. 264) for the ripple marks of the Mississippian Bedford and Berea formations in Ohio, and Kentucky. These two formations have remarkable, uniformly-oriented ripple marks over a distance of 125 miles (200 km). Figure 4-18 shows the uniform orientation of ripple marks in the Berea formation in a small area of Ohio. HYDE's (fig. 3) regional map has an essentially similar appearance. HYDE thought (p. 264) that wind waves could not produce such uniformity, although later a hypothesis of monsoon origin was proposed (KINDLE and BUCHER, 1932, p. 667—668).

KELLING (1958, fig. 2) and McIVER (1961, figs. 43 and 121) found ripple marks to be at right angles to flute and groove casts in flysch deposits and thus to parallel depositional strike.

Thus in marine shelf sands there is considerable evidence that oscillation ripple marks, if mapped for a sufficiently wide area, will outline depositional strike. KINDLE (1917, p. 53) early suggested this possibility as did SCOTT (1930, p. 53). KINDLE also thought (p. 53) that asymmetrical or current ripples indicate a current direction parallel to the strand line. Little evidence has subsequently been obtained in ancient shelf sandstones to evaluate this possibility, whereas there is some evidence in littoral deposits that it may not be true. In flysch deposits, ripple marks, where they have been measured, tend to parallel depositional strike.

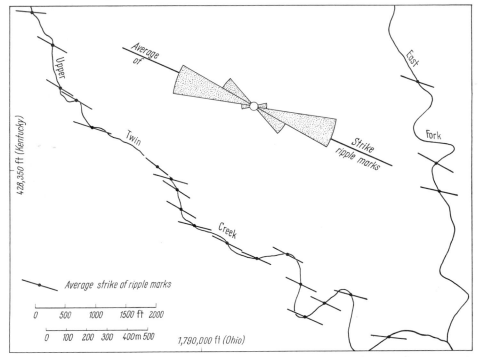

Fig. 4-18. Orientation of ripple marks (87 measurements) in sandstone of Berea formation (Mississippian) in a portion of Nile Township, Scioto County, Ohio, U.S.A. Note low variance

Ripple marks have been mapped on modern tidal and strand flats. HÄNTZSCHEL (1938, fig. 17) and VAN STRAATEN (1953, fig. 7) both show relatively complex patterns with asymmetrical crests making discordant angles with the shore. On the other hand, both MII (1955, figs. 2 and 3) and McKEE (1957b, fig. 28) show ripple marks subparallel to the shore in tidal flats. These studies indicate that asymmetric transverse ripple marks can have complicated relationships to strand line. There appear to be no published studies of ripple marks in ancient tidal strand flat deposits.

Ripple marks do not appear to have been mapped in either modern or ancient aeolian sands.

Variability

Comparatively little information is available on the variance of regional distributions of ripple marks with relatively straight crests. The distribution of

Ordovician ripple marks shown in figure 4-15 has a variance of 2088, the variance of a uniform distribution for line of movement data being 2704 (Chapter 10). The variance of the distribution shown in figure 4-16, the detailed map, is much less, 1050, reflecting its smaller and more homogeneous area. The ripple orientation of the Berea sandstone, as reported by HYDE (1911, fig. 3) and as shown in figure 4-18, probably represents a lower limit of variance for ripple marks and is only 148. This value is probably matched only by a comparatively few directional structures such as glacial striations (Chapter 5/I).

Sand Waves, Cross-Bedding and Ripple Marks

A *sand wave* has been defined as "a ridge on the bed of a stream formed by the movement of the bed material, which is usually approximately normal to the direction of flow, and has a shape somewhat resembling a wave" (LANE, 1947, p. 938). This definition is independent of size and obviously includes, at its lower limit, ripple mark. Although the definition refers to streams, it is equally applicable to other subaqueous environments and applies to wave forms on land such as eolian dunes and snow.

Other terms that have been used for sand waves are dune (GILBERT, 1914, p. 31) and *grève* by some French writers (GILBERT, 1914, p. 31). Large sand waves have been termed Großrippeln in the German literature. The term bar has also been used in America. Sand waves are termed *progressive*, if they migrate downstream, as they commonly do, and *regressive*, if they migrate upstream (BUCHER, 1919, p. 165—182). Regressive sand waves have also been called antidunes (GILBERT, 1914, p. 31).

At the present time there is no standard usage that separates small sand waves or ripples, from large ones that have variously been called bars, dunes, Großrippeln and so on. VANONI and KENNEDY (1961, Table 3) suggest, however, that when sand waves of two different sizes are present, the larger should be called dunes and the smaller ripples.

HIDER (1882, p. 84—85) early used the term sand wave to describe transverse ridges or bars in the bed of the Mississippi River. He found (p. 82—83 and pls. 3 to 7) sand waves ranging in height from several to 22 feet (7.1 m) and having variable wave lengths some up to 750 feet (150 m). OCKERSON (1884, cited in TWENHOFEL, 1932, p. 636) early described sand waves on a Mississippi point bar. Modern observation of the Mississippi River (CAREY and KELLER, 1957, p. 7—23) indicates that transverse sand waves occur on the river bottom in a systematic manner, that they migrate downstream and that amplitude and wavelength are related to stream discharge. They (1957, p. 7) found sand waves in the lower course of the Mississippi River to have amplitudes up to 30 feet (10 m). SUNDBORG (1956, p. 270—272) documents in detail transverse sand waves in the Klarälven River of Sweden. Large sand waves with wave lengths from a few to many feet can be observed on point bars and along the banks of many alluvial streams as well as under water, in favorable circumstances. Figure 4-19 shows a map of the crest pattern of large sand waves, as determined by fathogram, in the Columbia River. According to JORDAN (1962, p. 840), these sand waves have an average wave length of approximately 100 meters and amplitude of slightly over 2 meters and are asymmetrical. At point B the slip slope of a sand wave 5 meters in height is inclined 16° downstream while its back slope dips up stream at 2°.

Sand waves, called giant current ripples, have also been observed after catastrophic floods. Sand waves developed in gravel in a few hours when a modern lake was rapidly drained (THIEL, 1932, p. 455—458). BRETZ *et al.* (1956, pls. 4,

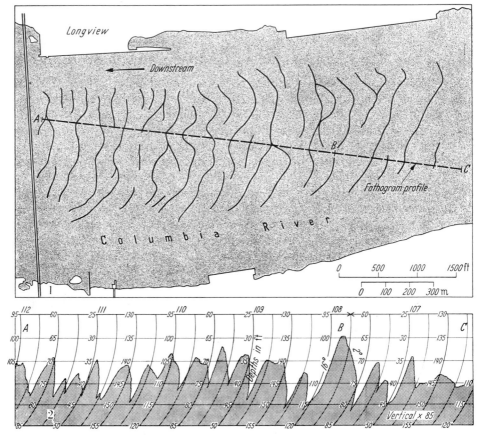

Fig. 4-19. (1) Map pattern of crests of large sand waves in Columbia River at Longview, Cowlitz County, Washington U.S.A. and (2) fathogram along line of section *A B C* (modified from JORDAN, 1962, fig. 2)

6, 9 and 13) also show sand waves developed in gravel. The sand waves they describe are usually transverse to direction of flow, as inferred from topographic evidence, in the channeled scablands of Washington and are believed related to rapid draining of a Pleistocene glacial lake. Heights of 10 feet (3.3 m) and wavelengths of 250 feet (83 m) are reported (p. 978).

Large sand waves are present in the other major environments. The term sand wave was early used by CORNISH (1901) to describe large ripples in tidal strand flats. Earlier, REYNOLDS (1889, 1890, and 1891) produced experimentally, in a series of carefully scaled experiments of a tidal estuary, sand waves of comparable size to those later found by CORNISH. KINDLE (1917, p. 21—22), LÜDERS and TRUSHEIM (1929), HÄNTZSCHEL (1938), VAN STRAATEN (1950), JORDAN (1962, fig. 3) and others describe large sand waves in tidal channels. Large sand waves can also be found on many shallow marine shelves (ILLING, 1954, pl. 7, figs. 3 and 4 and pl. 9; HARRINGTON and HAZLEWOOD, 1962, fig. 2a) as well as on the desert, of course, where they are usually called dunes. JORDAN (1962) describes sand waves on Georges Bank, a large bar at the entrance to the Gulf of Maine (fig. 4-20). He found (p. 843—846) orientation and asymmetry of the sand waves to be related to topography and tidal currents of the bank.

Thus observation of both modern and ancient sediments demonstrate that sand waves, both large and small, are practically an ubiquitous feature of sediment transport of granular material in every major environment and may even be the dominant means of bed load transport.

Fig. 4-20. Crest lines of sand waves and bottom topography of Georges Bank at entrance to Gulf of Maine (modified from JORDAN, 1962, fig. 1),

Plates 10 and 11 emphasize the range in size of sand waves: from small cuspate ripples on a point bar to large subaerial and submarine dunes. In a given reach of a river, size of subaqueous sand waves appears to be related to discharge; low discharge produces small sand waves and high discharge large sand waves (SUNDBORG, 1956, p. 207—208; CAREY and KELLER, 1957, p. 17). Thus, through variation in discharge, small sand waves may be superimposed on large ones (pl. 8B).

The mechanisms by which sand waves develop from flow over an initially smooth bed of granular material are not yet well understood. INGLIS (1949) suggested that sand waves are initiated by small irregularities at the sand-water interface even though the bed is, on the average, a plane surface. LUI (1957) proposed a theory of ripple formation based on the assumption that the upper part of a sediment-laden bed can be considered as a dense fluid which, in contact with the more rapidly moving and less dense water, is subject to boundary instability.

Although the mechanism is not yet well understood, flume experiments have established a well defined sequence of sand wave development from an initially flat bed (cf. BROOKS, 1958, p. 533—542; SIMONS et al., 1961, p. 36—58). The relations between bed form, water surface, and velocity are as shown in figure 4-21. Given an initial level surface, three major bed forms develop as velocity increases: dune, flat bed and antidune. Antidunes usually migrate upcurrent and are generally observed in flumes to be ephemeral features which are usually not preserved as velocity declines. Hydrodynamically, sand waves influence the frictional resistance of a granular bed to fluid flow and create bed roughness.

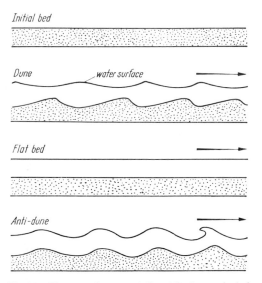

Disagreement exists concerning the extent to which different conditions of bed form are transitional to one another (cf. SUNDBORG, 1956, p. 208; SIMONS and RICHARDSON, 1961, fig. 4; VANONI and KENNEDY, 1961). Moreover, for given size characteristics of the granular material, it has not yet been possible to specify precisely the geometry of the bed forms as functions of flow conditions, such as discharge, depth, velocity and slope.

Because of differences in scale and limited field observations, laboratory studies cannot be applied to streams with complete confidence for it is at present virtually impossible to reconstruct the conditions of flow prevailing at the time of deposition from bed form alone. In addition, in most streams, bed forms are greatly altered as flow declines

Fig. 4-21. Diagrammatic representation of the three major bed forms: dune, flat bed, and antidune. Water surface is out of phase with dune bed form and in phase with antidune bed form. Arrows indicate direction of flow

following a rise or flood. In ancient sediments, the problem is made more difficult because of the difficulty of precisely specifying water depth.

Excluding oscillation and longitudinal ripples, the slip slopes of transverse sand waves, that migrate down stream, produce the foreset beds of most cross-bedding (pl. 8 B). This relation between sand waves and cross-bedding was early suggested by WALTHER (1893, p. 638), illustrated by BEADNELL (1910, fig. 11) for sand waves on the desert, and subsequently explicitly stated by KINDLE (1917, p. 55—56), SUNDBORG (1956, p. 207—208) and others. TRUSHEIM (1929, figs. 1 and 2) early illustrated the internal cross-bedding of sand waves along the strand.

Further evidence that the vast majority of cross-beds in ancient sediments is not of deltaic origin is provided by the configuration of the foresets. As noted by KNIGHT (1929, fig. 25), STOKES (1953, figs. 12 and 13), SHOTTON (1956, fig. 4), HAMBLIN (1958, figs. 46 and 48), NIEHOFF (1958, figs. 11 and 15), POTTER and GLASS (1958, pl. 1 a), BEUF et al. (1962, fig. 2) and others and as shown by figures 4-3, 4-4 and 4-5, the vast majority of foresets that have been figured in ancient sandstones are commonly concave in plan in the down-current direction. As shown by such observations in ancient sediments and by experiments such as those of BRUSH (1958), almost all cross-bedding is the product of sand wave migration at the sediment interface. Thus foresets are not generally deltaic in plan and the vast majority of cross-bedding is not related to micro-deltas as have been made by some experimenters (*cf.* McKEE, 1957a, p. 131—133, JOPLING, 1960).

Because of the direct dependence of sand wave size and form on the flow regime, it seems unwise to place much reliance on a direct environmental interpretation of type of cross-bedding or ripple marks, the two principal expressions of sand wave migration in ancient sandstones. However, hydrodynamic and environmental significance of cross-bedding should not be forgotten for it remains an important ultimate goal.

Summary

Cross-bedding and ripple marks are probably the two most widely recognized and described of all sedimentary structures. They occur in every major geologic environment, in sediments that range in size from fine silts to coarse gravels and conglomerates, and in carbonate as well as quartz sands. Cross-bedding, defined by the internal structure of a bed, and ripple marks, defined by external form, are both very largely the product of sand wave migration at the sediment interface. Widespread occurrence in both ancient and modern sands follows from the fact that sand waves are an almost ubiquitous feature of the traction transport of granular materials, by either air or water.

Both cross-bedding and ripple marks present problems of classification. The two principal types of cross-bedding, tabular and trough, are best thought of as end members; gradations between the two doubtless exist. Ripple marks with directional significance usually have either a relatively straight or markedly curved crest pattern. Ripple marks with relatively straight crests may have either an asymmetrical or symmetrical cross-section, whereas ripple marks with markedly curved crest patterns usually have asymmetrical cross-sections. Sand waves with a symmetrical crests have monoclinic symmetry, whereas ripple marks with symmetrical crests have orthorhombic symmetry.

Interpretation of the directional significance of both cross-bedding and ripple marks has developed largely from their systematic measurement and mapping in ancient sandstones, although both laboratory experiments and modern sediment study have made minor contributions. Interpretation of directional studies of cross-bedding and ripple marks in ancient sandstones has been facilitated by relating the inferred transport direction to other aspects of the deposit such as basin shape, down-current change in grain size or lithologic proportion and so on. The *internal consistency* of such different lines of evidence is usually essential for definitive interpretation in ancient sandstones.

Mapping reveals that both cross-bedding and ripple marks usually have preferred rather than random orientation in the majority of sandstones. Where both have been mapped, average cross-bedding direction tends to be perpendicular to average strike of transverse ripple marks in fluvial as well as marine sandstones.

Mapping confirms that cross-bedding in fluvial, deltaic, and turbidite deposits reflects the paleoslope. Measurement of cross-bedding in a modern littoral sand revealed a variable orientation that tends to subparallel strand line. Although less evidence is available, cross-bedding in a number of marine shelf sandstones also appears to reflect paleoslope, especially if sand input to the basin is high. Cross-bedding in sands of presumed aeolian origin bears no relation to paleoslope. Available data suggest that regional variability of cross-bedding is greater in marine shelf than in fluvial-deltaic sands, and that sands considered to be aeolian have a variability comparable to that of many fluvial-deltaic sands.

Present evidence indicates that transverse ripple marks in both fluvial and marine sandstones tend to parallel depositional strike. Ripple marks do not appear to have been mapped in either modern or ancient aeolian sands.

References

AGATSON, R. S., 1954: Pennsylvanian and lower Permian of northern and eastern Wyoming. Bull. Am. Assoc. Petrol. Geologists 38, 508—583.

ALLEN, J. R. L., 1962: Petrology, origin and deposition of the highest Lower Old Red Sandstone of Shropshire, England. J. Sediment Petrol. 32, 657—697.

ALLEN, P., 1949: Wealden petrology: The Top Ashdown pebble bed and the Top Ashdown sandstone. Quart. J. Geol. Soc. London 105, 257—321.

— 1960: Wealden environment: Anglo-Paris basin. Phil. Trans. Roy. Soc. London, Sec. B 242, 283—346.

ANDERSEN, S. A., 1931: The eskers and terraces in the basin of the River Susaa and their evidences of the process of the ice waning. Danish Geol. Surv., II Raekke No. 54, 201 p.

ANDRÉE, K., 1915: Wesen, Ursachen und Arten der Schichtung. Geol. Rundschau 6, 351—397.

ANDRESEN, M. J., 1961: Geology and petrology of the Trivoli sandstone in the Illinois basin. Illinois Geol. Survey, Cir. 316, 31 p.

BAARS, D. L., 1961: Permian blanket sandstone of Colorado Plateau in Geometry of sandstone bodies. Tulsa: Am. Assoc. Petrol. Geologists, p. 179—207.

BAGNOLD, R. A., 1941: The physics of blown sand and desert dunes. London: Methuen & Co. 265 p.

BALL, H. W., and D. L. DINELEY, 1961: The Old Red Sandstone of Brown Clee Hill and the adjacent area, Part I Stratigraphy. Bull. Brit. Museum, Geol. 5, 242 p.

BASSETT, D. A., and E. K. WALTON, 1960: The Hell's Mouth Grits: Cambrian greywackes in St. Tudwal's Peninsula, North Wales. Quart. J. Geol. Soc. London 116, 85—110.

BARTENSTEIN, H., and E. BRAND, 1938: Großrippel-Felder im oberen Muschelkalk von Gersheim (Saarland). Senckenbergiana 20, 115—121.

BAUSCH VAN BERTSBERGH, J. W. B., 1940: Richtungen der Sedimentation in der rheinischen Geosynkline. Geol. Rundschau 31, 328—364.

BEADNELL, H. J. H., 1910: The sand dunes of the Libyan desert. Geog. J. 35, 379.

BELL, W. A., 1929: Horton-Windsor district, Nova Scotia. Geol. Survey Canada, Memoir 155, 268 p.

BERRY, F. G., 1960: A new instrument for determining planar orientation. Geol. Mag. 97, 309—312.

BEUF, S., O. DE CHARPAL, J. DEBYSER, O. GARIEL and L. MONTADERT, 1962: Recherches Sédimentologiques, vol. 1, Le Cambro-Ordovicien du Tassili des Ajjers, Fas. 1. Institut Français du Pétrole, 68 p.

BIEBER, C. L., 1953: Current directions indicated by cross-bedding in deposits of early Mansfield age in southwestern Indiana. Proc. Indiana Acad. Sci. 62, 228—229.

BIGARELLA, J. J., and R. SALAMUNI, 1961: Early Mesozoic wind patterns as suggested by dune bedding in the Botucatú Sandstone of Brazil and Uruguay. Bull. Geol. Soc. Am. 72, 1089—1106.

BIRKENMAJER, K., 1959: Systematyka warstwowań w utworach fliszowych i podobnych. Polska Akad. Nauk, Studia Geol. Polonica 3, 113 [Polish, English-summary].

BOTVINKINA, L. N., 1959: Morphological classification of bedding in sedimentary rocks. Izvestiya Akad. Sciences USSR, Geologic Series, No. 6, 13—30 [English translation by Amer. Geol. Institute].

— A. P. FEOFILOV, and V. C. YABLOKOV, 1954: A study of the textures and conditions of deposition of recent alluvium and certain other deposits in the lower reaches of the Don River and the shore of the Azov Sea *in* Alluvial deposits in the Coal Measures in the Middle Carboniferous of the Donets Basin. Trans. Inst. Geol. Sci., USSR Acad. Sci. No. 151, 30—89 [Russian].

BRETT, G. W., 1955: Cross-bedding in the Baraboo quartzite of Wisconsin. J. Geol. 63, 143—148.

BRETZ, J H., H. T. U. SMITH and G. E. NEFF, 1956: Channeled scabland of Washington: new data and interpretations. Bull. Geol. Soc. Am. 67, 957—1049.

BRINKMANN, R., 1933: Überkreuzschichtung im deutschen Buntsandsteinbecken. Göttinger Nachr., Math.-physik. Kl. IV, Fachgr. IV, Nr. 32, 1—12.

BROOKS, N. H., 1958: Mechanics of streams with movable beds of fine sands. Trans. Amer. Soc. Civil Engrs. 123, paper 2931, 526—594.

BRUSH jr., L. N., 1958: Study of stratification in a large laboratory flume (abstract). Bull. Geol. Soc. Am. 69, 1542.

BUCHER, W. H., 1919: On ripples and related sedimentary surface forms and their paleographic interpretations. Am. J. Sci. [4] 47, 149—210, 241—269.

— 1944: The stereographic projection, a handy tool for the practical geologist. J. Geol. 52, 191—212.

BURACHEK, A. G., 1933: On the method of measurement of orientation of pebbles and cross-bedding. Repts. All-Russian Mineral. Soc. 62, 432—434 [Russian].

BURKALOW, A. VAN, 1945: Angle of repose and angle of sliding friction, an experimental study. Bull. Geol. Soc. Am. 56, 669—707.

CAREY, W. C., and M. D. KELLER, 1957: Systematic changes in the beds of alluvial rivers. J. Hydraulics Div., Proc. Amer. Soc. Civil Engrs., Hy. 4, paper 1331, 24 p.

CAZEAU, C. J., 1960: Cross-bedding directions in Upper Triassic sandstones of west Texas. J. Sediment. Petrol. 30, 459—465.

CHENOWETH, P. A., 1952: Statistical methods applied to Trentonian stratigraphy in New York. Bull. Geol. Soc. Am. 63, 521—560.

CLOOS, H., 1953: Conversation with the earth. New York: AA. Knopf, 413 p.

CORNISH, V., 1901: On sand-waves in tidal currents. Geog. J. 18, 170—202.

CRAIG, L. C., C. N. HOLMES, R. A. CADIGAN, V. L. FREEMANN, T. E. MULLENS and G. W. WEIR, 1955: Stratigraphy of Morrison and related formations, Colorado Plateau Region: A preliminary report. U.S. Geol. Survey Bull., 1009-E, 125—168.

CROOK, K. A. W., 1957: Cross-stratification and other sedimentary features of the Narrabeen group. Proc. Linnean Soc. N.S. Wales 82, 157—166.

CULEY, A. G., 1932: Ripple marks in the Narrabeen series along the coast of New South Wales. J. Roy. Soc. N. S. Wales 66, 248—272.

DAVIS, W. M., 1890: Structure and origin of glacial sand plains. Bull. Geol. Soc. Amer. 1, 195—202.

DOTT, R. H., 1955: Pennsylvanian stratigraphy of Elko and northern Diamond ranges northeastern Nevada. Bull. Am. Assoc. Petrol. Geologists 39, 2211—2305.

DZULYNSKI, S., and R. GRADZINSKI, 1960: Source of the lower Triassic clastics in the Tatra Mountains. Bull. Acad. polonaise sci. géol. et géogr. 8, 45—48.

EDWARDS, J. D., 1955: Studies of some early Tertiary red conglomerates of central Mexico. U. S. Geol. Soc., Prof. Paper 264, 153—185.

EMRICH, G. H., 1960: Cross-bedding and textural variations of the Miocene Hawthorne formation in northern Florida. J. Sediment. Petrol. 30, 561—567.

— 1962: Geology of the Ironton and Galesville sandstone of the upper Mississippi Valley. Unpublished Ph.D. thesis, University of Illinois, 109 p.

ERICSON, E. M., B. C. HEEZEN and G. WOLLIN, 1955: Sediment deposition in the deep Atlantic. Geol. Soc. Am., Spec. Paper No. 62, 205—220.

FAHRIG, W. F., 1961: The geology of the Athabasca formation. Geol. Survey Canada, Bull. 68, 41 p.

FARKAS, S. E., 1960: Cross-lamination analysis in the Upper Cambrian Franconia formation of Wisconsin. J. Sediment. Petrol. 30, 447—458.

FINKEL, H. J., 1959: The barchans of southern Peru. J. Geol. 67, 614—649.

FORCHE, F., 1935: Stratigraphie und Paläogeographie des Buntsandsteins im Umkreis der Vogesen. Mitt. geol. Staatsinst. Hamburg 15, 15—55.

FRANKS, P. C., G. L. COLEMAN, N. PLUMMER and W. K. HAMBLIN, 1959: Cross-stratification, Dakota sandstone (Cretaceous), Ottawa County, Kansas. Kansas Geol. Survey, Bull. 134, pt. 6, 223—238.

FRAZIER, D. E., and A. OSANIK, 1961: Point-bar deposits, Old River Locksite, Louisiana. Trans. Gulf Coast Assoc. Geol. Soc. 11, 121—137.

FREBOLD, H., 1928: Rippeln im Graptolithenschiefer. Z. Geschiebeforsch. 4, 60—65.

FULLER, J. O., 1955: Source of Sharon conglomerate of northeastern Ohio. Bull. Geol. Soc. Am. 66, 159—176.

GANGULY, S., 1960: Dimensional fabric of Barakar and Barren Measure sandstone in eastern part of Ramgarh coalfield, Hazaribagh. Quart. J. Geol. Mining Met. Soc. India 32, 39—47.

GEIKIE, A., 1882: Textbook of geology, 1st ed. London: Macmillan & Co. Ltd. 971 p.

GILBERT, G. K., 1914: The transportation of debris by running water. U.S. Geol. Survey., Prof. paper 86, 262 p.

GLASHOFF, H., 1935: Ein einfaches Gerät zur Messung von Schrägschüttungen. Mitt. geol. Staatsinst. Hamburg No. 24, 46—47.

GRADZINSKI, R., A. RADOMSKI and R. UNRUG, 1959: Preliminary results of sedimentological investigations in the Silesian Coal Basin. Bull. Acad. polan. sci., Ser. Sci. chim., géol. et géogr. 7, 433—440.

GREENSMITH, J. T., 1961: Cross-bedding in the Calciferous Sandstone Series of Fife and West Lothian. Geol. Mag. 98, 27—32.

— 1961: The petrology of the Oil-Shale Group sandstones of West Lothian and Southern Fifeshire. Geol. Assoc. Proc. 72, 49—71

GREINER, H. R., 1962: Facies and sedimentary environments of Albert shale, New Brunswick. Bull. Am. Assoc. Petrol. Geologists 46, 219—234.

GÜRICH, G., 1933: Schrägschichtungsbögen und zapfenförmige Fließwülste im „Flagstone" von Pretoria und ähnliche Vorkommnisse in Quartzit von Kuibis, S.W.A., dem Schilfsandstein von Maulbronn u. a. Z. deut. geol. Ges. 85, 652—664.

HAMBLIN, W. K., 1958: Cambrian sandstones of northern Michigan. Michigan Geol. Survey, Pub. 51, 149 p.

— 1961a: Paleogeographic evolution of the Lake Superior region from Late Keweenawan to Late Cambrian time: Bull. Geol. Soc. Am. 72, 1—18.

— 1961b: Micro-cross-lamination in Upper Keweenawan sediments of northern Michigan. J. Sediment Petrol. 31, 390—401.

— 1962: X-ray radiography in the study of structures in homogeneous sediments. J. Sediment. Petrol. 32, 201—210.

—, and W. J. HORNER, 1961: Sources of the Keweenawan conglomerates of northern Michigan. J. Geol. 69, 204—211.

HANNEMANN, M., 1961: Neue Beobachtungen zur Entstehung und Entwicklung des Berliner Urstromtals zwischen Fürstenwalde (Spree) und Fürstenberg (Oder). Geologie 10, 418—434.

HÄNTZSCHEL, W., 1935: Fossile Schrägschichtungs-Bogen, „Fließwülste" und Rieselmarken aus dem Nama-Transvaal-System (Südafrika) und ihre rezenten Gegenstücke. Senckenbergiana 17, 167—176.

— 1938: Bau und Bildung von Groß-Rippeln im Wattenmeer. Senckenbergiana 20, 1—42.

—, and A. SEIFERT, 1932: Groß- und Kleinrippeln im Elbsandsteingebirge: Ein Beitrag zur Paläogeographie des Oberkreidemeeres in Sachsen. Sitzber. Abh. nat. Ges. Isis Dresden 1931, 100—101.

HARRINGTON, J. W., and E. L. HAZLEWOOD, 1962: Comparison of Bahamian land forms with depositional topography of Nena Lucia dune-reef-knoll, Nolan County, Texas: Study in uniformitarianism. Bull. Am. Assoc. Petrol. Geologists 46, 354—373.

HIDER, A., 1882: Appendix D, Report of assistant engineer Arthur Hider upon observations at Lake Providence, November, 1879 to November 1880 *in* Report of the Mississippi River Commission, p. 80—98.

HOPKINS, M. E., 1958: Geology and petrology of the Anvil Rock sandstone of southern Illinois. Illinois Geol. Survey, Cir. 256, 48 p.

HÜLSEMANN, J., 1955: Großrippeln und Schrägschichtungs-Gefüge im Nordsee-Watt und in der Molasse. Senckenbergiana **36**, 359—388.

HUNTER, R. E., 1960: Iron sedimentation in the Clinton Group of the central Appalachian basin. Unpublished Ph.D. thesis, The Johns Hopkins University, 416 p.

HYDE, J. E., 1911: The ripples of the Bedford and Berea formations of central Ohio with notes on the paleogeography of that epoch. J. Geol. **19**, 257—269.

ILLIES, H., 1949: Die Schrägschichtung in fluviatilen und litoralen Sedimenten, ihre Ursachen, Messung und Auswertung. Mitt. geol. Staatsinst. Hamburg **19**, 89—109.

— 1952: Die eiszeitliche Fluß- und Formengeschichte des Unterelbe-Gebietes. Geol. Jahrb. **66**, 525—558.

— 1952: Eisrandlagen und eiszeitliche Entwässerung in der Umgebung von Bremen. Abhand. naturw. Verein. Bremen **33**, 19—56.

ILLING, L. V., 1954: Bahaman calcareous sands. Bull. Am. Assoc. Petrol. Geologists **38**, 1—95.

INGERSON, E., 1940: Fabric criteria for distinguishing pseudo-ripple marks from ripple marks. Bull. Geol. Soc. Am. **51**, 558—574.

INGLIS, C. C., 1949: The behaviour and control of rivers and canals, pt. 2. Central Water Power, Irrigation and Navigation Research Sta., Research Publ. 13, 459—467.

JONES, D. J., 1954: Sedimentary features and mineralization of the Salt Wash sandstone at Cove Mesa, Carrizo Mountains, Apache Co., Arizona. U.S. Atomic Energy Comm. Tech. Report RME-3093 (pt. 2), 40 p.

JONES, G. P., 1962: Deformed cross-stratification in Cretaceous Bima sandstone, Nigeria. J. Sediment. Petrol. **32**, 231—239.

JOPLING, A. V., 1960: An experimental study of the mechanics of bedding. Unpublished Ph.D. thesis, Harvard University, 358 p.

JORDAN, G. F., 1962: Large submarine sand waves. Science **136**, 839—848.

JÜNGST, H., 1928: Das Rhät, die Psilonoten- und Schlotheimienschichten im nördlichen Harzvorlande. Geol. u. Pal. Abh., n.s. **17**, 194.

— 1929: Zur Sedimentation des Meeressandes im Mainzer Becken. Zentr. Mineral., Geol. Paläont., B 65—84.

— 1938: Paläogeographische Auswertung der Kreuzschichtung. Geol. Meere Binnengewässer **2**, 229—277.

KANEKO, SHIRO, 1958: Some notes on cross-lamination. J. Geol. Soc. Japan **64**, 152.

KELLING, G., 1958: Ripple mark in the Rhinns of Galloway. Trans. Edinburgh Geol. Soc. **17**, pt. 2, 117—132.

KIDERLEN, H., 1931: Beiträge zur Stratigraphie und Paläogeographie des Süddeutschen Tertiärs. Neues Jahrb. Mineral. Geol. Paläontol., Beil-Bd. B **66**, 289—384.

KIERSCH, G. A., 1950: Small-scale structures and other features of Navajo sandstone, northern part of San Rafael Swell, Utah. Bull. Am. Assoc. Petrol. Geologists **34**, 923—942.

KINDLE, E. M., 1917: Recent and fossil ripple mark. Mus. Bull., Geol. Survey Canada **25**, 1—56.

—, and W. H. BUCHER, 1932: Ripple mark and its interpretation *in* Treatise on Sedimentation, 2nd ed., p. 632—668. Baltimore: Williams and Wilkins.

—, and E. M. EDWARDS, 1924: Literature of ripple mark. Pan Am. Geol. **41**, 191—203.

KING, P. B., 1948: Geology of the southern Guadaloupe Mountains, Texas. U.S. Geol. Survey Prof. paper 215, 183 p.

KNIGHT, S. H., 1929: The Fountain and the Casper formations of the Laramie basin: a study of the genesis of sediments. Wyoming Univ. Publ. Sci. Geol. **1**, 82 p.

KNILL, J. L., 1959: Paleocurrents and sedimentary facies of the Dalradian metasediments of the Craignish-Kilmelfort District. Proc. Geologist's Assoc. (Engl.) **70**, 273—284.

KONYUKHOV, I. A., 1951: Underwater slumps in the productive layer of the Apsheronsk peninsula. All-Union Petrol. Prospecting Scientific Res. Inst. Geol. Collection, No. 1 (4), 154—158 [Russian].

KUENEN, PH. H., 1950: Marine geology. New York: John Wiley and Sons. 568 p.

LAHEE, F. H., 1952: Field geology; 5th ed. New York: McGraw-Hill Book Co. 883 p.

LANDSBERG, S. Y., 1956: The orientation of dunes in Britain and Denmark in relation to wind. Geograph. Rev. 122, 176—189.

LANE, E. W., 1947: Report of the subcommittee on sediment terminology. Trans. Am. Geophys. Union 28, 937—938.

LEMCKE, K., W. F. v. ENGELHARDT and H. FÜCHTBAUER, 1953: Geologische und sedimentpetrographische Untersuchungen im Westteil der ungefalteten Molasse des Süddeutschen Alpenvorlandes. Geol. Jahrb., Beih. 11, 110 p.

LERBEKMO, J. F., 1961: Genetic relationships among Tertiary Blue sandstones in central California. J. Sediment. Petrol. 31, 594—602.

LIPPITT, L., 1959: Statistical analysis of regional facies change in Ordovician Cobourg limestone in northwestern New York and southern Ontario. Bull. Am. Assoc. Petrol. Geologists. 43, 807—816.

LOWELL, J. D., 1955: Applications of cross-stratification studies to problems of uranium exploration, Chuska Mountains, Arizona. Econ. Geol. 50, 177—185.

LÜDERS, K., and F. TRUSHEIM, 1929: Beiträge zur Ablagerung mariner Mollusken in der Flachsee. Senckenbergiana 11, 123—142.

LUI, HSIN-KUAN, 1957: Mechanics of sediment-ripple formation. Proc. Am. Soc. Civil Engrs., Hydraulics Div. 83, No. 2, 1—22.

MACKENZIE, F. T., and J. D. RYAN, 1962: Cloverly-Lakota and Fall River paleocurrents in the Wyoming Rockies. Wyoming Geol. Assoc. 17th Ann. Field Conf., p. 44—61.

MANOBAR, MADHAR, 1955: Mechanics of bottom sediment movement due to wave action: U.S. Beach Erosion Board, Corps. of Engineers, Tech. memo. No. 75, 121 p.

MAPEL, W. J., and C. L. PILLMORE, 1962: Stream directions in the Lakota formation (Cretaceous) in the northern Black Hills, Wyoming and South Dakota in Geologic Survey Research 1962: U.S. Geol. Survey, Prof. Paper 450-B, 35—37.

MARTIN, H., J. J. BIGARELLA and R. SALAMUNI, 1960: Ocorrência de arenitos ecólicos com estratificãco cruzada no grupo Itararé — Paraná. Bull. Univ. Paraná, Geologica, No. 2, 1—9.

McBRIDE, E. F., 1960: Martinsburg flysch of the central Appalachians. Ph.D. thesis, The Johns Hopkins University, 375 p.

—, and M. O. HAYES, 1962: Dune cross-bedding on Mustang Island, Texas. Bull. Am. Assoc. Petrol. Geologists 46, 546—552.

McDOWELL, J. P., 1957: The sedimentary petrology of the Mississagi Quartzite in the Blind River area. Ontario Dept. Mines, Geol. Circular 6, 31 p.

McIVER, N. L., 1961: Upper Devonian marine sedimentation in the Central Appalachians. Unpublished Ph.D. thesis, The Johns Hopkins University, 347 p.

McKEE, E. D., 1938: Original structures in Colorado River flood deposits of Grand Canyon. J. Sediment. Petrol. 8, 77—83.

— 1940: Three types of cross-lamination in Paleozoic rocks of northern Arizona. Am. J. Sci. 238, 811—824.

— 1945: Small-scale structures in the Coconino sandstone of northern Arizona. J. Geol. 53, 313—325.

— 1949: Facies changes in the Colorado Plateau in Sedimentary facies in geologic history. Geol. Soc. Am. Mem. 39, 35—48.

— 1957a: Flume experiments on the production of stratification and cross-stratification. J. Sediment. Petrol. 27, 129—134.

— 1957b: Primary structures in some Recent sediments: Bull. Am. Assoc. Petrol. Geologists 41, 1704—1747.

— C. G. EVENSEN and W. D. GRUNDY, 1953: Studies in sedimentology of the Shinarump conglomerate of northeastern Arizona. U.S. Atomic Energy Comm., Tech. Rept. RME-3089, 48 p.

McKee, E. D., M. A. Reynolds and C. H. Baker, Jr., 1962: Experiments on intraformational recumbent folds in cross-bedded sand. U. S. Geol. Sur. Prof. Paper 450 D, 155—160.
—, and G. W. Weir, 1953: Terminology of stratification and cross-stratification. Bull. Geol. Soc. Am. **64**, 381—390.
Mellen, J., 1956: Pre-Cambrian sedimentation in the northeast part of Cohutta Mountain Quadrangle, Georgia. Georgia Mineral Newsletter, Georgia Geol. Survey **9**, 46—61.
Mii, H., 1955: Ripple marks in Matsukawa-Ura *in* Studies in the ecology and sedimentation of Matsukawa-Ura, Soma City, Fukushima Prefecture, Part 2. Contributions from the Institute of Geology and Paleontology, Tohoku University, Sendai, Japan, p. 32—40 [Japanese].
Mikkola, T., 1955: Sedimentary transportation in Karelian quartzites. Bull. comm. géol. Finlande No. 168, 27—29.
—, 1960: Sedimentation of quartzite in the Kemi area, north Finland. 21st Internat. Geol. Congr., Session Norden, pt. 9, p. 154—161.
Miller, W. J., 1922: Intraformational corrugated rocks. J. Geol. **30**, 587—610.
Moberly jr., R., 1960: Morrison, Cloverly, and Skyes Mountain formations, northern Bighorn Basin, Wyoming and Montana. Bull. Geol. Soc. Am. **71**, 1137—1176.
Naha, K., 1961: Precambrian sedimentation around Ghatsila in East Singhbum, eastern India. Proc. Nat. Inst. Sci. India **27** A, 361—372.
Nickelsen, R. P., 1958: Cross-bedding in Pennsylvanian sandstones in central Pennsylvania: A preliminary study. Bull. Geol. Soc. Am. **69**, 791—796.
Niehoff, W., 1958: Die primär gerichteten Sedimentstrukturen, insbesondere die Schrägschichtung im Koblenzquartzit am Mittelrhein. Geol. Rundschau **47**, 252—321.
Norris, R. M., and K. S. Norris, 1961: Algodones dunes of southeastern California. Bull. Geol. Soc. **72**, 605—620.
Opdyke, N. D., and S. K. Runcorn, 1960: Wind direction in the western United States in the Late Paleozoic. Bull. Geol. Soc. Am. **71**, 959—972.
Osmond, J. C., 1954: Dolomites in Silurian and Devonian of east-central Nevada. Bull. Am. Assoc. Petrol. Geologists **38**, 1911—1956.
Otto, G. H., 1938: The sedimentation unit and its use in field sampling. J. Geol. **46**, 569—582.
Pelletier, B. E., 1958: Pocono paleocurrents in Pennsylvania and Maryland. Bull. Geol. Soc. Am. **69**, 1033—1064.
— 1960: Triassic stratigraphy, Rocky Mountain foothills, northeastern British Columbia 94 J and 94 K. Geol. Survey Canada, Paper 60-2, 32 p.
Pettijohn, F. J., 1957: Paleocurrents of Lake Superior Precambrian quartzites. Bull. Geol. Soc. Amer. **68**, 409—480.
— 1962: Paleocurrents and paleogeography. Bull. Am. Assoc. Petrol. Geologists **46**, 1468—1493.
Picard, K., 1950: Sedimentations-Verhältnisse des Hauptbuntsandsteins in der Bucht von Mechernich-Nideggen. Geol. Jahrb. **64**, 331—347.
— 1953: Zur Auswertung der Kreuzschichtung in fluviatilen Sedimenten. Geol. Rundschau **41**, 268—276.
Poole, F. G., 1961: Stream directions in Triassic rocks of the Colorado Plateau *in* Geological Survey Research, 1961. U.S. Geol. Sur. Prof. Paper 424 C, 139—141.
— 1962: Wind directions in late Paleozoic to middle Mesozoic time on the Colorado Plateaus *in* Geological Survey Research, 1962. U.S. Geol. Sur., Prof. Paper 450 D, 147—151.
Potter, P. E. 1955: The petrology and origin of the Lafayette gravel; Part I, Mineralogy and petrology. J. Geol. **63**, 1—38.
— 1962a: Sandbody shape and map pattern of Pennsylvania sandstones of Illinois. Illinois Geol. Survey, Cir. **339**, 36 p.
— 1962b: Late Mississippian sandstones of Illinois. Illinois Geol. Survey, Cir. **340**, 36 p.
— 1962c: Regional distribution patterns of Pennsylvanian sandstones in Illinois Basin. Bull. Am. Assoc. Petrol. Geologists **46**, 1890—1911.

POTTER, P. E., and H. D. GLASS, 1958: Petrology and sedimentation of the Pennsylvanian sediments in southern Illinois: a vertical profile. Illinois Geol. Survey, Rept. Inv. 204, 60 p.

— E. NOSOW, N. W. SMITH, D. H. SWANN and F. H. WALKER, 1958: Chester crossbedding and sandstone trends in Illinois basin. Bull. Am. Assoc. Petrol. Geologists 42, 1013—1046.

—, and J. S. OLSON, 1954: Variance components of cross-bedding direction in some basal Pennsylvanian sandstones of the Eastern Interior Basin: Geological application. J. Geol. 62, 50—73.

—, and W. L. PRYOR, 1961: Dispersal centers of Paleozoic and later clastics of the Upper Mississippi Valley and adjacent areas. Bull. Geol. Soc. Am. 72, 1195—1250.

—, and R. SIEVER, 1956: Sources of basal Pennsylvanian sediments in the Eastern Interior Basin: Part I, Cross-bedding. J. Geol. 64, 225—244.

POWER jr., W. R., 1961: Backset beds in the Coco formation, Inyo County, California. J. Sediment. Petrol. 31, 603—607.

PRYOR, W. A., 1958: Dip direction indicator. J. Sediment. Petrol. 28, 230.

— 1960: Cretaceous sedimentation in Upper Mississippi Embayment. Bull. Am. Assoc. Petrol. Geologists 44, 1473—1504.

RADOMSKI, A., 1958: Charakterystyka sedymentologiczna fliszu podhalanskiego. Acta Geol. Polon. 8, 335—409.

— 1961: On some sedimentological problems of the Swiss Flysch series. Eclogae Geol. Helv. 54, 451—459.

REICHE, P., 1938: An analysis of cross-lamination: The Coconino sandstone. J. Geol. 46, 905—932.

REINECK, H.-E., 1958: Longitudinale Schrägschichtung im Watt. Geol. Rundschau 47, 73—82.

— 1961: Sedimentbewegungen an Kleinrippeln im Watt. Senckenbergiana Lethaea 42, 51—67.

REINEMUND, J. A., 1955: Geology of the Deep River Coal field North Carolina. U.S. Geol. Survey Prof. paper 246, 159 p.

—, and W. DANILCHIK, 1957: Preliminary geologic map of the Waldron quadrangle and adjacent areas, Scott County, Arkansas. U.S. Geol. Survey Oil and Gas Invs. Map OM 192.

REYNOLDS, O., 1889: On model estuaries in Report of the committee appointed to investigate the action of waves and currents on the beds and fore shores of estuaries by means of working models. Rept. Brit. Assoc., p. 327—343.

— 1890: On model estuaries in Second report of the committee appointed to investigate the action of the waves and currents on the beds and fore shores of estuaries by means of working models. Rept. Brit. Assoc., p. 512—534.

— 1891: On model estuaries in Third report of the committee appointed to investigate the action of waves and currents on the beds and fore shores of estuaries by means of working models. Rept. Brit. Assoc., p. 386—404.

ROBSON, D. A., 1956: A sedimentary study of the Fell sandstone of the Coquet Valley, Northumberland. Quart. J. Geol. Soc. London 107, 241—262.

ROSS, J. V., 1962: Deposition and current direction within the Yellowknife group at Mesa Lake, N. W. T., Canada. Bull. Geol. Soc. Am. 73, 1159—1162.

RUBEY, W. W., and N. W. BASS, 1925: The geology of Russell County, Kansas, Part I. Kansas Geol. Survey, Bull. 10, 104 p.

RUKHIN, L. B., 1958: Grundzüge der Lithologie. Berlin: Akademie-Verlag. 806 p. [Translated from the Russian].

RUSNAK, G. A., 1957: A fabric and petrologic study of the Pleasantview sandstone. J. Sediment. Petrol. 27, 41—55.

SCHLEE, J. S., 1957: Upland gravels of southern Maryland. Bull. Geol. Soc. Am. 68, 1371—1410.

—, and R. H. MOENCH, 1961: Properties and genesis of "Jackpile" Sandstone, Laguna New Mexico in Geometry of sandstone bodies. Tulsa: Am. Assoc. Petrol. Geologists 134—150.

SCHMITT, PH., 1935: Zur Petrogenese des fränkischen Wellenkalkes. Chem. Erde 9, 321—364.

SCHWARZACHER, W., 1953: Cross-bedding and grain size in the Lower Cretaceous sands of East Anglia. Geol. Mag. 90, 322—320.

SCOTT, G., 1930: Ripple marks of large size in the Fredericksburg Rocks west of Fort Worth, Texas in Contributions to Geology, 1930. The University of Texas Bull. No. 3001, 53—56.

SEIFERT, A., 1935: Neue Beobachtungen über Großrippeln in den Turonsandsteinen der Sächsischen Schweiz. Sitzber. Abh. nat. Ges. Isis Dresden 1933/34, p. 122—135.

— 1942: Schrägschichtung im mittleren Buntsandstein des Saarlandes und angrenzender Gebiete. Z. deut. geol. Ges. 94, 489—510.

SHACKLETON, J. S., 1962: Cross-strata of the Rough Rock (Millstone Grit Series) in the Pennines. Liverpool Manchester Geol. J. 3, 109—118.

SHOTTON, F. W., 1937: The lower Bunter sandstones of North Worcestershire and East Shropshire. Geol. Mag. 74, 534—553.

— 1956: Some aspects of the New Red desert in Britain. Liverpool Manchester Geol. J. 1, 450—465.

SHROCK, R. R., 1948: Sequence in layered rocks. New York: McGraw-Hill Book Co. 507 p.

SIEVER, R., and P. E. POTTER, 1956: Sources of basal Pennsylvanian sediments in the Eastern Interior Basin: sedimentary petrology. J. Geol. 62, 317—335.

SIMONS, D. B., E. V. RICHARDSON and M. L. ALBERTSON, 1961: Flume studies using medium sand (0.45 mm). U.S. Geol. Survey Water-Supply Paper 1498-A, 76 p.

—, and E. V. RICHARDSON, 1961: Forms of bed roughness in alluvial channels. Am. Soc. Civil Engrs. 87, No. HY 3, 87—105.

SORBY, H. C., 1853: On the oscillation of the currents drifting sandstone beds of the southeast of Northumberland, and on their general direction in the coal field in the neighborhood of Edinburgh. Reports Proc. Geological and Polytechnic Soc. of the West Riding of Yorkshire, p. 225—231.

— 1858: On the ancient physical geography of the southeast of England. Edinburgh New Philosophical J., n.s. 7, 226—237.

SPURR, J. E., 1894: False bedding in stratified drift deposits. Am. Geologist 13, 43—47.

STEINLEIN, H., 1950: Eine einfache graphische Methode zur indirekten Bestimmung des Streichens und Fallens und der Schrägschichtungs-Richtung. Neues Jahrb. Mineral., Geol. Paläont. B 91, 149—160.

STEWART, A. D., 1962: On the Torridonian sediments of Colonsay and their relationship to the main outcrop in northwest Scotland. Liverpool Manchester Geol. J. 3, 121—155.

STEWART, J. H., F. G. POOLE and R. F. WILSON, 1957: Triassic studies in Geologic investigations of radioactive deposits; semi-annual report for 1 Dec. 1956 to 31 May 1957. U.S. Atomic Energy Comm., Oak Ridge, Tenn. TEI-690, book 2, p. 346—351.

STOKES, W. L., 1952: Uranium-vanadium deposits of the Thompson's area Grand County, Utah. Utah Geol. and Mineral. Survey, Bull. 46, 51.

— 1953: Primary sedimentary trend indicators as applied to ore finding in the Carrizo Mountains, Arizona and New Mexico. U.S. Atomic Energy Comm. RME-3043, 48 p.

— 1954a: Some stratigraphic, sedimentary and structural relations of uranium deposits in the Salt Wash sandstone. U.S. Atomic Energy Commission, RME-3102, 50 p.

— 1954b: Relation of sedimentary trends, tectonic features, and ore deposits in Blanding district, San Juan County, Utah. U.S. Atomic Energy Comm., Tech. Rept. RME-3093 (pt. 1), 33 p.

STRAATEN, L. M. J. U. VAN, 1950: Giant ripples in tidal channels. T. Kon. Nederl. Aardryksk, Genoot. 67, 76—81.

— 1951: Longitudinal ripple marks in mud and sand. J. Sediment. Petrol. 21, 47—54.

STRAATEN, L. M. J. U. VAN, 1953: Megaripples in the Dutch Wadden sea and in the basin of Arcachon (France). Geol. en Mijnbouw, n.s. **15**, 1—11.

SULLWOLD jr., W. H., 1960: Tarzana fan, deep submarine fan of late Miocene age, Los Angeles County, California. Bull. Am. Assoc. Petrol. Geologists **44**, 433—457.

SUNDBORG, Å., 1956: The river Klarälven, a study of fluvial processes. Geog. Annaler **48**, 316 p.

SUTTON, J., and J. WATSON, 1960: Sedimentary structures in the Epidotic Grits of Skye. Geol. Mag. **97**, 106—122.

SZÁDECZKY-KARDOSS, E., 1938: Geologie der Rumpfungarländischen kleinen Tiefebene mit Berücksichtigung der Donaugoldfrage. Kgl. ungar. Palatin-Joseph Univ. tech. u. Wirtschaftswiss., Fak. für Berg-, Hütten- u. Forstwes. zu Sopron (Hungary) **10**, 2, 444 p.

TANNER, W. F., 1955: Paleogeographic reconstructions from cross-bedding studies. Bull. Am. Assoc. Petrol. Geologists **39**, 247—248.

— 1959: The importance of modes in cross-bedding data. J. Sediment. Petrol. **29**, 221—226.

THIEL, G. A., 1932: Giant current ripples in coarse fluvial gravel. J. Geol. **40**, 452—458.

THOME, K. N., 1959: Die Begegnung des nordischen Inlandeises mit dem Rhein. Geol. Jahrb. **76**, 261—308.

TOWE, K. M., 1959: Petrology and source of sediments in the Narragansett basin of Rhode Island and Massachusetts. J. Sediment. Petrol. **29**, 503—512.

TRUSHEIM, F., 1929: Zur Bildungsgeschwindigkeit geschichteter Sedimente im Wattenmeer. Senckenbergiana **11**, 47—55.

TWENHOFEL, W. H., 1932: Treatise on sedimentation, 2nd ed. Baltimore: Williams and Wilkins Company. 926 p.

VALDYANADHAN, R., 1957: Significance of certain sedimentary structures in Cuddapah sandstones. Quart. J. Geol. Mining Met. Soc. India **29**, 191—198.

VAN HOUTEN, F. B., 1962: Frontier Formation, Big Horn basin, Wyoming *in* Symposium on early Cretaceous rocks of Wyoming and adjacent areas, Wyoming Geological Assoc., 17th Ann. Field Conf., p. 221—231.

VANONI, V. A., and J. G. KENNEDY, 1961: Discussion of "Forms of bed roughness in alluvial channels". J. Hydraulics Division, Proc. Am. Soc. Civil Engrs. **6**, 241—247.

VASSOEVICH, N. B., and V. A. GROSSGEYM, 1951: Paleogeography of the northeastern Caucasus in the middle Miocene Epoch. All-Union Petrol. Prospecting Scientific Res. Inst. Geological Collection No. 1 (4), 121—136 [Russian].

VAUSE, J. E., 1959: Underwater geology and analysis of Recent sediments off the northwest Florida coast. J. Sediment. Petrol. **29**, 555—563.

WALKER, C. T., 1955: Current-bedding directions in sandstones of lower *Reticuloceras* age in the Millstone Grit of Wharfdale, Yorkshire. Proc. Yorkshire Geol. Soc. **30**, 115—132.

WALTHER, J., 1893: Einleitung in die Geologie als historische Wissenschaft. Jena: Gustav Fischer. 1055 p.

WHITAKER, J. C., 1955: Direction of current flow in some Lower Cambrian clastics of Maryland. Bull. Geol. Soc. Am. **66**, 763—766.

WILLIAMS, E. G., 1959: Aspects of the paleogeography of the coal measures in western Pennsylvania. Mineral Industries, Coll. of Mineral Ind., Pennsylvania State Univ. **28**, 1—5.

WILSON, G., J. WATSON and J. SUTTON, 1953: Current-bedding in the Moine series of northwestern Scotland. Geol. Mag. **90**, 377—389.

WOOD, A., and A. J. SMITH, 1959: The sedimentation and sedimentary history of the Aberystwyth Grits (Upper Llandoverian). Quart. J. Geol. Soc. London **114**, 163—195.

WRIGHT jr., H. E., 1956: Origin of the Chuska sandstone, Arizona-New Mexico: a structural and petrographic study of a Tertiary eolian sediment. Bull. Geol. Soc. Am. **67**, 413—433.

WRIGHT, M. D., 1959: The formation of cross-bedding by a meandering or braided stream. J. Sediment. Petrol. **29**, 610—615.

WURSTER, P., 1958: Schüttung des Schilfsandsteins im mittleren Württemberg. Neues Jahrb. Geol. Paläontol., Monatsh. **1958**, 479—489.

— 1960: Kreuzschichtung im Buntsandstein von Helgoland. Mitt. geol. Staatsinst. Hamburg **29**, 61—65.

YEAKEL jr., L. S., 1959: Tuscarora, Juniata and Bald Eagle paleocurrents and paleogeography in the central Appalachians. Unpublished Ph.D. thesis, The Johns Hopkins University, 454 p.

— 1962: Tuscarora, Juniata, and Bald Eagle paleocurrents and paleogeography in the central Appalachians. Bull. Geol. Soc. Am. **73**, 1515—1540.

ZHEMCHUZHNIKOV, YU. A., 1926: Type of cross-bedding as criteria of origins of sediments. Ann. Inst. Mines (Leningrad) **7**, 35—69 [Russian].

Cross-Bedding and Ripple Marks (1963—1976)

Measurement of cross-bedding and ripple mark in terrigenous sandstones is now routine and has made great progress in the last thirteen years. There is some interest also in cross-bedding of carbonate and volcaniclastic sediments; in addition, there is strong emphasis on the paleohydraulic and environmental significance of cross-bedding and ripple mark in sandstones

The study of cross-bedding and ripple mark in ancient sediments and of bed-forms and sand-wave systems in flumes and modern environments is now commonplace; only the study of turbidite structures and processes is comparable. Certainly cross-bedding continues to be the most useful and widely studied of all the paleocurrent structures so that, for example, for each study of ripple mark, there are probably thirty or more of cross-bedding. Moreover, field studies now suggest that a cross-bedding map pattern has some environmental significance, an idea that in 1963 was not fully established. Several techniques have contributed significantly — the sonic depth finder primarily for rivers, photography from space for deserts (WOBBER, 1967), and side scan sonar for mapping sand-wave systems on continental shelves (BELDERSON *et al.*, 1972). In 1963 we were also just on the threshold of the flume studies of sand waves and their paleohydraulic significance. Although many have contributed to our better understanding of sand waves, the name of J. R. L. Allen easily stands foremost among sedimentologists.

Flow, Bedforms, and Their Internal Structures

ALLEN (1968, p. 1—4) and others have stressed that to understand cross-bedding and ripple mark fully, a better understanding of the origins of bedforms and sand-wave systems is essential, the logic being simply that flow conditions largely determine the external geometry of bedforms which in turn determine their preserved internal structures (see BRUSH *et al.*, 1966, for bedform nomenclature).

Below are ten statements summarizing much of the present knowledge of bedforms, sand waves and their internal structures with emphasis on their significance for paleocurrents. Most of this knowledge is a post-1963 development and comes from both sedimentology and hydraulic engineering.

1. An initial irregularity is needed as a nucleus to generate a sand-wave system on a granular, deformable bed

2. A sand-wave system migrates and evolves by grain to grain removal and deposition at the sediment interface

3. The fluid and bedforms of a sand-wave system interact in an as yet not fully understood way, although both flow intensity (as measured by either stream power or average velocity) and grain size go far to explain the sequential progression of bedforms with increasing flow intensity (fig. 4-22)

4. The individual sand waves may either be largely transverse to the flow (most ripples and dunes) or parallel to it (some of the larger dunes and erosional scours on the bottom)

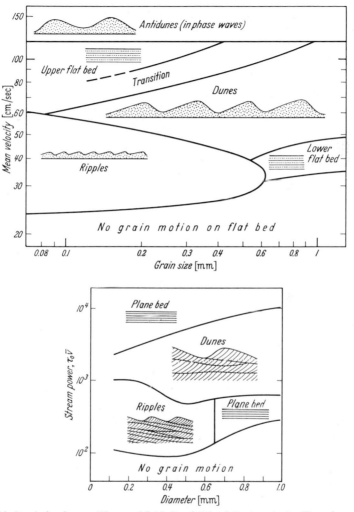

Fig. 4-22. Velocity, ripples, dunes, antidunes and flat beds and their relation to grain size. (Top redrawn from Southard and Boguchwal, 1973, fig. 1, J. Sediment. Petrol., Soc. Econ. Paleontol. Mineral.; botten redrawn from Allen, 1969, Fig. 1). The lower flat bed field (parting lineation) and that of ripples are not yet fully established

5. Transverse bedforms are related to *flow separation*, (the main flow separates from the boundary) whereas those that are parallel to the flow are a response to longitudinal, helical flow

6. Individual bedforms may migrate up or down current depending upon the relative rates of deposition on their lee or stoss sides

7. Sand-wave systems vary enormously in scale — from the smallest of ripples to those consisting of seif dunes over 100 km in length, although there seems to

be, nevertheless, a natural break between ripples and dunes at an amplitude of about 4 cm

8. Small sand waves can be superimposed on larger ones, the superposition resulting from marked change in flow intensity

9. Most types of ripple mark are not clearly related to water depth, whereas height of subaqueous dunes has a closer relationship

10. There is a hierarchy of bedforms that corresponds to a hierarchy of paleo-currents, the general rule being that the larger the structure the greater its paleocurrent significance and the less its variance, because it was formed by a stronger and larger current (fig. 4-23).

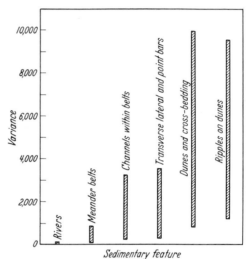

Fig. 4-23. Variance hierarchy. (Redrawn from MIALL, 1974, fig. 2, J. Sediment. Petrol., Soc. Econ. Paleontol. Mineral.)

Alluvial and estuarine bedforms and their flow conditions continue to be the most studied and best known, especially the alluvial environments followed by those of shelves and deserts. Typical is the work of WILLIAMS (1971) who related, in a beautifully illustrated study of flood deposits of ephemeral sand-bed streams in Australia, plane beds, large scale ripples, and longitudinal bars to high flow and transverse and lingulate bars to waning and low flows. He also found that trough stratification, which constitutes about 60 percent of the deposits, was formed by both small and large scale ripples whereas tabular crossbedding, which constitutes about 25 percent, resulted from the migration of transverse bars in channels. PICARD and HIGH (1973) published a monograph on ephemeral streams in the western United States. SMITH (1972a) studied stratification types in the braided Platte River of Nebraska as did RUST (1972) for a river in the Yukon of Canada and BLUCK (1974) for a sandur in Iceland. Although all of the above investigators related bedforms to flow, none, unfortunately, actually observed the bedforms during high flows. SMITH (1972b) also related bar and bedform morphology plus stratification type to grain size and distance downstream (over 960 km) in the braided Platte River. COLLINSON (1970) describes and documents the history of bedforms developed in a small river in Norway during and after the spring breakup of ice. MCGOWEN and GARNER (1970), although they do not emphasize paleocurrents, related stratification on a point bar to its physiographic development. Although nothing new has been added to their paleocurrent significance, there have been a number of papers on the paleohydraulic significance of climbing ripples (WALKER, 1969; ALLEN, 1970).

Bedforms, stratification types and paleocurrents have received much less attention in beaches, a notable exception being CLIFTON et al. (1971, Table 1), who studied bedforms of ripples and wave regime along the coast of Oregon and also measured cross-bedding orientation in analogously formed coastal terraces, the cross-bedding being poorly oriented.

BARTISH-WINKLER et al. (1975) made a somewhat similar study of a tidal bar in Cook Inlet, Alaska, where they mapped ripple marks, washouts and grain size. Here the tidal range is 10 m.

Marine sand-wave systems are now fairly well mapped in and near the North Sea (KENYON, 1970; McCAVE, 1971; LANGHORNE, 1973) and off parts of eastern North America (SWIFT et al., 1972), some effort being made to relate sand-wave type to surface currents and grain size (cf. McCAVE, 1971). These studies clearly show that superimposed on the large sand waves, some up to 15 km or more long, are dunes and ripples, the former producing the cross-bedding found within the sand waves. With box cores REINECK (1963, figs. 16 and 19) found similar dunes on large sand waves in an arm of the North Sea to have bipolar cross-bedding related to oscillatory tidal currents.

The most comprehensive process-oriented study of eolian bedforms is that of WILSON (1972, 1973), who suggested that there are four major, distinctly different bedform groups — draas, dunes, aerodynamic ripples, and impact ripples, draas being the largest. Each of the four types has transverse and longitudinal subtypes. GLENNIE (1970, Chap. 6) fully reviews the subject of dune morphology, origin, and bedding. SANDERSON (1974) has provided good documentation of cross-bedding geometry in the Navajo Sandstone, believed by him to be an ancient dune field. Sand-wave systems of gigantic size have also been observed on Mars (SAGAN, 1973).

For paleocurrents the key question is: "How well do cross-bedding and ripple mark correspond to *local* flow direction?" The answer is "very well", as demonstrated by a number of studies.

SMITH (1972a, figs. 11 and 12) found good correspondence between the vector mean of tabular cross-bedding of transverse bars and local channel orientation; he also found that 30 percent of all the foresets lie within 5° of the channel axis. RUST (1972, figs. 11 and 12) shows good agreement between bedform orientation on bars and channel axes as does BLUCK (1974). PICARD and HIGH (1973, p. 213) also found that the vector mean of tabular cross-bedding closely coincided with channel axes. LAND and HOYT (1966, fig. 2) illustrate good correspondence for estuarine point bars. KLEIN (1970), in an excellent study of intertidal sandbars, found very good correspondence between cross-bedding orientation and three different types of bedforms arising from ebb and flow tidal currents (fig. 4-24).

The cross-bedding studies of BIGARELLA (1970/1971) and BIGARELLA et al. (1970/1971) are exceptional in their scope and detail. They found good correspondence between regional wind pattern along the Brasilian coast and the orientation of cross-bedding in its coastal dunes. YAALON and LARONNE (1971) and GOLD-

SMITH (1973) also both report good correlation. Earlier McKEE (1966, p. 57—58) found good correspondence between wind and cross-bedding orientation for dome shaped, barchan and parabolic dunes in New Mexico, although an earlier study of seif dunes in Libya revealed more complicated relationships between external form and internal bedding (McKEE and TIBBITS, 1964). BIGARELLA (1972) provides, however, the most comprehensive review, whose main point is that dune cross-bedding is a good indicator of paleowinds.

Another aspect of paleocurrent analysis is the question of weighing paleocurrent measurements by some measure of their magnitude, say thickness or volume, to permit us to map a paleocurrent system for what it originally was — *a vector field having both a direction and magnitude at every point within its domain.* Volume, first suggested by IRIONDO (1973), appears to be better than thickness for cross-beds so that MIALL (1974, p. 1174—1184) suggested the *cube* of maximum

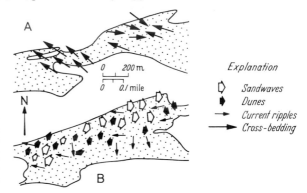

Fig. 4-24. Ebb and flow bedforms and their bimodal cross-bedding in some tidal sand-bodies. (Redrawn from KLEIN, 1970, fig. 16 and 17, J. Sediment. Petrol., Soc. Econ. Paleontol. Mineral.). Compare with bipolar crossbedding of Pennsylvanian and Ordovician calcarenites (figs. 4-16, 4-30 and 7-16)

observed thickness as a rough estimate of actual volume and found that by weighing both the vector mean and variance could be somewhat altered.

Underlying weighing is the idea that weaker currents are much more variable than stronger ones so that small structures should not count as much in a paleocurrent analysis as larger ones. In other words, *corresponding to a size-based hierarchy of bedforms, there is a hierarchy of paleocurrent magnitudes.* Bedform studies in alluvial, shelf and eolian environments all substantiate this idea, which was first proposed by ALLEN (1966) and more fully developed by MIALL (1974), even though PICARD and HIGH's study does provide some contrary evidence (1973, Table 16).

The above studies of bedforms and cross-bedding in modern environments clearly show not only the hierarchical structure of a paleocurrent system, but also that there is little foundation to ALLEN's (1967) philosophical discussion of the differential preservation potential of cross-bedding and ripple mark, at least as far as paleocurrents are concerned.

Classification and Measurement

In 1963 ALLEN proposed a very complete classification of cross-bedding with 15 different types to each of which he assigned a Greek letter. These types are based on their association (solitary or grouped), thickness, character of lower boundary (erosional or gradational), relation of foresets to lower boundary (flat or curved) and the lithologic homogeneity of the set. Subsequently, CROOK (1965) added to this classification. This exhaustive classification would be much more widely used, if it were possible in most outcrops, to see the full three-dimensional geometry of the units. Unfortunately, this is not commonly true, so most sedimentologists continue to specify only the maximum observed thickness, the azimuth of the maximum dip direction, possibly the dip angle, and classify the type into a few broad groups. An example of a more conservative classification is that of MICHELSON and DOTT (1973, fig. 2), who recognized four types — planar parallel, planar wedge, trough parallel and trough wedge (fig. 4-25).

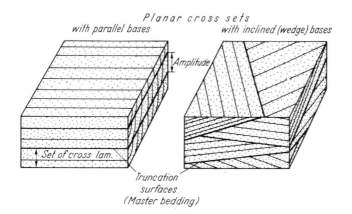

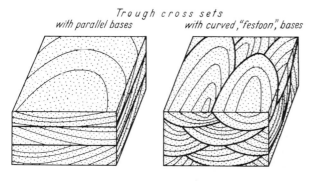

Fig. 4-25. Four major types of cross-bedding. (Redrawn from DALZIEL and DOTT, 1970, fig. 10). Because of possible gradation between types and difficulties of recognition in some essentially two dimensional outcrops, these four types are generally adequate for the vast majority of studies (cf. fig. 4-2)

Disagreement exists about the variability of maximum dip direction of planar versus concave cross-bedding. MECKEL (1967) found planar cross-beds to have a smaller variance than concave cross-bedding in closely spaced outcrops of Pennsylvanian sandstones. On the other hand, HIGH and PICARD (1974), who examined fluvial sands both ancient and modern, report exactly the opposite conclusion. Logically, it would seem that of the two, planar cross-beds would have the smaller variability. STEINMETZ (1975) investigated variability of cross-bedding orientation in a single, small sand body, his study being one of the most detailed in existence.

Still another aspect of the measurement problem, which has received some attention, is whether to measure the trough axis or a foreset. Most would agree with DOTT (1973) that measurement of a trough axis, should it exist and should it be exposed, is preferable, because its variance is usually less than that of random measurement of foresets from the same cross-bed. However, if only trough axes could be measured, the vast majority of the numerous cross-bedding studies would simply not exist.

Environmental Significance and Paleoslope

These two aspects of cross-bedding and ripple mark patterns are closely related.

It was very early suggested by BRINKMANN* (1933) that the variability of cross-bedding orientation has some environmental significance; he used the standard deviation as a measure of variability. Subsequently, this idea has been commented

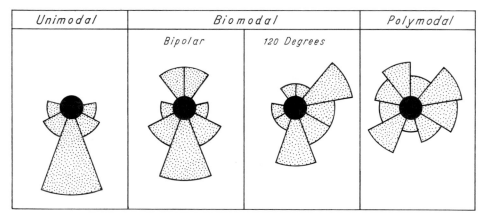

Fig. 4-26. Types of current roses

upon from time to time in the literature, most recently by PICARD and HIGH (1973, p. 213).

The regional current rose is more significant than variance, as observed by SELLY (1968) and ALLEN (1968, fig. 5-40). Is the current rose unimodal, bipolar (two modes 180° apart), bimodal (two modes separated by, say 90°) or polymodal-random (fig. 4-26)? Virtually all alluvial and deltaic environments as well as those believed to be eolian have unimodal current roses, tidal environments characteristically have bipolar ones, whereas some marine shelf sandstones may, in addition, have polymodal as well as unimodal cross-bedding orientation. The small-scale cross-bedding of current ripple origin in most turbidites also has a unimodal current rose. At the outcrop pervasive bipolar cross-bedding and ripple mark (fig. 4-24) is always a very strong indicator of tidal currents, (cf. DE RAAF and BOERSMA, 1971), one that certainly should not be ignored in any environmental analysis — another good reason always to *measure* paleocurrent structures. What can be achieved by study of sedimentary structures in outcrop is well illustrated by GIETELINK (1973) in the Cambro-Ordovician of Spain, who recognized thirteen different subfacies to which he related cross-bedding orientation.

How well do cross-bedding and ripple mark indicate the paleoslope? Cross-bedding and current ripples provide an unambiguous answer in alluvial–deltaic sandstones, as do oscillation and wave-formed ripples, which commonly strike parallel to the shorelines of lakes, seas, and oceans, even through very near shore and along the beach, locally more complicated patterns may exist. In turbidites ripple mark indicates paleoslope; but if the asymmetric current ripples are a product of contour currents, they should indicate a flow *parallel* to the slope (HOLLISTER and HEEZEN, 1972; BOUMA, 1972, Table 1). BEIN and WEILER (1976, fig. 15) provide a good example of a contour paleocurrent system on a carbonate slope as well as showing the paleocurrent systems of associated carbonate turbidites and those on the shelf in the Cretaceous of the Middle East (fig. 4-27).

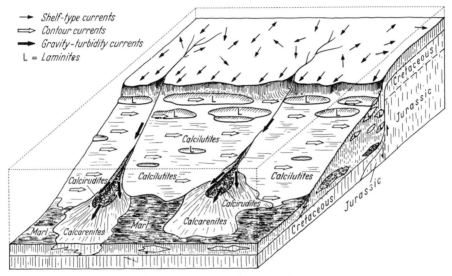

Fig. 4-27. Conceptual model of contour, gravity (turbidity), and shelf current systems on a Cretaceous carbonate shelf-to-basin transition in the Middle East. (Redrawn from BEIN and WEILER, 1976, Fig. 15.) The Talme Yafe Formation is over 3000 m thick.

The most complicated cross-bedding patterns are those of marine shelf sandstones and/or coastal barriers and beaches (cf. CLIFTON *et al.*, 1971, fig. 26). Some marine shelf sandstones have regional cross-bedding modes that point offshore and down dip (MICHELSON and DOTT, 1973), some oblique to it (ADAMS, 1970), and some possibly even parallel to it; however, most studies of ancient marine shelf sandstones, both terrigenous and carbonate, commonly suggest an overall net transport down paleoslope. SWIFT (1970, fig. 5) provided an explanation of this, when he pointed out that even random, possibly storm-generated wave motion, on a seaward-dipping shelf will result in a net slope or paleoslope transport of sand. Probably cross-bedding orientation within coastal barriers is much more complicated, because it may represent a mixture of beach, eolian, as well as possibly tidal environments.

By using the internal consistency of other evidence and by making an independent assessment of the environment of deposition, a rather definite interpretation with respect to paleoslope is nearly always possible, especially when one recognizes that in a large basin, paleoslope may systematically change along its strike or that in an elongate trough there may be both axial as well as lateral components to the slope. Mapping a sufficiently large area and/or integrating other evidence will nearly always resolve most uncertainties. Certainly it is safe to say that that studies of both modern and ancient paleocurrent systems give a certainty that far surpasses what has been surmised by theorizing (cf. KLEIN, 1967).

Terrigenous Sands and Sandstones

Cross-bedding studies are now so commonplace in terrigenous sands and sandstones that we have chosen to comment here only on a few aspects of their study,

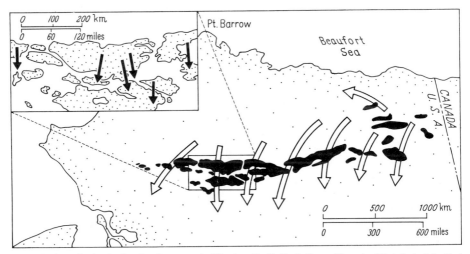

Fig. 4-28. Generalized paleoflow based on cross-bedding from the Endicott Group (Devonian-Mississippian) in Alaska. (Redrawn and simplified from DONOVAN and TAILLEUR, 1975). Paleocurrent map based only on 38 stations and a total of 177 measurements of which 116 were crossbeds and 40 were current lineations. Nonetheless, this sampling is sufficient to indicate clearly the regional paleocurrent system. In remote Arctic regions such as this, the helicopter is the chief field vehicle used to visit outcrops and, at a cost of about $300 an hour, must be used very efficiently

deferring until later the contribution of cross-bedding to the study of sandstone bodies and sedimentary basins.

There is a renewed interest in the sedimentology of Pleistocene sands and gravels, including their paleocurrents, five excellent studies being those of AARIO (1972) in Finland, HELM (1971) and SHAW (1972) in England, COSTELLO and WALKER (1972) in Ontario, and that of BAKER (1973), who studied Lake Missoula flood deposits in Washington. The latter is a comprehensive, very well done study of remarkably well-preserved deposits in which are found giant ripples, turbidites and a wide range cross-bedding. A common feature of many of these studies is their stress on the interrelationships of bedding facies, vertical profiles, paleoflow and paleocurrents as is, of course, increasingly true for pre-Pleistocene terrigenous sandstones. Commonly, the large literature on bedforms in flumes goes far to provide a basis for an interpretation, although it may be necessary to supplement this by special experiments as did McDONALD and VINCENT (1972) who, studying subglacial sedimentary environments, found it informative to duplicate them in a pipe!

A recent regional study of over 1,700 miles (2,720 km) in the Brooks Range of northern Alaska is that of DONOVAN and TAILLEUR (1975), who mapped cross-bedding and clast size and found a remarkably homogeneous pattern (fig. 4-28).

At the other end of the sedimentary realm there are a few papers about cross-bedding in metamorphic rocks at least one of which (MORTON, 1971) *measures its orientation* in sillimanite-rich, upper amphibolitic metasandstones and meta-conglomerates formed about 1,450 million years ago. Thus sedimentary structures cannot only be well preserved in metamorphic rocks, but even still used to

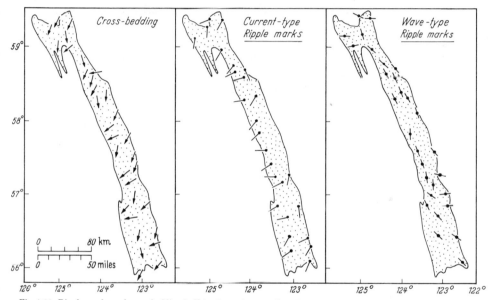

Fig. 4-29. Ripple marks and cross-bedding in Triassic sandstones of northeastern British Columbia. This map covers one of the widest areas ever studied for ripple marks. (Redrawn from PELLETIER, 1965, fig. 3, Soc. Econ. Paleontol. Mineral. Sp. Pub. 12). Strike of wave ripples gives depositional strike and thus net transport is seaward-downslope. See also the modern evidence on this conclusion in fig. 4-17

provide sedimentological information so that one might ask: "In how many metamorphic terrains is it now possible to infer the original shape of the sedimentary basin and better recreate both its fill and paleoslope?" In this age of speculation about plate tectonics answers to such questions may have much more relevance than they might have had 20 years ago.

Regional paleocurrent studies of ripple mark continue to be made, but are very uncommon. One notable recent study is that of BRUUN-PETERSEN and KRUMBEIN (1975), who systematically mapped oscillation ripples and inferred paleohydraulics in Helgoland. Earlier PICARD and HIGH (1968) studied a Triassic marine red bed in Wyoming and confirmed, in a general way, that ripple-mark orientation in marine rocks is dependent upon bathymetric slope and shoreline configuration. They also noted that complex patterns can occur nearshore so that the vector mean should not be used to summarize the resulting current use. The most comprehensive study, however, is that of PELLETIER (1965), who studied

Table 4-4. *Cross-Bedding in Carbonates*

Author and year	Remarks
SCHLEE, 1963	Early documentation of diverse orientation of cross-bedding in limestones of Mississippian age in the southern Appalachians
MACKENZIE, 1964	Well-defined cross-bedding in coastal skeletal grainstones dips onshore, formed by onshore winds at time of high sea level. One of the few studies of the cross-bedding of carbonate coastal dunes
KLEIN, 1965	The first paper to document fully the bipolar orientation of tidally influenced cross-bedding in carbonates
HOFMANN, 1966	Some cross-bedding, most of which shows a variable orientation, in a limestone–shale association. Mapped-over parts of two counties plus many observations on megaripples. Tabled data
Sedimentation Seminar, 1966	Bipolar cross-bedding indicates paleoslope in skeletal building stones. Tidal currents in shallow sea far onto the craton inferred
MACQUOWN, 1967	Measurement of ripple marks and cross-bedding in shelf limestones of a shallow shoal are an incidental part of water resource study
SELLEY, 1967	Cross-bedding orientation in shallow water, marine limestones is bimodal and dominantly onshore; tidal currents inferred. Good example of integration with other lithologies and environments
ERNST and WACHENDORF, 1968	An unspecified number of measurements in middle Triassic limestones in Germany show bipolar paleocurrents
HAMBLIN, 1969	Striking bipolar cross-bedding patterns in thin limestones on a craton. Local and regional cross-bedding maps, the latter outlining shape of depositional basin. Compare with CARR (1973)
HOFFMAN, 1969	Cross-bedding and ripple mark, 2,304 measurements, in Precambrian oolitic and other carbonate platform sands along the border of an alaucogen. Integration with much additional paleocurrent data from associated terrigenous sandstones provides massive documentation
KNEWTSON and HUBERT, 1969	Bipolar crossbeds with thicker sets (believed to indicate stronger currents) oriented to the northwest. Good pictures
ADAMS, 1970	Thickness and petrology of a carbonate–terrigenous sequence with carbonate paleocurrents, unimodal, bipolar, and polymodal. Paleocurrents not related to paleoslope
McBRIDE, 1970	Cross-beds in cherty calcareous conglomerates in a turbidite sequence support a paleoslope inferred from distribution of conglomerates within the formation
HRABAR et al., 1971	Bimodal cross-bedding in a skeletal bank located on a weak structural high. Discussion of how to sample bipolar cross-bedding (p. 112—113)
Sedimentation Seminar, 1972	Low angle cross-bedding measured in elongate carbonate bodies that accumulated along a slope break. Cross-bedding is oriented obliquely down dip of slope break and is largely parallel to grain fabric of an associated, elongate, coarsening upward sandstone body

Table 4-4 (Cont.)

Author and year	Remarks
CARR, 1973	Cross-bedding, shapes of oolitic banks, and their petrology. Both locally and regionally bimodal paleocurrents prevail; regionally these bimodal paleocurrents outline paleoslope of basin. Compare with HAMBLIN (1969)
SUGUIO et al., 1974	Integrated study of an oolite with 130 measurements of cross-bedding, randomly to moderately well oriented
PALMER and JENKYENS, 1975	Thirty seven cross-beds measured in an oolitic, island carbonate barrier of Jurassic age

the Triassic in British Columbia (fig. 4-29). He integrated lithologic facies and cross-bedding to find that asymmetric ripples are oriented down dip and oscillation ripples are parallel to depositional strike in this prograding marine depositional system. DAVIS (1965) also confirmed a general orientation parallel to the shoreline of Lake Michigan and also relates ripple dimension and grain size to water depth. STRICKLIN and AMSBURY (1974, fig. 23) report an exception to these generalizations in the Cretaceous of Texas, where they found ripples, in a calcisiltite, striking perpendicular to the old shoreline. They believed that this orientation resulted from wind driven marine currents flowing parallel to the shoreline in very shallow water. Ripple-mark orientation and stromatolite cenulations have also been compared (McCONNELL, 1975, fig. 5).

Ripple mark has also been observed in evaporite deposits (WEILER et al., 1974) and, as more sedimentologists recognize the detrital origin of many evaporites, more will undoubtedly be found.

TANNER (1967) has reviewed ripple-mark indices and their significance, while HARMS (1969) provides a good analysis of their hydraulics.

Carbonates

Cross-bedding studies in carbonates, virtually all skeletal limestones, have proceeded at a slow but rather constant pace of about one paper per year (Table 4-4). Of the 16 papers that we are aware of, all but three come from North America. Moreover, most are from the Paleozoic. The most striking feature of these studies is their bipolar current roses (fig. 4-30) that commonly have a close relationship to paleoslope (SEDIMENTATION SEMINAR, 1966; SELLEY, 1967; HAMBLIN, 1969; and CARR, 1973). Ebb and flood tidal currents essentially perpendicular to the shoreline, have been invoked to explain the bipolarity; supporting evidence comes from paleogeography and in some basins from parallelism between carbonate cross-bedding with that of associated deltaic sandstones. In Pleistocene eolianites in Bermuda onshore winds seem clearly responsible for persistent shoreward-dipping cross-bedding (MACKENZIE, 1964). Marine currents, either longshore or obliquely longshore, have also been inferred (ADAMS, 1970; HRABAR et al., 1971; SEDIMENTATION SEMINAR, 1972). More studies of cross-bedding in skeletal carbonates, which abound in the geologic record, certainly seem warranted, especially in dominantly carbonate-filled basins, where paleocurrents can be expected to outline effectively both the local and regional paleo-

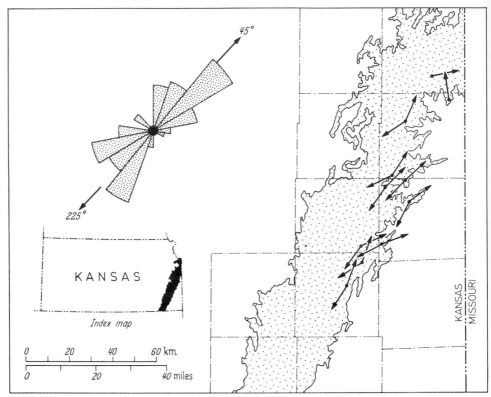

Fig. 4-30. Bimodal cross-bedding in shallow-water, detrital Pennsylvanian limestone in eastern Kansas. (Redrawn from HAMBLIN, 1969, fig. 10). Before cross-bedding was systématically mapped in limestones, many thought that tidal currents could not have existed in a shallow sea far onto a craton

current systems. One significant conclusion to be drawn from the existing studies is that *tidal currents have been operative in shallow epicontinental seas far onto the craton.*

We are not aware of any regional systematic post-1963 studies of ripple mark in carbonates, although it seems that their study should be very rewarding indeed.

Volcaniclastics

A small but growing paleocurrent literature, the authors almost all being sedimentologists, records the many diverse sedimentary structures found in airfall tuffs and base-surge deposits, the latter formed by a phreatic explosion near the topographic base of a volcanic cone.

CROWE and FISHER (1973, fig. 4) mapped cross-bedding formed by probable antidunes in Pleistocene base-surge deposits in California and found a radial pattern (fig. 4-31). MEYER *et al.* (1974) showed a similar radial pattern of fabric related to a Pleistocene volcano in Germany, the radial pattern helping to establish that the tuffs were not deposited as a normal ash fall — another example of the environmental contribution a paleocurrent map can provide. SCHMINCKE *et al.*

(1973) describe the cross-bedding in the same deposits. MATTSON and ALVAREZ (1973) also mapped antidune geometry in base-surge deposits. The inferred antidune cross-bedding of base-surge deposits is apparently analogous to the antidune cross-bedding reported in turbidites (SKIPPER, 1972) and indicates very rapid deposition in both environments. Earlier SCHMINCKE (1967) had mapped paleo-

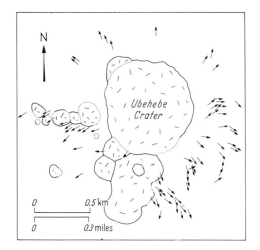

Fig. 4-31. Paleoflow from antidunes in Holocene ash flows. (Redrawn from CROWE and FISHER, 1973, fig. 4)

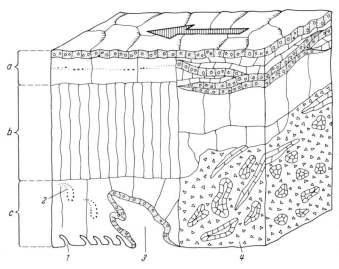

Fig. 4-32. Important mappable directional structures in basalt. (SCHMINCKE, 1957, fig. 3): *1:* pipe vesicles, *2:* vesicle cylinders, *3:* spiracles, and *4:* foreset bedding in pillow-palagonite-complex. Note elongation of pillows and lava tongues in longitudinal section. See also HOTCHKISS* (1923), FULLER* (1931) and WATERS* (1960). Transport from right to left.
The basalt flows of the Columbia River region are typically subdivided into these zones: (a) upper colonnade; (b) entablature; and (c) lower colonnade. These basalts commonly occur in flow units above pillow-palagonite-complexes. If interbedded sandstones are present, their paleocurrent directions should be compared with those of the flows

current structures in Columbia River basalts in Washington (fig. 4-32) as well as cross-bedding in a contemporaneous sandstone rich in volcaniclastic debris to make the first integrated "hot and cold" paleocurrent analysis of igneous flows and a water-laid sandstone. Schmincke's study contains a full list of the structures that he mapped in the basalt flows, one of which is a kind of Gilbert-type, delta cross-bedding formed when a basalt flow enters a body of standing water, cools and crystallizes to produce inclined foresets of granular debris that dip down paleoslope (JONES and NELSON, 1970). The current directions of some recent base-surge deposits have also been mapped from differential sandblasting, tilting of trees and houses, and from their differential coating of mud (MOORE, 1967, fig. 12).

Deformed Cross-Bedding

There is a vast literature on this subject even including its recognition in metamorphic terrains (TOBISCH, 1965). The most comprehensive review of field and experimental studies is that of ALLEN and BANKS (1972), who recognized three main types (a single recumbent fold, a series of folds with or without overturning and a complex type that includes faulting). They provide a model for these types, one that involves the geometry of the deformed beds, their original viscosity and settling velocity and also an estimate of the magnitude of the deforming force of the flowing fluid. They also suggest that the abundance of deformed cross-bedding — something that is not too easy to assess — may be related to seismic shocks. Earlier MCKEE et al. (1962a and b) also conducted experiments.

RITTENHOUSE (1972) has also commented on the difference in cross-bedding dip angle between the essentially flat-lying Navajo Sandstone of presumed eolian origin and that found in modern coastal dunes, the latter being appreciably greater. He concluded the lower dip of the Navajo resulted from a compaction of about 24 percent.

Future Problems

Although the vast number of cross-bedding studies attests to their utility and vigor, there are a number of areas where more effort should be rewarding.

Perhaps foremost is the mapping of cross-bedding in carbonates, which are widespread on many cratons and their margins. For example, in Europe many of the Jurassic carbonates are commonly cross-bedded, in places on a grand scale and over a substantial area, yet only KLEIN (1965) has mapped but a small part of them in England. Their systematic mapping would almost certainly, judging by North American experience, better outline basins and subbasins as well as establishing a regional general current system — that might ultimately be relatable to plate tectonics. And, of course, why not systematically map and collate in North America all the cross-bedding of the numerous Cretaceous sandstones that extend from New Mexico into the Arctic? Presently the widest-scale compilation of paleocurrents, based on a combination of eolian cross-bedding and dune orientation, is a world wide compilation for the Pleistocene based on six continents (BIGARELLA, 1972, fig. 23).

On a small scale, orientation of cross-beds in an oolite reservoir should also be useful in predicting the location of the next stepout well. More paleocurrent studies of Recent and Pleistocene skeletal carbonates could also go far to help establish paleocurrent systems in coastal and littoral environments and thus help extend the work of CLIFTON et al. (1971). Certainly paleocurrent systems based on cross-bedding in barrier and littoral sandstones are, of all environments, the least well-known.

Tidalites should also prove interesting to the paleocurrent mapper, be they carbonate or terrigenous. Certainly, it seems somewhat unusual to us that bipolar cross-bedding is so much more common in carbonates than in terrigenous sandstones. Is this because only a very small minority of terrigenous sandstones are of marine shelf origin?

More integrated studies of paleoflow in extrusives and their associated sandstones (beginning perhaps with the Deccan traps in India) seems most worthwhile, as does the extension of sedimentology and paleocurrent mapping, based on cross-bedding, onto the realm of metamorphic rocks.

Probably too, much more refined analyses will be obtained for cross-bedding, if its orientation is always weighted by some measure of volume and more closely related to its different depositional facies within a formation.

And finally, how long will our consciences allow us to let mapping of ripple mark remain the step-child of paleocurrent studies?

References

AARIO, R., 1972: Association of bed forms and paleocurrent patterns in an esker delta, Haapajärvi, Finland. Ann. Acad. Sci. Fennicae, Ser. A, III Geol.-Geogr. **111**, 55 p.

ADAMS, R. W., 1970: Loyalhanna Limestone — cross-bedding and provenance, *in* G. W. FISHER, F. J. PETTIJOHN, J. C. REED JR. and K. N. WEAVER, eds., Studies of Appalachian geology. New York: Interscience Publ., 83—100.

ALLEN, J. R. L., 1963: The classification of cross-stratified units with notes on their origin. Sedimentology **2**, 93—114.

—, 1966: On bed forms and paleocurrents. Sedimentology **6**, 153—190.

—, 1967: Notes on some fundamentals of paleocurrent analysis with reference to preservation potential and sources of variance. Sedimentology **9**, 75— 88.

—, 1968: Current ripples. Amsterdam: North Holland Publ. Co., 433 p.

—, 1969: Some recent advances in the physics of sedimentation. Geol. Assoc. Proc. Great Britain **80**, 1-42.

—, 1970: A quantitative model of climbing ripples and their cross-laminated deposits. Sedimentology **14**, 5—26.

—, and N. L. BANKS, 1972: An interpretation and analysis of recumbent-folded deformed cross-bedding. Sedimentology **19**, 257—283.

BAKER, V. R., 1973: Paleontology and sedimentology of Lake Missoula flooding in eastern Washington. Geol. Soc. Am., Sp. Paper **144**, 79 p.

BARTSCH-WINKLER, S., A. T. AVENSHINE and D. E. LAWSON, 1975: Sedimentological maps of the Girwood Bar, Turnagain Arm, Alaska for July–August 1973. U.S. Geol. Survey. Misc. Field Studies, Map MF **672**.

BEIN, A., and Y. WEILER, 1976: The Cretaceous Talme Yafe Formation: a contour current shaped sedimentary prism of calcareous detritus at the continental margin of the Arabian Craton. Sedimentology **23**, 511-532.

BELDERSON, R. H., N. H. KENYON, A. H. STRIDE and A. R. STUBBS, 1972: Sonographs of the sea floor. Amsterdam: Elsevier Publ. Co., 185 p.

BIGARELLA, J. J., 1970/1971: Wind pattern deduced from dune morphology and internal structures. Bol. Paranaense de Geociencias **28/29**, 73—113.

— 1972: Eolian environments: their characteristics, recognition and importance *in* J. K. RIGBY and W. K. HAMBLIN, eds., Recognition of ancient sedimentary environments. Soc. Econ. Palaeontol. Mineral. Sp. Pub. **16**, 12—62.

—, G. M. DUARTE and R. D. BECKER, 1970/71: Structural characteristics of the dune, foredune, beach, beach dune ridge and sand ridge deposits. Bol. Paranaense de Geociencias **28/29**, 9—72.

BLUCK, B. J., 1974: Structure and directional properties of some valley sand deposits in southern Iceland. Sedimentology 21, 533—554.

BOUMA, A. H., 1972: Fossil contourites in Lower Niesenflysch, Switzerland. J. Sediment. Petrol. 42, 917—921.

BRUUN-PETERSEN, J., and W. E. KRUMBEIN, 1975: Rippelmarken, Trockenrisse und andere Seichtwassermerkmale im Buntsandstein von Helgoland. Geol. Rundschau 64, 126—143.

BRUSH, L. M., H. A. EINSTEIN, D. B. SIMONS, V. A. VANONI and J. F. KENNEDY, 1966: Nomenclature for bed forms in alluvial channels. J. Hydraulics Div. Am. Soc. Civil Eng. 92, 51—64.

CARR, D. C., 1973: Geometry and origin of oolite bodies in the Ste. Genevieve Limestone (Mississippian) in the Illinois Basin. Indiana Geol. Survey, Bull. 48, 81 p.

CLIFTON, H. E., R. E. HUNTER and L. PHILLIPS, 1971: Depositional structures and processes in the non-barred high energy nearshore. J. Sediment. Petrol. 41, 651—670.

COASTAL RESEARCH GROUP, 1969: Field trip guidebook. Coastal environments of northeastern Massachusetts and New Hampshire, May 9—11. University Massachusetts Geol. Dept., Contrib. 1, 462 p.

COLLINSON, J. D., 1970: Bedforms of the Tana River, Norway. Geog. Annaler 52A, 31—56.

COSTELLO, W. R., and R. G. WALKER, 1972: Pleistocene sedimentology, Credit River, southern Ontario: a new component of the braided driver model. J. Sediment. Petrol. 42, 389—400.

CROOK, K. A. W., 1965: The classification of cross-stratified units, comment on a paper by J. R. L. ALLEN. Sedimentology 5, 249—254.

CROWE, B. M., and R. V. FISHER, 1973: Sedimentary structures in base-surge deposits with special reference to cross-bedding, Ubehebe Craters, Death Valley, California. Bull. Geol. Soc. Am. 84, 663—682.

DALZIEL, I. W. D., and R. DOTT Jr., 1970: Geology of the Baraboo district. Wisconsin Geol. Nat. History Survey, 163 p.

DAVIS, R. A., Jr., 1965: Underwater study of ripples, southeastern Lake Michigan. J. Sediment. Petrol. 35, 857—866.

DONOVAN, T. J., and I. L. TAILLEUR, 1975: Map showing paleocurrent and clast-size data from the Devonian-Mississippian Endicott Group, northern Alaska. U.S. Geol. Survey, Misc. Field Studies, Map MF-692.

DOTT, R. H., Jr., 1973: Paleocurrent analysis of trough cross-stratification. J. Sediment. Petrol. 43, 779—783.

ERNST, G., and H. WACHENDORF, 1968: Feinstratigraphisch-fazielle Analyse der „Schaumkalk-Serie" des Unteren Muschelkalkes im Elm (Ost-Niedersachsen). Beih. Ber. Naturh. Ges. 5, 165—206.

GIETELINK, G., 1973: Sedimentology of a linear prograding coastline followed by three high destructive delta complexes (Cambro-Ordovician, Cantabrian Mountains, NW Spain). Leidse Geol. Mededel. 49, 125—144.

GLENNIE, K. W., 1970: Desert sedimentary environments. Amsterdam: Elsevier Publ. Co., 222 p.

GOLDSMITH, V., 1973: Internal geometry and origin of vegetated coastal sand dunes. J. Sediment. Petrol. 43, 1128—1142.

HAMBLIN, W. K., 1969: Marine paleocurrent directions in limestones of the Kansas City Group (Upper Pennsylvanian) in eastern Kansas. Kansas Geol. Survey Bull. 194, Pt. 2, 25 p.

HARMS, J. C., 1969: Hydraulic significance of some sand ripples. Bull. Geol. Soc. Am. 80, 363—396.

HELM, D. G., 1971: Succession and sedimentation of glaciogenic deposits at Hendre, Anglesey. Geol. J. 7, 271—298.

HIGH, L. R., JR., and M. D. PICARD, 1974: Reliability of cross-stratification types as paleocurrent indicators in fluvial rocks. J. Sediment. Petrol. 44, 158—168.

HOFFMAN, P., 1969: Proterozoic paleocurrents and depositional history of the East Arm fold belt, Great Slave Lake, Northwest Territories. Canadian J. Earth Sci. **6**, 441—462.

HOFMANN, H. J., 1966: Ordovician paleocurrents near Cincinnati, Ohio. J. Geol. **74**, 868—890.

HOLLISTER, C. D., and B. C. HEEZEN, 1972: Geologic effects of ocean bottom currents: western North Atlantic *in* A. L. GORDON, ed., Studies in physical oceanography — A tribute to George West on his 80th birthday. New York: Gordon and Breach **2**, 37—66.

HRABAR, S. V., E. R. CRESSMAN and P. E. POTTER, 1971: Crossbedding of the Tanglewood Limestone Member of the Lexington Limestone (Ordovician) of the Blue Grass region of Kentucky. Brigham Young University, Geol. Studies **18**, 99—114.

IRIONDO, M. H., 1973: Volume factor in paleocurrent analysis. Bull. Am. Assoc. Petrol. Geologists **57**, 1341—1342.

JONES, J. G., and P. H. H. NELSON, 1970: The basalt lava from air into water — its structural expression and stratigraphic significance. Geol. Mag. **107**, 13—19.

KENYON, N. H., 1970: Sand ribbons of European tidal seas. Marine Geol. **9**, 25—39.

KLEIN, G. DEV., 1965: Dynamic significance of primary structures in Middle Jurassic Great Oolite Series, southern England *in* G. V. MIDDLETON, ed., Primary sedimentary structures and their interpretation. Soc. Econ. Paleontol. Mineral., Sp. Pub. **12**, 173—191.

—, 1967: Paleocurrent analysis in relation to modern marine dispersal patterns. Bull. Am. Assoc. Petrol. Geologists **51**, 366—382.

— 1970: Depositional and dispersal dynamics of intertidal sand bars. J. Sediment. Petrol. **40**, 1095—1127.

KNEWTSON, S. L., and J. F. HUBERT, 1969: Dispersal patterns and diagenesis of oolitic calcarenites in the Ste. Genevieve Limestone (Mississippian), Missouri. J. Sediment. Petrol. **39**, 954—968.

LAND, L. S., and J. H. HOYT, 1966: Sedimentation in a meandering estuary. Sedimentology **6**, 191—207.

LANGHORNE, D. N., 1973: A sand wave field in the outer Thames Estuary, Great Britain. Marine Geol. **14**, 129—143.

MACKENZIE, F. T., 1964: Bermuda Pleistocene eolianites and paleowinds. Sedimentology **3**, 52—64.

MACQUOWN, W. C., Jr., 1967: Factors controlling porosity and permeability in the Curdsville Member of the Lexington Limestone. University of Kentucky Water Resources Inst., Res. Rept. **7**, 80 p.

MATTSON, P. H., and W. ALVAREZ, 1973: Base surge deposits in Pleistocene volcanic ash near Rome. Bull. Volcanol. **37**, 553—571.

McBRIDE, E. F., 1970: Stratigraphy and origin of Maravillas Formation (Upper Ordovician), West Texas. Bull. Am. Assoc. Petrol. Geologists **54**, 1719—1745.

McCAVE, I. N., 1971: Sand waves in the North Sea off the coast of Holland. Marine Geol. **10**, 199—227.

McCONNELL, R. L., 1975: Biostratigraphy and depositional environment of algal stromatolites from the Mescal Limestone (Proterozoic) of central Arizona. Precambrian Res. **2**, 317—328.

McDONALD, B. C., and J. S. VINCENT, 1972: Fluvial sedimentary structures formed experimentally in a pipe, and their implications for interpretation of subglacial sedimentary environments. Geol. Survey Canada, Paper **72—27**, 30 p.

McGOWEN, J. H., and L. E. GARNER, 1970: Physiographic features and stratification types of coarse-grained point bars: modern and ancient examples. Sedimentology **14**, 77—111.

McKEE, E. D., 1966: Structures of dune at White Sands National Monument, New Mexico (and a comparison with structures of dunes from other selected areas). Sedimentology **7**, 1—69.

McKEE, E. H., M. A. REYNOLDS and C. H. BAKER JR., 1962a: Experiments on intraformational recumbent folds in cross-bedded sand. U.S. Geol. Survey Prof. Paper **450D**, 151—155.

—, M. A. REYNOLDS and C. H. BAKER JR., 1962b: Experiments on intraformational recumbent folds in cross-bedded sand. U.S. Geol. Survey Prof. Paper **450D**, 155—160.

—, and G. C. TIBBITTS JR., 1964: Primary structures of a seif dune and associated deposits in Libya. J. Sediment. Petrol. **34**, 5—17.

MECKEL, L. D., 1967: Tabular and trough cross-bedding: comparison of dip azimuth variability. J. Sediment. Petrol. **37**, 80—86.

MEYER, V. W., J. STETS and P. WURSTER, 1974: Gefüge und Entstehung der Ringdünen in den grauen Tuffen des Laacher Vulkans. Geol. Rundschau **63**, 1113—1132.

MIALL, A. D., 1974: Paleocurrent analysis of alluvial sediments: a discussion of directional variance and vector magnitude. J. Sediment. Petrol. **44**, 1174—1185.

MICHELSON, P. C., and R. H. DOTT JR., 1973: Orientation analysis of trough cross-stratification in Upper Cambrain sandstones of western Wisconsin. J. Sediment. Petrol. **43**, 784—794.

MOORE, J. G., 1967: Base surge in recent volcanic eruptions. Bull. Volcanol. **30**, 337—363.

MORTON, R. D., 1971: Geological investigations in the Bramble sector of the Fennoscandian shield, S. Norway. No. II. Norsk Geol. Tidsskrift **51**, 63—83.

PALMER, T. J., and H. C. JENKYENS, 1975: A carbonate island barrier from the Great Oolite (Middle Jurassic) of central England. Sedimentology **22**, 125—135.

PELLETIER, B. R., 1965: Paleocurrents in the Triassic of northeastern British Columbia *in* G. V. MIDDLETON, ed. Primary sedimentary structures and their hydrodynamic interpretation. Soc. Econ. Paleontol. Mineral. Sp. Pub. **12**, 233—245.

PICARD, M. D., and L. R. HIGH JR., 1968: Shallow marine currents on the early (?) Triassic Wyoming shelf. J. Sediment. Petrol. **38**, 411—423.

— 1973: Sedimentary structures of ephemeral streams. Amsterdam: Elsevier Publ. Co., 223 p.

RAAF, J. F. M. DE, and J. R. BOERSMA, 1971: Tidal deposits and their sedimentary structures. Geol. Mijnbouw **50**, 479—504.

REINECK, H. E., 1963: Sedimentegefüge im Bereich der südlichen Nordsee. Abh. Senckenberg naturf. Gesell. Nr. **505**, 1—138.

RITTENHOUSE, G., 1972: Cross-bedding dip as a measure of sandstone compaction. J. Sediment. Petrol. **42**, 682—683.

RUST, B. R., 1972: Structure and process in a braided river. Sedimentology **18**, 221—245.

SAGAN, C., 1973: Sandstones and eolian erosion on Mars. J. Geophys. Res. **78**, 4155—4161.

SANDERSON, I. D., 1974: Sedimentary structures and their environmental significance in the Navajo Sandstone, San Rafael Swell, Utah. Geol. Studies **21**, 215—246.

SCHLEE, J., 1963: Early Pennsylvanian currents in the southern Appalachian Mountains. Geol. Soc. Am., Bull. **74**, 1439—1452.

SCHMINCKE, H.-U., 1967: Flow direction in Columbia River basalt flow and paleocurrents of interbedded sedimentary rocks, south-central Washington. Geol. Rundschau **56**, 992—1020.

—, FISHER, R. V., and A. C. WATERS, 1973: Antidune and chute and pool structures in base surge deposits of the Laacher See area, Germany. Sedimentology **20**, 553—574.

SEDIMENTATION SEMINAR, 1966: Cross-bedding in the Salem Limestone of central Indiana. Sedimentology **6**, 95—114.

— 1972: Sedimentology of the Mississippian Knifley Sandstone and Cane Valley Limestone in south-central Kentucky. Kentucky Geol. Survey, Rept. Inv. **13**, 30 p.

SELLEY, R. C., 1967: Paleocurrents and sediment transport in nearshore sediments of the Sirte Basin, Libya. J. Geol. **75**, 215—223.

—, 1968: A classification of paleocurrent models. J. Geol. **76**, 99—110.

SHAW, J., 1972: Sedimentation in the ice-contact environment, with examples from Shropshire (England). Sedimentology **18**, 23—62.

SKIPPER, K., 1972: Antidune cross-stratification in a turbidite sequence, Cloridorme Formation, Gaspe, Quebec. Sedimentology 17, 15—68.

SMITH, N. D., 1972a: Some sedimentological aspects of planar cross-stratification in a sandy braided river. J. Sediment. Petrol. 42, 624—634.

— 1972b: The braided stream depositional environment: comparison of the Platte River with some Silurian clastic rocks, north-central Appalachians. Bull. Geol. Soc. Am. 81, 2993—3014.

SOUTHARD, J. B., and L. A. BOQUCHWAL, 1973: Flume experiments on the transition from ripples to lower flat bed with increasing sand size. J. Sediment. Petrol. 43, 1114—1121.

STEINMETZ, R., 1975: Cross-bed variability in a single sand body in E. H. TIMOTHY WHITTEN, ed., Quantitative studies in the geological sciences. Geol. Soc. Am., Mem. 142, 89—102.

STRICKLIN, F. L., JR., and D. L. AMSBURY, 1974: Depositional environments on a low-relief carbonate shelf, middle Geln Rose Limestone, Central Texas in B. F. PERKINS, ed., Aspects of Trinity Division geology. Geoscience and Man 8, 53—66.

SUGUIO, L., E. SALATI and J. H. BARCELOS, 1974: Calcarios ooliticos de Taquai (SP) e seu possivel significado paleoambiental na deposicão da Formacão Estrada Nova. Rivista Brasileira de Geociencias 4, 142—165.

SWIFT, D. J. P., 1970: Quaternary shelves and the return to grade. Marine Geol. 8, 5—30.

—, D. B. DUNAE and O. H. PILKEY, 1972: Self sediment transport: Process and pattern. Stroudsburg: Dowden Hutchinson and Ross, Inc., 656 p.

TANNER, W. F., 1967: Ripple mark indices and their uses. Sedimentology 9, 89—104.

TOBISCH, O. T., 1965: Observations on primary deformed sedimentary structures in some metamorphic rocks from Scotland. J. Sediment. Petrol. 35, 415—419.

WALKER, R. G., 1969: Geometrical analysis of ripple-drift cross-lamination. Canadian J. Earth Sci. 6, 383—391.

WEILER, Y., E. SASS and I. ZAK, 1974: Halite oolites and ripples in the Dead Sea, Israel. Sedimentology 21, 623—632.

WILLIAMS, G. E., 1971: Flood deposits of the sand-bed ephemeral streams of central Australia. Sedimentology 17, 1—40.

WILSON, I. G., 1972: Aeolian bedforms — their development and origins. Sedimentology 19, 173—210.

— 1973: Ergs. Sed. Geol. 10, 77—106.

WOBBER, F. J., 1967: Space photography: a new analytical tool for the sedimentologist. J. Sediment. Petrol. 37, 166—174.

YAALON, D. H., and J. LARONNE, 1971: Internal structures in eolianites and paleo-winds, Mediterranean coast, Israel. J. Sediment. Petrol. 6, 1059—1064.

Linear Structures up to 1963

Introduction

In the broadest sense, lineation is a descriptive, nongenetic term, for any kind of linear structure, on or within a rock (CLOOS, 1946, p. 1). Our concern here, however, is the primary lineation of sedimentary rocks acquired during the depositional process or imposed on the sediment while it was still in the environment of deposition. Lineations are of several types and include the preferred linear arrangement of elongate skeletal elements, plant fragments, or nonspherical clasts of any kind. Such preferential arrangement of the framework elements of a deposit is usually called the depositional "fabric" Lineation includes also the various features which are found at the interface between beds such as the slide marks, produced by movement of one bed over another — in reality a species of slickensides, and the various groovings and striations of mudstones, preserved only as casts on the sole of the overlying sandstone bed. Various other asymmetric sole markings on sandstone beds also impart a lineation to the bedding surface. Included here are the striations and groovings on those surfaces which underlie till or tillite. These glacial striations are in one sense tectonic — i.e. they are produced by an overthrust (of glacial ice) but, inasmuch as they have been systematically mapped to reconstruct the pattern of ice-movement, they are indeed a paleocurrent structure. Also included are those *internal* linear structures observed when certain sandstones are split or parted along bedding planes. Such parting lineation is a very common feature of some sand facies.

Linear structures are important records of movement and direction of sedimentary transport; others are records of penecontemporaneous, soft-sediment deformation. Excluded here are those sedimentary lineations which for various reasons are best treated in other sections of this book. Fabric lineation, bisectrices of trough cross-bedding, rib-and-furrow, ripple azimuths, and the axes of slump folds are all linear features which are discussed elsewhere. Despite this fragmentation of the subject, there is much merit in describing and attempting to analyze the various linear sole markings most characteristic of sandstone beds, especially those of the flysch facies, and internal parting lineation. A brief comparison will be made of glacial and the substratal striations and groovings.

Substratal Lineations (Sole Marks)

The undersides of many sandstone and siltstone beds are characterized by curious markings, most of which stand in relief on the bedding plane. They are in reality casts of structures formed in the underlying stratum. These structures, commonly referred to as "sole markings," are most abundant, most conspicuous, and most commonly reported from sandstones of the flysch facies, although similar structures are occasionally reported from sandstones of other sedimentary facies.

Sole markings have been known for a long time, among the earliest papers dealing with them being those of JAMES HALL (1843 a, b and c) in which the markings found in the Devonian of New York State are described and figured. These structures have been also called hieroglyphs and fucoids (FUCHS, 1895). Some are indeed of organic origin (biohieroglyphs) but most are now believed to be current produced and, because of their striking orientation, may be termed "substratal lineations."

The abundance and variety of these hieroglyphs has led to a large and confusing nomenclature. VASSOEVICH (1953) has proposed a comprehensive and logical system of nomenclature. In general, however, the system currently used has gradually evolved. Various nomenclatural problems have been treated in papers by GROSSGEYM (1946, 1955), CROWELL (1955), KUENEN (1957), TEN HAAF (1959), DZULYNSKI and SLACZKA (1959), and DZULYNSKI and SANDERS (1962). Some controversial problems of nomenclature have been discussed in various journals (see CROWELL, 1958; KUENEN and TEN HAAF, 1958; PRENTICE, 1960, 1961; KUENEN and PRENTICE, 1957).

Occurrence

Sole marks are most numerous and most prominant in sequences of rhythmically interbedded sands and shales, especially in the flysch facies. In fact, they have been identified by some with this facies and considered to be indicative of the activity of turbidity currents. This view is too restrictive inasmuch as some well-formed sole marks, especially load casts but also including flute casts, have been observed in fluviatile and shelf deposits (CUMMINS, 1958, p. 40) and some are even found on the soles of limestones interbedded with shales (fig. 5-3). Substratal lineations are a good illustration of the fact that few, if any, primary sedimentary structures *only* occur in a specific environment. In general, environmental contrasts are expressed by the differences in *abundance* rather than in *kinds* of primary sedimentary structures.

Sole marks are most prominent in the turbidite sequences and, when present in great profusion on the soles of graded sandstones, they can confidently be regarded as a product of turbidity currents. Owing to their occurrence on the underside of overhanging sandstone beds, they are commonly overlooked, especially in areas of flat-lying strata. In such regions the best examples are found on the overturned slabs found in the creek bottoms.

Sole marks are likely to be seen only in those strata that separate readily along the bedding. In the welded strata of the Precambrian and of metamorphic terranes, the parting is commonly along secondary cleavage planes so that sole markings, if present, are not seen. The poorly-consolidated younger strata also fail to yield bedding surfaces and hence substratal structures are overlooked. Some terranes, therefore, exhibit a profusion of sole markings; others display few or none, even where such markings are presumed to be present.

Methods of Investigation and Measurement

Substratal lineations include those which indicate a line of movement as well as a direction of movement (Chapter 3). For example, parting lineation indicates a line of movement, whereas flute marks indicate a direction of movement.

The measuring, recording, and mapping of linear features, especially sole markings, is less troublesome than of other directional structures (TEN HAAF, 1959, p. 68; McBRIDE, 1962, p. 76). The general orientation of sole marks on a particular bed is easily determined by inspection and only one compass reading per bed is required. In rare cases, where numerous markings deviate from the dominant orientation by more than 30 degrees, two readings per bed should be taken. At a given outcrop one may take readings on several beds but, as their dispersion is generally very low, only a few beds usually need to be measured. The arithmetic average is usually computed for the station.

Care must be taken not to confuse sedimentary and tectonic lineations. The intersection of cleavage and bedding for example, produces a bedding-plane lineation which might erroneously be thought primary.

In areas of tilted beds the azimuth of the linear structure must be corrected for the tilt of the strata. The azimuth error from failure to correct is negligible in strata with dips under 25 degrees but is appreciable in strata with high dips. The problem of tilt correction is discussed in Chapter 10.

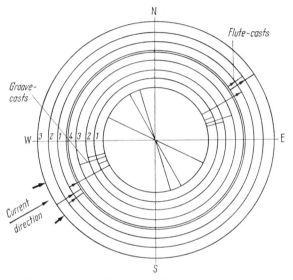

Fig. 5-1. Current-rose diagram (modified from KELLING, 1958, fig. 5). Observations plotted: 3 ripple marks (at center), 10 groove casts (intermediate circles), and 6 flute casts (outer circles)

Usually an average value for the directional structures is computed for each outcrop and, if plotted on a map, different symbols distinguish between direction and line of movement structures (Chapter 10).

A useful technique is the plotting of the individual measurements of different directional structures from an outcrop or an area on a circular diagram (fig. 5-1).

Classification and Origin of Substratal Lineations

The various substratal lineations are a product either of (1) the current, or (2) the objects or load propelled by the current (Table 5-1). The first type are the product of current scour such as flutes. The second type includes three kinds of markings, namely, those made by objects which function as engraving tools and are, therefore, in continuous contact with the bottom, those produced by objects following saltatory paths and which, therefore, impinge intermittently on the bottom, and finally those made by objects that progress by rolling or tumbling.

The first method produces striations and grooves, the second produces brush marks, impact or prod marks, and bounce marks, whereas the third produces a continuous or interrupted, but regularly repeated, signature.

In the section which follows we first consider the products of current action, most commonly seen as flute casts and current casts, and then discuss the lineations produced by the transported load, also seen mainly as casts — groove casts, prod casts and a multitude of other hieroglyphs. We will discuss to some extent the modification of these structures by load-casting, although deformational structures in general are treated in the next chapter.

Table 5-1. *Sole Markings and Related Structures*

Agent	Process	Name of structure
Produced by current	Current scour	Flute (casts)
	Engraved by moving objects a) Drag b) Saltation c) Rolling	Drag marks (Groove and striation casts) Bounce, brush and prod marks (and casts) Roll-marks
	Unknown	Channels
Produced by gravity	Unequal loading	Load pockets (load casts)
	Foundering (thixotropic transformation)	Ball- and pillow structure
	Slump or slide	Slide marks (and casts); slump folds and faults and breccias
Produced by "liquifaction"	Injection and other processes	Sandstone dikes and sills; sand volcanoes; sandstone cylinders; mud volcanoes
Complex current and gravitational interaction		Convolute laminations

Flute Casts

The structure to which the name *flute cast* is now commonly given has been referred to by earlier workers as "flow mark" (RICH, 1950, p. 722), "lobate rill mark" (SHROCK, 1948, p. 131), "Fließwülste" (HÄNTZSCHEL, 1935), "Zapfenwülste" (RÜCKLIN, 1938, p. 96, fig. 1), "scour cast" (KINGMA, 1958, p. 12, fig. 12), "scour finger" (BOKMAN, 1953, p. 158, pl. 1c), "vortex casts" (WOOD and SMITH, 1959, p. 192) and the like. The term "flute cast" is generally attributed to CROWELL (1955, p. 1359), although it appears to have been first used by RICH (1951, pl. 3) in a plate caption. It is now generally accepted*.

As described by CROWELL, flute casts are sharp, subconical welts, one end of which is rounded or bulbous whereas the other end flares out and merges gradually with the bottom surface of the sandstone layer (fig. 5-2). Despite general con-

---

* The reader should note that the term "cast" is here and elsewhere used in an informal sense. Cast is properly defined as the filling of a mold. The flute itself is not the mold; hence its filling is a mold from which the flute can be reproduced as a plaster cast. The sole marks seen are, therefore, flute molds. Common usage, however, considers the flute itself to be the mold and its filling a cast. We defer to usage.

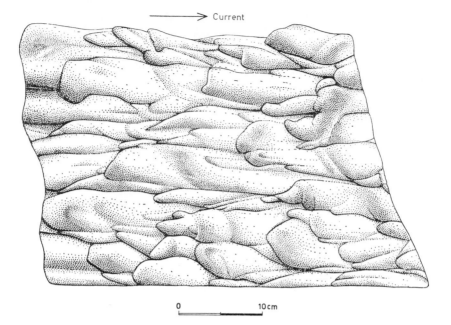

Fig. 5-2. Linguiform flute casts (TEN HAAF, 1959, fig. 12). Casts on base of sandstone bed, Marnoso-arenacea, Miocene, Apennines, Italy

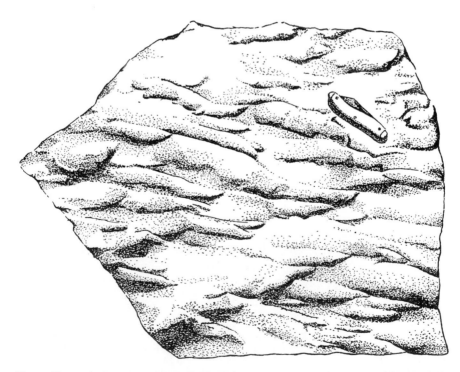

Fig. 5-3. Flute casts drawn from photograph (VASSOEVICH, 1932, pl. 1, fig. 5). Casts on base of limestone bed, Djorchy series, Kakhetia, Caucasus, U.S.S.R. (sketch by ANN ROMINGER)

formity in this pattern, flute casts are greatly varied in size, shape, and arrangement (pls. 12, 13, and 14). On a given sole the flute casts tend to be of the same size and much alike; they vary from sole to sole. They range in size from an inch (2—3 cm) to several feet (1 m) in length; most commonly they are two or three inches (5—8 cm) long. Most are bilaterally symmetrical; some are deltoid; others are more elongate. Some exhibit less symmetrical form with irregular stoss ends or terminate up-current in a twisted beak. A few have exaggerated relief, a result perhaps of modification by loadcasting. A few rare forms are sculptured and show a terracing which is not an expression of internal structure but is rather a cast of the differentially scoured laminations in the mud in which the flute was cut (TEN HAAF, 1959, p. 29; KSIAZKIEWICZ, 1961, p. 39).

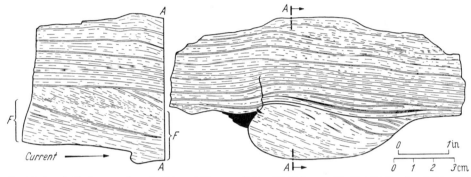

Fig. 5-4. Longitudinal cross-section of a flute cast (KUENEN, 1953, p. 25). Note current-bedded fill; flow from left to right

The flute casts themselves may appear structureless although if sectioned carefully, many show a small scale cross-stratification (fig. 5-4), which shows that the blunt end of the structure is indeed the up-current end (KUENEN, 1953, p. 25). Deformation of the internal structure may be seen, if the structure has been modified by load-casting. The coarsest materials of which a bed is composed are commonly collected in the flute casts.

Flute casts on a given sole are generally abundant, in some cases even overlapping, but on other soles they are more widely scattered. On some soles they appear to show a regular arrangement; they may be in rows parallel to the current or be arranged in diagonal rows or in still other cases closely crowded together in bunches with smooth surfaces in between.

Flute casts are not necessarily closely associated with groove and striation casts though there are many exceptions. Where there are several sets of groove casts, the associated flutes are found to be parallel with but one set of grooves, presumably the younger set. The relation between flutes and grooves is not wholly clear. Examples of fluting that has partially obliterated large groove casts, indicating a later age for the flutes, have been noted by KUENEN and SANDERS (1956, p. 662), KUENEN (1957, p. 346), and TEN HAAF (1959, p. 37). On the other hand, HSU (1959, p. 534), who studied these structures in the Alpine flysch, thought that flutes formed in an early turbulent phase of the current and that grooves were a product of laminar flow at a later stage of flow. MCBRIDE (1962, p. 58) concluded that in general the flutes of the Martinsburg flysch were earlier than the grooves. DZULYNSKI and SANDERS (1959) believed that relative age of flutes and grooves was dependent on the flow-load regimen. A thick, dense, basal saltation load ("saltation carpet") leads to formation of tool marks; a less dense basal layer permits scour.

The common orientation of a swarm of flute casts, their relation to other current structures, including their own internal cross-lamination, leaves no doubt of their current origin and their usefulness as a guide to direction of current flow. That

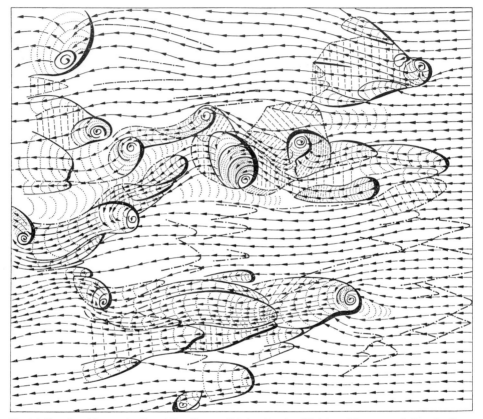

Fig. 5-5. Experimental reconstruction of flow lines on fluted surface (modified from RÜCKLIN, 1938, fig. 12). Sharp boundaries of fluted depressions shown by heavy black crescents; flow lines shown by arrows; depressed flutes shown by dotted concave lines and ridges by dotted convex lines

they are the product of filling of scour pits generated by current eddies is now generally accepted, despite early references to "flowing mud" or "flowing sand." RÜCKLIN (1938) clearly explained the origin of these features as a product of erosion in mud by vortices, such scour pits being preserved by subsequent filling with silt or sand (fig. 5-5). Not only did RÜCKLIN understand their origin, but he produced them experimentally. Several types, plain, cork-screw, flat or depressed, and horse-shoe or crescentic, were described by RÜCKLIN. The crescentic type is better known as "current crescents" than as flute casts.

Flute casts are one of the most widespread of sole marks and the most useful structures of this type as a guide to the direction of current flow. Although most characteristic of the flysch facies, from which most examples have been described, they do occur on the soles of sandstones from other facies (pl. 13 B), and even in some cases on the soles of limestones (fig. 5-3). It would be erroneous to attribute them solely to turbidity currents.

Current Crescents and "Sand Shadows"

Pebbles, shells, and similar objects form an obstacle to the flow of water. As noted by JAMES HALL (1843 a, p. 55) a semicircular depression or moat is commonly excavated on the up-current side of the obstacle and a little ridge of sand, narrowing and sloping down-current, is formed on the lee side (pl. 15 A). The crescentic depression is commonly preserved on the underside of a sand bed as a cast. The term "current crescent" has been applied to this structure (PEABODY, 1947, p. 73; MCKEE, 1954, p. 63, pl. 9 B). Commonly the object which diverted the flow of water was an intraformational shale fragment. The weathering out of such shale pebbles leaves a hole in the underside of the sandstone bed partially surrounded by the ridge or cast of the original crescentic furrow (pl. 15 B).

Current crescents seem to be most characteristic of fluvial sandstones with shale interbeds. Peabody, regarded those in the Moenkopi (Triassic) of Arizona as positive evidence of a thin sheet of water flowing over a mudflat. Those in the Ordovician Juniata sandstone of Pennsylvania seem to be of similar origin and significance as are perhaps those described from the Lower Triassic Buntsandstein of Germany, which were once interpreted as a reptile track (PEABODY, 1947, p. 73). They can and do form on tidal flats and beaches.

But current crescents are also known from some flysch sequences (DZULYNSKI and SLACZKA, 1959, pl. 27; KSIAZKIEWICZ, 1961, p. 39). See plate 20B. It is unlikely therefore, that they are proof of shallow water origin though they are most characteristic of such environments.

Current crescents may be related to "sand shadows." Just as the current scours the horseshoe-shaped moat on the upcurrent side of an obstacle so also it deposits a "tail" or leeside accumulation of coarser sand (HALL, 1843a, p. 52). These have been observed in some flysch sandstones (DZULYNSKI and SLACZKA, 1959, p. 237).

Groove and Striation Casts

The undersurface of many sand and siltstone beds which rest on shales, especially those in flysch sequences, are marked by numerous, straight, parallel ridges of slight relief (pls. 16, 17 and 18). Commonly these are raised but a fraction of a millimeter above the bedding surface; in a few cases, they have a relief of a centimeter or two. These structures are obviously the casts of striations or grooves formed in the underlying shale at the time it was a mud. They have, therefore, been called "groove casts" (SHROCK, 1948, p. 162). They have also been called "drag marks" (KUENEN, 1957, p. 243) and "Schleifmarken" (SEILACHER, 1960, p. 99).

The most striking characteristics of the groove casts are their straightness, uniformity of height, and great length. Curved and sinuous groove casts are very rare. They usually extend, without change of character, the full distance of the exposed bedding plane. Some groove casts are ornamented, that is, exhibit smaller superimposed longitudinal ridges and striations. They are also multiple and are in pairs, triplets, etc. They are generally numerous and closely spaced. Not uncommonly two sets are found on the same bedding plane, generally intersecting and making an acute angle with one another. One set, seemingly younger, has in part blotted out the earlier set. In a few cases groove casts show exaggerated relief and some even overhang — a feature believed due to loadcasting.

Groove or striations generally just fade out; only a few end abruptly and in rare cases some small object, such as a shale chip (McBride, 1962, p. 55; Dzulyn-

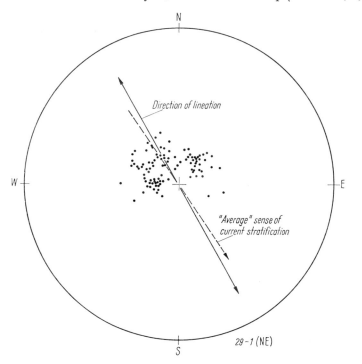

Fig. 5-6. Relation between current-bedding (cross-lamination) and sole marks (Crowell, 1955, fig. 7)

ski and Radomski, 1955) lies at the down-current end of the groove. Groove casts are not to be confused with prod casts which are shorter and have an abrupt *down*-current termination.

Groove and striation casts commonly cover an entire surface; they are generally not associated with flute casts though there are many exceptions. The age relations of groove and flute casts have been discussed above.

That grooves were current-produced is now generally conceded. A few similar markings have been interpreted as a species of slickensides made by sliding of a superincumbent bed. The general parallelism of groove and flute casts, the correlation between the mean direction of small-scale cross-bedding commonly formed in the upper part of a sandstone bed and the subjacent grooves (Crowell, 1955), figure 5-6, and the correlation also between the fabric (fig. 5-7) and the substratal groovings (McIver, 1961, p. 143) all support the view that the groovings are a current structure.

Clearly also the grooves and striations were cut in unconsolidated mud and preserved only as casts in the overlying sand bed. They were probably formed by the same current which deposited the sand though some (CROWELL, 1958, p. 334) have thought otherwise. The engraving tools are largely unknown. They have been thought to be algal anchor stones (SHROCK. 1948, p. 164), sand grains (CROWELL, 1955, p. 1358), or pieces of water-logged wood (RADOMSKI, 1958, pl. 38), the bodies of soft-bodied animals (TEN HAAF, 1959, p. 38), shale fragments (McBRIDE, 1962, p. 55), shells or other skeletal debris, or mats of plant debris

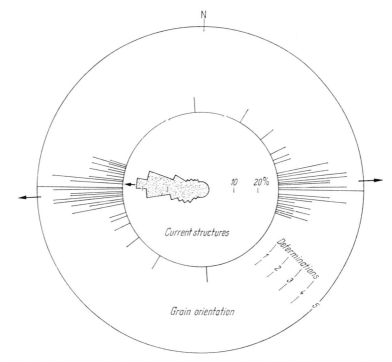

Fig. 5-7. Relation between fabric, lineation and sole markings, Devonian, New York, U.S.A. (McIVER, 1961, fig. 6)

(KÜLICK, 1960; TEICHMÜLLER, 1960). Seldom, however, is anything ever found to which the grooves could be attributed. Early workers such as CLARKE (1917, p. 206), who believed the strata exhibiting grooves were of shallow water origin, attributed the markings to drifting ice. Although CLARKE's concept of the origin of grooves has been generally abandoned, the stone tracks of some playa lakes seem to be produced by the shifting of large boulders by drifting ice (STANLEY, 1955, p. 1329).

Many problems remain unsolved. The straightness, continuity, and uniformity of height and structure, seems to require laminar flow (HSU, 1959, p. 534). In ordinary turbulent flow, the particles carried would perhaps follow saltatory paths and hence would lead to an interrupted and very incomplete trail. Some objects, fish vertebrae for example (DZULYNSKI and SLACZKA, 1958, p. 235), evidently did progress by rolling but it is unlikely that the hundreds of grooves could be formed by rolling or leaping objects. They could form by dragging of some objects, and have indeed been called "drag marks" (KUENEN, 1957, p. 243). It seems unlikely that so many grooved surfaces could be thus formed with so little evidence of such objects themselves.

If grooves were made by shell fragments or woody fragments or sticks, they should not occur in the Precambrian rocks, although the inadequacies of the Precambrian record do not permit a firm conclusion on this matter. The record is difficult to read because in the Precambrian, where such grooves might be expected, the rocks break along secondary cleavage planes instead of the bedding planes.

Grooves and related substratal lineations are most prominent and most abundant in the flysch sequences where, as noted, they have been attributed to the action of turbidity currents. They are, however, known from other types of deposits — deposits not presumed to be turbidites. In such strata, however, the grooves are few, less complete or continuous. Hence, although grooves cannot be said to indicate turbidity current action, such currents are more productive of this type of lineation.

Bounce, Brush, and Prod Casts

These structures are all related in that they appear to have formed by objects that intermittently impinged on the bottom as they were carried along by the current. The term "bounce cast" was proposed by WOOD and SMITH (1957, p. 182) for rather short ridges which fade out at either end. It is usually impossible to tell whether the striking object came from one end or the other. The origin of these markings was clearly stated by HALL (1843a, p. 236).

A brush cast (DZULYNSKI and SLACZKA, 1958, p. 231) is of similar form and origin and differs only in that it has a slight crescentic depression at one end. The depression is the cast of a mud ridge heaped up in front of the object as it momentarily slid along the bottom. The crescentic cast, therefore, points downcurrent. Where there is a row of brush casts, produced presumably by the striking body being alternately grounded and lifted up again by the current, the term "skip cast" (pls. 18 and 19) has been applied (DZULYNSKI and SLACZKA, 1958, p. 231).

Ctenoid casts (BEASLEY, 1908, 1914, p. opp. p. 35; CUMMINS, 1958, p. 39, fig. 4, pl. 2) are now interpreted as casts of impressions made by *Equisetites* stems intermittently touching a mud bottom as they were carried along by a current of water (LINCK, 1956). They are thus a species of brush cast but because of their resemblance to a large tortoise-shell comb they have been given a special name. They are very rare.

A "prod cast" (DZULYNSKI and SLACZKA, p. 232, pl. 25, fig. 1), the "impact cast" of RADOMSKI (1958, pl. 38, fig. 1), is a term applied to a short ridge which has one blunt end and fades out in the other direction. The abrupt termination is the down-current end of the structure (the opposite of flute casts). This structure is presumed to be formed by objects, perhaps a stick, grazing the bottom, plunging into it and then being lifted and taken away by the current. An impact cast is similar but is smaller and is without the extended "tail" of a typical prod cast.

Rolling and Tumbling Marks

These structures, the "Roll-Spuren" of KREJCI-GRAF (1932, p. 27), are a series of casts of rather small markings spaced at short and equal intervals. They were probably made by objects rolled along the bottom of the current. They grade into marks produced by saltation in which the intervals are widened and less regular.

DZULYNSKI and SLACZKA (1958, p. 235; 1960) describe such marks on the soles of the Krosno sandstones (Oligocene) of Poland, which they interpret as spinal joint imprints of fish vertebrae rolled along the bottom (pl. 18). A similar record in Oligocene beds in Switzerland is reported by PAVONI (1959). Other cylindrical objects, similarly, might be expected to produce such roll marks.

Miscellaneous Markings

There are various other current-produced sole markings, which though much less common than flute or groove casts, do indicate current direction and, perhaps more important, do yield some clues to the manner of origin of sole marks in general.

Of these minor structures, the V-shaped casts (fig. 5-8), variously designated chevron marks, vibration marks, etc., are the most intriguing. DZULYNSKI and SLACZKA (1958, p. 234) describe "vibration marks" as grooves modified by a succession of regularly spaced crescentic depressions (now raised as casts) with their convexities pointing down-current.

Fig. 5-8. Herringbone or chevron marks (drawn from photograph, KUENEN, 1957, pl. 2 C). Note distortion of earlier drag marks. Sherburn flags, Devonian, Taughannock Falls, New York, U.S.A. (sketch by R. W. B. DAVIS)

They appear to be the same as the "herringbone pattern" described by KUENEN (1957, p. 256, pl. 2). This structure has been likened to the "chatter marks" of glacial origin made by the vibratory action of the gouging tool. Although "chevron marks" (DUNBAR and RODGERS, 1957, p. 195) appear to be similar, their convexities are said to point up-current. A related feature are the "ruffled groove casts" (TEN HAAF, 1959, p. 32) which are grooves displaying a feather pattern, being accompanied by lateral wrinkles that join the main cast in the down-current direction (pl. 17B). The lateral wrinkles are presumed to have formed by tractional stress transmitted to the adjacent mud by the object cutting the groove.

Another sole mark of uncertain origin is that which McBRIDE (1962, p. 57) has described as "furrow flute casts" as "... casts of furrow-like depressions which have up-current noses similar to normal flute casts but differing in that they are much longer and are separated from adjacent furrows [casts] by narrow grooves which are filled with shale prior to weathering" (pl. 21 A). These seem to be the "delicate flute casts" of KUENEN (1957, fig. 9) and the "sludge casts" of WOOD and SMITH (1959, p. 169, pl. 7, fig. 4). These structures in turn grade into "furrow

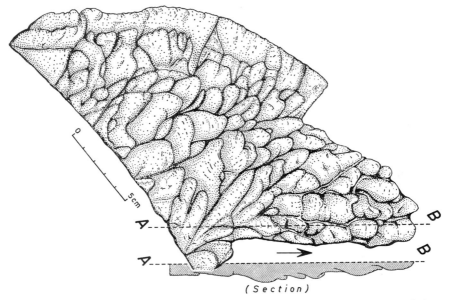

Fig. 5-9. Squamiform load casts (TEN HAAF, 1959, fig. 33). Picene flysch (Miocene), Cancelli, Apennines, Italy

casts" (McBRIDE, 1962, fig. 17) which lack the up-current "nose" and thus indicate current-orientation but not the direction of flow (pl. 21 B). The furrow casts are slightly raised casts separated by narrow shale septae which weather out as sharp grooves. They are commonly slightly sinuous, are more or less parallel, are 1 cm more or less wide, and cover the whole bedding surface. Furrows casts seem to have been called "load-casted longitudinal ripples" by KELLING (1959), "rill mark" or "rill casts" by DZULYNSKI and SLACZKA (1958, p. 230) and "load-cast striations" by TEN HAAF (1959, p. 46).

The furrow-mark pattern appears to grade into one showing up-current bifurcation and a somewhat dendritic character. To the dendritic mark, TEN HAAF (1959, p. 48) gave the name "syndromous load cast."

Other rare structures, little understood, include "squamiform casts" (TEN HAAF, 1959, p. 46), a fan-wise series of lobate casts overlapping down-current (fig. 5-9). This rather rare structure resembles some clusters of flute casts but, unlike flutes, the blunt ends are down-current.

"Frondescent casts," KUENEN's "cabbage leaf casts" (fig. 5-10), are casts resembling certain shrubs or large cabbage leaves (TEN HAAF, 1959, p. 30;

KUENEN, 1957, p. 255). The spreading "foliage" is said to be directed down-current. The frondescent casts of TEN HAAF bear a superficial resemblence to "frondescent 'furrow flute casting'" of MCIVER (1961, p. 168) in which, however, the "foliage" is directed up-current. Owing to the cross-cutting attitude of the frondescent casts, TEN HAAF ascribed them to erosion.

Channels and Channel Casts. In some turbidite sequences there are structures to which the terms "channel," "gouge channel," and "wash-out" have been applied. Unlike other sole markings, which exhibit such slight relief that some are seen only with strong, low angle illumination, channels are characterized both by their broad width and marked depth. They may be up to several meters wide and have a depth of several decimeters. They may cut entirely through

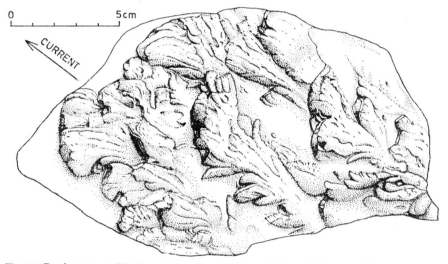

Fig. 5-10. Frondescent casts of "cabbage leaf" markings (TEN HAAF, 1959, fig. 15). Picene flysch (Miocene), Apennines, Italy

several thin beds. It is seldom possible to determine their length, although a channel over 100 feet (30 meters) long has been reported (WILLIAMS, 1881).

Channels are parallel to the current direction shown by other sole markings. Although the channel filling is generally structureless, some examples show horizontal laminations. The sides may be smooth; flute casts and striations have been reported in others (TEN HAAF, 1959, p. 40).

The origin of channels is obscure. Unlike stream or tidal channels, these channels are straight, not sinuous, and are all parallel to one another and to the associated groovings. They do not seem to be related either to the thickness or coarseness of the host strata.

Slide Marks (slide casts). Slide marks consist of broad areas of numerous parallel grooves or striations made by a large object sliding along the bottom surface. Larger clots of mud, sand, or perhaps colonial organisms, or water-logged masses of plant debris slipped, at times wavered or gyrated for a short distance along the bottom. At the down-current end the slide marks may fade out, as if the sliding mass had been swept away. They may also end abruptly with a lumpy cast as though the sliding body had disintegrated (pl. 20A).

Slide marks are characterized by the strict parallelism of the component grooves or striations, even where they display a curved or sinuous course; they are in some places at an angle to other sole markings. The divergent position may only be due to the pivoting of an asymmetric sliding body.

Slide marks are relatively rare and in many cases they are not easily distinguished from groove and striation casts. Although in theory drag marks are the product of the turbidity current and slide marks are the result of gravity alone, it may be that in many cases the two are combined and the distinction between them is meaningless. A slide mark may be just a drag mark produced by a larger rather than a small object.

Load-Casted Current Structures

Many current structures, including flutes and grooves, become modified by load casting — the process whereby soft hydroplastic mud, if unequally loaded with sand, yields to the weight of the superincumbent load by flowing. The initial depressions — flutes, grooves, *etc.* — become the loci of small overloads which sink into the underlying soft mud. Such deformation, even though slight, results in grotesque exaggerations of the original structures.

Flutes become load-casted and display heightened relief with bulbous and overhanging up-current terminations. Continued sinking of the flute pocket with its trapped coarser material and rise of the soft mud leads to an asymmetric sand "sack" hanging below the base of the bed (pl. 27A). Although the asymmetrical overhang, commonly displayed by all flute casts on a given sole, has been ascribed to the drag of the current, TEN HAAF points out that it may be only the result of accelerated load-casting on the deeper side of an initially asymmetric depression.

Groove casts may likewise show the effects of load casting. They tend to retain their characteristic length and rectilinearity. Load-casted grooves show greater than normal relief, and overhanging edges with squeezed-in or diapiric thin tongues of shale along their margins. Some of the furrow-cast structures, broad furrows separated by narrow pinched-in septae of shale, are a species of load-cast structure as is TEN HAAF's "squamiform load casts" with its fanwise pattern. A related form is the "syndromous load cast" a subparallel dendritic pattern of furrows converging down-current.

The distinction between true nondirectional load casts and strongly oriented load-casted current structures is not always clear as there are many transitions from one to the other. Probably all load-cast structures showing any kind of orientation are related to or derived from an initial current structure.

Interpretation and Analysis of Sole Marks

The early literature is largely a record of discovery and description of these features together with some speculation on their origin. The markings were indeed hieroglyphs and little understanding of their origin or significance was achieved.

There is a considerable German literature on these features in the thirties (KREJCI-GRAF, 1932; GÜRICH, 1933; FREUDENBERG 1934; RICHTER, 1935; HÄENTZSCHEL, 1935; PICKEL, 1937, p. 271: RÜCKLIN, 1938). Most workers seemed to think that they were of shallow water, if not intertidal, in origin. The structures, now referred to as flute casts were ascribed to the flow of wet mud or sand or to some type of rill action. The first interpretation of this structure as a product of vortex-scour seems to be that of RÜCKLIN (1938, p. 101). It was recognized quite early that many sole-markings could be used to distinguish top and bottom of beds in the flysch sequences (cf. VASSOEVICH, 1932, and GROSS-GEYM, 1946, 1955, cited by STRAKHOV, 1958, p. 112).

Interest in these features was stimulated by an important paper by RICH in 1950. RICH described what he called flow markings, now generally called flute casts, which he attributed to " ... density currents rendered abrasive with respect to the underlying mud by a heavy burden of silt" (p. 728). In a paper published a year later he described "flutings and striations" produced by " ... sheet flow of water (density currents) moving down the slope... seen usually as casts on the under sides of the overlying siltstone beds..." These structures were believed, therefore, to be formed in water of considerable depth.

RICH's interpretation has been generally accepted, in large part perhaps, because of the work of KUENEN and his students. Although sole markings are now generally believed to be due to turbidity currents, there are some important exceptions (DOTT and HOWARD, 1962; GLAESSNER, 1958). But as KUENEN (1957) wrote, sole markings " ... although ubiquitous in graded graywackes are not necessarily limited to beds deposited by turbidity currents (p. 232) ... However, ... it is apparent that the great majority of occurrences, especially those present in rock series showing a profusion of several types of sole markings, can confidently be attributed to the action of turbidity currents."

Evaluation and Utility. Much of the present interest in sole markings has to do with the relations between current flow and the subaqueous slope and with the paleogeographic value of these structures. In 1953, for example, KUENEN and CAROZZI (p. 371) wrote: "Current bedding, ripple mark, *flow markings*, distribution of grain size, and alignment of pebbles can be used to find the direction of the currents and hence of the slope. This, in turn, may give a clue to the shape of the sedimentary basins, the source area of the materials, and the age of the tectonic structures." Systematic measurements of the orientation of the sole markings did not become common, however, until KOPSTEIN's paper (1954) on the Cambrian of the Harlech dome in Wales and CROWELL's study (1955) on the Flysch of Switzerland. These papers contained excellent discussions of the directional structures of sediments, together with a large number of measurements which were used to construct paleocurrent maps. It soon became apparent to many workers that in the flysch facies, sole markings are the most common and most useful criterion of current direction.

The various types of sole marks are of very unequal utility for the mapping of current directions in the field. Although the relative frequency of specific features varies from one region to another, flute casts are by far the most common sole markings and, moreover, they generally occur in such numbers that a well-defined mean direction of flow is readily estimated. Moreover their dissimilar terminations indicate at once from which direction the current came.

Groove casts, though less helpful, are generally common and, if used in conjunction with other structures (prod casts, flutes, cross-lamination) which give the sense of direction of current flow, can be measured readily and used in paleocurrent analysis. Other sole markings are uncommon and hence useful only in special cases.

Variability. Sole markings, unlike other directional properties, such as cross-bedding, exhibit a surprising degree of uniformity of direction, both on the same sole, and in successive beds, although some bedding planes do indeed exhibit two sets of markings which intersect, generally at a small angle. The mean current direction indicated by the sole marks is the same as the mean determined from ripple cross-laminations commonly found in the upper parts of the same beds (CROWELL, 1955, p. 1362). The cross-bedding measurements have a markedly greater "spread" or "scatter" than do the substratal lineations.

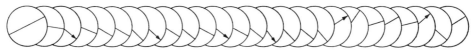

Fig. 5-11. Sketch showing uniformity of orientation of sole marks in Martinsburg (Ordovician) flysch, central Appalachians, U.S.A. (McBRIDE, 1962, fig. 34)

Bed-to-bed variation is commonly nil. A common sight is a display of the soles of a series of upturned turbidite beds all showing lineations pointing exactly in the same direction (fig. 5-11). In outcrops displaying many beds, the directions vary within a few degrees; a shift, if present, rarely exceeds 20 degrees. There are, however, a few beds with strongly divergent lineations. These do not seem to differ materially from their neighbors nor do they disturb the regularity of the sequence in any way.

Regions do differ somewhat in the homogeneity of the current pattern represented by the sole markings. The directions in the graded beds of the northern Apennines (TEN HAAF, 1959) and those in the Devonian "Portage" of New York State (SUTTON, 1959; McIVER, 1961) are remarkably uniform over large areas. Those of the Ordovician Martinsburg shale (McBRIDE, 1962, fig. 32) show much greater variability.

Relations to Other Structures: Beds exhibiting sole marks may also display other primary directional structures. CROWELL (1955, p. 1363) has explored the relations between the sole markings on a bed and the ripple cross-lamination seen at the top of the same bed. The direction of flow indicated by both structures is the same (fig. 5-6) and this suggests that the current responsible for the erosion of the underlying mud surface, the features of which are preserved as casts on the bottom of the sandstone bed, was the same current that deposited the bed itself.

Likewise the fabric of the sole-marked beds, both tilted and undeformed, was investigated by McIVER (1961, p. 141) and McBRIDE (1962, p. 68). The fabric vector was generally closely correlated with the substratal lineation (fig. 5-7).

The relations between the lineations and slump structures has been reported by several investigators. CUMMINS (1959, p. 177) and MURPHY and SCHLANGER (1962) state that the axes of slump folds in the cases studied by them, were more or less parallel to the substratal lineations and believed, therefore that the sub-aqueous slump moved transversely to the currents. Likewise TEN HAAF (1959, p. 50) in his work in the Apennines noted the perpendicularity between the presumed direction of slump and that of current flow.

Paleogeographic Significance. The interpretation of the graded flysch sandstones as products of turbidity currents led to the conclusion that the sole markings of these turbidite sands were also a product of such currents. If so, such density currents would be expected to behave as a heavy fluid, to underflow the clearer, less dense wa-ters, and to travel down the submarine slope. The sub-stratal markings, therefore, should indicate the direc-tion of greatest slope. Such a concept, however, meets several contradictions. As noted elsewhere, there are many examples of two sets of grooves or striations on one and the same sole. Per-haps these are analogous to the two sets of glacial striations commonly seen on the same glacial pave-ment which are produced by the advance of but a single glacial lobe (CHAM-BERLIN, 1888, p. 202). See figure 5-12. However, only one set can point down slope.

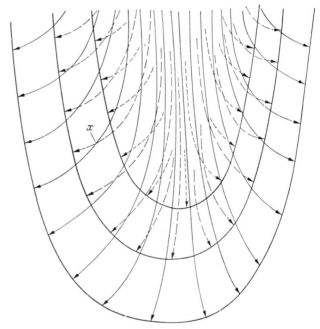

Fig. 5-12. Sketch showing pattern of flow at base of expanding ice lobe. The pattern within an expanding turbidity underflow would presumably be similar. Note that at a given point, X, two slightly divergent flow directions and hence two slightly divergent sole markings, are possible, (modified from CHAMBERLIN, 1888, fig. 25)

Although the orientation of markings on successive sand beds is commonly very uniform, there are cases where this is not so. One cannot presume the submarine slope to be tilted first one way and then another. If turbidity currents show a movement pattern analogous to that of a glacial lobe, such currents, ori-ginating at different places along the flanks of a deep trough, might exhibit divergent directions of flow at a given place situated in an "interlobate" position (fig. 5-13).

In some cases the direction of current flow, as shown by flutes and grooves, is at variance with the slope direction deduced from slump structures. Although it is notoriously difficult to estimate the direction of movement from most slump structures, one might expect the slump folds to trend at right angles to the direc-tion of movement. The fold axes, therefore, would be parallel to the strike of the submarine slope and, if the depositional slope controlled both the movement of the density current and the subaqueous slumping, the fold axes should be perpendicular to the current direction indicated by the sole marks. But as TEN HAAF (1959, p. 50) notes, commonly the slump direction, inferred from well-defined fold axes or slump scratches, lies athwart the current.

This anomalous relation, between the presumed direction of slumping inferred from slump folds and the direction of current flow inferred from the lineations, led MURPHY and SCHLANGER (1962) to reject the turbidite concept for the strata studied by them. If the slumps were down-slope, the currents were not — the latter ran parallel to the slope. CUMMINS (1959, p. 177), however, came to a different view, namely that the turbidity currents ran parallel to the axis of the trough whereas the slumps moved down the flanks. A similar relation between currents and slides is described from the marginal parts of the flysch basin in the central Carpathians (MARSCHALKO, 1961, fig. 11). See fig. 6-11.

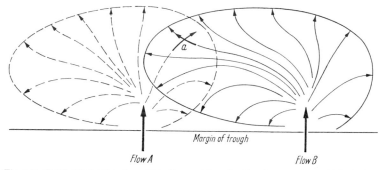

Fig. 5-13. Sketch showing how two widely divergent flow directions in successive turbidite beds are possible. Note divergent flow at *a*

Some doubt exists that many structures such as "flow rolls" commonly ascribed to slump are indeed slump structures. If they are not, their orientation, whatever its origin, is as yet unexplained, and cannot be used to delineate slopes.

Although density currents are responsive to relatively steep slopes, they can also travel over wide areas of negligible slope. A heavy liquid will underflow a light one even on a horizontal bottom. Seemingly, turbidity currents can travel long distances over nearly horizontal surfaces, such as the abyssal plains, with a slope of about 0.001 (MENARD, 1955, p. 247). The shape of the basin and location of the places of supply must be the main factors (fig. 5-13).

The systematic mapping of sole marks and the utilization of these structures in paleographic reconstruction is comparatively recent. Most of the paleocurrent maps are summarized in Table 5-2.

As can be seen by inspection of this table, most studies so far published, excepting those of SUTTON in the upstate New York area and of MURPHY and SCHLANGER in Brazil, have been made on sediments in Alpine-type orogenic belts. Nearly all are in flysch or flysch-like sequences of graded beds. A considerable number are from the central Carpathians or from the Alps (seven each); the Appalachians yield but two studies, the Caledonian geosyncline but three. The beds studied range in age from Precambrian to Miocene.

In general, these studies have yielded paleocurrent maps of great simplicity (fig. 5-14), although there are exceptions. In some cases the uniformity of orientation is so marked that any departure from the regional pattern was taken as

Table 5-2. *Published Maps of Sole Mark and Other Lineations*

Author and year	Age	Area	Remarks
KOPSTEIN, 1954	Cambrian	Harlech Dome, Wales	44 measurements
CROWELL, 1955	Paleocene-Eocene	Switzerland	337 measurements, diagrammatic map
DZULYNSKI and RADOMSKI, 1955	Tertiary	Polish Carpathians	
KUENEN and SANDERS, 1956	Carboniferous	Schiefergebirge and Harz, Germany	
CUMMINS, 1957	Silurian	Wales	67 flute casts; 17 groove casts; 10 miscellaneous linear structures
KUENEN, FAURE-MURET, LANTEAUME, and FALLOT, 1957	Tertiary	Maritime Alps, France and Italy	
REINEMUND and DANILCHIK, 1957	Carboniferous	Ouachita Mts., Arkansas, U.S.A.	
KSIAZKIEWICZ, 1958 . .	Upper Cretaceous, Upper Eocene-Oligocene	Carpathians, Poland	Several hundred measurements; diagrammatic map
RADOMSKI, 1958	Tertiary	Carpathians, Poland	
TEN HAAF, 1958	Tertiary	Apennines, Italy	
BOUMA, 1959a	Oligocene-Miocene	Maritime Alps, France	
BOUMA, 1959b	Oligocene	Maritime Alps, France	
CUMMINS, 1959	Silurian	Wales	204 measurements; diagrammatic map
DZULYNSKI, KSIAZKIEWICZ, and KUENEN, 1959	Cretaceous-Eocene	Carpathians, Poland	
DZULYNSKI and SLACZKA, 1958	Eocene-Oligocene	Carpathians, Poland	
TEN HAAF, 1959	Tertiary	Apennines, Italy	
KNILL, 1959	Precambrian	Scotland	
NEDERLOF, 1959	Upper Carboniferous	Cantabrian Mts., Spain	
SUTTON, 1959	Upper Devonian	New York	
WOOD and SMITH, 1959 .	Ordovician	Wales	
ALLEN (1960)	Cretaceous	England	3 measurements

Table 5-2 (Cont.)

Author and year	Age	Area	Remarks
Hsu, 1960	Cretaceous-Eocene	Alps, Switzerland	About 400 measurements
Marschalko and Radomski, 1960	Eocene-Oligocene	Carpathians Poland	
Frakes, 1961	Upper Devonian	Appalachians of Pennsylvania	
Marschalko, 1961 . . .	Eocene-Oligocene	Carpathians, Poland	Mapped current markings, slump structures and gravel fabric
Plessmann, 1961 . . .	Eocene	Maritime Alps, Italy	
Stanley, 1961	Tertiary	France	
Johnson, 1962	Mississippian	Marathon Basin, West Texas, U.S.A.	About 3400 measurements
McBride, 1962	Ordovician	Appalachians	
Murphy and Schlanger, 1962	Cretaceous	Brazil	Mapped current markings and slump folds

evidence of tectonic re-orientation of the strata with anomalous sole directions (ten Haaf, 1957).

The contrast between the remarkably uniform orientation of sole markings in some flysch basins and the disorder shown in others may reflect the difference between a single distant source and a diversity of local sources of supply.

As noted by Kuenen (1958, p. 329), in many elongate basins the dominant transport is parallel to the axis of the basin. This longitudinal transport does not rule out lateral supply. The sides of a trough must be steeper than the floor and a current entering from any direction must descend more or less transversely and then veer around to follow the axial slope towards the deepest part. The record of lateral transport might be absent — owing to its uplift and removal by erosion or the absence of deposition owing to the steepness of grade. Possibly the record commonly seen is that of axial movement of currents which alternate with slumps of slides derived from the steeper flank but which came to rest in the axial portion of the trough.

As noted by ten Haaf (1959, p. 82) and as postulated for the deep sea plains by Menard (1955, p. 239), turbidity currents modify the submarine topography by filling in low areas, filling these irregularities and building up of a "profile of equilibrium" adjusted to the average grain size of the sediment and the dynamics of the turbidity current. The regular stratification, great extent of turbidite beds, and general constancy of current direction suggest a near attainment of such equilibrium.

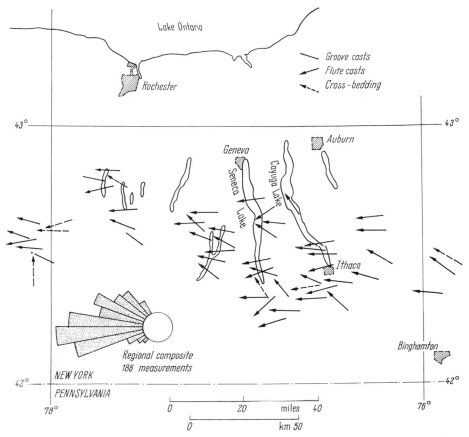

Fig. 5-14. Map showing orientation of sole markings in Upper Devonian strata of Finger Lakes area, New York, U.S.A. (modified from SUTTON, 1959, fig. 4)

Glacial Lineations

One of the earliest, perhaps the first, efforts to determine and map the pattern of sediment transport was the study of the movement pattern of past continental glaciers. Because the glacial striations are so conspicuous and widespread, it soon became clear that the mapping of their orientation over large areas would

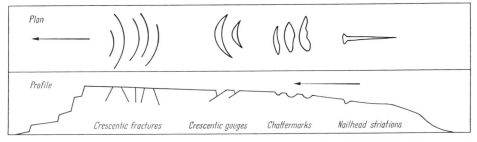

Fig. 5-15. Diagrammatic illustration of various glacial markings on roches moutonnées and their relation to ice movement (modified from DREIMANIS, 1956, fig. 5)

enable one to unravel the glacial transport plan and locate the centers of glaciation, delineate the various ice-lobes, and the like (fig. 5-16).

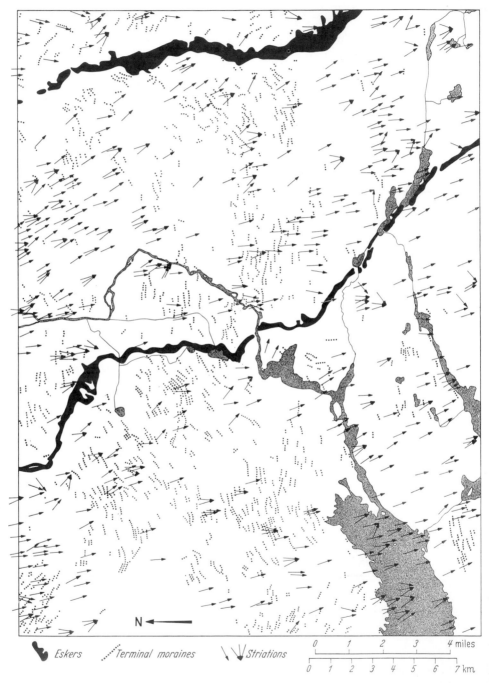

Fig. 5-16. Map showing orientation of glacial striations, terminal moraines and eskers (modified from Lundegärdh and Lundqvist, 1959, fig. 29)

Glacial grooves and striations are indeed sedimentary lineations and are analogous to the substratal lineations produced by turbidity currents. Unlike the
latter, preserved as casts on the soles of turbidite sandstones, the glacial markings
themselves are seen on the bedrock surface underlying till. Glacial markings vary
from faint striations to pronounced grooves. They commonly cover the whole
of the surface on which they occur (pl. 23). Two sets, generally intersecting at
a low angle, are present in many places. Associated with many striations are the
chattermarks, crescentic fractures, crescentic gouges, and nailhead striations
found in rows of two or more (fig. 5-15), produced by the rhythmic vibratory
action of the engraving tool moved by the ice (CHAMBERLIN, 1888; GILBERT
1905).

Inhomogeneities of the rock overridden by the ice are accentuated by glacial
scour. More resistant objects, such as a chert nodule, appear slightly raised
above the general surface and exhibit a ridge of less resistant rock on their pro-
tected lee side and thus form what the glacial geologist calls "knob and trail"
(CHAMBERLIN, 1888, fig. 48).

Glacial grooves and lineations and the other associated markings are only
one aspect of the record of ice movement. The interrelations between substratal
lineations, internal fabric of till and depositional form is nowhere better displayed
than in the glacial record. The various aspects of the glacial model are discussed
elsewhere.

Parting Lineation

Parting lineation is the term now generally applied to a lineation which char-
acterizes the bedding surfaces of most flagstones. It is a structure which is revealed by splitting or part-
ing of the stone along one
of its many very uniform
and closely-spaced bedding
planes. It is therefore, an
internal structure — not
a sole marking. Parting
lineation was apparently
first described by SORBY
(1856, p. 114); was also
described and figured by
HANS CLOOS (1938, fig. 2,
p. 358), rediscovered by
STOKES (1947, p. 21) who
called it "primary current
lineation," and renamed
"parting lineation" by CRO-
WELL (1955, p. 1361). It is
best seen in bright, low
angle illumination, appears
as "... a streamlining or

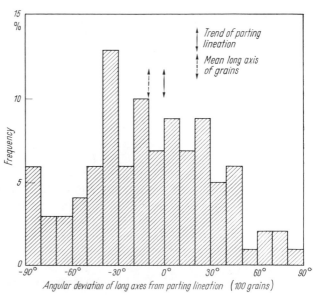

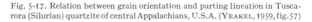

Fig. 5-17. Relation between grain orientation and parting lineation in Tuscarora (Silurian) quartzite of central Appalachians, U.S.A. (YEAKEL, 1959, fig. 57)

streaming effect of sand particles in relatively low poorly defined windrow-like ridges parallel with the current direction" (STOKES, 1947, p. 52). This lineation is commonly associated with an imperfection in the parting. A thin veneer, but a millimeter or two thick, is seen to cling in irregular elongate patches to the bedding surface — the residue of a once continuous layer now largely removed but having plaster-like remnants adhering to the bedding plane. These patchy, irregular areas are elongated in the direction of the parting lineation (pl. 22 B).

It is most characteristic and most abundant in those thin- and even-bedded sandstones most used as flagstones and hence is commonly seen in flagstone walks. Well-known flagstones, that display excellent parting lineation include the "Crab Orchard Stone" (Pennsylvanian), the Catskill flagstones (Devonian) and the flagstones of the Morrison (Jurassic). Although parting lineation is also found in some flysch sandstones, it is not characteristic of this facies.

The cause of the structure is obscure. Newly deposited sands display this feature. The lineation, although parallel to the current, indicates the line of movement but not the direction of the flow. The structure is usually closely correlated with grain orientation (fig. 5-17) as shown by YEAKEL (1959). The peculiar defects in parting associated with the lineation is perhaps an expression of some kind of imperfection in cementation, which is fabric controlled.

Summary

Excluding some depositional fabrics, sedimentary lineations are mainly of two kinds, namely substratal and internal. The former include a variety of markings or hieroglyphs which occur on the base or soles of many sandstones and some limestone beds. These markings are casts of depressions cut in the underlying bed by currents or objects propelled by the current and filled by sand. Included here are flute casts, groove and striation casts, and the less common bounce, prod, and brush casts as well as a few other marks of uncertain origin.

The substratal current markings are commonly ascribed to turbidity currents rendered abrasive by the load carried by them. Hence mapping of the sole marks of turbidity sands enables one to determine the direction of transport, sheds light on the source of supply of sediment, and enables one to draw some conclusions about the configuration of the basin and evolution of tectonic elements within it.

The internal lineation, designated parting lineation, characteristic of the flagstone facies is a fabric-controlled feature most characteristic of fluvial sands, though it is also known to be formed in other environments.

Glacial lineations, the bedrock striations in particular, are most nearly like the substratal lineations produced by turbidity currents. They were the first structures to be mapped, to be related to other glacial features in unravelling the pattern of ice-movement — the first true paleocurrent analysis.

What is known about sole markings is largely the product of field study of rocks bearing these markings. Our interpretation of their origin and significance is based on study of the features themselves, their relation to one another and to current and other structures in the rock. Our analysis is thus based on internal evidence. Study of modern sediments has provided no insight into the problems of sole markings. They are not amenable to study in present-day deposits inasmuch as the same current which produces them also covers them in a single short-lived event.

Although some experimental work has been done in an effort to make sole markings in the laboratory, further studies of this kind might be profitable. Experimental work in this field requires only patience and ingenuity on the part of the experimenter. No problems of scale, time, or physical conditions are encountered, as is usually the case in experimental geology. The structures were formed in a very short time, with ordinary sedimentary materials, at normal pressures and temperatures. Indeed, the little experimental work that has been done, by RÜCKLIN on flutes, for example, seems to indicate that all the features observed on the soles of sandstones could be reproduced in the laboratory.

References

ALLEN, P., 1960: The Wealden environment: Anglo-Paris basin. Phil. Trans. Roy. Soc., London, Ser. B **242**, 283—346.

BEASLEY, H. C., 1908: Some markings, other than footprints, in the Keuper sandstones and marls. Proc. Liverpool Geol. Soc. **10**, 262—275.

— 1914: Some fossils from the Keuper sandstone of Alton, Staffordshire. Proc. Liverpool Geol. Soc. **12**, 35—39.

BIRKENMAJER, K., 1958: Oriented flowage casts and marks on the Carpathian Flysch and their relation to flute and groove casts. Acta Geol. Polon. **8**, 117—118 [Polish, English summary].

BOKMAN, J., 1953: Lithology and petrology of the Stanley and Jackfork formations. J. Geol. **61**, 152—170.

BOUMA, A., 1959a: Some data on turbidites from the Alps Maritimes (France). Geol. en Mijnbouw **21**, 223—227.

— 1959b: Flysch oligocène de Pëira-Cava (Alpes-Maritimes, France). Eclogae Geol. Helv. **51**, 893—900.

CHAMBERLIN, T. C., 1888: Rock-scorings of the great ice invasions. 7th Ann. Rep. U.S. Geol. Survey, p. 218—223.

CLARKE, J. M., 1917: Strand and undertow markings of Upper Devonian time as indications of the prevailing climate. N.Y. State Museum Bull. **196**, 199—210.

CLOOS, E., 1946: Lineation, a critical review and annotated bibliography. Geol. Soc. Amer. Mem. **18**, 122 p.

CLOOS, H., 1938: Primäre Richtungen in Sedimenten der rheinischen Geosynkline. Geol. Rundschau **29**, 357—367.

CROWELL, J. C., 1955: Directional-current structures from the Prealpine Flysch, Switzerland. Bull. Geol. Soc. Am. **66**, 1351—1384.

— 1958: Sole markings of graded graywacke beds: a discussion. J. Geol. **66**, 333—335.

CUMMINS, W. A., 1957: The Denbigh grits; Wenlock greywackes in Wales. Geol. Mag. **94**, 433—452.

— 1958: Some sedimentary structures from the Lower Keuper sandstones. Liverpool, Manchester Geol. J. **2**, 37—43.

— 1959: The Lower Ludlow Grits in Wales. Liverpool, Manchester Geol. J. **2**, 168—179.

DOTT jr., R. H., and J. K. HOWARD, 1962: Convolute lamination in non-graded sequences. J. Geol. **70**, 114—120.

DREIMANIS, A., 1956: Steep Rock boulder train. Proc. Geol. Assoc. Can. **8**, 28—70.

DUNBAR, C. O., and J. RODGERS, 1957: Principles of stratigraphy. New York: John Wiley and Sons. 356 p.

DZULYNSKI, ST., M. KSIAZKIEWICZ and PH. H. KUENEN, 1959: Turbidites in flysch of the Polish Carpathian Mountains. Bull. Geol. Soc. Am. **70**, 1089—1118.

DZULYNSKI, ST., and A. RADOMSKI, 1955: Origin of groove casts in the light of turbidity current hypothesis. Acta. Geol. Polon. 5, 47—56.
—, and J. E., SANDERS, 1959: Bottom marks on firm lutite substratum underlying
turbidite beds (abstract). Bull. Geol. Soc. Am. 70, 1594.
—, and A. SLACZKA, 1958: Directional structures and sedimentation of the Krosno
beds (Carpathian flysch). Ann. soc. géol. Pologne 28, 205—260.
— — 1960: Sole markings produced by fish bones acting as tools. Ann. soc. géol.
Pologne 30, 249—255.
FRAKES, L. A., 1961: Sedimentary structures of the Upper Devonian of central
Pennsylvania. Proc. Penna Acad. Sci. 35, 116—123.
FREUDENBERG, W., 1934: Die Gürich-Berckhemerschen Wülste. Z. deut. geol.
Ges. 86, 59 (discussion of GÜRICH, 1933).
FUCHS, TH., 1895: Studien über Fucoiden und Hieroglyphen. Denkschr. Akad. Wiss.
Wien, Math.-nat. Kl. 62, 369—448.
GILBERT, G. K., 1905: Crescentic gouges on glaciated surfaces. Bull. Geol. Soc. Am.
17, 303—316.
GLAESSNER, M. P., 1958: Sedimentary flow structures on bedding planes. J. Geol.
66, 1—7.
GROSSGEYM, V. A., 1946: On the significance and methods of study of hieroglyphs.
(on the material of the Caucasian Flysch). Acad. Sci., U.S.S.R., B., Ser. Geol.,
No. 2, 111—120 [Russian, Engl. summ.].
— 1955: Experiment in the creation of a terminology for the morphological description
of hieroglyphics. Geol. Prospecting Inst. 3 (6), 314—325 [Russian].
GÜRICH, G., 1933: Schrägschichtungsbögen und zapfenförmige Fließwülste im
„Flagstone" von Pretoria und ähnliche Vorkommnisse im Quartzit von Kubis,
S.W.A., dem Schilfsandstein von Maulbronn u.a. Z. deut. geol. Ges. 85, 652—654.
HAAF, E. TEN, 1957: Tectonic utility of oriented resedimentation structures. Geol.
en Mijnbouw, n.s. 19, 33—35.
— 1959: Graded beds of the northern Apennines. Ph.D. thesis, Rijks University,
Groningen, 102 p.
HÄNTZSCHEL, W., 1935: Fossile Schrägschichtungs-Bogen, „Fließwülste" und Rieselmarken aus dem Nama-Transvaal-System (Süd-Afrika) und ihre rezenten Gegenstücke. Senckenbergiana 17, 167—177.
HALL, J., 1843a: Geology of New York. pt. IV, Survey of the 4th District. Albany:
Carroll and Cook. 683 p.
— 1843b: On wave lines and casts of mud furrows (Abstract): Am. J. Sci. 45, 148—149.
— 1843c: Remarks upon casts of mud furrows, wave lines, and other markings upon
rocks of the New York system. Rept. Assoc. Amer. Geologists and Naturalists,
p. 422—423.
HSU, K. J., 1959: Flute- and groove-casts in the Prealpine Flysch, Switzerland.
Am. J. Sci. 257, 529—536.
— 1960: Paleocurrent structures and paleogeography of the ultrahelvetic Flysch
basins, Switzerland. Bull. Geol. Soc. Am. 71, 577—610.
JOHNSON, K. E., 1962: Paleocurrent study of the Tesnus formation, Marathon Basin,
Texas. J. Sediment. Petrol. 32, 781—792.
KELLING, G., 1959: Ripple-mark in the Rhinns of Gallaway. Trans. Edinburgh Geol.
Soc. 17, 117—132.
—, and E. K. WALTON, 1961: Flow structures in sedimentary rocks: a discussion.
J. Geol. 69, 224.
KINGMA, J. T., 1958: The Tongaporutuan sedimentation in central Hawkes Bay.
New Zealand J. Geol. and Geophys. 1, 1—30.
KNILL, J. L., 1959: Paleocurrents and sedimentary facies of the Dalradian meta-
sediments of the Craignish-Kilmelfort District. Proc. Geol. Assoc. 70, 277—284.
KOPSTEIN, F. P. H. W., 1954: Graded bedding of the Harlech Dome. Ph.D. thesis,
Rijks University, Groningen, 97 p.
KREJCI-GRAF, J., 1932: Definition der Begriffe Marken, Spuren, Fährten, Bauten,
Hieroglyphen und Fucoiden. Senckenbergiana 14, 19—39.

KSIAZKIEWICZ, M., 1952: Graded and laminated bedding in the Carpathian flysch. Ann. soc. géol. Pologne **22**, 399—449.
— 1961: On some sedimentary structures of the Carpathian flysch. Ann. soc. géol. Pologne **31**, 37—46.
KUENEN, PH. H., 1953: Graded bedding, with observations on lower Paleozoic rocks of Britain. Verhandel. Koninkl. Ned. Akad. Wetenschap. Afdel. Natuurk. **20**, 1—47.
— 1957: Sole markings of graded graywacke beds. J. Geol. **65**, 231—258.
— 1958: Problems concerning source and transport of flysch sediments. Geol. en Mijnbouw, n.s. **20**, 329—339.
—, and A. CAROZZI, 1953: Turbidity currents and sliding in geosynclinal basins in the Alps. J. Geol. **61**, 363—372.
— A. FAURE-MURET, M. LANTEAUME and P. FALLOT, 1957: Observations sur les flysches des Alpes-Maritimes françaises et italiennes. Bull. soc. géol. France [6] **7**, 11—26.
—, and J. E. PRENTICE, 1957: Flow markings and load-casts. Geol. Mag. **94**, 173—174.
—, and J. E. SANDERS, 1956: Sedimentation phenomena in Kulm and Flözleeres graywackes, Sauerland and Oberharz, Germany. Am. J. Sci. **254**, 649—671.
—, and E. TEN HAAF, 1956: Graded bedding in limestones. Koninkl. Nederl. Akad. Wetenschap. Pr., s. B.**59**, 314—317.
KULICK, J., 1960: Driftmarken im Kulm des Edersee-Gebietes. Fortschr. Geol. Rheinld. u. Westf. **3**, 289—296.
LINCK, O., 1956: Driftmarken von Schachtelhalm-Gewächsen aus der mittleren Keuper (Trias). Senckenbergiana Lethaea **37**, 39—51.
LUNDEGARDH, P. H., and G. LUNDQVIST, 1959: Beskrivning till kartbladit Eskilstuna. Sveriges Geol. Undersökn., Ser. Aa, No. 200, 125.
MARSCHALKO, R., 1961: Sedimentologic investigation of marginal lithofacies in Flysch of central Carpathians. Geol. práce (Bratislava) **60**, 197—230.
—, and A. RADOMSKI, 1960: Preliminary results of investigations of current directions in the flysch basin of the central Carpathians. Ann. soc. géol. Pologne **30**, 259—261.
MCBRIDE, E. F., 1962: Flysch and associated beds of the Martinsburg formation (Ordovician), central Appalachians. J. Sediment. Petrol. **32**, 39—91.
MCIVER, N. L., 1961: Upper Devonian marine sedimentation in the central Appalachians. Unpublished Ph.D. thesis, The Johns Hopkins University, 347 p.
MCKEE, E. D., 1954: Stratigraphy and history of the Moenkopi formation of Triassic age. Geol. Soc. Am. Mem. **61**, 133 p.
MENARD, H. W., 1955: Deep-sea channels, topography and sedimentation. Bull. Am. Assoc. Petrol. Geologists **39**, 236—255.
MURPHY, M. A., and S. O. SCHLANGER, 1962: Sedimentary structures in Ihlas and São Sebastião formations (Cretaceous), Recôncavo Basin, Brazil. Bull. Am. Assoc. Petrol. Geologists **46**, 457—477.
NEDERLOF, M. H., 1959: Structure and sedimentology of the Upper Carboniferous of the upper Pisuerga Valleys, Cantabrian Mountains, Spain. Leidse Geol. Mededel. **24**, 603—703.
PAVONI, N., 1959: Rollmarken von Fischwirbeln aus den oligozänen Fischschiefern von Engi-Matt (Kt. Glarus). Eclogae Geol. Helv. **52**, 941—949.
PEABODY, F. E., 1947: Current crescents in the Triassic Moenkopi formation. J. Sediment. Petrol. **17**, 73—76.
PICKEL, W., 1937: Stratigraphie und Sedimentanalyse des Kulms an der Edertalsperre. Z. deut. geol. Ges. **89**, 233—280.
PLESSMANN, W., 1961: Strömungsmarken in klastischen Sedimenten und ihre geologische Auswertung. Geol. Jahrb. **78**, 503—566.
PRENTICE, J. E., 1960: Flow structures in sedimentary rocks. J. Geol. **68**, 217—225.
— 1961: Flow structures in sedimentary rocks: a reply. J. Geol. **69**, 225.
RADOMSKI, A., 1958: The sedimentological character of the Podhale flysch. Acta Geol. Polon. **8**, 335—408 [Polish, English summary].

REINEMUND, J. A., and W. DANILCHIK, 1958: Preliminary geologic map of the Waldron Quadrangle and adjacent areas, Scott County, Arkansas. U.S. Geol. Survey, Oil and Gas Investigation Map OM 192.

RICH, J. L., 1950: Flow markings, groovings, and intra-stratal crumplings as criteria for recognition of slope deposits with illustrations from Silurian rocks of Wales. Bull. Am. Assoc. Petrol. Geologists 34, 717—741.

RICH, J. L., 1951: Three critical environments of deposition, and criteria for recognition of rocks deposited in each of them. Bull. Geol. Soc. Am. 62, 1—20.

RICHTER, RUD., 1935: Marken und Spuren in Hunsrück-Schiefer. I. Gefließ-Marken. Senckenbergiana 17, 244—263.

RÜCKLIN, H., 1938: Strömungsmarken im unteren Muschelkalk des Saarlandes. Senckenbergiana 20, 94—114.

SEILACHER, A., 1960: Strömungsanzeichen im Hunsrückschiefer. Notizbl. hess. Landesamtes Bodenforsch. Wiesbaden 88, 88—106.

SHROCK, R. R., 1948: Sequence in layered rocks. New York: McGraw-Hill Book Co. 507 p.

SORBY, H. C., 1856: On the physical geography of the Old Red Sandstone sea of the central district of Scotland. Edinburgh New Philosoph. J., n.s. 3, 112—122.

STANLEY, D. J., 1961: Études sédimentologiques des grès d'Annot et leur équivalents latéraux. Inst. Français du Petrole, Ref. 6821, 158 p.

STANLEY, G. M., 1955: Origin of playa stone tracks, Racetrack Playa, Inyo County, California. Geol. Soc. Am. 66, 1329—1350.

STOKES, W. L., 1947: Primary lineation in fluvial sandstones. J. Geol. 55, 52—54.

STRAKHOV, N. M., ed., 1958: Méthodes d'étude des roches sédimentaires. Ann. Serv. Info. Geol., Bur. Recherches géol. geophys. et Min., No. 35, 1, 542 p. [Translated from the Russian].

SUTTON, R. G., 1959: Use of flute casts in stratigraphic correlation. Bull. Am. Assoc. Petrol. Geologists 43, 230—237.

TEICHMÜLLER, R., 1960: Ein rezentes Analogon zu Driftmarken im Kulm des Edersee-Gebietes. Fortschr. Geol. Rheinld. Westf. 3, 297—300.

VASSOEVICH, N. B., 1932: Some data allowing us to distinguish the overturned position of flysch sedimentary formations from the normal ones. Academy Sciences, U.S.S.R., Trudy Geol. Inst. 2, 47—63 [Russian].

— 1953: On some Flysch textures. Trans. Soc. geol. Lwow, ser. geol., 3, 17—85 [Russian].

WILLIAMS, H. S., 1881: Channel-fillings in the Upper Devonian shales. Am. J. Sci. 121, 318—320.

WOOD, A., and A. J. SMITH, 1959: The sedimentation and sedimentary history of the Aberystwyth grits (Upper Llandoverian). Quart. J. Geol. Soc. London 114, 163—195.

YEAKEL, L. S., 1959: Tuscarora, Juniata and Bald Eagle paleocurrents and paleography in the central Appalachians. Unpublished Ph.D. thesis, The Johns Hopkins University, 454 p.

Linear Structures (1963—1976)

Linear structures are the most important criteria for determination of current direction in turbidites and glacial deposits

Although linear structures can be found in sediments deposited in many environments, they are of primary importance in turbidites, where as sole marks, they are the chief guides to paleocurrent flow. Mapping of sole mark orientation, a novelty before 1963, is now commonplace (fig. 5-18) and turbidity current deposits and their sole marks have even now been recognized in lacustrine deposits, ancient (NEGENDANK, 1972) and Pleistocene (BAKER, 1973, p. 42—47).

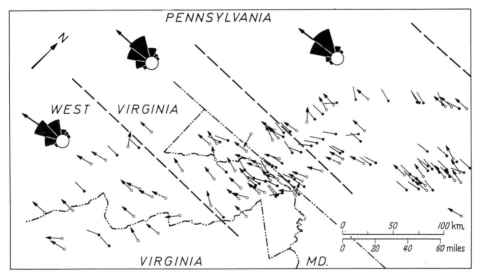

Fig. 5-18. Uniformly oriented paleocurrents of Upper Devonian age in part of Appalachian Basin. (Redrawn from McIVER, 1971, fig. 14. By permission of John Wiley and Sons, New York). In southwestern two thirds of area paleocurrents are inferred from sole marks on turbidites

There has been much interest in the relations between the dimensions and orientation of sole marks and other bed properties such as bed thickness, grain size, and fabric. These relations are best studied in the field. Interest in the hydrodynamics of sole mark generation has led to a number of experimental studies of their formation.

There are now many regional maps of paleoglacial flow based primarily on glacial striations. These are mainly of Pleistocene ice movement though some more ancient patterns have also been mapped. Paleoglacial flow patterns cover the widest area of any paleocurrent system, either ancient or modern.

New Structures

Only a few heretofore undescribed linear structures have been identified since 1963. It is of interest that many of these are either formed in snow or produced by ice, both ephemeral materials. Moreover nearly all have been recognized only in modern deposits, few in the rock record. These newer linear structures are listed in Table 5-3.

Tool Marks

Under tool marks we include all those features produced by objects moved by the current. These include grooves, impact-, prod-, bounce-, brush-, and roll-marks.

Table 5-3. *New Linear Structures*

Name	Characteristics	References
Setulf	Positive relief bedforms on modern tidal flat resembling flute molds	FRIEDMAN and SANDERS, 1974
Iceberg furrow marks	Large-scale furrows (3 km long, 30 m wide and 6 m deep) on continental shelf of Newfoundland	HARRIS and JOLLYMORE, 1974
Gutter casts	Scour-and-fill structures (large groove casts ?). A scour rather than a tool mark	WHITAKER, 1973
Scour-remnant ridges	Longitudinal scour features that stand up above the bedding plane — a ridge left by general deflation in the lee of an obstacle. Analogous to the "knob and trail" of glacial pavement (CHAMBERLIN, 1888*)	ALLEN, 1965
Obstacle shadow	Any flow-aligned features produced either by scour or deposition in the lee of obstacles; would include sand shadows (POTTER and PETTIJOHN, 1963, p. 121) and remnant ridges (ALLEN, 1965)	KARCZ, 1968
Harrow marks	Current-aligned, fine-grained ridges with intervening trough-like strips of coarser sediment in ephemeral stream beds	KARCZ, 1967
Longitudinal markings	A general term for various current-aligned longitudinal furrows and ridges including rill molds, dendritic ridge molds passing into scaly flute-like molds, fleur-de-lys patterns and frondescent casts	DZULYNSKI and WALTON, 1965

SEILACHER (1963) published a well-illustrated summary of the various roll- and impact-marks made by cephalopod shells. The varying forms of these marks are related to current velocity and the differing modes of shell movement. KELLING and WHITAKER (1970) describe bounce-, prod-, and roll-casts of longitudinally ribbed orthoconic nautiloids.

Both large and small grooves produced by ice have been reported. These range from the small ribbed grooves and striations produced by ice blocks carried seaward by ebb tide, described by DIONNE (1972), to gigantic iceberg furrows, 3 km long, 30 m wide, and 6 m deep observed on the continental shelf of Newfoundland reported by HARRIS (1974) and HARRIS and JOLLYMORE (1974). To date there is no rock record of marks made by these ephemeral tools.

Groove and striation casts are very common on the soles of turbidite sandstones (Plate 17* and PETTIJOHN and POTTER, 1964, Plate 62). These have been considered tool marks produced by the scribing action of current-swept debris over the mud substrate. No new light has been shed on their origin. ALLEN (1969, fig. 3), however, was able to produce experimentally very regular rectilinear longitudinal grooves, which in many respects resemble those found in nature and are generally attributed to tool action. The experimentally formed grooves, however, were presumed to form by erosion and their size and spacing controlled by "the flow structure present in the lower part of the stream" (ALLEN, 1969, p. 613). They are thus scour rather than tool marks and according to Allen (p. 613) "no structures similar to the grooves have yet been reported from the fossil record". At higher velocities, Allen produced longitudinal meandering grooves with an anastomosing pattern. These resemble the "rill molds" of DZULYNSKI and WALTON (1965, fig. 43) that are regarded by them as scour features rather than tool marks. They are not the usual dendritic rills of intertidal areas. In summary, we can say that the longitudinal features (furrows, ridges, striations and grooves) are manifestations of two dissimilar processes — engraving by current-propelled debris on the one hand and scour by current action without tools on the other.

Obstacle Marks

KARCZ (1968) uses the term "obstacle mark" for all those flow-aligned features associated with obstacles (shells, pebbles, mud lumps, etc.) which lie on the surface of sedimentation. A current crescent is such a mark. Included also are sand shadows and remnant ridges (Allen, 1968). Thus some are due to erosion or general deflation whereas others are due to deposition. All are formed by a secondary circulation generated by the obstacle. Karcz's paper contains many references to the engineering literature and analysis of flow patterns around bridge piers and other obstacles.

ALLEN (1965) distinguishes between current crescents and Hufeisen-Wülste (RÜCKLIN, 1938*), the latter being presumed to occur without a resistant object projecting above the bottom. Large scour marks in snow form a crescentic furrow related to a resistant object and are the analog of the current crescents of beaches and river bars. Allen describes the air-flow pattern in the snow furrow. KARCZ (1968) also describes several types of obstacle mark including current crescents.

Flutes and Other Scour Marks

Flute casts (flute molds), one of the most common and perhaps the most useful of the linear sole markings, continue to receive the attention of sedimentologists. Very good descriptions of various types of flute marks and a discussion of their origin is given by Dzulinski and Walton (1965, p. 40—55). A more quantitative study of flutes and flute parameters is contained in the paper by Pett and Walker (1971 a). The hydrodynamics of flute formation is presented by Allen (1968); flutes have also been made experimentally under closely controlled labora-

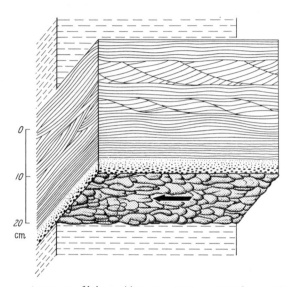

Fig. 5-19. Block diagram shows basal flute marks (current from right to left) and internal directional structures of a typical turbidite bed grading upward from granules at base into fine sandstone-siltstone at top with planar bedding followed by ripple bedding and again planar bedding. Relations such as these occur both in terrigenous as well as carbonate turbidite beds. (Redrawn from Richter, 1965, fig. 24.)

tory conditions (Allen, 1971a, 1971b, p. 287—320). The hydrodynamic conditions for flute formation vary with respect to grain size of sediment and a particular critical velocity of flow (Allen, 1968, fig. 8). However, the hydraulic origin of flutes and their related structures is most complex and still far from being fully understood.

In addition to the experimental studies of flutes and analyses of their formation, there have been many empirical field studies. Ricci Lucchi (1969), for example, concluded from his studies of the Marnoso-arenacea formation in the Apennines, that flutes form in the turbulent head of a turbidity current whereas the tail or depositional phase with little or no turbulence is responsible for grooves. Flutes are, therefore, earlier than grooves.

Pett and Walker (1971a) report on the morphology of flutes and its relation to other bed properties. The flutes of Bouma A beds (the graded interval) are more bulbous, wider, and are associated with coarser sediment than those strata which begin with the Bouma B and C intervals. The flutes of the latter are narrower and more apt to be associated with organic and tool markings. The relation between grain size and the Bouma interval postulated by Pett and Walker (1971 a, 1971 b) was also discussed by Allen (1971 c). The subject had also been investigated by Sestini and Curcio (1965) who found a close positive correlation between flute cast dimensions, bed thickness and grain size.

Parting Lineation

McBride and Yeakel (1963) discriminate between "parting-plane lineation" and "parting-step lineation", the former being essentially a plane surface with faint subparallel linear markings whereas the latter shows a series of subparallel, somewhat irregular, plaster-like remnants of an adjacent lamination clinging to the split surface. Fabric studies showed the long axes of the sand grains to be parallel to both lineation types.

Allen (1964) notes that primary current lineation (parting lineation), characteristic of the flat-bedded sandstone facies, is parallel to a preferred dimensional orientation of the sand grains and to a grain imbrication. Although all details of its origin are yet to be understood, parting lineation appears to form both at low intensity as well as high intensity flows (fig. 4-22). It occurs primarily on beaches, submerged bars in rivers and tidal channels.

Interpretative Problems

Several problems of interpretation are as yet not adequately solved. Some of these have been reviewed by Walker (1970). Turbidites whose sole markings show more than one trend are such a problem. Glennie (1963) presumed them to be due to several turbidity currents which crossed the same point within a short time period. Anderson (1965) accepted this concept, but believed it necessary to postulate complete removal of the deposit of the first current prior to erosion of the substrate and deposition by the second. Stanley (1965) presumed the diverse directions to be formed by a single current, which first ran transversely across the slope then became deflected along the maximum inclination of the sea floor.

Another problem is the relation of the orientation of the sole markings to other paleocurrent indicators in the same bed, such as grain fabric and ripple cross-bedding. Work to explore this problem includes that of Sestini and Pranzini (1965), Jipa (1968), Colburn (1968), Boccaletti and Micheli (1968), Tucker (1969), and Sestini (1970). In general, the long axis orientation of grains in a turbidite deposit deviates only slightly from the current direction indicated by flute casts on the base of the same bed (Sestini, 1970; Sestini and Pranzini, 1965; Boccaletti and Micheli, 1968). Essentially the same relation was observed by Colburn (1968). Tucker (1969, p. 288) has shown that the cross-bedding and the sole mark orientation in crinoidal turbidites are concordant. Although marked deviations are known, in general, the azimuths of the cross-bedding and that of the sole marks in sandstone turbidites are in good agreement (Jipa, 1968). These results are all consistent with earlier observations.

There has been considerable controversy concerning the nature of the current responsible for flutes and other sole markings. These structures, most commonly seen in the flysch deposits, have been most generally attributed to turbidity currents. They have been reported in lake deposits, where they are presumed also to be of turbidity current origin.

The presumption that turbidity currents flowed downslope leads to difficulties as noted in Chapter 5. In particular, the relations between directions of flow and

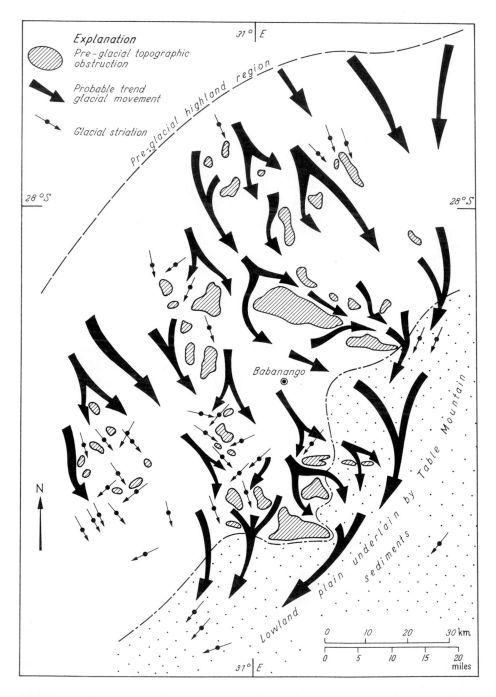

Fig. 5-20. Late Paleozoic glacial striae and preglacial topographic obstructions near Natal, Republic of South Africa. (Redrawn from CROWELL and FRAKES, 1972, fig. 14)

that of mass movement and slumping have created problems and even led some to presume that the currents responsible for the markings were not in fact turbidity currents at all but were the products of "contour currents", that is currents which flow parallel to the slope (HUBERT, 1966, p. 697 and 1968; BOUMA, 1972). This view has been challenged (KUENEN, 1967).

The relations between the direction of flow of the currents reponsible for the flutes, grooves, and other linear sole markings and the direction of slope deduced from slump structures, have been further explored by MCBRIDE and KIMBERLY (1963), SCOTT (1966), KUENEN (1966a, 1966b, 1967), and BAILEY (1967). The whole problem has been reviewed and summarized by WALKER (1970, p. 223 to 226), who stated that: "In none of the basins is there any evidence that an anomalous slope existed during turbidity current flow."

Certainly, some flutes are formed by ordinary non-marine suspension currents (SEN, 1967). The prevailing view, however, is that those found on the soles of the flysch sandstones are the product of turbidity current action. The problem of slump movement and paleoslope is explored further in the next section.

Paleoglacial Flow

Mapping of bedrock striations and other directional features to ascertain the movement pattern of now vanished glaciers is the oldest form of paleocurrent

Table 5-4. *Paleoglacial Flow Maps*

Age	Area	References
Precambrian	Ontario-Quebec Canada	LINDSEY, 1969
Precambrian	Scotland	SPENCER, 1971
Cambro-Ordovician	North Africa	BEUF *et al.*, 1971
Neopaleozoic	Brazil	FARJALLAT, 1970
Permo-Carboniferous	South Africa	CROWELL and FRAKES, 1972, STRATTEN, 1969
Devonian-Permian	Brazil	MARTIN, 1964, FRAKES and CROWELL, 1969
Permo-Carboniferous	India	FRAKES *et al.*, 1975
Permian (?)	Antarctica	FRAKES and CROWELL, 1968, FRAKES *et al*, 1971
Late Paleozoic	Africa	FRAKES and CROWELL, 1970, MATTHEWS, 1970
Late Paleozoic	Australia	CROWELL and FRAKES, 1971
Late Paleozoic	Asia	FRAKES *et al.*, 1975
Late Paleozoic	South America	FRAKES and CROWELL, 1969
Pleistocene	Northern Europe	GLÜCKERT, 1974
Pleistocene	North America	PREST, 1969; ANDREWS, 1973

analysis and even today continues to receive a great deal of attention (fig. 5-20). Whereas the pattern of paleoglacial flow of the Pleistocene ice sheets of northern Europe and North America has been known for a long time, we are now seeing comparable flow patterns of ice sheets of much older age, even Precambrian (Table 5-4). As noted elsewhere, some of the patterns of ancient times have been disrupted by continental breakup and the drifting apart of once contiguous land segments. There is likewise much interest in the relation between the centers of ice dispersal and pole position established by paleomagnetic studies as established from rocks of contemporary age.

Annotated References of Experimental Studies of Linear Structures

ALLEN, J. R. L., 1964: Primary current lineation in the Lower Old Red Sandstone (Devonian), Anglo-Welsh Basin. Sedimentology 3, 89—108.
Flume experiments suggest the lineation to be a stable equilibrium bed configuration only in the upper flow regime. This paper also contains notes on field occurrence, relation to grain fabric, and environmental significance.
—, 1966: Flow visualization using plaster of Paris. J. Sediment. Petrol. 36, 806—811.
The pattern of motion close to the surface of a bed of plaster of Paris beneath flowing water is made visible owing to differential solution. The flow patterns are useful in experimental studies of sedimentary structures.
—, 1969: Erosional current marks of weakly cohesive mud beds. J. Sediment. Petrol. 39, 607—623.
The sedimentary structure formed is dependent on the "severity" of the flow (primarily velocity). In order of increasing severity, the structures formed were (1) long, parallel longitudinal grooves, (2) long meandering grooves, (3) flute marks, and (4) transverse erosional markings.
—, 1971 a: Some techniques in experimental geology. J. Sediment. Petrol. 41, 695—702.
Describes briefly methods of preservation and reproduction of molds of experimentally produced sole marks, preservation of internal structures of uncemented sand beds with latex, and visualization of flow on models made of plaster of Paris.
—, 1971 b: Transverse erosional marks of mud and rock: their physical basis and geological significance. Sedimentary Geol. 5, 165—385.
The most comprehensive experimental studies ever made of flutes. Many beautiful pictures of rubber molds of flutes produced on cohesive mud surfaces plus some theory, although because the latter involves turbulence, it is not fully satisfactory. Many, many references.
DZULYNSKI, S., 1966: Influence of bottom irregularities and transported tools upon experimental scour markings. Ann. Soc. Geol. Pol. 36, 285—294.
The formation of structures resembling flutes is facilitated by the presence of tools in the generating current. Patterns of current-formed ridges and polygonal markings are shown to be related to original bed irregularities and obstacles.
DZULYNSKI, S., G. L. SHIDELER and A. SLACZKA, 1972: Impact-induced dendritic ridges in soft sediments. Ann. Soc. Geol. Pologne 42, 285—288.
Experiments showed that small dendritic ridge patterns associated with impact marks are the result of instantaneous loading and horizontal shear forces at the interface of the tool and the underlying sediment.
DZULYNSKI, S., and F. SIMPSON, 1966: Experiments on interfacial current markings. Geol. Romana 5, 197—214.
Deals with markings initially produced by suspensions flowing over soft bottom sediments. There is a gradation in morphology from tool markings which show no

trace of scouring, through structures with clear evidence of scouring and impingement of tools to flute marks. There is no discernable difference between structures formed by turbidity currents and those produced by normal suspensions. Well illustrated.

DZULYNSKI, S., and E. K. WALTON, 1963: Experimental production of sole markings. Trans. Edinburgh Geol. Soc. **19**, 279—305.
Experimentally produced sole markings including dendritic ridges, flute and scour structures, frondescent marks and various load structures. Mainly structures generated by turbidity flows.
See also annotated references on experimentally produced deformation structures in the next section.

References

ALLEN, J. R. L., 1964: Primary current lineation in the Lower Old Red Sandstone (Devonian), Anglo-Welsh Basin. Sedimentology **3**, 89—108.
—, 1965: Scour marks in snow. J. Sediment. Petrol. **35**, 331—338.
—, 1966: Flow visualization using plaster of Paris. J. Sediment. Petrol. **36**, 806—811.
—, 1968: On criteria for the continuance of flute marks, and their implications. Geol. Mijnbouw. **47**, 3—16.
—, 1969: Erosional current marks of weakly cohesive mud beds. J. Sediment. Petrol. **39**, 607—623.
—, 1971a: Some techniques in experimental geology. J. Sediment. Petrol. **41**, 695—702.
—, 1971b: Transverse erosional marks of mud and rock: their physical basis and geological significance. Sediment. Geol. **5**, 167—385.
—, 1971c: Relationship of flute-cast morphology to internal sedimentary structures in turbidites: a discussion. J. Sediment. Petrol. **41**, 1154—1156.
ANDERSON, T. B., 1965: Turbidites whose sole markings show more than one trend — a further interpretation. J. Geol. **75**, 812—814.
ANDREWS, J. T., 1973: The Wisconsin Laurentide ice sheet; dispersal centers, problems of rates of retreat, and climatic implications. Arctic Alpine Res. **5**, 185—199.
BAILEY, R. J., 1967: Paleocurrents and paleoslopes: a discussion. J. Sediment. Petrol. **37**, 1252—1255.
BAKER, V. R., 1973: Paleohydrology and sedimentology of Lake Missoula flooding in eastern Washington Geol. Soc. Am. Sp. Paper 144, 79 p.
BEUF, S., B. BIJU-DUVAL, O. DE CHARPAL, P. ROGNON, O. GARIEL and A. BENNACEF, 1971: Les grès de Paléozoique inférieur au Sahara. Paris, Éditions Technip, 464 p.
BOCCALETTI, M., and P. MICHELI, 1968: Analisi statistica dell'orientamento dei granuli in una torbidite della Marnoso-arenacea (Appennino settentrionale). Boll. Soc. Geol. Ital. **87**, 65—82.
BOUMA, A. H., 1972: Fossil contourites in Lower Neisen flysch, Switzerland. J. Sediment. Petrol. **42**, 917—921.
COLBURN, P. P., 1968: Grain fabrics in turbidite sandstone beds and their relationship to sole mark trends on the same bed. J. Sediment. Petrol. **38**, 146—158.
CROWELL, J. C., and L. FRAKES, 1971: Late Paleozoic glaciation: Part IV, Australia. Bull. Geol. Soc. Am. **82**, 2515—2540.
—, 1972: Late Paleozoic glaciation: Part V, Karroo Basin, South Africa. Bull. Geol. Soc. Am. **83**, 2887—2912.
DIONNE, J.-C., 1972: Ribbed grooves and tracks in mud tidal flats of cold regions. J. Sediment. Petrol. **42**, 848—851.
DZULYNSKI, S., and E. K. WALTON, 1965: Sedimentary features of flysch and greywackes. Develop. Sediment. **7**, Amsterdam: Elsevier Publ. Co. 274 p.
FARJALLAT, J. E. S., 1970: Diamictites Neopaleózoicos e sedimentos associados do sul de Mato Grosso. Republica Federativa do Brasil. Dept. Nac. Producão Mineral, Div. Geol. Mineral., Bol. **250**, 52 p.

FRAKES. L. A., and J. C. CROWELL, J. C., 1968: Late Paleozoic glacial geography of Antarctica. Earth and Planetary Sci. Letters 4, 253—256.

—, 1969: Late Paleozoic glaciation: I, South America. Bull. Geol. Soc. Am. 80, 1007—1042.

—, 1970: Late Paleozoic glaciation: II, Africa exclusive of the Karroo Basin. Bull. Geol. Soc. Am. 81, 2261—2286.

FRAKES, L., E. KEMP and J. C. CROWELL, 1975: Late Paleozoic glaciation: VI, Asia. Bull. Geol. Soc. Am. 86, 455—464.

FRAKES, L., J. MATTHEWS and J. C. CROWELL, 1971: Late Paleozoic glaciation: III, Antartica. Bull. Geol. Soc. Am. 82, 1581—1604.

FRIEDMAN, G. M., and J. E. SANDERS, 1974: Positive-relief bedforms on modern tidal flat that resemble molds of flutes and grooves; implications for geopetal criteria and for origin and classification of bedforms. J. Sediment. Petrol. 44, 181—189.

GLENNIE, K. W., 1963: An interpretation of turbidites whose sole markings show multiple directional trends. J. Geol. 71, 525—527.

GLÜCKERT, G., 1974: Map of glacial striation of the Scandinavian ice sheet during the last (Weichsel) glaciation in northern Europe. Bull. Geol. Soc. Finland 46, Pt. 1, 1—8.

HARRIS, I. McK., 1974: Iceberg marks on the Labrador shelf in B. R. PELLETIER, ed., Offshore geology of eastern Canada, vol. 1, Concepts and application of environmental marine geology. Geol. Surv. Canada Paper 74—30, 97—101.

HARRIS, I. McK., and P. G. JOLLYMORE, 1974: Iceberg furrow marks on the continental shelf northeast of Belle Isle, Newfoundland. Canadian J. Earth Sci. 11, 43—52.

HUBERT, J. F., 1966: Sedimentary history of Upper Ordovician geosynclinal rocks, Girvan, Scotland. J. Sediment. Petrol. 36, 677—699.

—, 1968: Currents and slopes in flysch basins: A discussion. J. Sediment. Petrol. 38, 1390.

JIPA, D., 1968: Azimuthal relationship between cross-stratification and current markings in flysch deposits: Upper Precambrian of central Dobrogea, Romania. J. Sediment. Petrol. 38, 192—199.

KARCZ, I., 1967: Harrow marks, current-aligned sedimentary structures. J. Geol. 75, 113—120.

—, 1968: Fluviatile obstacle marks from the wadis of the Negev (southern Israel). J. Sediment. Petrol. 38, 1000—1012.

KELLING, G., and J. H. McD. WHITAKER, 1970: Tool marks made by orthoconic nautiloids. J. Geol. 78, 371—374.

KUENEN, PH. H., 1966a: Light thrown on general problems by the Roumanian results. Sedimentology 7, 323—326.

— 1966b: Geosynclinal sedimentation. Geol. Rundschau 56, 1—19.

— 1967: Emplacement of flysch-type sand beds. Sedimentology 9, 203—243.

LINDSEY, D. A., 1969: Glacial sedimentology of the Precambrian Gowganda Formation, Ontario, Canada. Bull. Geol. Soc. Am. 80, 1685—1702.

MARTIN, H., 1964: The directions of flow of the Itararé ice sheets in the Paraná Basin, Brasil. Bol. Paranaense Geografia 10—15, 25—77.

MATTHEWS, P. E., 1970: Paleorelief and the Dwyka glaciation in the eastern region of South Africa in 2nd Gondwana symposium, Int. Union Geol. Sci., Comm. on Stratigraphy, Sub-Comm. on Gondwana stratigraphy and paleontology, 491—494.

McBRIDE, E. F., and L. YEAKEL, 1963: Relationship between parting lineation and rock fabric. J. Sediment. Petrol. 33, 779—782.

McBRIDE, E. F., and J. E. KIMBERLY, 1963: Sedimentology of Smithwick Shale (Pennsylvanian), eastern Llano region, Texas. Bull. Am. Assoc. Petrol. Geologists 47, 1840—1854.

McIVER, N. L., 1970: Appalachian turbidites in G. W. FISHER, F. J. PETTIJOHN, J. C. REED JR. and K. N. WEAVER, eds., Studies of Appalachian geology: central and southern. New York: John Wiley and Sons, 69—81.

NEGENDANK, JÖRG, F. W., 1972: Turbidite aus dem Unterrotliegenden des Saar-Nahe-Gebietes. Neues Jahrb. Geol. Paläontol. Mh. Jg 1972, 561—575.

PETT, J. W., and R. G. WALKER, 1971a: Relationship of flute cast morphology to internal sedimentary structures in turbidites. J. Sediment. Petrol. 41, 114—128.

—, 1971b: Relationship of flute cast morphology to internal sedimentary structures — reply. J. Sediment. Petrol. 41, 1156—1158.

PETTIJOHN, F. J., and P. E. POTTER, 1964: Atlas and glossary of primary sedimentary structures. Heidelberg-Berlin-New York: Springer, 370 p.

PREST, V. K., 1969: Retreat of Wisconsin and Recent ice in North America. Geol. Sur. Canada, Map 1257A, Scale 1 : 5,000,000.

RICCI LUCCHI, F., 1969: Considerazioni sulla formazione di alcune impronte da corrente. Giorn. Geol. Ser. 2, 36, 363—415.

RICHTER, D., 1965: Sedimentstrukturen, Ablagerungsart, und Transportrichtung im Flysch der baskischen Pyrenäen. Geol. Mitt. 4, 153—210.

SCOTT, K. M., 1966: Sedimentology and dispersal patterns of a Cretaceous flysch sequence, Patagonian Andes, southern Chile. Bull. Am. Assoc. Petrol. Geologists 50, 72—107.

SEILACHER, A., 1963: Umlagerung und Rolltransport von Cephalopoden-Gehäusen. Neues Jahrb. Geol. Paläontol. Mh. 11, 593—615.

SEN, D. P., 1967: Turbidite structures from Barakar rocks of Ramgarh Coalfield, Bihar, India. J. Sediment. Petrol. 67, 699—702.

SESTINI, G., 1970: Flysch facies and turbidite sedimentology. Sediment. Geol. 4, 559—597.

SESTINI, G., and M. CURCIO, 1965: Aspetti quantitativi delle impronte di fondo da corrente nelle torbidite dell'Appennino tosco-emiliano. Bull. Soc. Geol. Italiana 84, 1—126.

SESTINI, G., and G. PRANZINI, 1965: Correlation of sedimentary fabric and sole marks as current indicators in turbidites. J. Sediment. Petrol. 35, 100—108.

SPENCER, A. M., 1971: Late Precambrian glaciation in Scotland. Geol. Soc. London, Mem. 6, 98 p.

STANLEY, D. J., 1965: An interpretation of turbidites whose sole markings show multiple directional trends. J. Geol. 73, 670—671.

STRATTEN, T., 1969: A preliminary report of a directional study of the Dwyka Tillites in the Karoo Basin of South Africa in Gondwana stratigraphy 2, Paris: UNESCO, 741—762.

TUCKER, M. E., 1969: Crinoidal turbidites from the Devonian of Cornwall and their paleogeographic significance. Sedimentology 13, 281—290.

WALKER, R. G., 1970: Review of the geometry and facies organization of turbidites and turbidite-bearing basins in J. LAJOIE, ed., Flysch sedimentation in North America. Geol. Assoc. Canada Spec. Paper 7, 219—251.

WHITAKER, J. H. McD., 1973: "Gutter casts", a new name for scour-and-fill structures: with examples from the Llandoverian of Ringerike and Malmöya, southern Norway. J. Geol. Soc. London 129, 91 p.

Chapter 6/I

Deformational Structures up to 1963

Introduction

This chapter deals with disturbed, distorted or deformed bedding and with distortions of various primary sedimentary structures. Such deformation varies from mildly disturbed layers to those which are intricately crumpled and in extreme cases to those thoroughly broken and brecciated. In special cases the sediment is rendered incoherent and mobile and behaves as a semifluid and forms slurry slumps or is injected as sedimentary sills and dikes.

Our concern here is for deformation that took place at the time of or very shortly after deposition of the sediment. This penecontemporaneous deformation is to be distinguished from that which accompanies and is an expression of tectonic movement or from that imposed on the strata in a very late post-consolidation period. Excluded, therefore, is deformation resulting from chemical alteration and associated volume changes or from frost action or landslides and other earth movements in the present cycle.

Penecontemporaneous deformation is commonly ascribed to slumping or sliding down slope under the action of gravity at the time of or shortly following accumulation of the deposit. But deformation is also due to differential vertical movements unrelated to slope and without any lateral displacement (load-casting) and also to change of structure (transformation from loose packing to close packing) often accompanied by a momentary quasifluid condition. Perhaps in a few cases, it is due to forces arising from frictional drag of the current on the already-deposited sediment. It is not always easy to decide which of these several origins is to be assigned to a particular structure.

The criteria for discriminating between penecontemporaneous deformation and that due to tectonic action have been formulated by NEVIN (1942, p. 186), LEITH (1923, p. 228), RETTGER (1935) and others. Both types of deformation produce folds, faults, and breccias. In general "soft-sediment deformation" is characterized by its local character — its confinement to a single bed between undeformed beds. The folding is commonly chaotic and unlike drag folding it is unrelated to the larger structures of the area. Commonly also the deformed strata are truncated by a penecontemporaneous erosion surface. The faults are not sharp and neither in connection with faulting nor with brecciation are openings produced or maintained. Hence, unlike tectonic, post-consolidation fracturing, there are no vein fillings nor is there any associated rock cleavage or any other evidence of recrystallization and mineral orientation.

Disturbed and deformed bedding has been seen and described many times. The recognition of such disturbance as a product of penecontemporaneous processes probably was made rather early, perhaps first in glacial deposits, where the structures certainly could not be attributed to tectonic action. Folding in unconsolidated Quaternary clays in New York State was described by VANUXEM in 1842 (p. 214). GEIKE, in the first edition (1882, p. 480) of his well-known textbook, remarks on the disturbed bedding in some glacial clays which he attributed to the "... stranding of heavy masses of drift-ice upon still unconsolidated sand and mud," a conclusion reached as early as 1840 by LYELL (SORBY, 1859, p. 220). Recognition of the widespread nature and the importance of soft-sediment deformation and the role played by slumping, rather than stranded icebergs, grew out of the discovery and careful description of the distorted strata in the early Paleozoic rocks of Wales. As noted by JONES in discussion of a paper by KUENEN (1949, p. 380), prior to these studies, begun in the early part of this century, nobody "... had any conception how contorted and crumpled masses... intercalated among undisturbed sediments, were formed, although the process of slumping and some of its results, on a small scale had already been described." The great interest in turbidites in particular, has led to renewed study of primary sedimentary structures including both those modified by and produced by intraformational deformation.

The recognition of processes of deformation, other than subaqueous slump, and the criteria for differentiating between the several processes and the structures produced by them is quite recent. A few structures and processes were identified a very long time ago. Sandstone dikes, for example, were recognized in the first half of the 19th century; those of Patagonia were described by CHARLES DARWIN (1851, p. 150) and similar dikes in Oregon were reported by DANA (1849, p. 654). On the other hand, many structures once thought to be slump phenomena are now believed to form by other processes. Convolute bedding and pillow-form sandstones are examples.

The importance of soft-sediment deformation lies in (1) its modification, by load-casting and slumping, of current structures, (2) its relation, in the case of true slumps, to the direction of slope of the principal surface of accumulation, and (3) its bearing on the state of the material at the time of deposition.

Deformation by load-casting can alter the appearance of many current structures, particularly the substratal lineations; deformation can also distort cross-bedding, both large and small scale; and can greatly modify or obliterate sedimentary fabrics.

Except for load-casted current structures, most deformational features lack directional significance. Even the so-called slump structures are notoriously difficult to measure and interpret in terms of the paleoslope. Many penecontemporaneous folds are very disorderly and the fold axes are seemingly without any preferred orientation; even the directions of overturning may be contradictory.

The classification of structures due to intraformational disturbance is troublesome. These structures are difficult to define and describe independently of their origin. And the origin of many is even yet unknown or poorly understood. Yet a classification independent of origin is virtually meaningless. For our purposes we will group the structures according to the relative importance of vertical and lateral displacements and according also to the degree of disorganization of the original bed (see Table 5-1). Hence the structures may be arranged in a sequence:

1) Undeformed bed.

2) Load-casted bed, lower surface only (vertical movement).

3) Pillows and balls, due to partial to complete break-up of bed (vertical movement).

4) Asymmetric load-casts (combined vertical and lateral movement).

5) Convolute lamination, folding confined to a single layer (combined vertical and lateral movement?).

6) Slump folds involving several layers (lateral movement).

7) Slump breccias bedding disrupted (lateral movement).

8) Injected material such as dikes, sills, mud lumps and others.

Load Casts

The term "load cast" was proposed by KUENEN (1953a, p. 1048) for structures which had earlier been designated "flow cast" (SHROCK 1948, p. 156). Flow cast is too readily confused with "flow mark." The term "cast," moreover, implies the filling of a mold. Inasmuch as load casts are not the result of filling of a once empty mold, the term is not strictly correct. For these reasons SULLWOLD (1959, 1960) suggested the term "load pocket" to which some objections have been made (HOLLAND, 1960). The problems of terminology have been discussed further by others (KUENEN and PRENTICE, 1957; PRENTICE 1956; KUENEN 1957, p. 246). The terms load cast for the structure and load-casting for the process have become so generally used and so strongly fixed in the literature that we shall adopt them here despite their shortcomings. They have been referred to as "Belastungsmarken" in the recent German literature (PLESSMANN, 1961, p. 523).

Load casts occur on the base of sandstone beds that overlie shales or other sediments that were in a plastic condition at the time of deposition. They appear as swellings varying from slight bulges, deep or shallow rounded sacks, knobby excrescenses, or highly irregular proturberances (pls. 24A and 25). Load casts show an almost endless variety of form. They can be distinguished from flute casts by their much greater irregularity of form and absence of distinct up- and down-current ends. They generally show no alignment, although there are exceptions (fig. 6-1). However, flute casts can be transformed into load casts which may show some features inherited from the flute-cast form including orientation (pl. 26). This transformation is irreversible.

Load casts on a given bedding plane tend to be of the same general size and character. In some cases they are much flattened; in others they exhibit a striking mammillary or papilliform appearance. Some are highly irregular and of very large size. In a few cases they are notably asymmetrical.

Load casts vary from a centimeter or two to several decimeters in diameter and from a centimeter or less to several decimeters in relief.

The filling of these pockets is sand, commonly a little coarser than that found in the overlying layer itself (pl. 27A). The underlying shales are distorted and bent downward beneath the load bulge — not truncated by scour, as they are adjacent to flutes.

SHROCK (1948, p. 156) believed that "... soft hydroplastic sediments, if unequally loaded with sand or gravel, yield to the weight of the superincumbent load by flowing." This explanation has been accepted by all subsequent workers. Several questions remain. Why are some sand-shale interfaces so markedly load-casted whereas others in the same sequence are without the slightest hint of load casts? Perhaps the water content of the underlying mud was a factor as was also the rapidity with which the overlying sand was deposited. What is

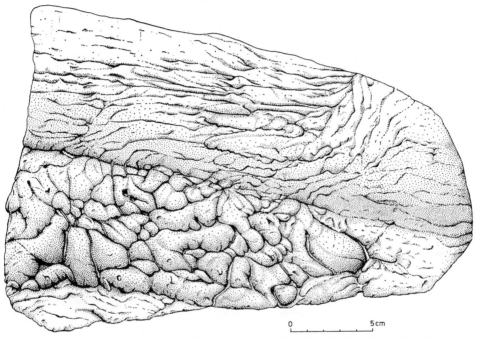

Fig. 6-1. Small load casts arranged in zones parallel to current on base of sandstone bed from Picene flysch (Miocene), Cancelli, Apennines, Italy (TEN HAAF, 1959, fig. 30)

the relation between load casts and the ball-and-pillow structure of some sandstones? Are these structures just very large load casts? Or is it that load casts grow slowly during the sedimentation process and that the pillow structure is produced by a later catastrophic foundering of the sand deposit? How do the highly asymmetrical load pockets (pl. 27A) form? To some, the direction of overturning of the tongues of shale involved suggest gravitational creep down slope or drag by the current. Others have thought the asymmetry to be inherited from the asymmetrical form of the flutes from which they were presumably derived (TEN HAAF, 1959, p. 45).

Closely related to load casts are the curved, pointed tongues of shale which project up into the overlying sandstone (pl. 24B; fig. 6-2). These structures, perhaps first described by SORBY (1908, p. 196, pl. 14), were designated "antidunes" by LAMONT (1938), "flame structures" by WALTON (1956), and "load waves" by SULLWOLD (1959, 1960). Flame structures tend to point in one direction. KELLING and WALTON (1957, p. 487) attribute this to the upward migration of ripple-mark crests into the overlying sand. This interpretation was challenged (HILLS, 1958; PRENTICE, 1958). Ripples do not commonly occur in shales. KUENEN und MENARD (1952, p. 90) ascribed the structure to two factors, namely, "(1) drag exerted by the turbidity current on the watery clay film of its bed, (2) local settling and squeezing caused by the rapid accumulation

Fig. 6-2. Flame structures on base of siltstone beds. Note that overturning of mud plumes indicates current flow from right to left as does the small-scale cross-bedding. Wilhite formation, Walden Creek group, Ocoee series (Precambrian), State Highway 73, Kinzel Springs, Blount County, Tennessee, U.S.A. (drawn by R.W. B. DAVIS from U.S.Geol. Survey photograph)

of overburden on the highly mobile foundation." Flame structure is best seen in cross sections of rocks in which the sand and shale layers are welded together, as they commonly are in Precambrian rocks.

Load casts are indicative of no particular environment. The only requirement for their formation is deposition of a bed of sand on a water-saturated hydroplastic layer. Load casts occur most commonly in turbidite sequences but they are also abundant in other facies.

A structure, strongly resembling load casts, is produced by cryoturbation, a process which operates in permafrost regions. The structures are common in

Pleistocene glacial deposits. The term "involution" has been applied to the local deformation and penetration produced in stratified materials by frost action (SHARP, 1942, p. 115). Involutions consists of intensely deformed, involved, and haphazardly penetrating masses of silt and sand originally deposited in horizontal beds. There is a lack of linear trend or continuity to the structure. Involutions are largely a product of vertical movement and have been attributed to "... local vertical movements caused by differential freezing and the formation of ground-ice" (SCHAFER, 1949, p. 164). To the extent that the movement is largely vertical and that diapiric relations are common, involutions are akin to large-scale load casts and to the ball-and-pillow structure of some sandstones.

Ball-and-Pillow Structure

The structure designated "sandstone balls" or "pillows" is that commonly displayed by a sand bed overlying a silt or shale. The same bed, or at least the basal portion of it, is broken up into numerous, in many cases closely packed, ball- or pillow-form masses of sandstone. SMITH (1916) suggested the designation "ball- or pillow-form structure." It is graphic, descriptive and noncommital as to origin. Hence we have modified it slightly and adopted it. The appearance has led to the designation "hassock structure," "pseudo-nodules" (MACAR, 1948) and the like. Some writers have called ball-and-pillow structures "flow rolls" (PEPPER et al., 1954, p. 88), "flow structures" (COOPER, 1943), "intrastratal flowage phenomena" (RICH, 1951, p. 13, pl. 2) or "slump balls" (DESTOMBES and JEANNETTE, 1955, pl. 1) based on presumed mechanics of formation. The term "flow rolls" was also used by RICH (1950, p. 727) for what are now called "flute casts." Others have designated the ball-and-pillow structures "storm rollers" (CHADWICK, 1931) or used other equally uninformative terms. The structure is in many ways the analogue of pillow lava, — just as a lava flow may be broken up into closely packed pillow-form or ellipsoidal bodies so also a sandstone, or in some cases a clastic limestone, is broken up into a larger number of pillow-shaped masses (see pl. 29B). The character and meaning of this "muddled" structure is the subject of this section.

Characteristically the structure is confined to one bed — being present neither in the overlying or underlying beds. Several such disturbed beds, however, may be present in the same outcrop. The bed is broken into a series of hassock- or pillow-shaped masses. They are rarely spherical, more generally they are hemispherical or kidney-shaped, ranging from a few inches (0.1—0.2 m) to very large bodies several feet (1 m) in size. The bed involved will thus vary from less than a foot (0.3 m) to 8 or 10 feet (3 m) in thickness. The pillow structure is commonly found only in the lower part of the bed and grades upward into ordinary undisturbed sandstone (pl. 28B; fig. 6-3). The bed, therefore, appears to have a sharp but highly undulatory base and a flat top. In extreme cases the pillows become isolated in a matrix of shale and fail to form a continuous bed.

The underlying shale is very much deformed and appears to be wrapped around the pillows, to be squeezed in between them and to extend as thin tongues up into the sandstone bed.

The pillows themselves vary from seemingly structureless to laminated. The laminations are deformed and conform to the boundaries of the pillow structure (pls. 27 B and 28 A). In some cases they are crumpled in the central area but

Fig. 6-3. Ball-and-pillow structure in sandstone of Caseyville formation (Pennsylvanian), Interstate 57, SE 1/4 NE 1/4, sec. 27, T. 11 S., R. 1 E., Union County, Illinois, U.S.A. (Illinois Geological Survey)

generally the pillow structure is cup-shaped in cross-section and the laminations therefore show a synclinal concave-upward structure (fig. 6-4). The margins of the basin structure are curved inward in a manner reminiscent of inverted mushrooms (DESTOMBES and JEANNETTE, 1955, p. 78). This vertically-directed asym-

metry has led some workers to conclude that the cup-shaped or basin-like structures are the synclinal depressions left after truncation of the anticlinal portion
of a folded bed. As noted later, this is an unlikely explanation, especially as it
fails to account for the inward-curved margins of these basins. The basins were
formed as we now see them; there were no anticlinal counterparts. Some are
tilted at an angle to the principal bedding.

Pillow structure is most common in certain sandstones but it is equally well
displayed by some limestones. The limestones which show it are themselves
finely-laminated carbonate sands or silt resting on shale. The structures they
exhibit match in every detail those found in non-carbonate sands.

The pillow structure is most characteristic of rather fine sands where such
materials are interbedded with shales. Best known occurrences in the United
States are those in the Berea (Mississippian) sandstone of Ohio (COOPER, 1943;
PEPPER *et al.*, 1954, p. 88), the Chemung sandstone (Upper Devonian) in the

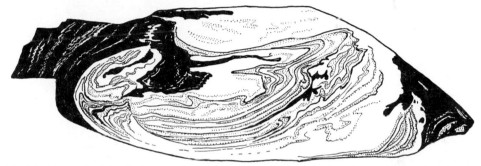

Fig. 6-4. Sketch showing internal structure of "pseudo-nodule", Devonian of Luxemburg (MACAR and ANTUN, 1950, fig. 10)

central Appalachian area (McIVER, 1961, p. 219) and the limestones of the Cynthiana formation (Ordovician) of southern Ohio (CONATSER, 1958). Well-known
European examples occur in the Psammites du Condroz of Devonian age (MACAR,
1948; MACAR and ANTUN, 1949; VAN STRAATEN, 1954). The structure has also
been found in recent deposits (EMERY, 1950, p. 111; MACAR, 1951; KAYE and
POWER, 1954). The structure seems not to be characteristic of a particular environment but rather of a particular material and condition at the time of deposition.

The ball-and-pillow structure of sandstones has been explained in various
ways. It has been mistaken for concretions and has also been considered to be
a concentric or spheroidal weathering phenomenon. The true pillow-form
structure is clearly neither concretionary nor a product of weathering.

A great many authors have attributed it to slumping or subaqueous sliding.
The disrupted bedding, the intermingled shale and sandstone, the folded laminations within some pillows all clearly signify movement and deformation. Deformation can occur by downslope sliding, and hence the conclusion that the
structure is due to subaqueous slide or slump, is not unreasonable. The absence
of anticlinal structures and the preservation of synclines could be explained by
erosional planation of a folded sequence.

However, careful study of the structure suggests another process, namely
foundering of a sand bed. It can hardly be supposed that erosion always truncates

the folds; the absence of anticlines, the basinal structure and recurved margins
of the sandstone segments and the passage upward into undisturbed sandstone
require an explanation.   Moreover, it seems unlikely that an unconsolidated
sand could be broken into fragments.   Such a sand would yield by disintegration
and flowage during slumping.   On the other hand, if the sand were consolidated
it could not be expected to fold.   Hence the pillow structure requires some special
conditions — not just subaqueous slump.

KAYE and POWER (1954) describe a "flow cast" structure in modern unconsolidated deposits.   This structure appeared as a series of "... sharply convoluted waves of clayey silt..." between which were "pockets of medium-
grained sand whose stratification, like that of the silt, was disturbed to conform

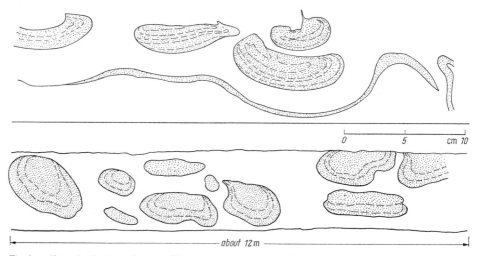

Fig. 6-5. *Above*, sketch of experiment by KUENEN showing production of pseudonodules by foundering of sand bed into
underlying clay rendered semi-fluid by shock (drawn from KUENEN, 1958, pl. 1).   *Below*, sketch of pseudonodules from
Devonian of Luxemburg (after MACAR and ANTUN, 1949, fig. 8).   Note similarity of structure and orientation to structures
produced by KUENEN despite difference in scale

to the outline of the convolutions."   The sand-silt complex was sharply terminated
above by a horizontal bedding plane.   The sand pockets appear to be about
4 to 8 inches (10 to 20 cm) in diameter.   There is little doubt that the structure
described and figured by KAYE and POWER is a recent example of the ball-and-
pillow structure.

They state that the "... sand layer foundered into the mud, breaking up into
isolated units or cells in so doing, and squeezing of the mud, which is now displaced, as complicated intrusions along breaks and weak places in the foundering
sand layer."   Moreover KUENEN (1958, p. 17) was able to reproduce the structure
experimentally.   A layer of sand was deposited over clay.   A shock applied to the
deposit caused the sand layer to founder, to break up into the saucer and kidney-
shaped segments which sank into the clay rendered semifluid by the thixotropic
transformation.   The result was strikingly similar to the natural phenomenon
(fig. 6-5).   TEN HAAF had come to a similar conclusion (KUENEN, 1958, p. 18)
as had VAN STRAATEN (1954, p. 34).   MACAR (1948) had likewise emphasized the
vertical descent of the sand into the mud as an essential part of the process.

MACAR, however, supposed some lateral displacement also accompanied the ball formation, although in a later study (1951) of the structure in recent sediments of the Zuider Zee area, lateral displacement was proved improbable. Similar structures found in Pleistocene sediments of southern California have been described by EMERY (1950), who likewise attributed the structure to down-sinking of the sand into a yielding substratum. Moreover, the occurrence of the ball-and-pillow structure in Ordovician limestones in southwestern Ohio (CONATSER, 1958) in beds which even now dip but a few feet per mile makes the postulate of down-slope movement or slide most unlikely. The downward sinking and encasement of the sand, even though unconsolidated, into the mud explains why disintegration did not occur.

Unlike ordinary load casts, the structure did not form during the accumulation of the sand layer. The internal structure of some pillows, the absence of coarser material in these structures — unlike load casts — all indicate a later and sudden formation. Abrupt thixotropic transformation of the clay substrate seems most probable cause. Hence KUENEN suggested (1958, p. 20) that an earthquake shock might be the triggering mechanism and that the resultant ball-and-pillow bed be called a "quake sheet."

Clearly, if the pillow structures originated in the manner outlined and were not formed by down-slope movement of sediment, they have no directional significance. CONATSER (1958) measured and plotted the long axes of the pillow structures in the Ordovician limestones of Ohio. No preferred orientation was found. Some preferential arrangement, however, seems to be present in other examples (MACAR and ANTUN, 1949, p. 141; pl. 29 B). Whether such orientation is controlled by a pre-existing structure in the sand or not is uncertain. Its value in directional studies is, therefore, unknown, although MACAR and ANTUN believed the orientation was due to slump movement.

Convolute Lamination

Convolute lamination, first designated "convolute bedding" by KUENEN (1953 b), is a structure characterized by marked crumpling or intricate folding of the laminations within a well-defined, undeformed sedimentation unit (fig. 6-6). It has been variously termed "curled bedding" (FEARNSIDES, 1910, p. 150), "crinkled bedding" (MIGLIORINI, 1950, cited by KUENEN, 1953 b, p. 15), "intrastratal contortions" (RICH, 1950, p. 729, fig. 9), "slip bedding" (KSIAZKIEWICZ, 1951), "convolute lamination" (TEN HAAF, 1956, p. 190), and "Wulstbänke" (PLESSMANN, 1961, p. 528). Convolute lamination is a structure that is difficult to define (see HOLLAND, 1959, 1960; SULLWOLD, 1959, 1960; DOTT and HOWARD, 1962, p. 114).

Convolute lamination is readily confused with the crumpling formed by gravitative slumping. Unlike convolute lamination, true slump structures do not have continuous corrugated laminations as far as the bed can be traced. Most slumps result in disruptive, piling up of rolled or crinkled masses or in a chaotic mixture of fragments and matrix. Slump structures are commonly faulted. They generally involve more than one sedimentation unit. Moreover the thickness of the disturbed zone is uneven; in the down-slope direction the beds may be superimposed, whereas in the up-slope direction they may be thinned or even absent. Clearly slump structure and convolute laminations are two distinct phenomena.

In the field, convolute lamination is best seen on weathered surfaces, where slight contrasts in color and relief render it visible; it might otherwise be easily overlooked. It is found most commonly in the finest grained sand or coarsest siltstone beds.

The convolutions do not involve the external bedding planes. The bed itself remains remarkably constant in thickness and extends unchanged as far as exposed in a given outcrop. The convolutions may have their greatest amplitude near the center of the bed and die out both upward and downward or they may be confined to the upper part of the bed. They commonly pass downward into ripple cross-laminations with which they may be associated. The thickness of the bed and the magnitude of the convolutions both increase with increase in the coarseness of the sand. Most beds containing convoluted laminations are one to ten inches (2.5 to 25.0 cm) thick.

The convolutions are remarkably continuous. Faulting does not occur even in the most complicated structures, although some laminations are cut out by small internal unconformities which are themselves involved in convolutions. The convolutions are characterized by sharp-crested anticlines and wider rounded to box-shaped synclines. The downfolds are basin-shaped and the anticlines are domical (pl. 30). Involutions of recumbent folds result in closed, more or less ellipsoidal, concentrically-laminated "convolutional balls" which may be mistaken for concretionary bodies.

The directional significance of the orientation, if any, of the axes of the convolutions is uncertain. In some cases, they show a definite trend perpendicular to the depositing current and display down-current overturning (KUENEN, 1953b; KUENEN and SANDERS, 1956, p.661). KÜHN-VELTEN (1955, p.18)

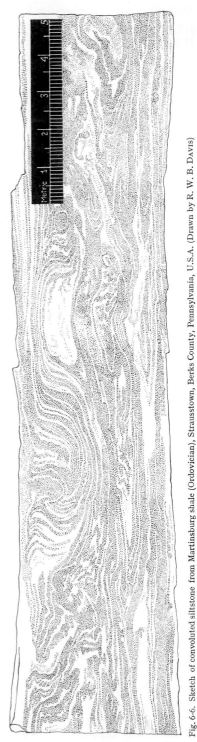

Fig. 6-6. Sketch of convoluted siltstone from Martinsburg shale (Ordovician), Strausstown, Berks County, Pennsylvania, U.S.A. (Drawn by R. W. B. DAVIS)

measured, on bedding surfaces, the normals to the fold axes of a structure, termed "Gleitfältelung" that appears to be convolute bedding. The measurements, when plotted on a map, disclosed a consistent pattern thought to be related to regional anticlinal fold axes and to indicate movement down the flanks of rising structures. More commonly, however, convolutions show only a disarray of equidimensional peaks and basins and have no close correlation either with current direction (TEN HAAF, 1956, p.162) or with submarine slopes. Only in exceptional cases can the trend of convolute structures be used to establish current direction.

Convolute lamination has been believed to occur only in turbidite sequences (SANDERS, 1960, p.419). DOTT and HOWARD (1962, p.119), however, believe it also occurs in non-turbidites and is not, therefore, an infallible criterion of turbidity current action. It is said not to occur in modern sediments (SANDERS, 1960, p.419). It is however, found in some Pleistocene glaciolacustrine varved silts. Convoluted beds are most common in flysch sequences from the Precambrian onwards. Like other sedimentary structures, it can form in any situation where the right materials and conditions prevail but may be common only in certain environments.

Convolute bedding is not only one of the most difficult structures to define but it is also the most difficult to explain. It has been attributed to slumping, although RICH (1950, p.730) suggested that the sliding took place *after* the superjacent beds had been deposited. But, as noted by TEN HAAF (1956, p.192), some convolutions are beheaded by intrastratal surfaces of erosion showing that the structures were formed before the deposition of the cover bed. The disorganization of the folds, the absence of regularity as might be expected if the structures were due to down-slope movement, and the absence of faulting further seem to make sliding improbable. Seemingly localized slight differential forces acting on a very weak hydroplastic deposit during its process of accumulation are needed to explain the structure.

MIGLIORINI (1950, cited by KUENEN, 1953, p.16) thought that compaction of the lower part of the bed would lead to expulsion of the contained pore water; the laminations of higher levels would be dragged upward into convolutions by concentration of the upwelling at definite points. It is difficult to prove or disprove this hypothesis.

KUENEN (1953b, p.18) believed convolute lamination is produced by plastic deformation of current ripples. According to him, water flowing over a rippled surface exerts a vertical suction over the crests and a downward pressure on the troughs. Hence weakly hydroplastic materials would yield and produce the peaked anticlines and broad rounded synclines seen. Moreover, as noted by KUENEN (1953b, fig.8), some anticlines show erosional truncation whereas the adjacent synclines contain sets of inclined laminations resembling small-scale foreset beds. In some cases both the ripple lamination and the foreset filling of the synclines are overturned. TEN HAAF (1956, p.193) accepts KUENEN's hypothesis, although he observed a lack of correlation between the azimuths of the corrugations and the known current direction in the turbidites of the Apennines investigated by him.

HOLLAND (1959, p.233) attributed convolute laminations to an "... irregular distribution of pressure or suction, drawing up the anticlinal convolute folds and depressing the laminations between them." The necessary conditions are attributed to turbidity currents, in a manner not fully explained, and to "... slight readjustments... of the structures by bulk movement of the sediment..." down-slope.

WILLIAMS (1960, p. 208) postulated the formation of convolute laminations by "lateral intrastratal flow of liquified beds, according to flow patterns mainly determined by the distribution of parts of the layers that have remained solid..." The structure, therefore, would be formed only *after* the deposition of the super-jacent layers.

Current drag during deposition is believed to be responsible for the commonly observed deformation of ripple cross-lamination (pl. 4A). The deformed "ripple-drift bedding" of PRENTICE (1960), however, may not be the same as the con-voluted bedding so closely associated with some rippled sands or silts. Nevertheless SANDERS (1960, p. 409) believes there is a common origin for both. According to SANDERS, "convolutions arise when formerly cohesionless sand grains become cohesive after deposition and respond to increased shearing due to higher current velocity by a décollement type of adjustment..."

Thus, although convolute laminations have been attributed to current action and also to slumping, the disorganized character of the structure and general lack of correlation between established current and/or slope direction suggests that it is formed by neither. The broad box-shaped downfolds and the sharp-peaked anticlines are suggestive only of load deformation, concommitant with sedimen-tation, perhaps localized by rippling.

Slump Structures

The term "slump structure" has been applied rather uncritically to many structures, some quite unlike one another, and to many that are almost certainly not due to slumping. "Slump" is the term generally applied to gravity-generated movement of small lateral displacements; "slides" are similar movement of greater lateral extent. The ball-and-pillow structure of some sandstones and lime-stones and convolute laminations are examples of structures inappropriately attributed to slumping. In many published reports, any disturbed bedding has been designated a "slump structure."

True slump structures were early recognized in Pleistocene glacial sediments, especially in the lacustrine varved silts and clays which seem especially prone to sliding and which characteristically yield décollement folds. Many such struc-tures have been explained by the drag of grounded icebergs or even by a readvance of the glacier itself which indeed might produce disturbed zones in places. Although many disturbed glacial beds are of slump origin, some may be produced by cryoturbation, a process active in the permafrost zone. See, for example, papers by SHARP (1942), TROLL (1944), SCHAFER (1949) and others.

Occurrence

Slumps are widespread in the geologic record. They are ubiquitous in glacio-lacustrine deposits (FAIRBRIDGE, 1947; VAN STRAATEN, 1949); are present in a small measure in every cross-bedded sand deposit (see KIERSCH, 1950, p. 939;

ROBSON, 1956, and JONES, 1962) and are especially common in the flysch basins. The most famous examples of the latter are found in the lower Paleozoic deposits of the Caledonian geosyncline in Wales.

This occurrence has been described by JONES (1937, 1939, 1953). In this example, the sediments became folded or even corrugated; in places they were rolled over and became balled-up or otherwise greatly disturbed. In many places they disintegrated into a slurry and all original structure was lost. The slump folds and strikes of the slump beds bear no relation to those structures which are known to be tectonic (fig. 6-9). The uppermost contact of the disturbed bed is sharp; the overlying undeformed strata lie directly on the deformed beds without any signs of thrusting between; irregularities in the upper surface of the slump sheet have been filled in by the overlying bed which is graded. The slump interpretation of the disturbed beds in Wales is relatively recent and has not yet been accepted by all students of these structures (BOSWELL, 1949a, p. 111; 1953).

Other classic examples of structures attributed to subaqueous slumping include those in the Ordovician at Trenton Falls, New York (HAHN, 1913), near Girvan, Scotland (HENDERSON, 1935), and near Dublin, Ireland (LAMONT, 1938), in the Devonian of Gaspé, Canada (LOGAN, 1863, p. 391), in the Jurassic of Sweden (HADDING, 1931), in the Solnhofen limestone of Germany (WALTHER, 1904), in the Tertiary of the Caucasus (KUGLER, 1939), in the Miocene of Öhningen, Germany (HEIM, 1908) and in Tertiary sediments near Florence, Italy (BEETS, 1946). Beets has given a rather lengthy bibliography on these and still other occurrences of reported slump structures as has LIPPERT (1937). VOIGT (1962) describes in detail a slump and associated early diagenetic deformation in calcareous Turonian sediments of Cretaceous age in northwest Germany. Well documented also are the structures found in the Polish Carpathians (KSIAZKIEWICZ, 1958).

Character of Slump Structures

The effects of slumping are many and varied. The effects are the products of either (1) pervasive movement involving the interior of the transported mass, producing a chaotic mixture of incompetent matrix, generally shale, with diverse fragments of more resistant materials, such as sandstone, or (2) a décollement type of movement in which the lateral displacement is concentrated along a sole. In the latter case, the superincumbent beds are tightly folded and piled up into nappe-like structures accompanied by attenuation or even a hiatus in the up-slope detachment area. Many sedimentation units may be involved in either case.

Where the movement is distributed throughout the sliding mass, the thin, more competent beds, are broken into irregular, small to large, slabby fragments of variable shapes. Some are folded over into hook-like structures (fig. 6-8), termed "slump overfolds" by CROWELL (1957, p. 998). These slump overfolds and the spiral slump balls or "snowball structure" (HADDING, 1931, p. 386) may give a clue to the direction of the slide movement (KSIAZKIEWICZ, 1958, p. 143—144). The result is a chaotic mixture, which with a higher water content, might acquire greater mobility and become in fact a mud flow and lead to the formation of a "pebbly mudstone" (CROWELL, 1957). If the sand beds are very thin and alto-

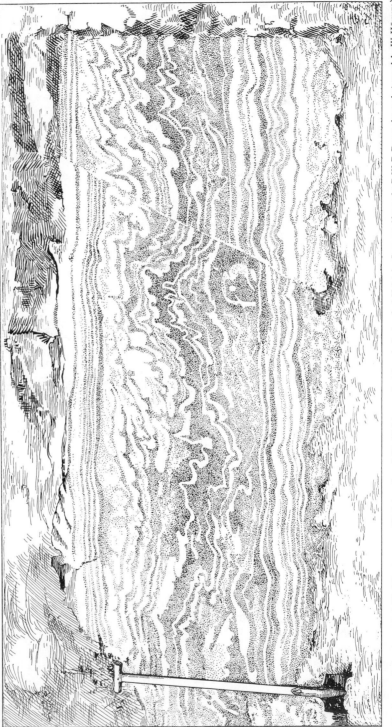

Fig. 6-7. Sketch showing contorted bedding in unconsolidated sands and silts, Pleistocene glacial sands, Lechendorf, Holstein, Germany (drawn by R. W. B. DAVIS, from photograph by W. HÜHNEL)

gether lacking cement, they and the interbedded mudstone may develop a "flow-age structure" (HADDING, 1931, p. 386) resembling that seen in some migmatites.

Slump deposits may be both very thick and widespread. KSIAZKIEWICZ (1958, p.135) describes accumulations 55 meters in thickness. Those described by CROWELL are of similar magnitude. On the other hand, the entire slumped body may be less than a meter in thickness. Some slump sheets are thick and extensive enough to be mapped (JONES, 1937). KUENEN (1949, p.373) traced a slump sheet between 5 and 15 feet (1.7 and 5 m) thick over a mile (1.6 km).

Under certain conditions the slumped strata retain their integrity and apparently slid along a sole and were thrown into disharmonic folds (fig.6-7). Glacio-lacustrine varved clays seem especially prone to this type of décollement displacement (VAN STRAATEN, 1949; FAIRBRIDGE, 1947). The folds are very complex; even the axial planes are folded. Thrusting and the formation of nappe-like structures are common.

Although slump structures seem to occur mainly in mudstones and sandy shales with thin interbedded sands, they occur also in pure calcareous sediments, and, more surprising perhaps, in pure sandstones (see p.80). The slump structure of sandstones is most commonly associated with cross-bedding, both small (pl.4A) and large scale (pl.6B). The cross-laminations may be only abnormally steepened or they may be crumpled or in rare but interesting cases they may be recumbent. As noted by ROBSON (1956), the deformation in pure sand is mainly by folding an observation which, in light of RETTGER's experiments (1935), suggests that the deformation is subaqueous rather than subaerial.

Slumping in calcareous sediments differs in no important way from that in other kinds of deposits. In the Permian limestones of the Guadaloupe area, New Mexico, for example, slide structures vary from small crumplings between undisturbed beds to rather large-scale folds with amplitudes of 40 to 50 feet (13—17m), and coarse breccias, 30 to 40 feet thick (10—13 m), extending over many square miles (NEWELL et al., 1953, p.69, 86; RIGBY, 1958). Limestone breccias associated with graded limestone beds in the Alps are also attributed to slides and slumps generated along a reef front (KUENEN and CAROZZI, 1953, p.369). VOIGT (1962, p. 187—228) describes slump overfolds and snowball structures in Cretaceous carbonates of a large slump on the flanks of a broad anticline in northwest Germany.

Origin

Slumping can presumably take place in any environment where unstable deposits of sediment accumulate. Instability is greatest where bottom slopes are abnormally steep, although there is evidence that movement can occur on slopes of but one or two degrees (ARKHANGUELSKY, 1930, p.80; HEIM, 1908, p.139). KUENEN (1956, p.136) estimates the most suitable slope for slumping of marine clays to be 5 to 10 degrees. Slumping is most prone to occur in any over-steepened area such as found on the margins of tectonic structures that were active during sedimentation, on the front of a delta, on reef fronts, on foresets of subaerial and subaqueous dunes, along the margins of erosional channels, at heads of submarine canyons, and especially in those glacial lakes where silts and clays were draped over an irregular topography. Slumps are said to be characteristic of the borders of geosynclinal basins, especially the flysch basins (FAIR-BRIDGE, 1946, p. 91). Certainly some facies display a greater abundance of slump structures than others.

Slumping masses may travel many miles. A deposit attributed to sliding, the Paleocene Babice conglomerate in the Polish Carpathians, has been traced

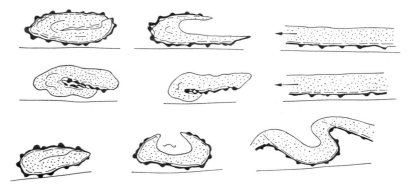

Fig. 6-8. Sketch of hook-shaped and roll-up structures produced by slumping (KSIAZKIEWICZ, 1958, fig. 19)

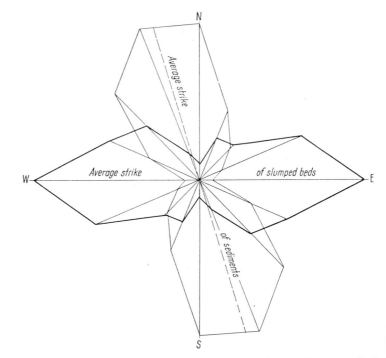

Fig. 6-9. Diagram showing relation of strike of slumped beds to normal tectonic strike (JONES, 1939, fig. 1)

30 kilometers in the direction of movement (DZULYNSKI et al., 1959, p. 1113). A slide may become a subaqueous mud flow and travel even greater distances or grade insensibly into slurries and turbidity currents which are reported to travel hundreds of miles.

The mechanisms of underwater gliding and the generation of slumps is not well known. Modern subaqueous slides are largely hidden from view, although interesting accounts of some known slides have been published (HEIM, 1908; ARKHANGUELSKY, 1930). Still less is known about the internal structure and organization of these deposits. As is the case with sole markings, most of what we know about these phenomena has been learned from the study of the now exposed ancient examples and to a lesser extent, perhaps, by laboratory experiments.

Directional Significance of Slump Structures

Commonly there is an absence of any well-defined direction of strike in slumped beds. However, some organization may be discerned. KSIAZKEWICZ (1958, p.134) noted that the hook-shaped overfolds of sandstone slabs generally closed in a preferred direction. He believes that during sliding the frontal part of a sliding sheet bends upward or downward, and in this way folded lumps of sandstone are formed (fig.6-8). This predominance of one direction of closure in sandstone lumps may be helpful in determining the direction of slumping.

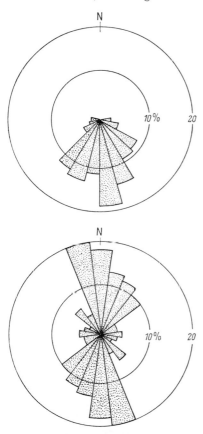

As noted earlier, there is commonly a discordance between the direction of movement of slumped materials and the direction of travel of the associated turbidity currents. The slides studied by KSIAZKIEWICZ took place obliquely and nearly in a direction opposite to the direction of flow shown by cross-bedding and sole markings.

JONES (1939, fig. 1. p. 338) made over 100 observations on the strike of slumped beds. The average turned out to be about perpendicular to the tectonic strike (fig.6-9). JONES believed that the plastic mass, in sliding down a slope, would tend to acquire folds and corrugations roughly at right angles to the down-slope movement. He concluded, therefore, that the average strike of the slump structure was the strike of the subaqueous slope. The same conclusion was reached by MURPHY and SCHLANGER (1962), who showed a parallelism between strike of slump folds and the direction of flow of the bottom currents (fig. 6-10).

Fig. 6-10. Rose diagrams of paleocurrent directions (above) and orientation of axes of slump folds (below) in Ilhas and São Sebastião formations (Cretaceous), Recôncavo basin, Brazil (MURPHY and SCHLANGER, 1962, fig. 7)

Perhaps owing to the difficulty of measurement in the more chaotic slump deposits, only a few systematic measurements of their directional structures have been made (Table 6-1). An exception is the study of the décollement type structure in Pleistocene varved clays by HANSEN *et al.* (1961). Careful measurement of the fold axes and the fabric of the associated overlying till showed the two to be closely correlated and led to the conclusion, in this case, that the structure was indeed produced by the overriding glacier and not by slump.

Careful and detailed mapping of slump occurrences in a particular stratigraphic unit may make possible not only determination of the index of slumping, i.e. the proportion of slumped to undisturbed layers, but also its areal variability.

Table 6-1. *Measurements of Slump Folds and Related Structures*

Author and year	Age	Place	Remarks
JONES, 1939	Ordovician	Wales	Over 100 measurements of strike of slumped beds; strike approximately normal to tectonic strike.
CHENOWETH, 1952 . .	Ordovician	New York	Measured and plotted 18 fold axes; parallel to seaway and shoreline.
KÜHN-VELTEN, 1955 .	Upper Devonian	Sauerland, Germany	Plotted normals to fold axes of convolute structure (Gleitfältelung).
CONATSER, 1958 . . .	Ordovician	Ohio and Kentucky	Plotted long axes of pillow-structures in limestone; orientation random.
KSIAZKIEWICZ, 1958 .	Cretaceous	Carpathians, Poland	Summarized criteria for direction of slump; measured pebble fabric of slumped beds.
MARSCHALKO, 1961 . .	Tertiary	Carpathians, Poland	Mapped "oriented slide deformation."
HANSEN *et al.*, 1961 .	Pleistocene	New York	Plotted axis of décollement folds in glacial clays; axes correlated with till fabric; due to overriding glacier.
MURPHY and SCHLANGER, 1962	Cretaceous	Brazil	Measured and mapped fold axes; axes parallel to slope and to direction of paleocurrents.

Some areas seem more characterized by slumping than others. Such knowledge, plus that concerning the direction of sliding, may give us a better insight into the conditions and shape of flysch basins at various periods in their history and, in particular, may shed light on the "Wild-flysch" and the exotic block problem.

It may be that slumping, small-scale soft-sediment faulting, and related phenomena are most concentrated at the line between the shelf and trough accumulations. If so, the orientation and direction of movement should help estimate trend and position of this important structural line. MARSCHALKO's work (1961, fig. 11) on the marginal facies of a Tertiary turbidite basin in Poland seems to bear this out. The slumps in this area moved down the sides of the basin whereas the main system of bottom currents was longitudinal (fig. 6-11).

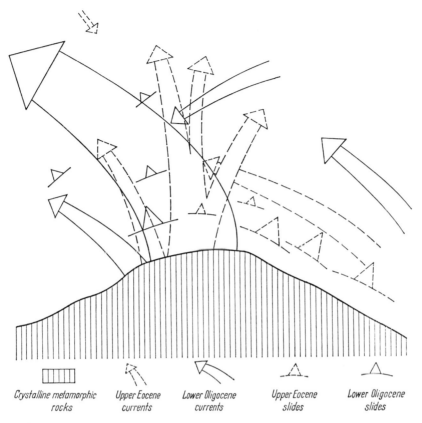

| Crystalline metamorphic rocks | Upper Eocene currents | Lower Oligocene currents | Upper Eocene slides | Lower Oligocene slides |

Fig. 6-11. Schematic map showing relations between direction of slump movement and presumed turbidity currents (modified from Marschalko, 1961, fig. 11). Marginal facies, Upper Eocene and Lower Oligocene, central Carpathians Poland

Clastic Dikes and Related Structures

Sandstone Dikes and Sills

Sandstone dikes are tabular cross-cutting bodies of sandstone. They are to be distinguished from fissures which have been filled by downward movement of sand, grain by grain, under the influence of gravity. They are, instead, true "neptunian dikes" (Smith and Rast, 1958, p. 234) filled by forceful injection of quicksand, either from above or below. Sandstone sills are similar bodies more or less concordant with the enclosing strata.

Sandstone dikes have long attracted the attention of geologists. Charles Darwin saw them in his travels in Patagonia in 1833 (Darwin, 1851, p. 150) and J. D. Dana discovered a series of sandstone dikes at Astoria, near the mouth of the Columbia River (Dana, 1849, p. 654) in 1841. Sandstone dikes have been seen many times, cutting rocks from Precambrian to Pleistocene in age. Among the better known examples are those cutting Cretaceous rocks in the Sacramento Valley in California described by Diller (1890). Diller's description of sandstone dikes can scarcely be improved upon. Other well-known examples include the dikes cutting the White River beds (Oligocene) in Nebraska (HAY, 1892, LAWLER, 1923), those cutting the Mowry shale (Cretaceous) in the Black Hills of South Dakota (RUSSELL, 1927), and the Woodbine (Cretaceous) of Texas (MONROE, 1950), the dikes cutting the Precambrian Pikes Peak granite in Colorado (CROSS, 1894; CROSBY, 1897; ROY, 1946; VITANAGE, 1954), and the dikes cutting Pleistocene deposits in the state of Washington (JENKINS, 1925; LUPHER, 1944). Of special interest are the dikes in the Carpathian flysch beds (Tertiary) of Poland described by DZULYNSKI and RADOMSKI (1956), the dikes and sills which cut Kimmeridgian (Jurassic) beds near Cromarty (WATERSON, 1950), and the quartzite dikes that cut the Espanola formation (Huronian, Precambrian) near Lake Huron in Ontario, Canada (QUIRKE, 1917, p. 37, pl. 11 B; COLLINS, 1925, p. 53, pl. 7). The early work on clastic dikes has been reviewed by NEWSOM (1903) who added some observations of his own on dikes in California and elsewhere.

Sandstone dikes are greatly varied in their size, attitude, and shape. Most are sharp-walled, vertical, tabular bodies, varying from an inch or so (2 to 5 cm) to great bodies up to 35 feet (11 m) in thickness. Generally they can be traced only a short distance but, under favorable circumstances, they have been followed several miles. VITANAGE (1954), for example, mapped the Colorado dikes in granite for distances up to about 8 miles (12.9 km) and DILLER (1890) traced the "Great Dike" for about 9 miles (14.5 km). See figure 6-12.

Normally the dikes are regular tabular bodies but in some cases, such as those in the Polish Carpathians, and in the Numidian (Paleogene) of Tunisia (GOTTIS, 1953), they are very irregular, sinuous, or are but a series of interrupted pod-shaped masses. The sinuousities of those dikes, which are perpendicular to the bedding, are strongly reminiscent of ptygmatic pegmatites. Their irregularity was attributed to reduction in thickness of the host shales by compaction and consequent adjustment of the enclosed dikes (DZULYNSKI and RADOMSKI, 1956, p. 261). If this explanation is correct, injection of the dikes at an early, pre-compaction stage is implied.

On the other hand, many dikes, such as those described by DILLER, are not deformed. Because they are vertical, even though the host strata are tilted, DILLER regarded them as being injected after tilting and hence at a late rather than early stage.

The normal filling of the dikes is massive sandstone. In some, worn pebbles are common, including granite pebbles six inches (11 cm) in diameter (COLLINS, 1925, p. 53). In some dikes the pebbles are markedly concentrated near the center of the dike. DILLER (1890, p. 425) has observed a faint alignment of mica flakes and shale fragments parallel to the dike walls. A similar arrangement of the micas and the long axes of quartz grains was reported by VITANAGE (1954, p. 498). This alignment, rather than a transverse arrangement, supports the notion that the dike filling was injected and that the fabric is a product of the wall influence on the flow.

Related to sandstone dikes are sandstone sills. These are tabular bodies of similar character which are parallel to the bedding of the host rock and which closely resemble and are, no doubt, commonly mistaken for beds. Unlike many ordinary beds, they lack internal stratification, grading, and the sole marks so characteristic of the turbidite facies where they are most commonly found. Moreover, if followed far enough, they can be seen to veer across the structure of

Fig. 6-12. Large sandstone dike and subsidiary smaller dike cutting Cretaceous shales, Sacramento Valley, on Roaring River, California, U.S.A. (drawn by R.W. B. DAVIS from DILLER, 1890, pl. 8)

the host rock to a different stratigraphic level. They also are connected to dikes and have dike offshoots.

The source of the sand filling the dikes and sills can in some cases be determined. Proof of injection from a subjacent source is indicated in some cases; in others the injection was downward from above (Pavlow, 1896; Vitanage, 1954). The mechanism most commonly invoked is earthquake shock, momentary lique-faction of water-saturated sand, and injection into fissures opened by the shock. The driving force is the pressure of the overlying strata. The formation of sills suggests that the injected material is capable of making space for itself.

The varied manner of occurrence of sandstone dikes makes it difficult to gene-ralize about them or to attribute them to any particular environment or situation. They are most common, perhaps, in thick geosynclinal sand-shale accumulations and have, in fact, been attributed to an unstable environment (Fairbridge, 1946) characterized also by extensive slumping. But there are many exceptions.

Dzulynski and Radomski (1956, p.258) believed there was a close relation between submarine slumping and dike formation. The slides may produce a local tension in the subjacent strata which favors quicksand injection. The dikes, however, showed variable strikes and quite a random orientation relative to the main direction of the slide movement. In other cases, as in the dikes in the Pikes Peak granite, the dikes are related to faulting and must have a very different history and significance. The dike swarm described by Diller (1890) occupies joints, not faults, and is remarkable for the parallelism of the many dikes.

Sandstone dikes, therefore, seem not to have a particular meaning other than the mobility and injection of quicksand, perhaps produced by liquefaction. They are but one additional manifestation of this transformation and are thus akin to the pillow sandstones or "quake sheets" and related also to other structures produced by hydroplastic deformation. They have no proved directional signi-ficance.

The reader interested further in the subject is referred to the extensive literature briefly reviewed by Shrock (1948, p.212).

Sand Volcanoes and Related Structures

There are a number of rare structures related or attributed to sand injection including sand volcanoes (Gill and Kuenen, 1958). These structures are closely associated with slump sheets and commonly occur on the eroded upper surfaces of these deposits. The volcanoes, which range from a few inches to over ten feet (3 m) in diameter, resemble, in miniature, igneous volcanoes. They are circular in plan with well-defined central crater or vent and were clearly built up by sand and silt ejected from this vent. The experiments and field relations observed by Kuenen led to the conclusion that the sand volcanoes were associated with the extrusion of water from the sediment resulting either from slump and the piling up of the uncompacted sediment to abnormal thickness or from the agitation of freshly deposited sediment without change in thickness, resulting in closer packing of sands and silts and concomitant expulsion of water.

Phenomena, possibly in part of related origin, are the pit-and-mound structure (Shrock, 1948, p.132) ascribed to currents of water or gas bubbles rising through mud, the larger spring pits (QUIRKE, 1930) formed on sandy beaches, gas pits (MAXON, 1940) and mud volcanoes, and certain cylindrical structures found in ancient sandstones some of which have been attributed to rising waters (GABEL-MAN, 1955; HAWLEY and HART, 1934; ARAI, 1959). The reader interested in

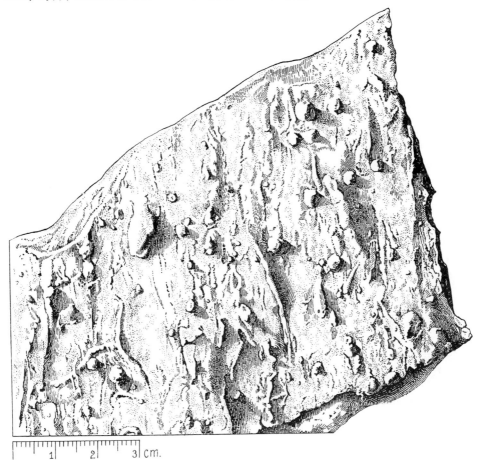

Fig. 6-13. Burrow-fillings and cast of tracks on bottom of bed, Gurnigel sandstone, La Mortivue, Switzerland (drawing by R. W. B. DAVIS after CROWELL, 1955, pl. 1, fig. 4)

these structures is referred to the original literature inasmuch as they seemingly have no directional value and perhaps have no clear relation to deformational movements.

A phenomenon an order of magnitude larger than most of the sedimentary structures described elsewhere in this volume, are the "mud lumps" of the delta of the Mississippi River. These are large swellings of upheavals of tough bluish-gray clay found within a mile or two of the mouths of the Mississippi. They occur just offshore and form islands an acre or two in extent and rise 5 to 10 feet (1.5 to 3 m) above the water surface. They are presumed to form by flow of semifluid clay under pressure of overburden and by upward buckling wherever this seaward flow meets resistance (SHAW, 1913).

Little is known of the internal structure of mud lumps. It is not even known whether or not they occur elsewhere nor is it likely that they would be identified properly, if seen in ancient sediments. The interested reader is referred to the more recent comprehensive studies of MORGAN (1961).

Any discussion of disturbed bedding would be incomplete without mention of the deformation produced by organisms, principally those of the burrowing type (fig. 6-13). Although the action of organisms is on a small scale, the results may be so extensive as to destroy all traces of bedding just as effectively as a large-scale slump.

Summary

Preconsolidation deformation of sediments leads to distortion of ordinary current structures, such as cross-bedding and sole markings. Such deformation also leads to the production of new structures of distinctive character.

Deformation structures include those due to vertical displacements under load and those due to lateral displacement as a result of sliding, slump, or, in rare cases, overriding by ice, or by drag of the superjacent current on the watery substratum. The structures related to load have a distinctive asymmetry which reflects the vertical displacement. Included here are load casts and the ball-and-pillow structure. Both are characterized by downward-facing convexities and both may be a manifestation of the same process. Load casting, however, appears to be contemporaneous with sedimentation; the ball-and-pillow structure may be a product of catastrophic foundering of a sand sheet following loading beyond a critical point, of an unstable substratum. Convolute lamination is a structure of more uncertain origin but its symmetry pattern suggests a kinship to load structures.

Structures due to lateral displacement display internal patterns indicative of such displacement including flame structures with overturned wisps and tongues of shale, spiral balls, the hook-like slump overfolds, and migmatite-like flowage structures.

A by-product of sediment deformation are the accessory phenomena of injection such as sandstone dikes and sills, and volcanoes and perhaps sand pipes.

The deformational structures of soft-sediment origin seem not to be indicative of any particular environment. They are found in sediments of diverse origins and form anywhere that the requisite conditions of stress and material are met. If is possible, however, that some structures may be more common in certain facies than others.

The directional value of deformational structures is highly varied. The pure load structures have no value in delineation of paleoslope; their symmetry, however, may be a guide to stratigraphic order in vertical or overturned beds. An exception should be made of those whose form and orientation is inherited from current structures that localized the loading. The slump overfolds and related features, if due to sliding, may enable one to determine direction of slope of the subaqueous bottom. Directions thus inferred are in some cases at variance with the direction of flow of currents as recorded by bottom markings and other structures. Sandstone dikes and related phenomena have little or no directional value.

The investigation of soft-sediment deformation is best made in the field where the structures formed can be clearly seen in the better outcrops. Examination of cores, however, will disclose some of the small-scale features such as convolute bedding, oversteepened and overturned ripple cross-lamination, and small sandstone dikes and sills. The sills are likely to be mistaken for beds.

A full understanding of the structures produced by deformation of unconsolidated sediments will come in part from a consideration of the principles of soil mechanics. The behaviour of water-saturated sediments under stress is the basic problem. The liquefaction which water-saturated silts and some sands undergo upon application of stress due to shock or to overloading is probably the underlying cause of sand injection as dikes, sills, etc. and the cause also of the ball-and-pillow structure of the founder or "quake" sheets. The thixotropic transformation of some clays is probably involved in many mass movements such as subaqueous mudflows, slumps, or slides and related phenomena. Comparatively few efforts have been made to understand these phenomena and to explain them in terms of soil mechanics. The importance of soil mechanics in this connection has been recognized and a few papers with geological applications have been published (BOSWELL, 1949b, 1950, 1952, 1961; WILLIAMS, 1960, p. 209). Further exploration of this subject and carefully controlled experimentation would seem worthwhile.

References

ARAI, J., 1959: Cylindrical structures in the Tertiary sediments of the Chichibu Basin, Saitama Prefecture, Japan. Bull. Chichibu Museum Nat. Hist. No. 9, 61—68.

ARKHANGELSKY, A. D., 1930: Slides of sediments on the Black Sea bottom and the importance of this phenomenon for geology. Bull. Soc. Naturalistes, Moscou (Sci. Geol.) 38, n.s., 38—80.

BEETS, C., 1946: Miocene submarine disturbances of strata in northern Italy. J. Geol. 54, 229—245.

BOSWELL, P. H. G., 1949a: The Middle Silurian rocks of Wales. London: Edward Arnold Co. 448 p.

— 1949b: A preliminary examination of the thixotropy of some sedimentary rocks. Quart. J. Geol. Soc. London 104, 499—526.

— 1950: Thixotropic and allied phenomena in geological deposits. Proc. Liverpool Geol. Soc. 20, 86—105.

— 1952: The determination of the thixotropic limits of sediments. Liverpool Manchester Geol. J. 1, 1—22.

— 1953: The alleged subaqueous sliding of large sheets of sediment in the Silurian rocks of north Wales. Liverpool Manchester Geol. J. 1, 148—152.

— 1961: Muddy sediments. Cambridge: W. Heffer and Sons Ltd. 140 p.

CHADWICK, G. H., 1931: Storm rollers (abstract). Bull. Geol. Soc. Am. 42, 242.

CHENOWITH, P. A., 1952: Statistical methods applied to Trentonian stratigraphy in New York. Bull. Geol. Soc. Am. 63, 521—560.

COLLINS, W. H., 1925: North Shore of Lake Huron. Geol. Survey Canada, Memoir 143, 160 p.

CONATSER, W. E., 1958: The contorted strata of the Cynthiana limestone. Unpublished MS thesis, University of Cincinnati, 112 p.

COOPER, J. R., 1943: Flow structure in the Berea sandstone and Bedford shale of central Ohio. J. Geol. 51, 190—203.

CROSBY, W. O., 1897: The great fault and accompanying sandstone dikes of Ute Pass, Colorado. Science, n.s., **5**, 604—607.

CROSS, C. W., 1894: Intrusive sandstone dikes in granite. Bull. Geol. Soc. Am. **5**, 225—230.

CROWELL, J. C., 1957: Origin of pebbly mudstones. Bull. Geol. Soc. Am. **63**, 993—1010.

DANA, J. D., 1849: Geology, United States exploring expedition during the years 1838, 1839, 1840, 1841, 1842, under the command of Charles Wilkes, U.S.N., vol. 10, Philadelphia, 756 p.

DARWIN, CHARLES, 1851: Geological observations on coral reefs, volcanic islands and on South America, Part III. London: Smith, Elder & Co.

DESTOMBES, J., and A. JEANNETTE, 1955: Étude pétrographique et sédimentologique de la série Acadienne de Casablanca présence de glissements sous-marins (slumpings). Notes du Service géol. du Maroc **11**, 75—98.

DILLER, J. S., 1890: Sandstone dikes: Bull. Geol. Soc. Am. **1**, 411—442.

DOTT jr., R. H., and J. K. HOWARD, 1962: Convolute lamination in non-graded sequences. J. Geol. **70**, 114—120.

DZULYNSKI, ST., M. KSIAZKIEWICZ and PH. H. KUENEN, 1959: Turbidites in flysch of the Polish Carpathian Mountains. Bull. Geol. Soc. Am. **70**, 1089—1118.

— and A. RADOMSKI, 1956: Clastic dikes in the Carpathian Flysch. Ann. soc. géol. Pologne **26**, 225—264.

EMERY, K. O., 1950: Contorted strata at Newport Beach, California. J. Sediment. Petrol. **20**, 111—115.

FAIRBRIDGE, R. W., 1946: Submarine slumping and the location of oil bodies. Bull. Am. Assoc. Petrol. Geologists **30**, 84—92.

— 1947: Possible causes of intraformational disturbances in the Carboniferous varve rocks of Australia. J. Proc. Roy. Soc. N.S. Wales **81**, 99—121.

FEARNSIDES, W. G., 1910: Tremadoc Slates and associated rocks of southeast Carnarvonshire. Quart. J. Geol. Soc. London **66**, 142—187.

GABELMAN, J. W., 1955: Cylindrical structures in Permian (?) siltstone, Eagle County, Colorado. J. Geol. **63**, 214—227.

GEIKE, A., 1882: Text-book of Geology, 1st ed. London: Macmillan Company. 971 p.

GILL, W. D., and PH. H. KUENEN, 1958: Sand volcanoes on slumps in the Carboniferous of County Clare, Ireland. Quart. J. Geol. Soc. London **113**, 441—460.

GOTTIS, CH., 1953: Les filons clastiques "intra formationnels" du flysch numidien tunisien. Bull. soc. géol. France [6] **3**, 775—784.

HAAF, E., TEN, 1956: Significance of convolute lamination. Geol. en Mijnbouw **18**, 188—194.

— 1959: Graded beds of the northern Apennines. Ph.D. thesis, Rijks University of Groningen, 102 p.

HADDING, A., 1931: On subaqueous slides. Geol. Fören. i Stockholm Förh. **53**, 377—393.

HAHN, F. F., 1913: Untermeerische Gleitung bei Trenton Falls (Nordamerika) und ihr Verhältnis zu ähnlichen Störungsbildern. Neues Jahrb. Mineral., Geol. Paläont., Beil.-Bd. **36**, 1—41.

HANSEN, EDWARD, S. C. PORTER, B. A. HALL and ALLAN HILLS, 1961: Décollement structures in glacial-lake sediments. Bull. Geol. Soc. Am. **72**, 1415—1418.

HAWLEY, J. E., and R. C. HART, 1934: Cylindrical structures in sandstones. Bull. Geol. Soc. Am. **45**, 1017—1034.

HAY, R., 1892: Sandstone dikes in northwestern Nebraska. Bull. Geol. Soc. Am. **3**, 50—55.

HEIM, ARNOLD, 1908: Über rezente und fossile subaquatische Rutschungen und deren lithologische Bedeutung. Neues Jahrb. Mineral., Geol., Paläont. **2**, 136—157.

HENDERSON, S. M. K., 1935: Ordovician submarine disturbances in the Girvan district. Trans. Roy. Soc. Edinburgh **58**, 487—507.

HILLS, E. S., 1958: Load-casts and flame structures. Geol. Mag. **95**, 171—172.

HOLLAND, C. H., 1959: On convolute bedding in the Lower Ludlovian rocks of north-east Radnorshire. Geol. Mag. **96**, 230—236.
— 1960: Load-cast terminology and origin of convolute bedding. Bull. Geol. Soc. Am. **71**, 633—634.
JENKINS, O. P., 1925: Clastic dikes of eastern Washington and their geologic significance. Am. J. Sci. [5] **10**, 234—246.
JONES, G. P., 1962: Deformed cross-stratification in Cretaceous Bima sandstone, Nigeria. J. Sediment. Petrol. **32**, 231—239.
JONES, O. T., 1937: On the sliding or slumping of submarine sediments in Denbighshire, north Wales, during the Ludlow period. Quart. J. Geol. Soc. London **95**, 335—382.
— 1939: The geology of the Colwyn Bay district; a study of submarine slumping during the Salopian period. Quart. J. Geol. Soc. London **95**, 335—376.
— 1953: On submarine slumping in the Lower Ludlow rocks of north Wales. Geol. Mag. **90**, 220—221.
KAYE, C. A., and W. R. POWER, 1954: A flow cast of very recent date from north-eastern Washington. Am. J. Sci. **252**, 309—310.
KELLING, G., and E. K. WALTON, 1957: Load cast structures: their relationship to upper surface structures and their mode of formation. Geol. Mag. **94**, 481—491.
KIERSCH, G. A., 1950: Small-scale structures and other features of Navajo sandstone, northern part of San Rafael Swell, Utah. Bull. Am. Assoc. Petrol. Geologists **34**, 923—942.
KSIAZKIEWICZ, M., 1951: Slip bedding in the Carpathian Flysch. Ann. soc. géol. Pologne **19**, 493—501 [Polish, English summary].
— 1958: Submarine slumping in the Carpathian Flysch. Ann. soc. géol. Pologne **28**, 123—150.
KUENEN, PH. H., 1949: Slumping in the Carboniferous rocks of Pembrokeshire. Quart. J. Geol. Soc. London **104**, 365—385.
— 1953a: Significant features of graded bedding. Bull. Am. Assoc. Petrol. Geologists **37**, 1044—1066.
— 1953b: Graded bedding with observations on Lower Paleozoic rocks of Britain. Verhandl. Koninkl. Ned. Akad. Wetenschap. Amsterdam, Afd. Nat. **20**, 1—47.
— 1956: The difference between sliding and turbidity flow. Deep-Sea Research **3**, 134—139.
— 1957: Sole markings of graded graywacke beds. J. Geol. **65**, 231—258.
— 1958: Experiments in geology. Trans. Geol. Soc. Glasgow **23**, 1—28.
—, and A. CAROZZI, 1953: Turbidity currents and sliding in geosynclinal basins in the Alps. J. Geol. **61**, 363—372.
—, and H. W. MENARD, 1952: Turbidity currents, graded and non-graded deposits. J. Sediment. Petrol. **22**, 83—96.
—, and J. W. PRENTICE, 1957: Flow-markings and load-casts. Geol. Mag. **94**, 173—174.
—, and J. E. SANDERS, 1956: Kulm and Flozleeres graywackes, Sauerland and Oberharz, Germany. Am. J. Sci. **254**, 649—671.
KÜHN-VELTEN, H., 1955: Subaquatische Rutschungen in höheren Oberdevon des Sauerlandes. Geol. Rundschau **44**, 3—25.
KUGLER, H. G., 1939: A visit to the Russian oil districts. J. Inst. Petrol. Technologists **25**, 68—88.
LAMONT, A., 1938: Contemporaneous slumping and other problems at Bray series, Ordovician and Lower Carboniferous horizons, in County Dublin. Proc. Roy. Irish Acad. B **45**, 1—32.
LAWLER, T. B., 1923: On the occurrence of sandstone dikes and chalcedony veins in the White River Oligocene. Am. J. Sci. [5] **5**, 160—172.
LEITH, C. K., 1923: Structural geology. New York: Henry Holt & Co. 390 p.
LIPPERT, H., 1937: Gleit-Faltung in subaquatischen und subaerischen Gesteinen. Senckenbergiana **19**, 355—375.
LOGAN, WM., 1863: Geology of Canada. Geol. Survey Canada, Rept. Prog., 983 p.

LUPHER, R. L., 1944: Clastic dikes of the Columbia Basin region, Washington and Idaho. Bull. Geol. Soc. Am. **55**, 1431—1462.

MACAR, P., 1948: Les pseudonodules du Famémien et leur origine. Ann. soc. géol. Belg. **72**, 47—74.

— 1951: Pseudo-nodules en terrains meubles. Ann. soc. géol. Belg. **75**, 111—115.

—, and P. ANTUN, 1950: Pseudo-nodules et glissement sousaquatique dans l'Emsien inferieur de l'Oesling. Ann. soc. géol. Belg. **73**-B, 121—150.

MARSCHALKO, ROBERT, 1961: Sedimentologic investigation of marginal lithofacies in Flysch of central Carpathians. Geol. práce (Bratislava) **60**, 197—230.

MAXON, J. H., 1940: Gas pits in non-marine sediments. J. Sediment. Petrol. **10**, 142—145.

MCIVER, N. L., 1961: Upper Devonian marine sedimentation in the central Appalachians. Unpublished Ph.D. thesis, The Johns Hopkins University, 346 p.

MIGLIORINI, C. I., 1950: Dati a conferma della risedimentazione della arenarie del macigno. Mem. Soc. Toscana Sci. Nat. **57**, 1—15.

MONROE, J. N., 1950: Origin of the clastic dikes of the Rockwell area, Texas. Field and Lab. **18**, 133—143.

MORGAN, J. P., 1961: Genesis and paleontology of Mississippi mud lumps, Part 1. Louisiana Geol. Survey Bull. **35**, 116 p.

MURPHY, M. A., and S. O. SCHLANGER, 1962: Sedimentary structures in Ilhas and São Sabastião formations (Cretaceous), Recôncavo Basin, Brazil. Bull. Am. Assoc. Petrol. Geologists **46**, 457—477.

NEVIN, C. M., 1942: Principles of structural geology, 3rd ed. New York: John Wiley and Sons. 320 p.

NEWELL, N. D., and others, 1953: The Permian reef complex of the Guadaloupe Mountains region, Texas and New Mexico. San Francisco: W. H. Freeman & Co. 236 p.

NEWSOM, J. F., 1903: Clastic dikes. Bull. Geol. Soc. Am. **14**, 227—268.

PAVLOW, A. D., 1896: On dikes of Oligocene sandstone in Russia. Geol. Mag., n.s. **3**, 49—53.

PEPPER, J. F., W. DeWITT jr. and D. F. DEMAREST, 1954: Geology of the Bedford shale and Berea sandstone in the Appalachian basin. U.S. Geol. Survey, Prof. Paper 259, 111 p.

PLESSMANN, W., 1961: Strömungsmarken in klastischen Sedimenten und ihre geologische Auswertung. Geol. Jahrb. **78**, 503—566.

PRENTICE, J. E., 1956: The interpretation of flow markings and load casts. Geol. Mag. **93**, 393—400.

— 1958: Anti-dunes and flame structures. Geol. Mag. **95**, 169—171.

— 1960: Flow structures in sedimentary rocks. Jour. Geol. **68**, 217—225.

QUIRKE, T. T., 1917: Espanola district, Ontario. Geol. Survey Canada, Memoir 102, 92 p.

— 1930: Spring pits, sedimentation phenomena. J. Geol. **38**, 88—91.

RETTGER, R. E., 1935: Experiments on soft-rock deformation. Bull. Am. Assoc. Petrol. Geologists **19**, 271—292.

RICH, J. L., 1950: Flow markings, groovings, and intrastratal crumplings as criteria for the recognition of slope deposits, with illustrations from Silurian rocks of Wales. Bull. Am. Assoc. Petrol. Geologists **34**, 717—741.

— 1951: Three critical environments of deposition and criteria for recognition of rocks deposited in each of them. Bull. Geol. Soc. Am. **62**, 1—20.

RIGBY, J. K., 1958: Mass movements in Permian rocks of Trans-Pecos Texas. J. Sediment. Petrol. **28**, 298—315.

ROBSON, D. A., 1956: A sedimentary study of the Fell sandstones of the Coquet Valley, Northumberland. Quart. J. Geol. Soc. London **112**, 241—262.

ROY, C. J., 1946: Clastic dikes of the Pikes Peak region (abstract). Bull. Geol. Soc. Am. **57**, 1226—1227.

RUSSELL, W. L., 1927: The origin of the sandstone dikes of the Black Hills region. Am. J. Sci. [5] **14**, 402—408.

SANDERS, J. E., 1960: Origin of convoluted laminae. Geol. Mag. **97**, 409—421.

SCHAFER, J. P., 1949: Some periglacial features in central Montana. J. Geol. **57**, 154—174.

SHARP, R. P., 1942: Periglacial involutions in northeastern Illinois. J. Geol. **50**, 113—133.

SHAW, E. W., 1913: The mud lumps at the mouths of the Mississippi. U.S. Geol. Survey Prof. Paper 85, p. 11—28.

SHROCK, R. R., 1948: Sequence in layered rocks. New York: McGraw-Hill Book Co. 507 p.

SMITH, A. J., and N. RAST, 1958: Sedimentary dykes in the Dalradian of Scotland. Geol. Mag. **95**, 234—240.

SMITH, BERNARD, 1916: Ball- or pillow-form structure in sandstone. Geol. Mag. **3**, n.s., 146—156.

SORBY, H. C., 1859: On the contorted stratification of the drift of the coast of Yorkshire. Proc. Geol. and Polytechnic Soc. West Riding of Yorkshire, 1849—1859, p. 220—224.

— 1908: On the application of quantitative methods to the study of the structure and history of rocks. Quart. J. Geol. Soc. London **64**, 171—233.

STRAATEN, L. M. J. U., VAN, 1949: Occurrence in Finland of structures due to subaqueous sliding of sediments. Bull. comm. géol. Finlande No. 144, 9—18.

— 1954: Sedimentology of recent tidal flat deposits and the Psammites du Condroz (Devonian). Geol. en Mijnbouw **16**, 25—47.

SULLWOLD jr., H. H., 1959: Nomenclature of load deformation in turbidites. Bull. Geol. Soc. Am. **70**, 1247—1248.

— 1960: Load cast terminology and origin of convolute bedding, further comments. Bull. Geol. Soc. Am. **71**, 635—636.

TROLL, CARL, 1944: Strukturboden, Solifluktion und Frostklimate der Erde. Geol. Rundschau **34**, 545—694.

VANUXEM, LARDNER, 1842: Geology of New York, Pt. III, Surv. of the 3rd District. ALBANY: W. and A. WHITE and J. VISSCHER, 306 p.

VITANAGE, P. W., 1954: Sandstone dikes in the South Platte area, Colorado. J. Geol. **62**, 493—500.

VOIGT, EHRHARD, 1962: Frühdiagenetische Deformation der turonen Plänerkalke bei Halle/Westf., Mitt. geol. Staatsinst. Hamburg **31**, 147—275.

WALTHER, J., 1904: Die Fauna der Solnhofener Plattenkalke bionomisch betrachtet. Festschr. zum 70. Geburtstage Ernst Haeckel, Jena.

WALTON, E. K., 1956: Limitations of graded bedding and alternative criteria of upward sequence in the rocks of the Southern Uplands. Trans. Geol. Soc. Edinburgh **16**, pt. 3, 262—271.

WATERSON, C. B., 1950: Note on the sandstone injections of Eathie Haven, Cromarty. Geol. Mag. **87**, 133—139.

WILLIAMS, E., 1960: Intra-stratal flow and convolute folding. Geol. Mag. **97**, 208—214.

Deformational Structures (1963—1976)

Much better understanding of the largest of all deformational structures — olistostromes — plus recognition of dewatering during compaction as an agent for much internal deformation of silts and sands.

Structures due to deformation contemporaneous with sedimentation continue to attract attention. Interest has focused on structures produced by those vertical or convective movements due to inverse density stratification and on those structures related to dewatering of sediments and momentary "liquification" and, also, on the largest deformational features — olistostromes — structures

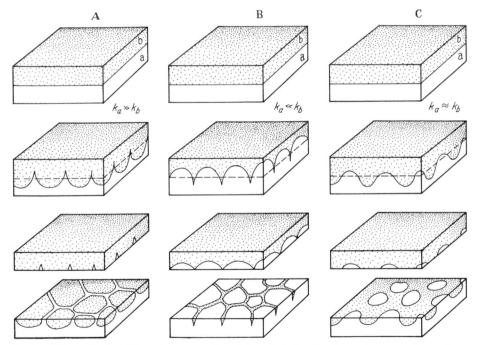

Fig. 6-13. Different idealized deformation patterns of two beds — typically either fine sand or coarse selt — with contrasting viscosities, k_a and k_b (ANKETELL *et al.*, 1970, fig. 1). Deformation is by viscous or plastic flow

difficult to distinguish from true tectonic mélanges. The relations between olistostromes, turbidites, and tectonics are explored by ABBATE (1970b). The deformational structures generally have limited paleocurrent value except perhaps those produced by slump or other lateral movements which are presumed to define the paleoslope.

Many papers have appeared since 1963 that describe deformational structures. The beautifully illustrated paper by DIONNE (1971) on unconsolidated Quaternary deposits is worthy of special note.

Most notable, perhaps, is the effort to relate various deformational structures to a process. An important group is related to excess pore water and to the momentary liquification or fluidization of the bed. In some sands a convective system is set up due to inverse density stratification (fig. 6-13). "Flow rolls" (ball-and-pillow structures) and periglacial involutions seem to be an expression of this process (DZULYNSKI and RADOMSKI, 1966; ANKETELL *et al.*, 1970; AR-TYUSHKOV, 1960a, 1960b). Other structures seem largely to be produced during the escape of excess pore waters. These include dish and pillar structures, perhaps also sandstone dikes and sills, and convolute bedding. The linkage between these and the dewatering processes has been explored by SELLEY (1964), SHEARMAN (1964), and LOWE (1975). Some experimental work has been done on this problem (SELLEY and SHEARMAN, 1962).

Over-steepening of foresets and overturning of cross-bedding forming recumbent folds, is a special aspect of soft-sediment deformation that was discussed in the section on cross-bedding.

New Structures

Few, if any, new structures have been observed but several, previously seen but not well described or understood, have been better defined and named. Perhaps the most important of these are the dish and pillar structures.

Dish structure (fig. 6-14) is defined by concave-upward laminations in siltstone or sandstone and is commonly associated with vertical or near-vertical cross-

Fig. 6-14. Dish structure — long seen but not explicitly recognized until after 1963 — not only tells the way up, but also indicates rapid deposition and consequent rapid dewatering. (Redrawn from CHIPPING, 1972, fig. 4, J. Sediment. Petrol., Soc. Econ. Paleontol. Mineral.)

0   5   10   15 cm

0   2   4   6 ins.

cutting columns of massive sand termed *pillars* (LOWE and LoPICCOLO, 1974, p. 487). Dish structure is commonly overlooked as it is weakly defined in many sediments. It occurs *within* sandstone beds, and if the sand is only faintly laminated, the structure is almost undetectable. It is best defined by thin laminations of heavy minerals or by faint more argillaceous laminations. The structure was named by WENTWORTH (1967) though apparently seen and described earlier by SELLEY (1964). It has been observed by many others (STAUFFER, 1967; CHIPPING, 1972; CORBETT, 1972; KRUIT et al., 1972; STANLEY, 1974; NAGAHAMA et al., 1975; LOWE, 1975).

The structure itself appears as one or many shallow concave-upward dishlike forms one to 50 cm in diameter. The structures may be superimposed one on another. The edges of each "dish" turn sharply upward. Some beds displaying dish structure pass upward into convolute laminations, an observation that suggests a common origin. LOWE and LoPICCOLO (1974, p. 493), believe the structures form by upward flow of pore-waters during sediment consolidation, a view not unlike that earlier postulated by SELLEY (1964). Dish structures occur in turbidite sands and in other sands which were rapidly deposited with entrapment of excess pore water.

Convolute Lamination

Convolute lamination is, perhaps, the most enigmatic of all deformational structures. It has been attributed to a diversity of causes. Recent work on the structure includes that of DZULYNSKI and SMITH (1963), EINSELE (1963), NAGTE-GAAL (1963), DAVIES (1965), SUTTON and LEWIS (1966), KUENEN (1968), ANKE-TELL et al. (1969), and LOWE (1975). The close association of the structure with ripple-bedding showing deformation and oversteepening of the ripple laminations have led many to conclude that "convolutions are ... the fossil indicators of the forces which existed within a freshly deposited group of laminae still under the influence of a moving current" (DZULYNSKI and SMITH, 1963, p. 622). The structure is thus a "frozen" record of sediment deformation by *current* action. The close correlation between the orientation of some convolute structures and the paleocurrent direction lends some support to this idea (SUTTON and LEWIS, 1966, fig. 1). On the other hand, there is convincing evidence that some convolutions, if not all, are related to the dewatering process during sediment consolidation. LOWE (1975), and also DAVIES (1965), ascribe the structure "to the rising pressure of trapped fluids, reduced strength and density of liquified layers, and fluid drag" (LOWE, p. 188) — a combination of thixotropic transformation and loading. The relation with rippling is, perhaps, fortuitous and the orientation observed in some cases is presumed due to control of the dewatering process and loading by pre-existing current structures or the inhomogeneities related to such structures.

Ball-and-Pillow Structures

Ball-and-pillow structures (sometimes called "flow rolls" or "pseudonodules") have been described in the Devonian of New York by SORAUF (1965), who cites eleven references to similar structures elsewhere, and by HOWARD and LOHRENGEL (1969) in rocks of Upper Cretaceous age in Utah. The latter are unusually large, some exceeding 20 feet (6 m). They are presumably related to foundering of a sand stratum into an underlying fluidized layer.

The involutions commonly described in sandy periglacial sediments are also attributed to convective movement due to a reversed density gradient (BUTRYM et al., 1964; CEGLA and DZULYNSKI, 1970).

Both theoretical (ARTYUSHKOV, 1960a, 1960b) and experimental studies (AN-KETELL et al., 1970; DZULYNSKI and RADOMSKI, 1966) support the concept of a convective origin of the ball-and-pillow related structures.

Rather large ball-and-pillow structures also occur in limestones as has been noted (see Plate 27B*). KELLING and WILLIAMS (1966) describe similar structures from Wales, but note their preferred orientation which they attribute to bottom configuration or possibly bottom currents thus precluding in some measure an origin by foundering. Orientation of ball-and-pillow structures is not unknown; it may be "inherited from pre-existing channels elongated in the current direction" (SORAUF, 1965, p. 562 and fig. 6) or, as has been shown from field studies by HUBERT (1972, p. 111—114), the long axes of pillows parallel the depositional dip of a delta front.

Sandstone Dikes

Where fluidization is complete, conditions for the formation of sandstone dikes and volcanoes exist. These structures continue to attract attention. Recent work on sandstone dikes includes that of BORRELLO (1962), HARMS (1965), CASTELLARIN (1966), HAYASHI (1966), MARSCHALKO (1965), OOMKENS (1966), STRAUCH (1966), PETERSON (1968), EISBACHER (1970), SMYERS and PETERSON (1971), and TRUSWELL (1972). Sand volcanoes were reported by BURNE (1970). Of special interest, per-haps, are the clastic limestone dikes, less easy to recognize because of the similarity to their host rock (PRAY, 1964; SUNDERMAN and MATHEWS, 1975).

The later work on dikes has not disclosed much that is new. Some dikes are injected into shales at an early stage and undergo deformation due to compaction; others are injected into late fractures of tectonic origin. Some are injected upward, others downward (even into subjacent crystalline basement). Some authors de-scribe new occurrences of dikes; others provide a better description of well-known occurrences. An example of the latter is EISBACHER's (1970) description of the Precambrian dikes on the north shore of Lake Huron. These sandstone intrusions appear to be fault related, but the associated conglomeratic dikes were attributed to a "slow, plug-like injection" presumed to be a "sub-permafrost" phenomenon.

Two of the most comprehensive papers on dikes are those by HAYASHI (1966) and STRAUCH (1966). The first summarizes what is known about over 10,000 dikes in Japan, ranging from Permian to Pleistocene in age. Hayashi covers all aspects of the subject and presents classifications based on geometry and on genesis. Strauch describes a swarm of Plio-Pleistocene dikes in northern Iceland that cut both sediments and basalt. He reviews the literature on clastic dikes going back to Werner (!) and includes the most complete bibliography known to us.

Slumping and Olistostromes

It has gradually become apparent that all soft-sediment deformation is not related to slumping as once thought. But after we eliminate that deformation due to current drag, to dewatering during compaction, and that due to convective transfer related to inverse density gradients, there remains some deformation due to lateral displacement, generally related to movement down slope, albiet in many cases a very gentle one. Such movement gives rise to some of the largest sedimentary structures known. The largest and most spectacular are the olistostromes (slide layers) with their contained olistoliths (slide blocks). Olistostromes are sometimes difficult to distinguish from *tectonic structures*. And in some

Table 6-2. *Conceptual Difference Between Mélanges and Olistostromes* (after Hsu, 1974, p. 326)

Item	Mélange	Olistostromes
Underlain by	A formation not pervasively sheared	A formation slightly older than the olistostrome
Overlain by	A formation not pervasively sheared	A formation slightly younger than the olistostrome
Time of chaotic displacement	Tectonic emplacement is under overburden of hundreds or thousands of meters	Sedimentary emplacement of upper surface of olistostrome was either subaerial or submarine
Site of chaotic emplacement	Is a major shear zone, in some cases the consuming plate margin	Is a major submarine topographic depression, a suitable receptacle for a large olistostrome
Duration of chaotic emplacement of one unit	Is duration of shearing movement, which may have been short or long	Is geologically very short, such as deposition of a sedimentary layer

cases, where rupturing of stratification and mixing occur, it is difficult to distinguish between the chaotic deposits thus generated and mélanges related to subduction of crustal plates. Because slumping is related to a slope, we can use the directional features of slump structures to define the paleoslope.

Slump and slide deposits have been the subject of many recent papers (AARIO, 1971; CORBETT, 1973; GREGORY, 1969; GRUMBT, 1966). Slumping also occurs in carbonate deposits (KENNEDY and JUIGNET, 1974; SPRENG, 1967). The deposits formed by slides and slumps show the greatest range in size, from relatively minor features to great chaotic bodies — the olistostromes. The term olistostrome, first commonly used in Europe (see ABBATE *et al.*, 1970a, for history of term), is applied to very large-scale chaotic deposits which form a part of an otherwise normal sedimentary sequence. In their scale, chaotic internal structure, and large size of many enclosed blocks (olistoliths), olistostromes resemble mélanges — the sheared chaotic tectonic breccias associated with subduction zones. Distinction between olistostromes and mélanges may be difficult or even impossible to make in some cases. HSU (1974) has, however, addressed himself to this problem and has summarized the differences between the two (Table 6-2). The reader is referred to Hsu's masterful paper for careful discussion of the criteria for discriminating between the two types of deposit. HOEDEMAEKER (1973) has also summarized the characteristics of olistostromes, and discussed their origin and classification. He termed them "delapsional" deposits.

In general, it is agreed that slumps are slope controlled and the direction of movement can, therefore, be used to ascertain paleoslope. Usually the orientation of the slump fold axes and their direction of overturning have been employed for this purpose. LAJOIE (1972, p. 584), however, has noted that in snow slumps, many fold axes were as much as 45° to the strike of the slump surface. Caution needs to be exercised in interpreting slump fold measurements, a view expressed earlier by CROWELL et al. (1966, p. 36), and MARSCHALKO (1963).

Experimental Studies

Much effort has been expended to produce those structures attributed to soft-sediment deformation. Most noteworthy are those related to reverse density stratification (high-density sediment resting on a low-density substrate). These studies show that deformation by plastic or viscous flow produces regular patterns similar to those of convective currents in fluids, that various "intraformational" folds may originate on flat surfaces and without lateral movement, that where brittle layers are involved sedimentary breccias may form, and where deformation occurs near the sediment–water interface, unconformable surfaces truncating disturbed laminations may be formed. Though they resemble unconformities, no erosion is implied.

Rapid sedimentation may lead to the entrapment of excess pore water. The escape of this fluid produces small-scale deformational structures. There have been a few experimental studies of the process of dewatering.

Annotated References of Experimental and Theoretical Studies

ANKETELL, J. M., J. CEGLA and S. DZULYNSKI, 1969: Unconformable surfaces formed in the absence of current erosion. Geol. Romana 8, 4—46.
Experiments showed that in stratified materials with inverted density gradient, rising diapirs, upon reaching the sediment-water interface, formed a thin layer deposited on upturned edges of the deformed strata — a false unconformity formed without current erosion.
—, J. CEGLA and S. DZULYNSKI, 1970: On the deformational structures in systems with reversed density gradients: Ann. Soc. Geol. Pol. 40, 3—30.
Systems deforming by plastic or viscous flow take the form of ascending or descending columns (convective cells). The deformation of the interface between two deforming layers takes one or the other of three configurations which depend on the ratio of their kinematic viscosities, K_a/K_b. This is perhaps the best paper on experimental deformation; many references; in English.
DZULYNSKI, S., and A. RADOMSKI, 1966: Experiments on bedding disturbances produced by the impact of heavy suspensions upon horizontal sedimentary layers: Bull. Acad. Sci. Pol., Geol. Geog. Ser. 14, 227—230.

The moving suspension caused considerable deformation of the previously deposited turbidite layers. Features similar to "slump overfolds" and "pseudonodules" were formed. The results are products of both shear and "settling convection."

— and E. K. WALTON, 1963: Experimental production of sole markings: Trans. Edinburgh Geol. Soc. **19**, 279—305.
Experimental studies of a large variety of sole markings, but including some related to load structures and other deformation phenomena.

EINSELE, G., R. OVERBECK, H. U. SCHWARZ and G. UNSÖLD, 1974: Mass physical properties, sliding and erodibility of experimentally deposited and differentially consolidated clayey muds. Sedimentology **21**, 339—372.
A fundamental background paper for understanding many, if not all, of the sole marks plus gravitational mass movements on slope. Key ideas: critical sedimentation rates, shallow sediment flow, creep, intrastratal flow and erodibility.

McKEE, E. D., and M. GOLDBERG, 1969: Experiments on formation of contorted structures in mud. Bull. Geol. Soc. Am. **80**, 231—243.
Fifteen experiments showing effects of differential loading of freshly deposited muds and sands; largely directed toward understanding of convolute and flame structures. Contains a review of what is known about convolute-type structures and ideas concerning their origin.

—, M. A. REYNOLDS and C. H. BAKER JR., 1962: Laboratory studies on deformation in unconsolidated sediment. U.S. Geol. Survey Prof. Paper **450-D**, 151—155.
Saturated mud layers alternating with sand were experimentally deformed by both vertical and progressive loading. Structures resembling convolute folds and the recumbent structure of cross-bedding were produced.

RAMBERG, H., 1967: Gravity, deformation, and the earth's crust. London-New York: Academic Press, 214 p.
Most complete and comprehensive experimental studies of "gravity tectonics", i.e., deformation related to vertical or convective transfer of materials owing to inverted density gradients. Although main focus is understanding of large-scale tectonic structures, the experiments are readily applicable to analysis of soft-sediment deformation.

References

AARIO, R., 1971: Syndepositional deformation in the Kurkiselkä esker, Kliminki, Finland. Bull. Geol. Soc. Finland **43**, 163—170.

ABBATE, E., V. BORTOLOTTI and P. PASSERINI, 1970a: Olistostromes and olistoliths. Sediment. Geol. **4**, 521—557.

ABBATE, E., et al., 1970b: The geosyncline concept and the northern Apennines. Sediment. Geol. **4**, 625—636.

ANKETELL, J. M., J. CEGLA and S. DZULYNSKI, 1969: Unconformable surfaces formed in the absence of current erosion. Geol. Romana **8**, 41—46.

—, J. CEGLA and S. DZULYNSKI, 1970: On the deformational structures in systems with reversed density gradients. Ann. Soc. Geol. Pol. **40**, 3—30.

ARTYUSHKOV, YE. V., 1960a: Possibility of convective instability in sedimentary rocks and the general laws of its development. Dokl. Akad. Nauk. S. S. S. R., Geol. Ser., **153**, 26—28.

—, 1960b: Principal forms of convective structures in sedimentary rocks. Dokl. Akad. Nauk. S. S. S. R. **153**, 43—45.

BORRELLO, A. V., 1962: Sobre los diques clasticos. Rev. Museo La Plata (New Ser.) sec. geol. **5**, 155—191.

BURNE, R. V., 1970: The origin and significance of sand volcanoes in the Bude formation, Cornwall. Sedimentology **15**, 211—228.

BUTRYM, J., J. CEGLA, S. DZULYNSKI and S. NAKONIENZNY, 1964: New interpretation of periglacial structures. Folia Quaternaria **17**, 1—34.

CASTELLARIN, A., 1966: Filon sedimentari nel Giurese di Loppio. Giorn. Geol., 2, **33**, 527—546.

CEGLA, J., and S. DZULYNSKI, 1970: Uklady niestatecznie warstwowne l ich wysepowanie w srodowisku peryglacjalnym (Systems with reversed density gradient and their occurrence in periglacial zones). Acta Universitatis Wratislaviensis **124**, 17—43 (Extended English summary).

CHIPPING, D. H., 1972: Sedimentary structures and environment of some thick sandstone beds of turbidite type. J. Sediment. Petrol. **42**, 587—595.

CORBETT, K. D., 1972: Features of thick-bedded sandstones in a proximal flysch sequence, Upper Cambrian, Southwest Tasmania. Sedimentology **19**, 99—114.

—, 1973: Open-cast slump sheets and their relationship to sandstone beds in an Upper Cambrain flysch sequence, Tasmania. J. Sediment. Petrol. **43**, 147—159.

CROWELL, J. C., R. A. HOPE, J. E. KAHLE, A. T. OVENSHINE and R. H. SAMS, 1966: Deep-water sedimentary structures, Pliocene Pico Formation, Santa Paula Creek, Ventura Basin, California. California Div. Mines Geol., Spec. Rept. **89**, 40 p.

DAVIES, H. G., 1965: Convolute lamination and other structures from the Lower Coal Measures of Yorkshire. Sedimentology **5**, 305—325.

DIONNE, J.-C., 1971: Contorted structures in unconsolidated Quaternary deposits, Lake Saint-Jean and Saguenay regions, Quebec. Rev. Géog. Montreal **25**, 5—33.

DOTT, R. II., 1963: Dynamics of subaqueous gravity depositional processes. Bull. Am. Assoc. Petrol. Geologists **47**, 104—128.

DZULYNSKI, S., and A. RADOMSKI, 1966: Experiments on bedding disturbances produced by the impact of heavy suspensions upon horizontal sedimentary layers. Bull. Acad. Sci. Pol., Geol. Geog. Ser. **14**, 227—230.

—, and A. J. SMITH, 1963: Convolute lamination, its origin, preservation, and directional significance. J. Sediment. Petrol. **33**, 616—627.

—, and E. K. WALTON, 1965: Sedimentary features of flysch and greywacke. Devel. Sedimentol. **7**, 274 p.

EINSELE, G., 1963: "Convolute bedding" und ähnliche Sedimentstrukturen im rheinischen Oberdevon und anderen Ablagerungen. Neues Jahrb. Geol. Paläontol., Abhandl. **116**, 162—198.

EISBACHER, G. H., 1970: Contemporaneous faulting and clastic intrusions in the Quirke Lake Group, Elliot Lake, Ontario. Canadian J. Earth Sci. **7**, 215—225.

GREGORY, M. R., 1969: Sedimentary features and penecontemporaneous slumping in the Waitemata Group, Whangaparaoa Peninsula, North Auckland, New Zealand. New Zealand J. Geol. Geophys. **12**, 248—282.

GRUMBT, E., 1966: Schichtungstypen, Marken und synsedimentäre Deformationsgetüge im Buntsandstein Südthüringens. Berlin Deut. Gesell. geol. Wiss. A., Geol. und Paläontol. **11**, 217—234.

HARMS, J. C., 1965: Sandstone dikes in relation to Laramide faults and stress distribution in the southern Front Range, Colorado. Bull. Geol. Soc. Am. **76**, 981—1001.

HAYASHI, T., 1966: Clastic dykes in Japan. Trans. Japanese J. Geol. Geog. **37**, 1—20.

HOEDEMAEKER, PH. J., 1973: Olistostromes and other depolapsional deposits, and their occurrence in the region of Moratalla (Prov. of Murcia, Spain). Scripta Geol. **19**, 207 p.

HOWARD, J. D., and C. F. LOHRENGEL II, 1969: Large non-tectonic deformational structures from Upper Cretaceous rocks of Utah. J. Sediment. Petrol. **39**, 1032—1039.

HSU, K. J., 1974: Mélanges and their distinction from olistostromes in R. H. DOTT JR., and R. H. SHAVER, eds., Modern and ancient geosynclinal sedimentation. Soc. Econ. Paleontol. Mineral. Sp. Publ. **19**, 321—333.

HUBERT, J. F., 1972: "Shallow water" prodelta flysch-like sequence in Upper Cretaceous deltaic rocks, Wyoming, and the problem of the origin of graded sandstones. Int. Geol. Congr. 24th Sess. Sec. 6, 107—114.

KELLING, G., and B. P. J. WILLIAMS, 1966: Deformation structures of sedimentary origin in the Lower Limestone Shales (basal Carboniferous) of South Wales. J. Sediment. Petrol. **36**, 927—939.

KENNEDY, W. J., and P. JUIGNET, 1974: Carbonate banks and slump beds in the Upper Cretaceous (Upper Turonian-Santonian) of Haute Normandie, France. Sedimentology 21, 1—42.

KRUIT, C., J. BROUWER and P. EALEY, 1972: A deep-water sand fan in the Eocene Bay of Biscay. Nature Phys. Sci. 240, 59—61.

KUENEN, PH. H., 1968: So-called turbidite structures. J. Sediment. Petrol. 38, 943—944.

LAJOIE, J., 1972: Slump fold axis orientation: an indication of paleoslope? J. Sediment. Petrol. 42, 584—586.

LOWE, D. R., 1975: Water escape structures in coarse-grained sediments. Sedimentology 22, 157—204.

—, and R. D. LoPICCOLO, 1974: The characteristics and origins of dish and pillar structures. J. Sediment. Petrol. 44, 484—501.

MARSCHALKO, RUDOLF, 1963: Sedimentary slump folds and the deposition slope (Flysch of central Carpathians). Geol. Práce 28, 161—168.

MARSCHALKO, ROBERT, 1965: Clastic dykes and their relations to syn-sedimentary movements (Flysch of central Carpathians). Geol. Práce 36, 139—148.

NAGAHAMA, H., R. OTA and H. AOYAMA, 1975: Dish structure in the Nichinon Group, Kyushu, Japan. Bull. Geol. Survey Japan 26, 217—225.

NAGTEGAAL, P. T. C., 1963: Convolute lamination, metadepositional ruptures and slumping in an exposure near Pobla de Segur (Spain). Geol. Mijnbouw 42, 363—374.

OOMKENS, E., 1966: Environmental significance of sand dikes. Sedimentology 7, 145—148.

PETERSON, G. L., 1968: Flow structures in sandstone dikes. Sediment. Geol. 2, 177—190.

PRAY, L. C., 1965: Limestone clastic dikes in Mississippian bioherms, New Mexico (Abs.). Geol. Soc. Am. Sp. Paper 82, 154—155.

SELLEY, R. C., 1964: The penecontemporaneous deformation of heavy mineral bands in the Torridonian Sandstone of northwest Scotland in L. M. J. U. VAN STRAATEN, ed. Deltaic and shallow marine deposits, Amsterdam: Elsevier Publ. Co., 362—367.

—, and D. J. SHEARMAN, 1962: Experimental production of sedimentary structures in quicksands. Proc. Geol. Soc. London 1599, 101—102.

SHEARMAN, D. J., 1964: On the penecontemporaneous disturbance of bedding by "quicksand" movement in the Devonian rocks of North Devon in L. M. J. U. VAN STRAATEN, ed. Deltaic and shallow marine deposits, Amsterdam: Elsevier Publ. Co., 368—370.

SMYERS, N. B., and G. L. PETERSON, 1971: Sandstone dikes and sills in the Moreno Shale, Panoche Hills, California. Bull. Geol. Soc. Am. 82, 3201—3208.

SORAUF, J. E., 1965: Flow rolls of Upper Devonian rocks of south-central New York State. J. Sediment. Petrol. 35, 553—563.

SPRENG, A. C., 1967: Slump features, Fayetteville Formation, northwestern Arkansas. J. Sediment. Petrol. 37, 804—817.

STANLEY, D. J., 1974: Dish structures and sand flow in ancient submarine valleys, French Maritime Alps. Bull. Centre Rech., Pau-S. N. P. A. 8, 351—371.

STAUFFER, P. H., 1967: Grain-flow deposits and their implications, Santa Yuez Mountains, California. J. Sediment. Petrol. 37, 487—508.

STRAUCH, F., 1966: Sedimentgänge von Tjörnes (Nord-Island) und ihre geologische Bedeutung. Neues Jahrb. Geol. Paläontol., Abhandl. 124, 259—288.

SUNDERMAN, J. A., and G. W. MATHEWS, 1975: Clastic dikes and sills in Silurian carbonate rocks of northeastern Indiana (Abs.). Ann. Meeting Abstr. 2, 72—73, Am. Assoc. Petrol. Geol. and Soc. Paleontol. Mineral.

SUTTON, R. G., and T. L. LEWIS, 1966: Regional patterns of cross-laminae and convolutions in a single bed. J. Sediment. Petrol. 36, 225—229.

TRUSWELL, J. F., 1972: Sandstone sheets and related intrusions from Coffee Bay, Transkei, South Africa. J. Sediment. Petrol. 42, 578—583.

WENTWORTH, C. M., 1967: Dish structure, a primary sedimentary structure in coarse turbidites (abstr.). Bull. Am. Assoc. Petrol. Geologists 51, 485.

Internal Directional Structures and Shape of Sedimentary Bodies up to 1963

Introduction

Internal directional structures and the shapes of most, if not all, sedimentary deposits are the joint response to the depositing medium, especially its prevailing direction of transport.

This hypothesis is the basis for the economic interest in using internal directional structures to predict the trend or orientation of deposits such as sand or gravel bodies in ancient sediments. Such economic applications are numerous. Improved prediction of sand-body trend can help minimize costs of exploration for petroleum and ground water, for some types of sedimentary mineral deposits such as gold placers and uraniferous sandstones, can improve prediction of the direction of washouts or cutouts in coal seams, and can contribute to fluid reservoir development, where performance may be linked to the primary, internal directional anistropy of the reservoir.

WILLIAM MORRIS DAVIS, in a study made in 1890, of a glacial sand plain, considered the relationship between internal directional structures and external form of deltaic land forms. Since this time, comparatively few data have been published on the relations between the shape of sedimentary bodies and their internal directional structures. In ancient sediments it has been difficult to relate directional structures, usually measured only in outcrop, to the three-dimensional shape of the sand bodies, for as yet subsurface directional data are not readily obtainable. Moreover, little is known about directional structures and the three-dimensional form of modern sand bodies, either quartz or carbonate. Some information is available, however, on the internal fabric of sedimentary bodies of glacial origin, such as drumlins and moraines.

Analysis of the Problem

Sand bodies with length several to many times greater than width, occur in all major sedimentary environments. In ancient sediments, such elongate sand bodies have been called shoestrings (RICH, 1923, p. 103), channels, and bars as well as pods, ribbons, dendroids and belts (POTTER, 1962a, p. 3). Modern equivalents of such elongate sands include beaches or cheniers, delta and bar-finger sands (FISK, 1962, fig. 16; 1961) and various bars such as meander or point bars in streams or offshore bars associated with the strand line. On other hand, blanket, tabular, and prism (KRYNINE, 1948, fig. 9) and sheet are terms applied to non-elongate sand bodies. Nomenclature and classification of sand-body shape is, at the present time, far from standardized.

To what extent do internal directional structures of elongate sand bodies correlate with their direction of elongation? Figure 7-1 shows diagrammatically various possible orientations of both directional structures and elongate sand bodies. Obviously, the preferred orientation of directional structures shown by most formations, each of which consists of many individual sand bodies, would not exist, if directional structures were randomly oriented within sand bodies (figs. 7-1a and 7-1b). Nor could it exist, if directional structures had a fixed relation to randomly oriented sand bodies within a formation (fig. 7-1c).

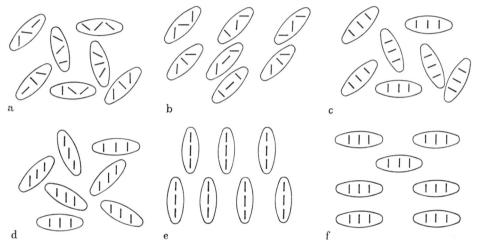

Fig. 7-1. Diagrammatic relations between internal directional structures and orientation of elongate sand bodies

Direct rather than indirect evidence is required to distinguish between the remaining possibilities of figures 7-1 each of which has preferred orientation of directional structures. Such evidence indicates that the uniform orientation of directional structures within randomly oriented sand bodies, figure 7-1d, does not exist. Hence there must be a preferred orientation of directional structures having a constant relationship to the direction of elongation in most sand bodies (figs. 7-1e and 7-1f). This conclusion also follows from the correlation of both with current orientation. Direct observation is required to distinguish between directional structures parallel with from those perpendicular to direction of elongation.

Actual measurements show that directional structures, even in thick formations vary but little in vertical profiles. PELLETIER (1958, fig. 8) and YEAKEL (1959, fig. 23) show good vertical homogeneity of cross-bedding; BASSETT and WALTON (1960, fig. 8) and McIVER (1961, figs. 115 and 116) demonstrate similar vertical uniformity in sole mark orientation.

It is likely, therefore, that within the time span required for the accumulation of most individual sand bodies, the direction of sediment transport will change slowly, if at all. Hence most elongate sand bodies probably have a preferred transport direction, generally unidirectional, perhaps bi-directional in some cases;

those with a random transport direction are probably exceptional. TEN HAAF (1959, fig. 50) and McBRIDE (1960, fig. 123) found small bed-to-bed variation in turbidites. EMRICH (1962, fig. 23), and MAST and POTTER (1963, fig. 11) demonstrated good homogeneity of cross-bedding in individual sand bodies. Figure 7-2 shows a vertical profile of cross-bedding azimuths in a thin sand body of the Pocono formation in which the cross-bedded layers have an average thickness of only 3 to 4 inches (7.5 to 10.0 cm). The cross-bedding of many sand bodies appears to have even smaller variability as shown by detailed published maps (KNIGHT, 1929, figs. 26 and 27; STOKES, 1953, figs. 14, 15 and 16; FRAZIER and OSANIK, 1961, fig. 11; POTTER, 1962a, fig. 7; 1962b, fig. 6).

In sandstones that have been considered to be regionally uniform, thick blankets, preferred orientation of directional structure has been found also (cf. BAARS, 1961, figs. 12 and 15). However, such sandstones may consist of many individual, coalescent sand bodies.

Little information is available for individual thin sheet or blanket sands.

Actual observation shows that, in addition to particular directional structures having preferred orientation, the different directional structures of sand bodies are correlated. BRETT (1955, p. 137) found good correlation between cross-bedding and ripple mark orientation in the Baraboo quartzite. HSU (1960, figs. 4, 6, 7, 8, 12 and 13) shows good agreement between different structures, mostly sole marks, in a turbidite. In outcrop sections, CROWELL (1955, figs. 8 to 12) shows general consistency between cross-bedding, flute casts and parting lineation in turbidites. KELLING (1958, figs. 2 and 5) found agreement between strike of cross-bedding and ripple mark azimuth.

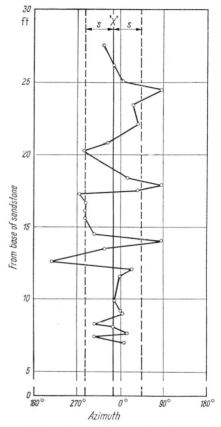

Fig. 7-2. Vertical profile of cross-bedding orientation in sandstone of Pocono formation (Mississippian) at crest of Sideling Hill on U.S. Route 40, Washington County, Maryland, U.S.A. Twenty-nine observations have a vector mean of 344° and standard deviation, s, of 57° (dashed lines)

Good correspondence exists between cross-bedding and parting lineation (fig. 4-3). However, the correlation between plant debris and various other directional structures is not always consistent. Thus most of the evidence indicates that the different directional structures are the product of a common current system.

This current system is also expressed in grain fabric. Several studies show that consistent, rational relations exist between directional structures such as cross-bedding, parting lineation and flute marks and the associated grain fabric (SMOOR, 1960, fig. 25; McIVER, 1961, fig. 35; POTTER and MAST, 1963, fig. 6). FRAZIER and OSANIK (1961, fig. 11) also found that photoelectric anisotropy correlated well with cross-bedding direction in a Mississippi River deposit. The correlation between cross-bedding and fabric implies that *individual dunes, either subaqueous or aeolian, have a relatively homogeneous grain fabric that is related to the shape of the dune.* Because many important directional structures thus appear consist of aggregates of like-oriented grains, they can be considered as forming small sub-domains or fields within the sand body.

A sand body consists of beds with directional structures and of massive beds. Both types of beds have grain fabric. Figure 7-3 shows the relations between directional structures, massive beds and grain fabric of a sand body and some dependent geophysical properties. This figure emphasizes that a sand body, from its smallest detrital grain to large aggregates of grains in directional structures

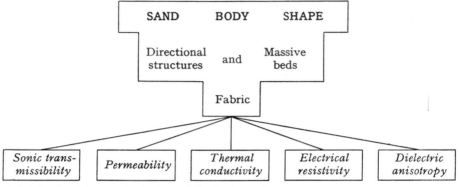

Fig. 7-3. Relations between sand-body shape and directional properties

and massive beds, is a *unified response to the local transport direction and symmetry of its depositing currents.* These orienting currents may be just as pervasive, in their effect on orientation of both grain and directional structures, as is glacial motion which produces an orientation of all sized particles in till.

A unified response of a sand body and its components to the direction and symmetry of its depositing currents implies that the sand body will have a directional anisotropy that can be reconstructed from various internal structures and fabric-dependent geophysical properties. The hierarchy shown in figure 7-3 is useful in evaluating the relative effectiveness of these different predictors of sand-body elongation. They include dimensional grain fabric measured in thin sections, aggregate measures of fabric based on geophysical properties such as optical transmissibility (MARTINEZ, 1958), sonic transmissibility (NANZ, 1960), and dielectric constant (ARBOGAST et al., 1960), and directional structures such as cross-bedding and sole marks.

The effectiveness of these different predictors is dependent on the number of grains involved. The orientation of each grain in a sand body can be considered as the outcome of an experiment where the variables are: current and gravity vectors, grain shape, and the roughness of the sediment interface. Assuming an average grain diameter of 0.12 mm and a porosity of 15 per cent, the number of grains in a core, a sand wave and a sand body of the dimensions was calculated (fig. 7-4). A geophysical measurement on a core is the aggregate effect of 6×10^8 grains. A sand wave, on the other hand, has 2.4×10^{30} grains and even a comparatively small sand body has as many as 1.1×10^{35} sand grains. Thus, either a geophysical measurement on a core or a single measurement of a directional structure is many times superior to a count of 200 grains in a thin section. Obviously, directional structures are the best predictors of current direction. In both outcrops

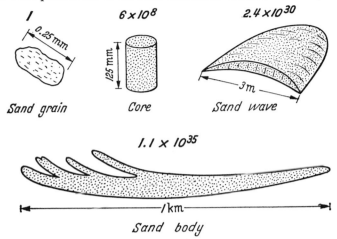

Fig. 7-4. Number of sand grains in a core, sand wave and sand body of indicated dimensions assuming an average grain diameter of 0.12 mm and a porosity of 15 per cent

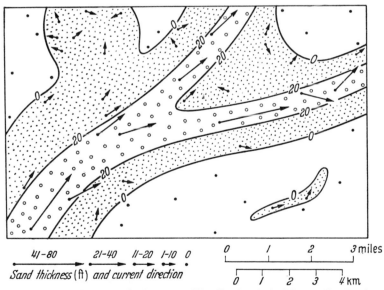

Fig. 7-5. Hypothetical example of a sand body contoured using cross-bedding direction obtained from well wall and weighted according to sand thickness. Areas of thick sand show more persistent current direction than proximal areas of thin sand. Stippling outlines area of sand deposition

and oriented cores, ease and rapidity of measurement also favors directional structures over either particulate or aggregate fabrics measured on separate samples. However, the superiority of directional structures over fabric would be less, if continuously recording, remote sensing devices were available to measure a fabric-dependent property in the well. Moreover as directional structures, such as cross-bedding, are usually present only in a part of a sand body, and even may be absent (cf. BRICAUD and POUPON, 1959, NANZ, 1960), some reliance on fabric may be necessary.

The contouring of sand body thickness can be done more rationally if the current directions are known. Figure 7-5 shows a hypothetical example of a sand-

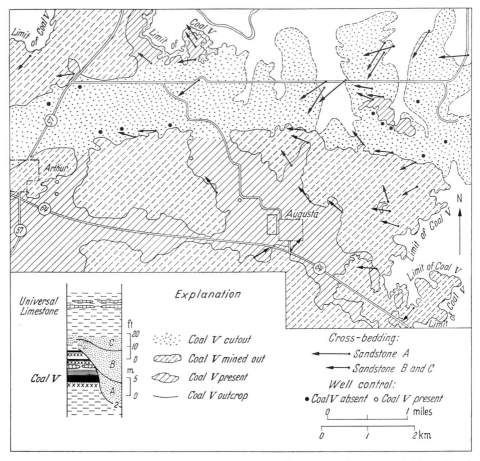

Fig. 7-6. Cross-bedding and sand-body trend in coal cutout (washout). Note agreement between cross-bedding and trend of cutout. Sandstone above Coal V, Petersburgh formation (Pennsylvanian), Pike County, Indiana, U.S.A. (modified from WIER, 1953)

thickness contour map made utilizing current direction as well as sand thickness. This example assumes sand-body elongation to be greatest parallel to transport direction. Actual case histories substantiate this assumption for fluvial sands.

Elongate Sand Bodies and Directional Structures

Elongate sand bodies with preferred orientation of directional structures are of principal interest. They appear to be of two major types: those that are elongate parallel to transport direction and those that are elongate perpendicular to it. Some variable relationships probably also occur.

Sand bodies of fluvial origin provide good examples of elongation parallel or nearly so to transport direction. The map, by RUBEY and BASS (1925, pl. 3), of cross-bedding in Cretaceous channel sand in Kansas supports this view as do the qualitative observations of EVANS (1949, p. 87) on the Permian Verden sandstone of Oklahoma. Later TANNER (1959, fig. 2) found cross-bedding to be subparallel to the trend of this channel sand. Figure 7-6 shows an example of parallelism of cross-bedding azimuths and the shape of a Pennsylvanian sand body in the Illinois basin. The boundaries of the sand body were established by the absence of an underlying mineable coal as determined by both mining and well data. The sand body is considered to be of fluvial origin (FRIEDMAN, 1960, p. 34). Good agreement between cross-bedding and sand body trend is also shown by the Pennsylvanian Warrensburg sandstone in Missouri (fig. 7-7). Other examples of good agreement between cross-bedding direction and sand body elongation have also been documented (POTTER and OLSON, 1954, fig. 6; LOWELL, 1955, fig. 5; TANNER, 1959, figs. 2 and 3; SCHLEE and MOENCH, 1961, fig. 4; POTTER and MAST, 1963, fig. 11). MOBERLY (1960, fig. 8) also shows a generalized map of channel trend and current direction in the Himes member of the Cloverly formation (Cretaceous). Unusually detailed correspondence between orientation of directional structures (cross-bedding, current parting, and ripple mark) and

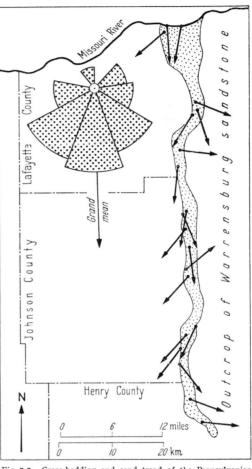

Fig. 7-7. Cross-bedding and sand trend of the Pennsylvanian Warrensburg sandstone (modified from DOTY and HUBERT, 1962, fig. 1)

the sand-shale ratio is shown by the fluvial Salt Wash sandstone at Cove Mesa, Arizona (fig. 7-8). In detail and throughout the mesa, there is good agreement. Directional structures have been used routinely as predictors of channel trends in exploration of uranium in fluvial sand bodies of the Jurassic Morrison formation of the Colorado Plateau (cf. LOWELL, 1955). Although few published data are available, the correlation between grain fabric and directional structures indicates that both particulate and aggregate measures of fabric could effectively predict direction of elongation (cf. POTTER and MAST, 1963, fig. 11).

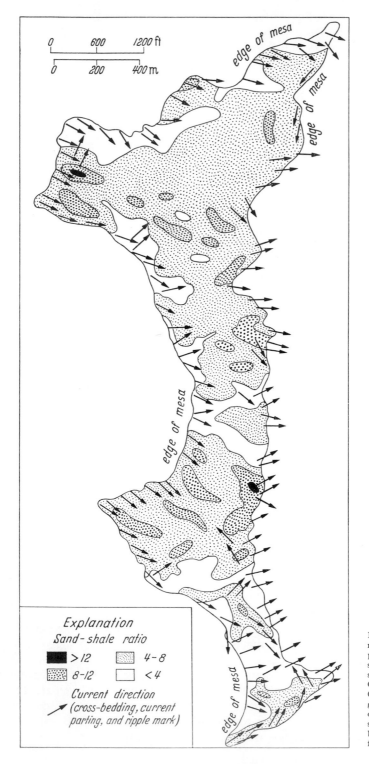

Fig. 7-8. Directional structures (cross-bedding, current parting, and ripple mark) and sand-shale ratio of Salt Wash sandstone at Cove Mesa, Carrizo Mountains, Apache County, Arizona, U.S.A. Note general agreement between directional structures and sand trends. Sand-shale ratio based on 256 drill holes (modified from JONES, 1954, fig. 5)

Descriptions of fluvial channel systems in modern and Recent sediments that may be helpful as guides to interpreting ancient fluvial sand bodies have been made by FISK (1944, p. 21—37; 1959; 1961) and others. Figure 7-9 shows the

Explanation

Former river beds of braided system, partly filled with peat

Late-glacial fluvial loam on fluvial sand and gravel

Thin cover of fluvial loam

Inland dunes, partly covering the fluvial loam

Fluvio-glacial outwash

Water

Fig. 7-9. Fluvial loam landscape, showing a portion of an abandoned channel system. Southwest of Nijmegen, Gelderland, Netherlands (modified from PANNEKOEK, 1956, fig. 42)

surface pattern of abandoned channels of a braided, Holocene river in the Netherlands. NANZ (1954, figs. 6 and 16), BUSCH (1959, fig. 12), FRIEDMAN (1960, figs. 4, 7, 9, 10 and 12) and POTTER (1962a, figs. 8 to 11) show maps of ancient fluvial and deltaic sands based on detailed subsurface studies. Elongate fluvial sand bodies, as part of a regional fluvial or deltaic system, are commonly perpendicular to depositional strike, although many local deviations are known.

Gravel fabric has also been used to predict direction of elongation. BECKER (1893, p. 54) long ago cited the use of gravel imbrication as a guide to exploration of Tertiary channel deposits by the placer miners in the Sierra Nevada region of California. REINECKE (1928, pl. 4) early mapped long axes of pebbles and related them to current direction in the gold-bearing Witwatersrand deposits of South Africa, which he considered to be fluvial origin (p. 114—115). Gravel fabric has been reported to show good agreement with erosional channels in a Permian delta (APRODOV, 1949, cited in ZHEMCHUZHNIKOV, 1954, fig. 5) and in a Pleistocene

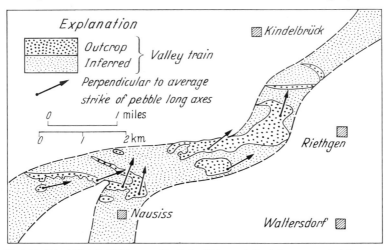

Fig. 7-10. Gravel fabric in Pleistocene valley train. Note correlation between normals to long axes of pebbles and trend of valley fill (modified from UNGER and ZIEGENHARDT, 1961, fig. 1)

valley train (UNGER and ZIEGENHARDT, 1961, fig. 1). Figure 7-10 shows the relations reported by UNGER and ZIEGENHARDT, who found that long axes of pebbles were transverse to channel direction. However, because long axes of gravel pebbles have been reported both transverse and parallel to current direction in streams (cf. Chapter 3), the best procedure is to specify the dip and strike of the maximum projection ($A\,B$) plane of the pebble.

Thus, in fluvial deposits, substantial evidence indicates good agreement between direction of channel elongation and both cross-bedding and gravel fabric, especially the former.

Beach deposits are the best known examples of sand bodies that are elongate perpendicular to their internal directional structures. NANZ (1955) found the long axes of sand grains on the foreshore of beaches to be parallel to the direction of ebb and flow and thus be perpendicular to the elongation of the sand body. However, NANZ (1955, p. 130) emphasized that his findings about beach deposits do not necessarily give information about orientation of sand grains, which form the major part of barrier islands. The regional studies of CURRAY (1956, figs. 1 and 3) on the Gulf Coast further documented NANZ's results. CURRAY further concluded (p. 2447) that backwash was the principal orienting current on the forebeach. Of course, other directional structures of a beach such as rill mark and current crescents also indicate a transport direction usually transverse to elongation, inasmuch as they are but different expressions of an integrated, local transport system. The more complex-shaped land forms along the strand line display more complex relations between internal grain fabric and sand body shape (CURRAY, figs. 6, 7, and 8).

Detailed studies of the relation between gravel fabric and shape of the marine beach gravel bodies seem not to have been made. However, disk-shaped pebbles are said to dip seaward on marine beaches (RUKHIN, 1958, p. 416). Both FRASER (1935, Table 8) and RUKHIN (1958, fig. 161) also state that the long axes of pebbles tend to be parallel to the strand line.

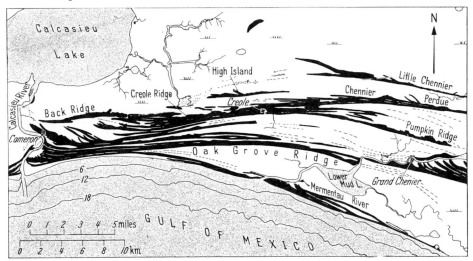

Fig. 7-11. Raised beaches or cheniers along western Gulf Coast. Note parallelism with strand line (modified from BYRNE *et al*, 1959, pl. 1)

Descriptions that emphasize the distribution and development of recent littoral sand bodies and barrier islands, both of which are largely parallel to the shore line include those of PRICE (1955), BYRNE *et al.* (1959), and SHEPARD (1960). Such depositional-strike sand bodies may also include aeolian deposits, which may complicate relations between direction of elongation and internal directional structures. Figure 7-11 shows a map of littoral sand bodies along the Gulf Coast.

BASS *et al.* (1937, p. 55—65), DILLARD (1941, p. 464), FETTKE (1941, p. 492) and BUSCH (1959, p. 2829—2832) report what they believe to be examples of strand-line deposits in ancient sediments. In general, however, strand-line sands do not appear to have been as commonly incorporated into the sedimentary record, especially in pre-Tertiary sediments, as have fluvial, deltaic, and marine sand bodies.

Obviously, distinction between elongate sand bodies of strand line origin that are parallel to depositional strike and those of fluvial origin that are perpendicular to it is vital for successful use of directional properties as predictors of sand-body elongation. Internal characteristics of the sand body as well as possible environmental contrasts in the associated muds and shales may be helpful in making the needed distinction. Independent evidence of depositional strike can also be very valuable.

Virtually no data have been published on the relations between internal directional properties and the shapes of either modern or ancient marine shelf sand bodies. In the absence of asymmetrical boundary conditions, such as on a beach, elongation may be parallel to transport direction. However, discrete marine shelf sand bodies may also be the submarine analogues of isolated dunes in the aeolian environment, in which case, elongation may be transverse as well as parallel to current direction. Large sand waves have been described from many marine shelves. JORDAN (1962) summarizes these studies and notes that height of sand waves up to 30 meters have been reported (p. 848). Air photographs of carbonate sands of the Bahamian banks show large transverse, underwater sand waves (PURDY, 1961, figs. 5, 6 and 7a; HARRINGTON and HAZLEWOOD, 1962, fig. 2b) that are, in many respects, the analogue of subaerial dune fields. RICH (1948, fig. 8) sketched the elongate patterns of sand accumulation on portions of the Bahamian shelf.

Unlike strand-line deposits, marine shelf sand bodies are commonly preserved in ancient marine sediments. Unfortunately, comparatively little systematic information is available on their relations to depositional strike in ancient sediments. However, OFF (1963) calls attention to modern "tidal current ridges" that are 25 to 100 feet (8 to 33 m) high and 5 to 40 miles (8.0 to 64 km) long. These ridges parallel tidal currents and are therefore perpendicular to shore line. According to OFF (1963, Appendix), such ridges develop along coasts, wherever tidal velocities range between 1 and 5 knots and there is a plentiful supply of sand. POTTER (1962a, fig. 13) shows sand bodies of similar size and orientation in the Chesterian sediments of the Illinois Basin.

Published data are also lacking on the relations between internal directional properties and the shapes of discrete turbidite sand bodies. However, like fluvial sand bodies, directional structures of turbidites should be parallel to their elongation for both are a response to down-slope, gravity transport. GORSLINE and EMERY (1959, figs. 3 and 5) show the actual distribution of modern turbidite sands off southern California. The sand trends are approximately normal to strand line. SULLWOLD (1960, figs. 2 and 3) also indicates how turbidites may form discrete, elongate sand bodies on deep sea fans as well as thinner more blanket-like sand bodies on basin floors. In elongate basins turbidite sands might be expected to be elongated parallel to the basin axis. EMERY (1960, p. 32—60) described a modern turbidite basin off southern California.

Some information is available on the internal structure of aeolian sand dunes and external shape. These relations probably differ little, if at all, from subaqueous dunes. In general, transverse dunes, either subaerial or aeolian have cross-bedding parallel to lee-side slip face. Aeolian dunes have a lee slope up to 30° or more; JORDAN (1962, p. 848) reports lee slopes of large subaqueous sand waves of at least 20°. Longitudinal dunes may show more complex relations between internal cross-bedding and external form (cf. SHOTTEN, 1956, fig. 3; MCBRIDE and HAYES, 1962, figs. 5 and 6). Actual identification of individual dunes, of either subaerial or subaqueous origin, as discrete sand bodies in ancient sediments is probably unusual unless they are large and subsurface control is unusually good.

Reef Form and Structure

There are few published studies of inclined bedding and fabric of reefs and reef-flank deposits. Ideally the inclined bedding of reef flanks should dip away from the reef core. In large compound reefs, however, more complex relations probably exist. MAXWELL *et al.* (1961, fig. 4) show a relatively complicated pattern of grain fabric around a modern atoll of the Great Barrier Reef. However, SCHWARZ-ACHER (1961, fig. 1) found that the fabric of crinoid debris indicated a uniform current direction in the non-reef facies whereas inclined bedding was clearly related to the reef core. He also found (p. 1492—1495) that cavity pattern was related to primary slope of the reef flank.

The shape of the reef may itself, under certain conditions, be a response to currents for currents supply nourishment and thus influence growth pattern. Such an interpretation may be difficult to establish, however, because other factors such as water depth and local sedimentary strike can also affect the pattern of reef growth.

Glacial Deposits

Some studies have been made of the relations of till fabric to constructional land forms such as moraines and drumlins.

HARRISON (1957) found that commonly disk-like particles were imbricated up-current perpendicular to end-moraines in Illinois (fig. 3); in some places a more variable orientation prevailed (fig. 4). In Sweden, GILLBERG (1961, fig. 6) found

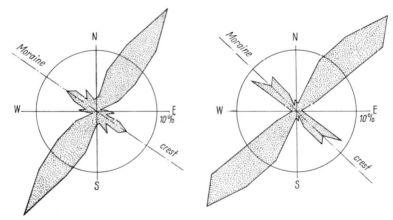

Fig. 7-12. Relations between long axes of till pebbles and strike of lodgement moraines in Dalsland, Sweden (modified from GILLBERG, 1961, fig. 6)

most of the long axes of pebbles to be perpendicular to moraine crest although a secondary mode was parallel to it (fig. 7-12). WRIGHT (1962, fig. 10) shows that most of the long axes of till stones are parallel to long axes of drumlins and are imbricated up-current. Thus morainic landforms of lodgement origin show close relations between internal fabric and external shape.

LUNDQVIST (1949, 1951, p. 77—81) and HARRISON (1957, p. 284) found that mud flow and solifluxion deposits have fabrics similar to those of ground moraine in which long axes are chiefly parallel to direction of glacial movement. Hence there should be good agreement between the shapes of mud flows and like deposits and their internal fabric.

Most fluvial deposits of glacial origin have relations between internal structures and external form similar to those of most nonglacial fluvial deposits. In ice contact deposits, such as some kame terraces and eskers, more complex relations may exist.

RAPP (1960, fig.63) found that long axes of boulders in avalanche tongues were parallel to the line of movement and hence to the elongation of the deposit.

Summary

Elongate deposits of clastic sediments occur in all major sedimentary environments.

Regardless of origin, most such bodies appear to be an integrated response to the direction of flow of the depositing currents. Evidence from both modern and ancient sediments suggests that most elongate sand bodies probably have persistent, rational relations between their direction of elongation and their internal directional structures. Such relations are of great practical value because knowledge of them would make possible more successful prediction of sand body trend in exploration and development.

The best predictors are those that correlate well with the direction of elongation and that can be rapidly and objectively measured. Published studies, most of which have been made on ancient sand bodies in outcrop, have been based on either cross-bedding or grain fabric, although some use has also been made of geophysical properties. Lack of oriented cores or a remote sensing device that could routinely determine directional anisotropy in the well has retarded knowledge about the relations between shape of sand bodies and their internal directional properties. Available data, however, indicate a good correlation between the two.

In glacial deposits, such as drumlins and some moraines, there is a good correlation between external shape and internal till fabric. Long axes of till pebbles tend to be parallel to the long axes of drumlins and to be at right angles to crests of lodgement moraines.

Long axes of boulders have also been reported to be parallel to the line of movement, and hence elongation, of avalanche debris tongues.

Comparatively few detailed studies have been published on the internal directional structures of reef-flank deposits. Reef growth pattern may, in part, also be a response to current direction.

Cross-bedding is usually parallel to the direction of elongation of fluvial sand bodies. Fabric of both fluvial sands and gravels has also been related to direction of sand body elongation, which, on the average, tends to be perpendicular to depositional strike.

In contrast, beach deposits commonly lie parallel to depositional strike. The internal directional properties of the beach slope deposits show transport perpendicular to the strike of the beach. Unlike fluvial, deltaic and marine shelf sands, strand-line sands do not appear to be as commonly preserved in the pre-Tertiary record.

Little has been published about the internal directional structures of elongate marine shelf sand bodies, although such sand bodies are commonly preserved. Nor has much information about orientation of sand bodies to depositional strike been systematically tabulated. Directional properties of turbidite sand bodies are probably parallel to the direction of elongation, as in fluvial sands, for both are a response to down-hill, gravity transport.

Successful prediction of sand body elongation obviously requires, in addition to data on directional structures, independent knowledge of the environment of deposition and of regional depositional strike. Moreover, experience with local characteristics of sand bodies, such as their size and local variability, is usually essential. Knowledge of the size of the sand bodies is especially important. Internal directional structures can help predict orientation of a sand body but they generally cannot predict its length. Successful prediction thus demands a wide range of information ranging from detailed knowledge of the local variability and characteristics of sand bodies gained from actual working experience to the broadest aspects of basin analysis.

References

ARBOGAST, J. L., C. H. FAY and S. KAUFMAN, 1960: Method and apparatus for determining direction of dielectric anisotropy in solids. U.S. Patent 2,963,642.

BAARS, D. L., 1961: Permian blanket sandstone of Colorado Plateau *in* Geometry of sandstone bodies. Tulsa: Am. Assoc. Petrol. Geologists, p. 179—207.

BASS, N. W., CONSTANCE LEATHEROCK, W. R. DILLARD and L. E. KENNEDY, 1937: Origin and distribution of Bartlesville and Burbank shoestring sands in parts of Oklahoma and Kansas. Bull. Am. Assoc. Petrol. Geologists **21**, 30—66.

BASSETT, D. A., and E. K. WALTON, 1960: The Hell's Mouth Grits: Cambrian greywackes in St. Tudwal's Peninsula, North Wales. Quart. J. Geol. Soc. London **116**, 85—110.

BECKER, G. F., 1893: Finite homogeneous strain, flow and rupture in rocks. Bull. Geol. Soc. Am. **4**, 13—90.

BRETT, G. W., 1955: Cross-bedding in the Baraboo quartzite of Wisconsin. J. Geol. **63**, 143—148.

BRICAUD, J. M., and A. POUPON, 1959: Continuous dip meter survey, a new intrument: The poteclinometer and micro-focused devices. 5th World Petroleum Congr., New York, 9 p.

BUSCH, D. A., 1959: Prospecting for stratigraphic traps. Bull. Am. Assoc. Petrol. Geologists **43**, 2829—2843.

BYRNE, J. W., D. O. LeROY and C. M. RILEY, 1959: The chenier plain and its stratigraphy, southwestern Louisiana. Gulf Coast Assoc. Geol. Soc. **9**, 23 p.

CROWELL, J. C., 1955: Directional-current structures from the prealpine flysch, Switzerland. Bull. Geol. Soc. Am. **66**, 1351—1384.

CURRAY, J. H., 1956: Dimensional orientation studies of Recent coastal sands. Bull. Am. Assoc. Petrol. Geologists **40**, 2440—2456.

DAVIS, W. M., 1890: Structure and origin of glacial sand plains. Bull. Geol. Soc. Am. **1**, 195—202.

DILLARD, W. R., 1941: Olympic pool, Hughes and Okfuskee Counties, Oklahoma, *in* Stratigraphic type oil fields symposium. Tulsa: Am. Assoc. Petrol. Geologists, p. 456—472.

DOTY, R. W., and J. F. HUBERT, 1962: Petrology and paleogeography of the Warrensburg channel sandstone, western Missouri. Sedimentology **1**, 7—39.

EMERY, K. O., 1960: The sea off southern California. New York: John Wiley and Son. 366 p.

250   Internal Directional Structures and Shape of Sedimentary Bodies

EMRICH, G. H., 1962: Geology of the Ironton and Galesville sandstones of the Upper Mississippi Valley. Unpublished Ph.D. thesis, University of Illinois, 109 p.

EVANS, O. F., 1949: The origin of the Verden sandstone of Oklahoma: J. Sediment. Petrol. **19**, 87—94.

FETTKE, C. R., 1941: Music Mountain oil pool, McKean County, Pennsylvania, *in* Stratigraphic type oil fields. Tulsa: Am. Assoc. Petrol. Geologists, p.492—506.

FISK, H. N., 1944: Geological investigations of alluvial valley of the lower Mississippi. U.S. Corps of Engineers, Mississippi River Commission, 78 p.

— 1952: Geological investigations of the Atchafalaya Basin and the problem of Mississippi River diversion. U.S. Corps of Engineers, Mississippi River Commission, vol. 1, 145 p.

— 1959: Sand facies of Recent Mississippi delta deposits. Proc. 4th World Petroleum Congr., Rome, Sec. I/c, Reprint 3, p. 377—397.

— 1961: Bar-finger sands of the Mississippi delta *in* Geometry of Sandstone Bodies. Tulsa: Am. Assoc. Petrol. Geologists, p. 29—52.

FRASER, H. J., 1935: Experimental study of the porosity and permeability of clastic sediments. J. Geol. **43**, 910—1010.

FRAZIER, D. E., and A. OSANIK, 1961: Point-bar deposits, Old River lock site, Louisiana. Trans. Gulf Coast Assoc. Geol. Soc. **11**, 121—137.

FRIEDMAN, S. A., 1960: Channel-fill sandstones in the Middle Pennsylvanian rocks of Indiana. Indiana Geol. Survey, Rept. Prog. No. 23, 59 p.

GILLBERG, GUNNAR, 1961: The Middle-Swedish moraines in the province of Dalsland, W. Sweden. Geol. Fören. i. Stockholm Förrh. **83**, p. 335—369.

GORSLINE, D. S., and K. O. EMERY, 1959: Turbidity-current deposits in San Pedro and Santa Monica basins off southern California. Bull. Geol. Soc. Am. **70**, 279—290.

HAAF, ERNST TEN, 1959: Graded beds of the northern Apennines. Rijks University of Groningen, 102 p.

HARRINGTON, J. W., and E. L. HAZLEWOOD, 1962: Comparison of Bahamian land forms with depositional topography of Nena Lucia dune-reef knoll, Nolan County, Texas. Study in uniformitarianism. Bull. Am. Assoc. Petrol. Geologists **46**, 354—373.

HARRISON, P. W., 1957: A clay-till fabric: its character and origin. J. Geol. **65**, 275—303.

HSU, K. JINGHWA, 1960: Paleocurrent structures and paleogeography of the ultrahelvetic flysch basins, Switzerland. Bull. Geol. Soc. Am. **71**, 577—610.

JONES, D. J., 1954: Sedimentary features and mineralization of the Salt Wash sandstone at Cove Mesa, Carrizo Mountains, Apache County, Arizona. U.S. Atomic Energy Comm., RME-3093 (pt.2), 40 p.

JORDAN, G. F., 1962: Large submarine sand waves. Science **136**, 839—848.

KELLING, GILBERT, 1958: Ripple-mark in the Rhinns of Galloway. Trans. Geol. Soc. Edinburgh **17**, pt.2, 117—132.

KNIGHT, S. H., 1929: The Fountain and Casper formations of the Laramie basin: A study of the genesis of sediments. Univ. Wyoming Publ. Sci., Geol. **1**, 82 p.

KRYNINE, P. D., 1948: The megascopic study and field classification of sedimentary rocks. J. Geol. **56**, 130—165.

LOWELL, J. D., 1955: Applications of cross-stratification studies to problems of uranium exploration, Chuska Mountains, Arizona. Econ. Geol. **50**, 177—185.

LUNDQVIST, G., 1949: The orientation of block material in certain species of flow earth *in* Glaciers and Climate. Geog. Annaler HI-2, p. 335—347.

LUNDQVIST, G., 1951: Beskrivning till jordartskarta över Kopparbergs län. Sveriges Geol. Undersökn, Ser. Ca, No. 21, 213 p.

MARTINEZ, J. D., 1958: Photometer method for studying quartz grain orientation. Bull. Am. Assoc. Petrol. Geologists **42**, 588—608.

MAST, R. F., and P. E. POTTER, 1963: Sedimentary structures, sand shape fabrics, and permeability, pt. II. J.Geol. **71**, 548—565.

MAXWELL, W. G. H., R. W. DAY and P. J. G. FLEMING, 1961: Carbonate sedimentation on the Heron Island Reef, Great Barrier Reef. J. Sediment. Petrol. **31**, 215—230.

McBride, E. F., 1960: Martinsburg flysch of the central Appalachians. Ph.D. thesis, The Johns Hopkins University, 375 p.
—, and M. O. Hayes, 1962: Dune cross-bedding on Mustang Island, Texas. Bull. Am. Assoc. Petrol. Geologists 46, 546—552.
McIver, N. L., 1961: Upper Devonian marine sedimentation in the central Appalachians. Unpublished Ph.D. thesis, The Johns Hopkins University, 347 p.
Moberly jr., Ralph, 1960: Morrison, Cloverly, and Sykes Mountain formations, northern Bighorn Basin, Wyoming and Montana. Bull. Geol. Soc. Am. 71, 1137—1176.
Nanz jr., R. H., 1954: Genesis of Oligocene sandstone reservoir, Seeligson Field, Jim Wells and Kleberg Counties, Texas. Bull. Am. Assoc. Petrol. Geologists 38, 96—117.
Nanz, R. H., 1955: Grain orientation in beach sands: a possible means for predicting reservoir trend (abstract). J. Sediment. Petrol. 25, 130.
— 1960: Exploration of earth formations associated with petroleum deposits. U.S. Patent, 2,963,641.
Off, Theodore, 1963: Rhythmic linear sand bodies caused by tidal currents. Bull. Am. Assoc. Petrol. Geologists 47, 324—341.
Pannekoek, A. J., ed., 1956: Geological history of the Netherlands. The Hague: Geological Foundation, Government Printing and Publishing Office. 147 p.
Pelletier, B. R., 1958: Pocono paleocurrents in Pennsylvania and Maryland. Bull. Geol. Soc. Am. 69, 1033—1064.
Potter, P. E., 1962a: Late Mississippian sandstones of Illinois. Illinois Geol. Survey, Cir. 340, 36 p.
— 1962b: Sand body shape and map pattern of Pennsylvanian sandstones of Illinois. Illinois Geol. Survey, Cir. 339, 36 p.
—, and R. F. Mast, 1963: Sedimentary structures, sand shape fabrics, and permeability, pt. I. J. Geol. 71.
—, and J. S. Olson, 1954: Variance components of cross-bedding direction in some basal Pennsylvanian sandstones of the Eastern Interior Basin: Geological application. J. Geol. 62, 50—73.
Price, W. A., 1955: Environment and formation of the chenier plain. Quaternaria 2, 75—86.
Purdy, E. G., 1961: Bahamian oolite sands in Geometry of sandstone bodies. Tulsa: Am. Assoc. Petrol. Geologists, p. 53—62.
Rapp, A., 1960: Recent development of mountain slopes in Kärhevagge and surroundings, northern Scandinavia. Geog. Annaler 42, 71—200.
Reinecke, Leopold, 1928: The location of payable ore-bodies in the gold-bearing reefs of the Witwatersrand. Trans. Geol. Soc. S. Africa 30, 89—119.
Rich, J. L., 1923: Shoestring sands of eastern Kansas. Bull. Am. Assoc. Petrol. Geologists 7, 103—113.
— 1948: Submarine sedimentary features on Bahama Banks and their bearing on distribution patterns of lenticular oil sands. Bull. Am. Assoc. Petrol. Geologists 32, 767—779.
Rubey, W. M., and N. W. Bass, 1925: The geology of Russell County, Kansas, pt. I. Kansas Geol. Survey, Bull. 10, 104 p.
Rukhin, L. B., 1958: Grundzüge der Lithologie. Berlin: Akademie-Verlag. 806 p. [Translated from the Russian].
Schlee, J. S., and R. H. Moench, 1961: Properties and genesis of "Jackpile" sandstone Laguna, New Mexico in Geometry of sandstone bodies. Tulsa: Am. Assoc. Petrol. Geologists, p. 134—150.
Schwarzacher, W., 1961: Petrology and structure of some Lower Carboniferous reefs in northwestern Ireland. Bull. Am. Assoc. Petrol. Geologists 45, 1481—1503.
Shepard, F. P., 1960: Gulf Coast barriers in Recent sediments, northwest Gulf of Mexico, 1951—1958. Tulsa: Am. Assoc. Petrol. Geologists, p. 192—381.
Shotton, F. W., 1956: Some aspects of the New Red desert in Britain. Liverpool Manchester Geol. J. 1, 450—456.

SMOOR, P. B., 1960: Dimensional grain orientation studies of turbidite graywackes. Unpublished M.Sc. thesis, McMaster University, 97 p.

STOKES, W. L., 1953: Primary sedimentary trend indicators as applied to ore finding in the Carrizo Mountains, Arizona and New Mexico. U.S. Atomic Energy Comm. RME-3043, 48 p.

SULLWOLD jr., H. H., 1961: Turbidites in oil exploration *in* Geometry of sandstone bodies. Tulsa: Am. Assoc. Petrol. Geologists, p.63—81.

TANNER, W. F., 1959: The importance of modes in cross-bedding data. J. Sediment. Petrol. **29**, 221—226.

UNGER, K. P., u. W. ZIEGENHARDT, 1961: Periglaziale Schotterzüge und glazigene Bildungen der Mindel-(Elster-)Eiszeit im Zentralen Thüringer Becken. Geologie **10**, 469—479.

WIER, CHARLES, 1953: Distribution, structure, and mined areas of coals in Pike County, Indiana. Indiana Geol. Survey, Prelim. Coal Map 3.

WRIGHT jr., H. E., 1962: Role of the Wadena lobe in the Wisconsin glaciation of Minnesota. Bull. Geol. Soc. Am. **72**, 73—100.

YEAKEL, L. S., 1959: Tuscarora, Juniata and Bald Eagle paleocurrents and paleogeography in the central Appalachians. Unpublished Ph.D. thesis, The Johns Hopkins University, 454 p.

ZHEMCHUZHNIKOV, YU. A., 1954: The possibility and conditions of burial of alluvial sediments in fossil strata *in* Alluvial deposits in coal measures in the Middle Carboniferous of the Donets Basin. Trans. Inst. Geol. Sci. U.S.S.R. No.151, 9—29 [Russian].

Internal Directional Structures and Shape of Sedimentary Bodies (1963—1976)

Although there are many studies on the paleoenvironments and shape of terrigenous and carbonate bodies, data about shape and paleocurrents is still largely lacking and almost totally so in carbonates.

In 1963 there were relatively few comprehensive studies on the paleoenviron-ments of sandstone and carbonate bodies. Today, this is a major area of investigation both by industry and by academics, and there is a very substantial literature, a very small part of which is summarized in Table 7-1. Underlying this effort is the idea that knowledge of the environment of deposition can help predict the size, shape, and orientation of a terrigenous and/or carbonate body as well as provide a guide to the internal distribution of its porosity and permeability. We emphasize only that part of this vast literature that relates shape to paleo-currents.

The basic philosophy underlying all these studies, terrigenous or carbonate, is that there are fairly constant relationships between the position of a discrete lithologic body in a basin, its depositional environment and its shape and internal organization. Hopefully, these relationships even extend to grain size, fossils, porosity and permeability so that possibly there is a very well-defined hierarchy (fig. 7-13). Since 1963, considerable progress has been made toward this goal, although very much still remains to be done.

Terrigenous Sandstones and Carbonates

Most of our knowledge about shape and paleocurrents concerns alluvial, deltaic, coastal and turbidite sandstone bodies and is based on study of both ancient and modern environments, with more contributions from the former than the latter. The best general reviews are provided by MACKENZIE (1972) and by LE BLANC (1972), MacKenzie's summary table being the best composite source of information (Table 7-2). This table combines information from vertical sections with general sandstone body characteristics plus a strong emphasis on directional structures and is, in our mind, one of the best available, because it recognizes eleven depositional environments. Another very useful table is that by MECKEL (1975, Table 2), although his table only concerns modern tidal, fluvial, and barrier sandbodies. HAYS and KANA (1976) also provide much information about current and paleocurrent systems and shape in modern sands, primarily in the fluvial, deltaic, and coastal environments. It is the strong emphasis upon vertical sequence that developed in the mid 1960s (VISHER, 1965; SELLEY, 1968) that has provided sedimentologists with a *general integrating framework* to which the details of grain size, kinds and abundance of sedimentary structures, fossil content and associated lithologies can be all related. Certainly, this is one of sedimentol-ogy's major developments, since World War II.

Table 7-1. *Major Sources for the Study of Depositional Environments and Systems*

Author and year	Remarks
FRIEDMAN, 1969	Seventeen papers with considerable emphasis on geometry and composition, but only one on paleocurrents — the Dimple Limestone, a shelf to basin transition
SELLEY, 1970	Brief effective overview with much emphasis on vertical profiles. Conclusions include an interesting table (p. 218—224) summarizing diagnostic features of major environments, including paleocurrents
MULTER, 1971	This guidebook, organized into 7 major parts and two appendices, gives a good overview of some modern carbonate environments. See also PURSER (1973)
FISHER and BROWN, 1972	An outstanding compendium of knowledge with emphasis upon the different sedimentary environments as integral parts of several major depositional systems. Well referenced. Don't miss it — even though there is little on paleocurrents!
LeBLANC, 1972	Well-organized and well-written discussion of the geometry of sandstone reservoirs by a staff sedimentologist of Shell. Thirty-five schematic illustrations and over 600 references. One of the best papers on geometry, but see also SHELTON (1973)
MacKENZIE, 1972	Good overview with Table 2, a summary of environmental characteristics, being outstanding. Thirteen summary papers on all the major sedimentary environments. Well illustrated with many photographs of hand specimens and outcrops, air photographs, and line drawings. About 1,200 references, and many different points of view on both terrigenous and carbonate environments. Some paleocurrent maps
REINECK and SINGH, 1973	Probably the best single book on the depositional environments of terrigenous sediments. Two parts: primary structures and textures (158 p.) and modern environments (277 p.). Over 1,300 references and 579 informative figures — truly the place to start an inquiry into depositional environments of terrigenous sediments. Discussion of hydro-dynamic controls on sedimentary structures and of bedding is particularly lucid, although paleocurrent data is limited
SHELTON, 1973	Well-described case histories by an ex-Shell sedimentologist. Paleocurrent data is emphasized
WEIMER, 1973	Nineteen diverse papers from the world's foremost petroleum society for the period 1961—1972 provide an excellent source book for both the student and professional. Highly recommended
SHAWA, ed., 1974	Five parts in all with brief text and informative illustrations. Emphasis on vertical sequences
BROUSSARD, ed., 1975	Three parts: general studies (6 papers), modern deltas (8 papers) and ancient deltas (9 papers) give a wide coverage of just about all you might wish to know about deltas — including a little on their paleocurrents
GINSBURG, ed., 1975	An outstanding, well-illustrated collection of diverse, short papers on tidal deposits, both terrigenous and clastic. Also contains a list of symbols for use on columnar sections. When

Table 7-1 (Cont.)

Author and years	Remarks
	you study shallow-water sedimentation in the field, be sure and take this volume with you!
KLEIN, 1975	A privately printed syllabus on depositional environments. Many diagrams
WILSON, 1975	The first text to emphasise the many diverse aspects of carbonate facies with plentiful tables, cross-sections, and maps. The notable absence of paleocurrent data can hopefully be rectified by carbonate workers in the future

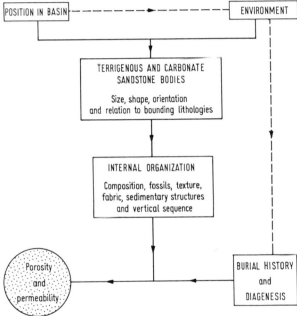

Fig. 7-13. In this schematic flow diagram porosity and permeability are seen as related to position in basin and depositional environment, subject to some modification by burial history which in some lithologies, especially carbonate ones, may themselves be strongly influenced by the original depositional environment

Attempts have been made to interpret the depositional environment directly from the curves of wire line logs, principally for terrigenous fluvial, deltaic, and barrier sandstone bodies (fig. 7-14). Electric, gamma ray and neutron logs reflect vertical change in lithology through variation in clay content and grain size, POUPON *et al.* (1970) giving the full technical details. A useful reference is that of GLAZE (1973) which provides environmental interpretation to core descriptions and their associated wire line logs. Additional reference articles to the sedimentological interpretation of wire line logs include those by KRUEGER (1968), PIRSON (1970, p. 1—58 and 153—179), JAGELER and MATUSZAK (1972) and HOLT (1973). After such profiles are confirmed in a basin by drilling a core and carefully correlating its properties with wire line logs, it should be possible to directly interpret the wire line logs environmentally, map the results and thus obtain a combined *thickness-environmental map*, normally the best type to use in exploration. Just exactly how vertical profiles of turbidite sandstone bodies, as assessed by wire

Table 7–2. *Characteristics of Sandstone Bodie*

| | General lithology | Characteristics of entire sediment body | | | | | |
		Thickness (m)	Shape, horizontal dimensions	Distribution trend	Relationship to adjacent or enclosing facies	Lithology, composition, texture, fauna	Bounding contacts
Subaerial migrated dune sands	Sands, no muds; homogeneous	3 to > 300	Elongate, or sheets up to 1,000 s of sq. km in area	Downwind from source of sand	Commonly the end stage of a regressive sequence	Well-sorted sands; pebbles and clasts rare	Variable
Alluvial sands	Sands, muds, some gravels	Usually 10—24 but 60—90 possible	Continuous bodies; usually 0.5 to 8 km wide, 10 s to 100 s of km	Make large angles with shoreline trends	Lower contacts erosional; lateral contacts erosional or indeterminate	Pebbles and clasts common; proportion or mud variable	Base erosional; top usually transitional
Deltaic deposits — Distributary channel fills	Sands, muds	Up to 60	Continuous sinuous bodies usually < 2 km wide		Commonly enclosed in non-marine or blackish muds		
Deltaic deposits — Delta-front sheet sands	Sands	6—24	Sheets		Underlain by marine pro-delta muds; overlain by marsh muds		←
Deltaic deposits — Reworked transgressive sands	Sands	0.3—12	Sheets		Underlain by or adjacent to deltaic deposits	Well-sorted; may contain coarse sand lag	Both sharp
Regressive shoreline sands mostly barrier islands	Sands; rare muds	6—18	Elongate or sheets; up to 5—10 km wide and 10 s of km long	Parallel to shoreline where elongate	Transitional downward and seaward into muds, landward into lagoonal or deltaic deposits	Well-sorted; pebbles and clasts rare; marine fauna, if any	Base transitional; top sharp
Offshore bars	Sands with mud partings	Several to 10—20	Elliptical lenses, less than a few sq. km in size	Scattered; orientation variable	Enclosed in and intertongues laterally with marine muds and silts		Sharp, or narrowly transitional
Strike-valley sands	Fine to coarse sands and muds; heterogeneous	3—15	Elongate; up to several km wide, 10 s of km long	Parallel to pre-unconformity paleostrike	Fills erosional strike valleys; intertongues with marine muds seaward; onlaps landward	Pebbles, clasts, glauconite, phosphate, marine fauna	
Deep-water sands — Proximal turbidites	Interbedded sands, silts and muds	100s—1,000s	Fans or sheets up to 1,000 s of sq. km in area	High flanks of deep basins near sand source	May be middle part of regressive sequence from deep- to shallow-water deposits	Graded bedding; displaced shallow water fauna; proximal turbidites often with interbedded debris beds	Variable
Deep-water sands — Distal turbidites				Sumps of deep basins	Sands interbedded with deep-water muds		

Slightly modified from MacKenzie 1972, *Table 2)*

Characteristics of individual vertical sections							Remarks
Overall vertical grain-size change	Primary sedimentary structures					Deformational and organic sedimentary structures	
	Stratification	Cross-stratification					
		Contact, set thicknesses	Nature of laminae	Shape of sets	Ripples		
Not systematic	Conspicuous high angle cross-bedding	Erosional, horiz. or sloping; sets up to 10—20 m thick	Lee dips 25°—34° commonly tangential to lower boundary	Tabular; sometimes enormous troughs	High indices; crests often parallel to dip of lee beds	Slumps not uncommon; vertebrate tracks	Still poorly known
Upward decrease	Many beds lenticular; abundant cross-bedding	Erosional, planar or concave up; usually 0.2—0.6 m thick	Maximum dips usually 20°—25°, inclined or tangential to lower boundary	Trough	Short-crested; linguoid; microtrough in cross-section; abundant	Slumps common; burrows uncommon	Many now attempt to distinguish braided from meandering stream deposits
						Slumps, burrows not uncommon	Several types possible depending upon ratio of stream input to onshore wave power

Similar to barrier sand bodies ⟶

Not systematic	High-angle cross-bedding; orientation diverse	Erosional, planar; 0.2—0.6 m thick	Maximum dips 20°—25°, tangential to lower boundary	Wedge or tabular	Not conspicuous; microtroughs present	Slumps, burrows uncommon	Lateral facies changes may provide proximity indicators
Upward increase, (but middle may have coarsest beds)	Upper and lower: subhorizontal stratification with low-angle truncations, exp. near base; sets < 0.3 m thick				Most abundant near base; symm. long-crested	Load structures and burrows common at base	
	Middle: high-angle cross-bedding, thick, tangential laminae; cross-laminae dip obliquely shoreward; local scours			Wedge or trough	Uncommon	Uncommon	
Not systematic	Low-angle cross-bedding	Erosional, planar; typically 0.3 m thick	Most dips < 10°; laminae parallel to lower set boundary	Wedge?	Common; some symm., long-crested	Burrows abundant only in marginal facies	Gradual outward decrease in sand/clay may provide proximity indicator
	Tabular units with high-angle cross-bedding dipping parallel to sand body elongation	Erosional, planar; 0.2—2 m thick	Max. dips 25°—30°; tangential to lower boundary	Tabular; sets straight and continuous for 20—30 m	Common locally esp. at toes of cross-sets; some are long-crested wave ripples	Burrowing common	Paleogeologic and paleotopographic maps effective in exploration
	Parallel stratified or structureless; may have large mud-lined scours		Trough-shaped sets found rarely		Asymmetric ripples, both short and long-crested, found at tops of individual beds	Burrows uncommon; bedding plane tracks and trails often present	Proximal beds are thicker, coarser grained, less well graded, less regular, more deformed, and more porous and permeable than distal beds
	1—3 continuous beds; parallel or ripple stratified						

line geophysical logs, compare with some of those from shallow marine environments is not yet fully known, although careful outcrop studies are well under way (RICCI LUCCHI, 1975). It seems possible, however, that some wire line profiles of turbidites may be almost identical so that it will be necessary to use paleontology and paleogeographic knowledge to determine first whether sand deposition occurred in a deep or shallow marine basin. Certainly more studies relating log characteristics of turbidites to their vertical sequence and relation to the sand dispersal systems are needed. In addition, virtually nothing is published about shape and paleocurrents of sheet sandstones that are eolian composites.

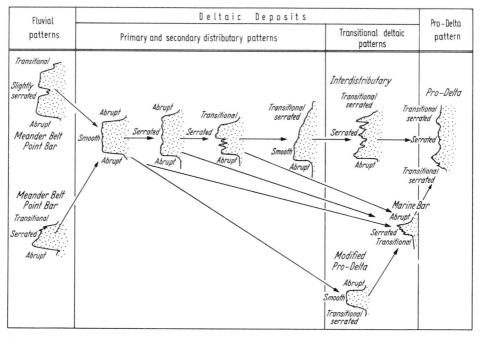

Fig. 7-14. Idealized self-potential log patterns representing some fluvial and deltaic environments of sandstone deposition (Redrawn from SAITTA and VISHER, 1968, fig. 8). More diagrams such as this are needed for the different environments summarized in Table 7-2

Paleocurrent studies of carbonate bodies and their shape are very, very rare. SEDIMENTATION SEMINAR (1972) found low angle cross-bedding in bryozoa-crinoidal packstones deposited along a break in paleoslope to dip roughly downslope and thus essentially have paleocurrents perpendicular to the main, long axis of the body (fig. 7-15). In an Ordovician example there are bipolar and bimodal current patterns in calcarenites that formed a low bank along the crest of the Cincinnati Arch (fig. 7-16). Although there are a modest number of paleocurrent studies in carbonates (cf. Table 4-4), few relate paleocurrent orientation to shape as do the above. Opportunities for such study seem especially plentiful on the dipping flanks of reefs and bioherms, where it may be possible to distinguish a gravity-induced fabric from a current-produced one. Certainly, the flanks of a reef, bioherm or other type of carbonate buildup deserve much more attention from students of paleocurrents than they have received, for the processes that actually form such sloping beds, usually much over-steepened in the ancient by differential compaction, are not very well understood.

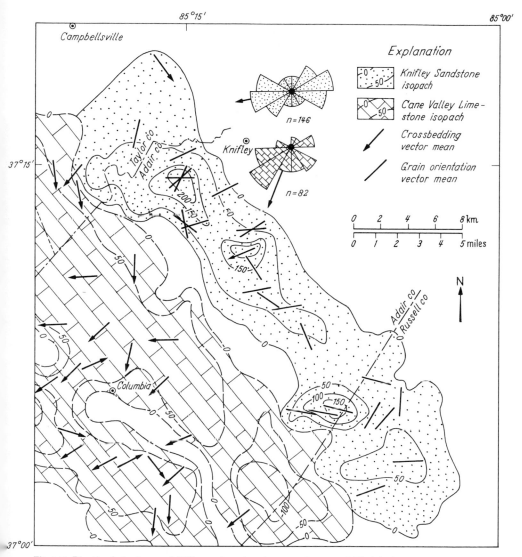

Fig. 7-15. Directional structures and thickness of an elongate, coarsening upward Mississippian sandstone body and in packstone–grainstone carbonates subsequently deposited in front of it in southcentral Kentucky. (Redrawn from SEDI-MENTATION SEMINAR, 1972, fig. 6). The carbonates are clinoform deposits that may in part have accumulated in channels. Note relation of paleocurrents to long axes of bodies

Although comparatively few studies have been published, carbonate bodies can be mapped by careful stratigraphic and petrologic study, the needed require-ments being (1) availability of samples (cuttings, chips and/or cores) as well as geophysical logs, (2) ability to correlate thin as well as thick units, and (3) absence of extreme diageneses that can destroy original textures.

Dipmeter Logs

In addition to using electric and radioactivity logs to estimate vertical sequence and thus depositional environments, dipmeter logs are also being widely used. The dipmeter log measures the orientation (inclination) of lithologically contrasting beds in the borehole. Ideally, it is fairly easy to distinguish between regional structural dip and deviations from it caused by faults and unconformities

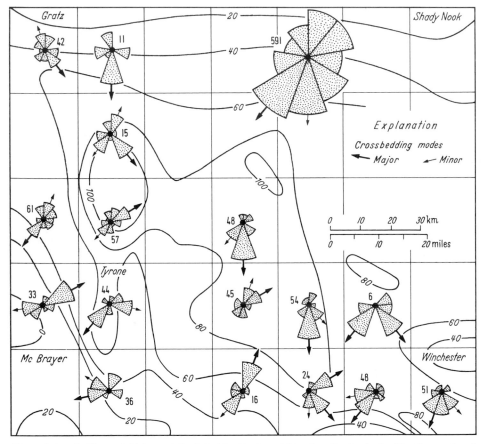

Fig. 7-16. Cross-bedding and thickness of the Ordovician Tanglewood Limestone Member in the Blue Grass region of Kentucky. Small current roses summarize paleoflow for each 7½ minute geologic quadrangle; large current rose is composite for entire area. Note tendency for bipolar flow to be perpendicular to lithologic trends, the latter suggesting shoaling on a regional high. (From Hrabar et al., 1971, fig. 8)

as well as differential compaction and/or primary dips associated with channel fills and reef structures (fig. 7-17). Articles by Gilreath and Maricelli (1964), Campbell (1968), Gilreath et al. (1969), Jageler and Matuszak (1972), Holland et al. (1974) and Gilreath and Stephens (1975) describe how to use the dipmeter log and give some examples of its application.

In terrigenous sediments the measured dip is chiefly that of sandstone–shale contacts, for example, the clay drape over a fluvial point bar. The dipmeter has also been used to infer cross-bedding orientation in sandstones, especially if they are not fully compacted and cemented as is commonly true of most of late Mesozoic and Tertiary age in nondeformed basins. PERRIN (1975) has performed a valuable service by presenting vertical profiles of cross-bedding orientation, measured in outcrop, in the style of a dipmeter log. Before completely relying on a

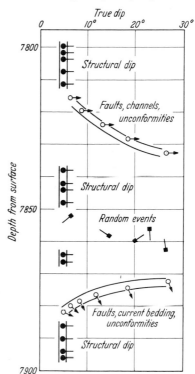

Fig. 7-17. Schematic dipmeter patterns and their geologic interpretation. (Redrawn from CAMPBELL, 1968, fig. 2)

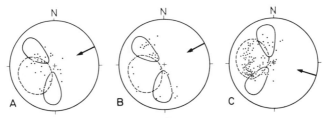

Fig. 7-18. Stereographic polar upper hemisphere plots of poles to Rotliegendes (Permian) cross-bedding of eolian origin (Redrawn from Glennie, 1972, fig. 13). A and B are based on outcrop studies in northern England and C is from a well with a dipmeter log in the southern North Sea. Arrow indicates average wind direction; solid line includes pole positions typical of seif dunes whereas dashed line includes those typical of barchans. Although stereographic plots are little used in paleocurrent studies today, the above suggested distinction between seif and barchan cross-bedding points to their continued utility.

dipmeter log to infer cross-bedding orientation and thus, in conjunction with other evidence, the shape and orientation of the sandstone body, cross-bedding measured in an oriented core should be compared with that inferred from the dipmeter log. A good example of cross-bedding measured by a dipmeter log is in the eolian sandstones of the Rotliegendes of the North Sea (Fig. 7-18).

The vertical pattern and magnitude of primary dips has also been used to infer the type and orientation of different types of delta front sandstone bodies (GILREATH and STEPHENS, 1975) and even to directly estimate possible ranges of water depth (GILREATH et al., 1969)!

Table 7-3. *Methods for Obtaining Oriented Cores*

BOUMA, 1964	Self-locking compass used with a box core sampler permits oriented samples to be obtained from shallow water. Accuracy claimed to within 5°
HÜGEL, 1965	Core oriented by movable marking knives and an inner tube that will not rotate
YOUNG, 1965	Gives technical details for a slim hole, oriented core
WINKEL, 1965	Core sample removed from hole after having been embedded with an epoxy resin containing magnetic particles; at surface core is reoriented so that its magnetic orientation is similar to that in the hole. Contains a formula for use when drilling directional holes
ROARK and SCHWEISBERGER, 1966	A piezo-electric system of orientation for both the core and its wall rock permits the core to be correlated with the latter when it is brought to the surface
WINKEL, 1966	Similar to WINKEL (1965)
HARRISON et al., 1967	Photographic compass for use of deep sea cores
KAZANTSEV, 1967	Extols the benefits of oriented cores and gives 9 additional references mostly to the Russian literature
ROSFELDER and MARSHALL, 1967	A free floating spherical compass is proposed for use at sea
EUSTACE, 1968	Brief expository account
SHEWAKE, 1968	Special outer and inner core barrels claimed to improve efficiency of core orientation
VOLOSHIN et al., 1968	In this engineering application oriented cores were used to map direction of potential planes of failure in alluvium and its underlying bedrock. Interesting application and informative description
FULLER, 1969	Proposes that a weak viscous component of magnetism related to the earth's present magnetic field can be recovered from a core and thus used to reorient its original orientation
MORRISON and CARSON, 1971	A gyrocompass is used for continuous monitoring of the orientation of a core barrel in deep sea drilling. Fourteen earlier references to oriented cores

In carbonates the dipmeter log measures the orientation of carbonate–shale contacts and is primarily used to infer reef position via differential compaction and/or primary depositional dips. Cox (1968) has one of the few papers discussing the use of the dipmeter in carbonates.

Oriented Cores

There is scattered literature on the use of oriented cores to estimate paleo-current currents and/or structural dip in exploration, some of the best examples of its use being in coal mining in Australia (DIESSEL and MOELLE, 1967; MOELLE and YOUNG, 1970). Oriented cores have been obtained both in the shallow near-shore and in the deep sea and, of course, onshore. However, their use onshore is rather rare, because of their much greater cost and because of possible lingering doubts by some explorationists as to their reliability. The methods of core orientation are diverse, although use of a compass and scribbers in the core barrel is probably most common (Table 7-3).

Annotated References

ASSERETO, R., and F. BENELLI, 1971: Sedimentology of the pre-Cenomanian Formations of the Jebel Ghorian, Libya in CARLYLE GRAY, ed. Symposium on the Geology of Libya. Tripoli, University of Libya, Fac. Sci., 37—83.
Integrated environmental study featuring sedimentary structures, paleocurrents and petrology of a carbonate-terrigenous platform sequence of Mesozoic age. This paper is a good example of the now routine use of paleocurrents in environmental reconstruction.

BAARS, D. L., and W. R. SEAGER, 1970: Stratigraphic control of petroleum in White Rim Sandstone (Permian) in and near Canyonlands National Park, Utah. Bull. Am. Assoc. Petrol. Geologists 54, 709—718.
One of the few studies of a large marine bar that can be seen in outcrop as well as mapped in subsurface. Megaripples super-imposed on a 200-foot high, 10-mile long bar. Some cross-bedding is up to 50 feet thick and is oriented transverse to trend of offshore bar.

BALL, M. M., 1967: Carbonate sand bodies of Florida and the Bahamas. J. Sediment Petrology 37, 556—591.
Geometry of tidal oolite bodies in relation to sediment transport directions and basin shape.

BANKS, N. L., 1973: Tide-dominated offshore sedimentation, Lower Cambrian, north Norway. Sedimentology 20, 213—228.
Four facies of tidally influenced, shallow marine shelf sandstones defined on the basis of grain size, bed thickness and sedimentary structures. Sand deposition believed to be a response to storm-enhanced, tidal currents. Author does not believe sandstone body shape can be predicted. Figure 9 is useful model for possible interpretation of offshore, tidally influenced sandstone bodies.

BENNACEF, A., S. BEUF, B. BIJU-DUVAL, O. DECHARPAL, O. GARIEL and P. ROGNON, 1971: Example of cratonic sedimentation, Lower Paleozoic of Algerian Sahara. Bull. Am. Assoc. Petrol. Geologists 55, 2225—2245.
Remarkable exposure and geologic history of several proglacial and periglacial sandstone bodies related to Cambrian-Ordovician glaciation plus raised beaches.

BERG, R., 1966: Point-bar origin of Fall River sandstone reservoirs, northeastern Wyoming. Bull. Am. Assoc. Petrol. Geologists 52, 2116—2122.
Classic subsurface study which uses the modern to interpret a Cretaceous alluvial deposit. Short and very sweet.

BROWN, L. F., JR., A. W. CLEAVES II and A. W. ERXLEBEN, 1973: Pennsylvanian depositional systems in north-central Texas. Geol. Soc. Am., Guidebook 14, 122 p.
Outstanding illustrations for geometry of sandstones on the shelf and slope.

CHISHOLM, J. I., and J. M. DEAN, 1974: The Upper Old Red Sandstone of Fife and Kinross: a fluviatile sequence with evidence of marine incursion. Scottish J. Geol. 10, 1—30.

Comprehensive paleocurrent study of cross-bedding displayed by histograms at outcrops, maps and histograms of dip angle plus an interesting illustration showing relationships between outcrops, their stratigraphic position, and their dominant sedimentary structures (Fig. 8).

CARR, D. C., 1973: Geometry and origin of oolite bodies in the Ste. Genevieve Limestone (Mississippian) in the Illinois Basin. Indiana Geol. Survey Bull. 48, 81 p.
One of the few studies relating orientation of oolitic bodies to paleoslope of basin plus a summary of geometry of oolitic bodies, modern and ancient (Table 6). Key ideas: oolitic bars, geometry, paleocurrents, and paleoslope.

DERR, M. E., 1974: Sedimentary structure and depositional environment of paleochannels in the Jurassic Morrison Formation near Green River, Utah. Geological studies, Brigham Young University 21, (3), 3—40.
Exhumed fluvial-channel segments provide opportunity to relate paleocurrents, textures, and vertical sequences to geometries. Point-bar and avulsion channel sequences compared.

EVANS, W. E., 1970: Imbricate linear sandstone bodies of Viking Formation in Dosland-Hoosier area of southwestern Saskatchewan, Canada. Bull. Am. Assoc. Petrol. Geologists 54, 469—486.
An apparent blanket sandstone really consists of three imbricate, linear members believed to have been deposited by tidal currents on a shallow sea. Some widespread chert pebble beds are present.

EXUM, F. A., and J. C. HARMS, 1968: Comparison of marine-bar with valley-fill stratigraphic traps, western Nebraska. Bull. Am. Assoc. Petrol. Geologists 52, 1851—1868.
Comparison and contrast of geometry of sandstone bodies, the dip of their cross-stratification, degree of bioturbation, petrology and reservoir characteristics. Classic.

GLAZE, R. E., ed., 1973: Core book. Wyoming Geol. Assoc., 25th Ann. Symp. and Core Seminar, 178 p.
Descriptions, photographs, some wire line logs, and environmental interpretations of 24 cores from the Rocky Mountain Region plus a brief section on core processing. A useful book for all those who must describe and interpret diamond drill core. No paleocurrent data, but still very worthwhile.

GOODWIN, P. W., and E. J. ANDERSON, 1974: Associated physical and biogenic structures in environmental subdivision of a Cambrian tidal sandbody. J. Geol. 82, 779—794.
This paper, although lacking geometry, is a good example of environmental analysis using sedimentary structures, both physical and biologic.

HARMS, J. C., 1966: Stratigraphic traps in a valley fill. Bull. Am. Assoc. Petrol. Geologists 50, 2119—2149.
Comprehensive core description combined with careful electric log study distinguishes an alluvial valley sandstone from marine sheet sandstones. Good pictures of cores.

HEWITT, C., and J. T. MORGAN, 1965: The Fry in situ combustion test — reservoir characteristics. J. Petrol. Technol. 17, 337—353.
An outstanding subsurface study including vertical sequence, sandstone body shape, directional permeability and reservoir characteristics. Far ahead of its time.

HOBDAY, D. K., and H. G. READING, 1972: Fair weather versus storm processes in shallow marine sand bar sequences in the Late Pre-Cambrian of Finnmark, north Norway. J. Sediment. Petrol. 42, 318—324.
Little directly on shape of sandstone bodies, but useful because it relates cross-bedding to inclined lateral accretion surfaces and distinguishes between foul and fair weather growth.

HOUBOLT, J. J. H. C., 1968: Recent sediments in the southern bight of the North Sea. Geol. Mijnbouw 47, (4), 245—273.
Good data on sand ridges and megaripples. Very impressive maps.

HRABAR, S. V., and P. E. POTTER, 1969: Lower West Baden (Mississippian) sandstone body of Owen and Greene Counties, Indiana. Bull. Am. Assoc. Petrol. Geologists 53, 2150—2160.

Detailed study of cross-bedding in a delta finger sandstone body, which has an upper and lower part: lower body was partially filled by tidal currents; upper body by downdip, fluvial currents. Key words: geometry ,paleocurrents, environmental reconstruction, and permeability.

KENDALL, C. G. ST. C., and Sir P. A. D'E. SKIPWITH, 1969: Geomorphology of a recent shallow-water carbonate province: Khar Al Bazam, Trucial Coast, southwest Persian Gulf. Geol. Soc. Am. Bull. **80**, 865—892.
Plates 5 and 6 map orientation of sand ribbons, longitudinal and transverse megaripples, sand waves, and gullies and runnels on a shallow carbonate shelf.

KING, R. E., ed., 1972: Stratigraphic oil and gas fields — classification, exploration methods and case histories. Am. Assoc. Petrol. Geologists Mem. **16**, Soc. Exploration Geophys. Sp. Pub. **10**, 687 p.
Many useful articles arranged into geologic and geophysical exploration methods and case histories of which there are 36.

LUDWICK, J. E., 1970: Sand waves and tidal channels in the entrance to Chesapeake Bay. Virginia Acad. Sci. **21**, 178—184.
Sand bodies 5—12 feet high and 200—1200 feet long occur in water 20—25 feet deep. Surely they must have many ancient equivalents.
See also NEWTON and WERNER (1971).

—, 1974: Tidal currents and zig-zag sand shoals in a wide estuary entrance: Bull. Geol. Soc. Am. **85**, 717—726.
An important paper that applies to both terrigenous and carbonate sand bodies; contains many relevant references to their dynamics. Good source of references.

McGOWEN, J. H., 1971: Gum Hollow fan delta, Nueces Bay, Texas: University of Texas Austin, Bur. Econ. Geol., Rept. Inv. **69**, 91 p.
Excellent source for maps of paleocurrent indicators and their relationship to sand distribution patterns. Case history of a 30-year old, man-made fan deposited at sea level.

McCUBBIN, D. G., 1969: Cretaceous strike-valley sandstone reservoirs, northwestern New Mexico. Bull. Am. Assoc. Petrol. Geologists **52**, 2114—2140.
Individual sandstone bodies localized on seaward side of buried cuestas having local relief of more than 30 m. Sand was transported along shore parallel to the orientation of the buried cuestas. Key words: paleocurrents, sandbody geometry, paleotopography, paleogeology, and longshore transport.

McDONNELL, K. L., 1974: Depositional environments of the Triassic Gosford Formation, Sydney Basin. J. Geol. Soc. Australia **21**, 107—132.
A matrix of primary sedimentary structures is used to interpret the depositional environment of a Triassic sand-shale sequence. Paleocurrents and many informative illustrations.

MILTON, D. J., 1973: Water and processes of degradation in the Martian landscape. J. Geophys. Res. **78**, 4037—4047.
Alluvial (?) sandbodies on the Martian landscape.

NEWTON, R. S., and F. WERNER, 1971: Form und Schichtungsgefüge periodischer Sandkörper im Strömungsfeld des Außenelbewatts. Geol. Rundschau **60**, 321—330.
Mapping of sandwaves on tidal flats gives good indication of current orientation. Discussion of external form and internal oriented structures. Compare with KENDALL and SKIPWITH (1969).

PARK, Y. A., 1974: Migration and textural parameters of intertidal channel sand bars in the tidal environments near Sylt, Schleswig-Holstein (F. R. Germany): Meyniana **24**, 73—89.
Bedforms, sandbody shapes and grain size all very well done and illustrated.

SCHMITZ, U., 1971: Stratigraphie und Sedimentologie im Kambrium und Tremadoc der westlichen Iberischen Ketten nördlich von Ateca (Zaragoza) NE-Spanien. Münster/Westf. **22**, 123 p.
A good example of an integrated field study wherein paleocurrents (cross-bedding, ripples, and fossil orientation) are routinely used along with other sedimentary

structures, trace fossils and vertical sequences to help establish the environment of deposition (English and French summaries).

SEDIMENTATION SEMINAR, 1972: Bethel Sandstone (Mississippian) of western Kentucky and south-central Indiana, a submarine-channel fill. Kentucky Geol. Survey, Ser. X, 24 p.
Excellent correlation between orientation of cross-bedding of marine sandstone and a sharply defined, narrow and very long paleovalley on the North American craton.

SELLEY, R. C., 1976: Subsurface environmental analysis of North Sea sediments: Am. Assoc. Petroleum Geologists. Bull. 60, 184—195.
Argues that modern sedimentary environments tend to generate characteristic vertical sequences of grain size and sedimentary structures (and to some degree composition) which can be perceived by examining a combination of wire line logs plus cuttings and/or cores. Informative schematic drawings, especially Figure 1.

SESTINI, G., 1971: Sedimentology of a paleoplacer: The gold-bearing Tarkwaian of Ghana *in* G. C. AMSTULZ and A. J. BERNARD, eds., Ores in sediments. Berlin-Heidelberg-New York, Springer, 275—305.
Precambrian gold, conglomerates, paleocurrents plus ore body geometry and trends — a very practical application of paleocurrents.

SPEARING, D. R., Compiler, 1974: Summary sheets of sedimentary deposits with bibliographies. Geol. Soc. Am., Misc. Chart 8.
Seven sheets, one each for alluvial fans, alluvial, eolian, regressive shoreline, coastal barriers, tidal and turbidity deposits. Handy and informative with a total of over 500 references, mostly to the English literature. An exceptionally *good and compact* source of information, which should not be overlooked.

SWETT, K., G. deVRIES KLEIN and D. E. SMIT, 1970: A Cambrian tidal sandbody — the Eriboll Sandstone of Northwest Scotland; an ancient recent analog. J. Geol. 79, 400—415.
Regional measurement of cross-bedding and inventory of other structures in a thick, 670 ft, sandstone. Paleocurrents markedly polymodal with weak regional trend. Table 1 is useful example of environmental matching.

THOMPSON, D. B., 1970: Sedimentation of the Triassic (Scythian) red, pebbly sandstones in the Cheshire Basin and its margins. Geol. J. 7, 183—216.
Paleocurrents are part of an environmental analysis with Figs. 7 and 8 good.

VEEN, F. R. VAN, 1971: Depositional environments of the Eocene Miador and Misoa Formations, Maracaibo Basin, Venezuela. Geol. Mijnbouw 50, 527—546.
Outcrops and cores used to recognize fluvial and deltaic sandstones, the latter comparable to the Mississippi. Point-bar, distributary channel, and barrier bar sandstone bodies recognized and profiled. Application to subsurface.

VOS, R. G., 1975: An alluvial plain and lacustrine model for the pre-Cambrian Witwatersrand deposits of South Africa. J. Sediment. Petrol. 45, 480—493.
Trends of gold-enriched placers and their similarity to the channels of a braided stream.

WEBER, K. J., 1971: Sedimentological aspects of oil fields in the Niger delta. Geol. Mijnbouw 50, 559—576.
Sandstone bodies and their depositional environments related to general delta development with good documentation of shape, grain size, and log characteristics. Point and barrier bar plus tidal sandstone bodies as well as discussion of on- and off-lap cycles, each about 50 feet thick. Good discussion of sedimentation and growth faults is an extra dividend.

References

BOUMA, A. H., 1964: Self locking compass. Marine Geol. 1, 181—186.
BROUSSARD, MARTHA LOU, ed., 1975: Deltas. Houston Geol. Soc., 555 p.
CAMPBELL, R. L., JR., 1968: Stratigraphic applications of dipmeter data in Mid-Continent. Bull. Am. Assoc. Petrol. Geologists 52, 1700—1719.

Cox, J. W., 1968: Interpretation of dipmeter data in the Devonian carbonates and evaporites of the Rainbow and Zama areas. J. Canadian Petrol. Tech. **7**, 164—171.

Diessel, C. F. K., and K. H. R. Moelle, 1967: The application of analysis of the sedimentary and structural features of a coal seam and its surrounding strata to the operations of mining *in* J. T. Woodcock, R. T. Madigan and R. G. Thomas, eds., Proc.-gen. 8th Commonwealth Mining and Metallurgical Congr., Australia and New Zealand, 1965, **6**, Paper **36**, 837—859.

Eustace, W. R., 1968: Sophisticated coring. Drill Bit **15**, 16—19.

Fisher, W. L., and L. F. Brown Jr., 1972: Clastic depositional systems — a genetic approach to facies analysis: Annotated outline and bibliography. University of Texas, Austin, Bur. Econ. Geol., 211 p.

Friedman, G. M., ed., 1969: Depositional environments in carbonate rocks. Soc. Econ. Paleontol. Mineral. Sp. Pub. **14**, 209 p.

Fuller, M., 1969: Magnetic orientation of borehole cores. Geophysics **34**, 772—774.

Gilreath, J. A., and J. J. Maricelli, 1964: Detailed stratigraphic control through dip computations. Bull. Am. Assoc. Petrol. Geologists **48**, 1902—1910.

—, J. S. Healy and J. N. Yelverton, 1969: Depositional environments defined by dipmeter interpretation. Trans. Gulf Coast Assoc. Geol. Soc. **19**, 101—111.

—, and R. W. Stephens, 1975: Interpretation of log responses in a deltaic environment *in* D. M. Curtis, J. R. Duncan and D. S. Gorsline, eds., Finding and exploring ancient deltas in the subsurface. Am. Assoc. Petrol. Geologists, Marine Geol. Comm. Workshop, Dallas, 31 p.

Ginsburg, R. N., ed., 1975: Tidal deposits: A casebook of recent examples and fossil counterparts. New York-Heidelberg-Berlin, Springer, 428 p.

Glaze, R. E., ed., 1973: Core book. Wyoming Geol. Assoc., 25th Ann. Symp. and Core Seminar, 178 p.

Glennie, K. W., 1972: Permian Rotliegendes of northwest Europe interpreted in light of modern desert sedimentation studies. Bull. Am. Assoc. Petrol. Geologists **56**, 1048—1071.

Harrison, C. G. A., J. C. Belshe, A. S. Dunlap, J. D. Mudie and A. I. Rees, 1967: A photographic compass inclinometer for the orientation of deep-sea sediment samples. J. Ocean. Technol. **1**, 37—39.

Hayes, M. O., and T. W. Kana, eds., 1976: Terrigenous clastic depositional environments. Univ. S. Carolina Coastal Research Div.-Dept. Geology, Tech. Rept. **11**-CRD, Pts. I and II, 131 and 171 p.

Holland, D. S., C. E. Sutley, R. E. Berlitz and J. A. Gilreath, 1974: East Cameron Block 270, a Pleistocene field. Trans. Gulf Coast Assoc. Geol. Soc. **24**, 89—106.

Holt, O. R., 1973: Some problems in the stratigraphic analysis of diplogs. Trans. Gulf Coast Assoc. Geol. Soc. **23**, 68—73.

Hrabar, S. V., E. R. Cressman and P. E. Potter, 1971: Cross-bedding of the Tanglewood Limestone Member of the Lexington Limestone (Ordovician) of the Blue Grass region of Kentucky. Brigham Young University. Geol. Studies **18**, 99—114.

Hügel, H., 1965: Apparatus for marking and for recovering oriented drill cores. U.S. Patent Office, 3, 207, 239.

Jageler, A. H., and D. R. Matuszak, 1972: Use of well logs and dipmeters in stratigraphic-trap exploration *in* R. E. King, ed. Stratigraphic oil and gas fields. Am. Assoc. Petrol. Geologists Mem. **16**, and Soc. Explor. Geophysicists Sp. Pub. **10**, 107—135.

Kazantsev, M. I., 1967: Spezialuntersuchungen an orientierten Kernen. Z. angew. Geologie **13**, 153—155.

Klein, G. de Vries, 1975: Sandstone depositional models for exploration for fossil fuels. Champaign, Illinois, Continuing Education Publ. Co., 109 p.

Krueger, W. C., Jr., 1968: Depositional environments as interpreted from electrical measurements. Gulf Coast Assoc. Geol. Soc. Trans. **18**, 226—241.

LeBlanc, R. J., 1972: Geometry of sandstone reservoir bodies *in* T. D. Cook, ed., Underground waste management and environmental implications. Am. Assoc. Petrol. Geologists Mem. **18**, 133—189.

MacKenzie, D. B., 1972: Primary stratigraphic traps in sandstones *in* R. E. King, ed., Stratigraphic oil and gas fields. Am. Assoc. Petrol. Geologists Mem. 16, and Soc. Explor. Geophysicists, Sp. Pub. 10, 47—63.

Meckel, L. D., 1975: Holocene sand bodies in the Colorado delta area, Northern Gulf of California *in* B. M. L. Broussard, ed., Deltas: Models for exploration. Houston Geol. Soc., 239—265.

Moelle, K. H. R., and J. D. Young, 1970: On geological and technological aspects of oriented N-size diamond drilling core. Eng. Geol. 4, 65—72.

Morrison, D. R., and B. Carson, 1971: A gyrocompass for measurement of core orientation and core behavior. Deep Sea Res. 18, 935—939.

Multer, H. G., 1971: Field guide to some carbonate rock environments, Florida Keys and western Bahamas. Madison, N. J., Fairleigh Dickinson University, various paging.

Perrin, G., 1975: Comparaison entre des structures sédimentaries à l'affleurement et les pendagemétries de sondage. Bull. Centre Rech. Pau SNPA 9, 147—181.

Pirson, S. J., 1970: Geologic well log analysis. Houston, Gulf Pub. Co., 370 p.

Poupon, A., C. Clavier, J. Dumanoir, R. Gaymard and A. Misk, 1970: Log analysis of sand-shale sequences — a systematic approach. J. Petrol. Technology, July, 867—881.

Reineck, H. E., and I. B. Singh, 1973: Depositional sedimentary environments. New York-Heidelberg-Berlin, Springer, 439 p.

Ricci Lucchi, F., 1975: Miocene paleogeography and basin analysis in the Periadriatic Apennines *in* Coy Squyres, ed. Geology of Italy. Tripoli, Petroleum Exploration Soc. Libya, 111 p.

Rigby, J. K., and W. K. Hamblin, 1972: Recognition of ancient sedimentary environments. Soc. Econ. Paleontol. Mineral. Sp. Pub. 16, 340 p.

Roark, J. J., and R. T. Schweisberger, 1966: Method of formation logging and core orientation by measuring the piezoelectric potential produced in response to an elastic pulse introduced into a formation and core. U.S. Patent Office 3, 243, 695.

Rosfelder, A. M., and N. F. Marshall, 1967: Obtaining large undisturbed and oriented samples in deep water *in* A. F. Richards, ed., Marine geotechnique. Urbana, University of Illinois Press, p. 243—263.

Saitta, S. B., and G. S. Visher, 1968: Subsurface study of the southern portion of the Bluejacket-Bartlesville Sandstone, Oklahoma: Oklahoma City Geol. Soc., 52—68.

Sedimentation Seminar, 1972: Sedimentology of the Mississippian Knifley Sandstone and Cane Valley Limestone in south-central Kentucky. Kentucky Geol. Survey, Ser. X, Rept. Invs. 13, 30 p.

Selley, R. C., 1968: Facies profile and other new methods of graphic data presentation: Application in a quantitative study of Libyan Tertiary shoreline deposits. J. Sediment. Petrol. 35, 363—372.

—, 1970: Ancient sedimentary environments. Ithaca, Cornell University Press, 237 p.

Shawa, M. S., ed., 1974: Use of sedimentary structures for recognition of clastic environments. Canadian Soc. Petrol. Geologists, 66 p.

Shelton, J. W., 1973: Models of sand and sandstone deposits: A methodology for determining sand genesis and trend. Oklahoma Geol. Survey Bull. 118, 122 p.

Shewake, P., 1968: Orientation coring tool. U.S. Patent Office 3, 363, 703.

Visher, G. S., 1965: Use of vertical profile in environmental reconstruction. Bull. Am. Assoc. Petrol. Geologists 49, 41—61.

Voloshin, V., D. D. Nixon and L. L. Timberlake, 1968: Oriented core: A new technique in engineering geology. Bull. Assoc. Eng. Geologists 5, 37—48.

Weimer, R. J., ed., 1973: Sandstone reservoirs and stratigraphic concepts. Am. Assoc. Petrol. Geologists, Reprint Ser. 7 and 8, 212 and 216 p.

Wilson, J. L., 1975: Carbonate facies in geologic history. New York-Heidelberg-Berlin, Springer, 471 p.

Winkel, D. E., 1965: Core orientation. U.S. Patent Office 3,209,823.

—, 1966: Apparatus and material for core orientation. U.S. Patent Office 3,291,226.

Young, J. O., 1965: Diamond drilling core orientation. Broken Hill Proprietary Co., Tech. Bull. 9, 29—32.

Dispersal and Current Systems up to 1963

Introduction

Current systems that deposited ancient clastics can be reconstructed by the mapping of dispersal patterns of clastic particles and unconformities as well as by mapping directional structures.

A dispersal pattern is the map pattern or areal extent of one or more variables whose distribution was controlled by a current system, in either air, water or ice. Dispersal patterns may be displayed, for example, by distribution of boulders, or by pollen concentration, by particle size and roundness, by bed or formation thickness, and by lithologic proportions in a formation or sequence of formations. Such patterns are particularly useful in economic exploration and in situations where directional structures are absent or cannot be readily measured, as for example, in well cuttings. Maximum information is obtained by the combined study of directional structures and dispersal patterns.

Students of glacial drift made the first dispersal studies. In 1740 TILAS described a boulder train of Rapakivi granite in what is now Finland. He recognized that size and concentration of boulders were clues to the location of the source ledges. Mapping of glacial boulder trains is now commonplace and has been successfully employed in prospecting.

Unlike directional structures, dispersal patterns cannot be seen and interpreted at a single outcrop. Instead, dispersal patterns can only be perceived by *systematic areal mapping* of one or more variables. Dispersal patterns may be local or they may be regional and cover many thousands of square miles and pervade a large volume of sediment. Such patterns depend on the direction and competence of the depositing currents.

Many types of dispersal patterns have been recognized. They are defined by two principal sedimentary properties: attributes and scalars (Table 1-1).

Attributes are those properties that are specified only by a *quality* such as a color. The presence or absence of a distinctive type of boulder in glacial drift or the presence or absence of a given heavy mineral or suite in a sandstone, are examples of attributes which, if systematically mapped, may be related to current systems.

Scalar quantities or scalars are specified by magnitude alone. Some, such as, mineral proportion or percentage, grain roundness, and lithologic proportions are examples of dimensionless scalar quantities. Others, such as grain size or formation thickness are dimensional scalar properties. Scalars, like attributes, acquire directional significance only if mapped. By dividing their range of variation into mutually exclusive classes, scalar quantities can be considered as attributes as, for example, the classification of pebble roundness in two classes: round and angular.

Usually attributes are quicker and easier to determine than scalars, for the latter always require *measurement*. Scalar properties, on the other hand, nearly always yield more information than attributes. Most dispersal patterns have been defined by scalar properties. PETTIJOHN (1957, p. 525—587) summarizes much of what is geologically known relating to the causes of areal variations of scalar properties.

The interpretation of dispersal patterns in ancient sediments has developed, like that of directional structures, along two principal paths. The one is largely empirical and depends on the internal relations between different variables. The other depends on direct analogy with either a laboratory experiment or controlled observation of modern sediments. The interpretation of maximum pebble size illustrates both paths. For example, in ancient sediments maximum pebble size was found to decrease systematically in the direction of transport as independently determined by cross-bedding. This observation alone, indicates that maximum pebble size is a useful guide to reconstructing current direction. The other path depends on knowledge of size decrease as observed experimentally in abrasion experiments and in modern streams. Both paths have made important contributions to our understanding of dispersal patterns.

Dispersal Patterns Defined by Composition

Introduction

Dispersal patterns, defined by mineral or pebble composition, may be based solely on the presence or absence of a single component. Usually, however, more information is obtained by estimating the *proportions* of the significant components in each sample, in which case, the pattern can be contoured. Although one mineral species or pebble lithology may suffice, a dispersal pattern usually will be based on more than one component and is defined by appropriate combinations of detrital minerals or rock fragments and is, then, a *multicomponent* or *multivariate system*.

Methodology

The mapping of a dispersal pattern is complicated by the fact that mineral composition and rock proportions in sediments vary with size grade. DAVIS (1951, fig. 2) and POTTER (1955, fig. 2) demonstrate this for gravels. HORBERG and POTTER (1955, fig. 3) and HARRISON (1960, fig. 2) show that the composition of gravel, sand, and silt fractions of tills is, in part, size dependent. Heavy mineral composition also varies with size grade and with sorting of the light minerals (*cf.* HAWKES and SMYTHE, 1931, fig. 1; RUSSELL, 1937, fig. 2; RITTENHOUSE, 1943, fig. 2; VAN ANDEL, 1950, fig. 13). This relationship has been called granular variation (VAN ANDEL, 1950, p. 18).

Several methods have been used to cope with this problem. Most glacial geologists have used a convenient size grade, 8 to 16 or 16 to 32 mm for pebble counts of glacial drift. On the other hand, usually no effort has been made to control granular variation, other than perhaps a selection of samples of approximately equal grain size, in thin-section studies of consolidated sandstones. The major, and even many minor, compositional contrasts of the light minerals, usually transcend all but the most extreme size contrasts.

Considerable effort has been directed at coping with the effects of granular variation on heavy mineral frequencies. VAN ANDEL (1950, p. 18—43) reviews this problem at length. Some investigators have studied the heavy minerals of either a single size grade or of a variable combination of size grades related to the size of the light minerals. Although such procedures minimize the effect of granular variation, it may be difficult to find one or two size grades common to samples of widely varying grain size. Moreover, significant heavy minerals may escape notice, if some size grades are rejected. RITTENHOUSE (1943, p. 1741—1947) suggested the use of hydraulic ratios. Such ratios, although effective, involve considerable work and depend upon good separation of the heavy and light minerals, not always achieved in consolidated sands.

An alternative procedure is to examine the whole heavy mineral-bearing size fraction, usually that portion between 0.062 and 0.50 mm in diameter. This fraction will include all significant heavy minerals and examination of it is rapid making possible the study of many samples. Many samples are usually desirable in order to more accurately delimit mineral province. Matters are facilitated, if the samples examined are of approximately equal grain size. Moreover, a number of studies, based on the entire sand fraction, have shown that contrasts of heavy mineral composition in different mineral province usually transcend differences in size and sorting, especially if many samples are used (VAN ANDEL, 1950, fig. 11 and 15; POTTER, 1955, fig. 6; VAN ANDEL, 1960, fig. 4). Thus, both the entire heavy-mineral and the light-mineral separates of the sand fraction of a sediment should be examined. In special cases it many be desirable to study the light and heavy mineral content of each size grade.

Most workers have expressed the abundance of mineral species in terms of number rather than weight per cent. Number of particles counted per sample has varied widely. Some investigators have counted as many as 500 grains per thin section or slide. More commonly, two hundred point counts per thin section have been used to estimate modal composition and counts of 100 to 200 grains have been deemed adequate for determination of mineral frequencies. Between 50 to 100 pebbles have commonly been counted per outcrop in studies of pebble lithology.

It is generally better to have many samples with a modest number of counts per sample rather than a few samples with a large count per sample.

Samples should, of course, be as unweathered as possible.

Relations Between Mineral Dispersal and Current Systems

Uniformity of current direction and number and location of different sources of supply are the basic factors controlling mineral disperal patterns. Two other factors, abrasion during transport and post-depositional alteration, should also be considered.

Studies of modern fluvial sediments (RUSSELL, 1937, p. 1332—1344; VAN ANDEL, 1950, p. 94—104) show that differential abrasion of sand-sized light and heavy minerals during transport is not of great importance, especially in streams with moderate to low gradients. In contrast, abrasion in the beach environment has been considered to be more intense. Even such abrasion, however, may not be as important as generally believed. Hsu (1960, p. 402), for example, considered that dilution by more mature detritus, rather than abrasion in the present cycle, is responsible for the feldspar-deficient, quartz-rich beach sands of the western Gulf Coast. Thus mineral province boundaries in sands appear to depend little or not at all on abrasion but may be primarily boundaries between detritus from differing source regions. In gravels and in till, however, abrasion of coarse mate-

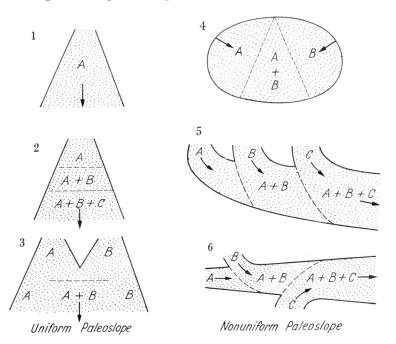

Fig. 8-1. Some common relations between current system, indicated by arrows, and dispersal provinces, indicated by letter

tials can be appreciable. In ancient sediments compositional provinces related ro primary deposition can be better understood, if knowledge of the current system based on directional structures is available.

In ancient sandstones both heavy mineral frequencies and abundance may be affected by post-depositional intrastratal solution (PETTIJOHN, 1941; 1957, p. 514—520), although some investigators doubt that intrastratal solution is of major importance (cf. VAN ANDEL, 1959, p. 156—160). Although there are some well-documented examples of vertical mineral variations attributed to intrastratal solution, there are very few examples of lateral variations within the same formation that have been assigned to this cause.

In any case, the correlation between several petrographic attributes makes possible evaluation of the importance of abrasional changes during transport as well as those due to post-depositional solution. If, for example, variations in the abundance and kind of feldspar and stable light minerals, such as chert and polycrystalline quartz, and the variations in grain roundness are closely correlated with content of specific heavy minerals or varietal types of particular heavy mineral, the role of solution and/or abrasion can be assessed. The possible effects of intrastratal solution may be minimized by relying only on the more stable species (PETTIJOHN, 1957, Table 95). Thus a number of means are available to link firmly most mineral provinces in ancient sediments to the current systems which governed their distribution.

Figure 8-1 shows some basic patterns of mineral dispersal and current systems, assuming that neither new sources of supply are unroofed nor that climatic

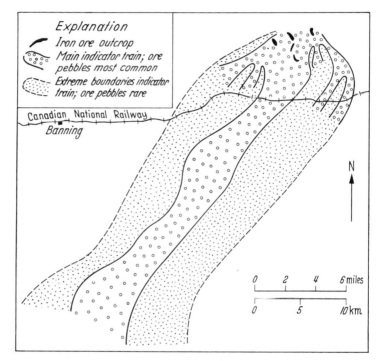

Fig. 8-2. Steep Rock boulder train, Rainy River District, Ontario, Canada (modified from DREIMANIS, 1956, fig. 2)

change produced an altered suite. These patterns are quite independent of agent of deposition and scale and may range from a few to hundreds of miles in size.

Dispersal by Glaciers. Dispersal by glacial ice is probably the best documented of all dispersal systems for its effects have been recognized and studied for well over one hundred years, principally by observation of transported boulders and more recently through systematic pebble counts. Glacially transported boulders may be called *indicators*, if their source ledges are known or, if not, they are called *erratics* (FLINT, 1957, p. 123), although the latter term is often employed for any foreign component of the drift.

Boulder trains are striking examples of glacial dispersal from point sources. Since TILAS first described a boulder train in 1740, many have been identified and mapped. LUNDQVIST (1935) gives a detailed summary of the history and methods of study of glacial dispersion in Scandinavia. CHARLESWORTH (1957, p. 362—375) and FLINT (1957, p. 122—130) summarize much of the English literature on this subject.

Boulder trains may be essentially linear (KNECHTEL, 1942) but more commonly they are fan-shaped and have a pattern such as that shown in figure 8-1-1. They may be mapped by recording the presence of a particular type of boulder on a map. The Iron Hill boulder train of Connecticut (SHALER, 1893), the Alsia Craig boulder train of Great Britain (WRIGHT, 1914, fig. 29) and many others have been delineated in this manner. Figure 8-2 shows the Steep Rock boulder

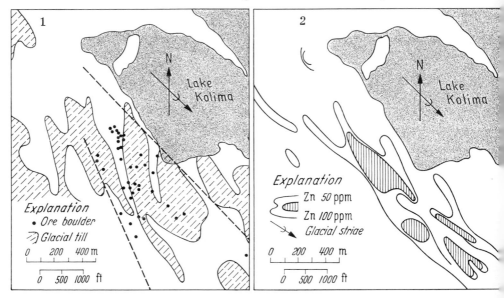

Fig. 8-3. Distribution of ore boulders (1) and trace element concentration of Zn in less than 0.05 mm fraction of C horizon of glacial till (modified from KAURANNE et al, 1961, figs.4 and 5). Note control of distribution patterns by average direction of ice movement

train of Ontario mapped to indicate two qualitative classes of abundance. This mapping clearly illustrates the southwestward dispersion of the boulders from the source ledges. An example that combines both attribute and scalar mapping of abundance is shown by figure 8-3. One map, figure 8-3-1, shows the observed distribution of ore boulders; whereas the other, figure 8-3-2, shows the trace element concentration of Zn. Both map patterns are parallel to the mean direction of glacial movement as recorded by striations and illustrate glacial dispersion on a small scale. Dispersion on a much larger scale is shown in figure 8-4, a map depicting the abundance of Jotnian sandstone pebbles in till. The principal dispersal pattern is controlled by the large area of outcrop of the Jotnian sandstone and direction of ice movement, indicated by striation pattern. The abundance of sandstone pebbles decreases rapidly away from the source ledges. Other recent studies of boulder trains include those by HYYPPÄ (1948), MÖLDER (1948) and SAKSELA (1950). Systematic study of glacial markings and perhaps even till fabric should be recorded when boulder trains are being systematically mapped.

KRUMBEIN (1937, p. 590—594) made a quantitative analysis of glacial dispersal from a point source. He found that boulders per unit area plotted against distance of transport followed the distribution function

$$y = y_0 e^{-ax}$$

where y is concentration in a given area, y_0 the concentration in a unit area at the source, a is a constant for a particular boulder train and x the distance of

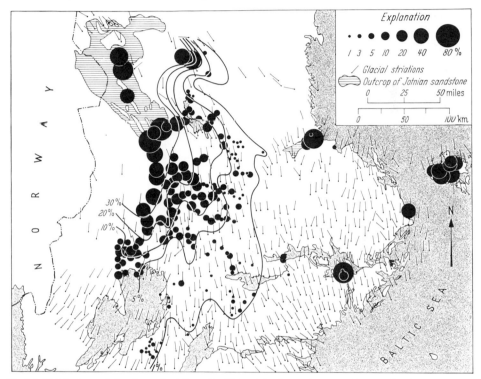

Fig. 8-4. Distribution and abundance of pebbles and cobbles of Jotnian sandstone, southeastern Sweden. Note rapid nonlinear decrease of concentration as ice moved away from source ledges (modified from LUNDQVIST, 1935, fig. 14)

transport. Concentration of a given constituent can be also expressed in per cent based on a pebble count. This equation will plot as a straight line if the logarithm of the concentration is plotted against distance. Down-current decrease in concentration is due to progressive *dispersion*, or scatter, to *dilution* and to *abrasion*, three independent processes, which if they could be independently estimated could be expressed as

$$y_0 e^{-(a_1 + a_2 + a_3)x}$$

where the a_1, a_2 and a_3 are constants for areal dispersion, dilution and abrasion and where $a_1 + a_2 + a_3 = a$. Similar negative exponential functions appear to characterize many dispersal systems.

Dispersal by ice moving across a linear outcrop belt produces a comparable down-current change in abundance of a particular constituent and a mineral association of the type shown by figure 8-1-2. HOLMES' (1952, figs. 6 and 7) work in New York State illustrates such patterns.

Indicator stones or pebbles of distinctive lithology have long been used to identify major source regions or also used to identify the deposits of major glacial lobes. The relations between erratics and their sources were shown by HITCHCOCK (1843, pl. 7) on an early map; the method has been used since by MILTHERS (1909), SEDERHOLM (1910), SLAWSON (1933), MILTHERS (1942), ANDERSON (1955) and many others. ANDERSON (1957) investigated the pebble content in the major glacial lobes of the American Midwest. He found (p. 1431) that far-traveled

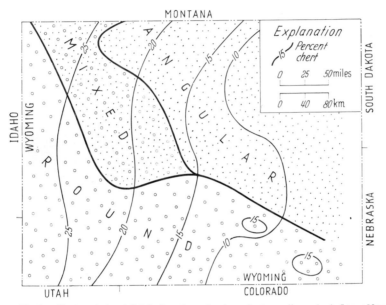

Fig. 8-5. Chert content and distribution of round and angular tourmaline grains in Lower Mesaverde (Cretaceous) sandstone of Wyoming, U.S.A. (modified from PRYOR, 1961, fig. 11)

Precambrian lithologies usually were uniformly distributed within a given lobe and were useful in distinguishing between them. Paleozoic and Mesozoic pebbles, on the other hand, displayed greater variations both within and between lobes. In a related study, HARRISON (1960) reconstructed a hypothetical lobe of glacial ice extending from Quebec to Indiana and compared observed composition of till found in central Indiana with that calculated on theoretical grounds.

The principles and concepts of glacial dispersion have been successfully applied to ore-finding in Fennoscandia (SAURAMO, 1924; GRIP, 1953) but have not been as widely used elsewhere. Maps showing distribution of boulders, their abundance and size, and maps of directional properties together with knowledge of the exponential down-current decrease in particle concentration are the keys to successful ore-finding. Knowledge of direction of ice movement is also vital for successful geochemical prospecting in glaciated areas.

Dispersal by Water and Air. The principal studies of dispersal by water and air have been made on sands. Nearly all date from the development of sedimentary petrology in the 1930's.

Actual examples of dispersal patterns in modern and ancient sands match most of those shown in figure 8-1.

The pattern shown in figure 8-1-1, produced by a uniformly oriented current system and a single source of supply, may be found along the edge of a depositional basin. A subaerial fan, a submarine turbidite fan, and a volcanic ash bed derived from a single vent are other examples. On a larger scale, an intracratonic basin with the source of supply far removed is still another example as is shown by the Cretaceous McNairy sandstone of the upper Mississippi Embayment (PRYOR, 1960, fig. 13 and Table 8).

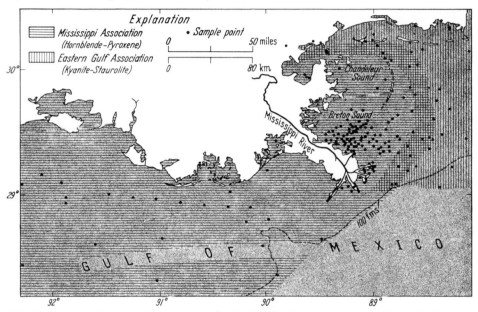

Fig. 8-6. Two mineral provinces with narrow zone of mixing (modified from VAN ANDEL, 1960, fig. 12). Radial currents from Mississippi delta (Mississippi province) have sharp boundary with longshore currents from east (Eastern Gulf province)

The pattern shown in figure 8-1-2 combines a uniformly oriented current system with progressive contamination in the down-current direction, as might occur in a basal transgressive deposit above a major unconformity. The initial deposits would be characterized by the association $A + B + C$ but, as deposition migrated up-dip in the direction of the source area, the association would change to $A + B$ and finally A, as successive sources of supply were progressively overlapped and covered.

A uniformly-oriented current system combined with two different sources of sediment produces the dispersal pattern of figure 8-1-3. Such a pattern could occur in two coalescent fans, in volcanic debris derived from two vents or in the fill of a broad depositional basin. FÜCHTBAUER (1958, fig. 6a) illustrates this basic pattern in the Tertiary Chatt sands of the foreland Molasse of the Alps. PRYOR (1961, figs. 7 and 11) also describes a striking example of this pattern (fig. 8-5). Two distinct mineral provinces and a zone of overlap are present in the Lower Mesaverde sandstones of Cretaceous age in Wyoming. According to PRYOR (p. 45) the source area was to the west. If the pattern of figure 8-1-3 were as wide as 1,000 km, the mixed province $A + B$, might not be recognized in a small basin and would be considered as simply as a single one, C.

Good examples of opposing current systems combined with contrasts in supply (fig. 8-1-4) have been mapped in the Pliocene Lafayette gravel in western Kentucky and adjacent areas (POTTER, 1955, figs. 13 and 17), in the basal Pennsylvanian sandstones of the Illinois basin (SIEVER and POTTER, 1956, fig. 3), and in Pliocene molasse gravels in the Caucasus region (GROSSGEYM, 1959, fig. 3). VAN ANDEL (1960, fig. 12) shows mineral provinces derived from opposing current systems in the modern sands of the Gulf Coast east of the Mississippi delta (fig. 8-6).

Longitudinal transport with asymmetrical supply, as might occur in an oblong basin, is illustrated by figure 8-1-5. FÜCHTBAUER (1958, fig. 2) shows this arrangement in the Tertiary Baustein beds of the Alpine foreland. In figure 8-1-6 longi-

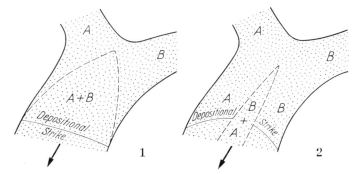

Fig. 8-7. Two oblong basins with uniform longitudinal transport each with two "up-dip" sources of supply. Good mixing in (1) implies little curvature of depositional strike and small mixing of (2) implies stronger curvature

tudinal transport prevails but supply is from either side. Modern streams also illustrate this situation.

Variants of the patterns of figure 8-1 can be obtained if more than three provinces are recognized or by combining two or more patterns. Complexities will multiply rapidly if the character of the supplying source does not remain constant owing to climatic changes or unroofing of new formations. Pleistocene changes in sea level can also produce complicated patterns of mineral provinces on modern marine shelves (cf. VAN ANDEL, 1960, fig. 8; KAGAMI, 1961, fig. 3).

Mineral dispersion patterns can provide information that may not be obtainable from mapping of directional structures. Figure 8-7 indicates, by hypothetical example, how knowledge of mineral provinces can delineate paleoslope more effectively than directional structures alone.

Mineral provinces can also be used to date the emergence of major geologic structures, especially in post-Paleozoic basins. Mineral provinces may be clearly related to positive tectonic elements or their boundaries may be independent of them (cf. PRYOR, 1961, p. 46).

Dispersal Patterns Defined by Scalar Properties

Unlike most compositional aspects which are commonly treated as attributes rather than as scalar properties, pollen concentration, bed thickness, formation thickness, grain size, and lithologic proportions are almost universally treated as scalars. Their areal variation is represented by contours on a map. The resulting contour pattern is a function of the configuration of the source area, particle supply, distance from the source of supply and of the current systems which transported the constituent particles. Identification and interpretation of these contour patterns in ancient sediments is important for the analysis of sedimentary basins for such patterns can yield information about distance to shore lines and source areas that usually cannot be obtained from directional structures alone. Such inferences usually apply to regional rather than local situations.

Mathematically, contour orientation and spacing depend on the scalar distribution function, $F(x, y)$, of a particular variable, say sand-shale ratio. When such a function is known, one can calculate a vector quantity, ∇F, called the gradient,

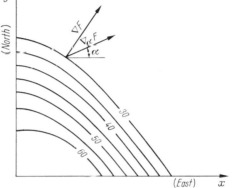

$$\nabla F = \operatorname{grad} F = \frac{\partial F}{\partial x} i + \frac{\partial F}{\partial y} j$$

which gives the direction, at an arbitrarily chosen point in the formation, of maximum rate of change of the scalar variable where x and y are map coordinates and i and j are the unit base vectors along the x and y axes. Knowledge of gradient is usually the best single measure of those systematic changes in mapped data over large areas that define regional sedimentary trends.

Fig. 8-8. Gradient, ∇F, and directional derivative, $\nabla_v F$, of function $F(x, y)$

The directional derivative, $\nabla_v F$, of the function $F(x, y)$ is simply the gradient of an arbitrarily chosen direction from a particular point and is

$$\nabla_v F = \operatorname{comp}_v \nabla F = \frac{\partial F}{\partial x} \cos \alpha + \frac{\partial F}{\partial y} \sin \alpha$$

as is illustrated in figure 8-8.

What information is available about such scalar distribution functions? Typically, they are nonlinear and may have the form of second or higher order polynomials or may be negative exponential functions. Such empirical functions usually are the result of the joint effects of abrasion, of selective sorting due perhaps to down-current decrease in competency, and of progressive dilution.

With the possible exception of down-current decrease in pebble size and bed thickness, theoretical derivation of such distribution functions has not yet generally been obtained. In practice, the investigator has usually been content to infer current systems and direction of source directly from the contour map (higher concentration or magnitude usually indicating proximity to the source) or to estimate empirically regional gradients graphically by vector resolution (cf. SWINEFORD and FRYE, 1951, fig. 5) or by the method of trend surface analysis based on least squares theory (see Chapter 10). Examples of the use of trend surface analysis are given by LIPPITT (1959) and GOODELL and GORSLINE (1961, figs. 2 and 4).

Regional sedimentary gradients of primary origin in a basin can be "telescoped" by subsequent tectonic deformation as in folded mountains of the Jura or central Appalachian type, especially if structural dips commonly exceed 30°, or by large scale overthrusting as in the Alps. If primary interest is the reconstruction of such gradients or in the true spatial relations of directional structures, tectonic shortening must be taken into account and independently estimated as was done by BAUSCH VAN BERTSBERG (1940). However, under certain conditions, it may be possible to estimate the tectonic shortening, from the displaced sedimentary gradient. Variation of maximum pebble size is particularly useful in making such estimates (PETTIJOHN, 1960, p. 449—452). In exploration, however, the present distribution pattern is of greater interest than the predeformation pattern.

Grain Size and Roundness

Areal variation of grain size has received considerable attention since the 1920's. It has been found that size and roundness change markedly with distance of transport, in the case of pebbles, but that sand grains show little or no change in roundness and in some cases no change in size.

Declining competency of water currents, primarily linked to declining gradients, appears to be the principal cause of variation in sands. Abrasion is of no importance. Consequently a change of size with distance can occur in fluvial and turbidite sands (cf. CAROZZI, 1957, fig. 10) but such change is unlikely in the sands of either marine shelf or aeolian dune regimens. However, CURRAY (1961) believed he found some size and sorting gradients in the modern marine shelf sands of the Gulf of Mexico. Poorly defined size-sorting gradients will also be likely in markedly regressive and transgressive deposits. In contrast, air dispersed particles such as volcanic ash falls (THORARINSSON, 1954, fig. 6) and loess (SWINEFORD and FRYE, 1951, fig. 4) show excellent size-distance relationships (cf. U.S. Dept. Commerce, Weather Bureau, 1955, p. 76—87).

Roundness variations have proved to be a guide to dispersal patterns in some ancient sands. Usually, however, in such cases, the presence or absence of rounded material was related to supply rather than to changes in roundness during transport.

Abrasion experiments on the wearing of gravels have been made by DAUBRÉE (1879, p. 249—259), ERDMANN (1879), MARSHALL (1929, p. 325—330), KRUMBEIN (1941), KUENEN (1956) and others. All demonstrate conclusively a decrease in size and increase in roundness with distance of transport. Size decrease of pebbles and cobbles and increase of roundness downstream has been observed and measured by PLUMLEY (1948, figs. 9, 18, 19 and 20), UNRUG (1957, figs. 3 and 9) and others. Much of the earlier European data on size-distance relationships are summarized by SCHOKLITSCH (1930, Table 25). Unlike experimental studies, however, downstream decrease in size of gravel is the result of change in competence as well as abrasion. Both rate of change of size and roundness depend, of course, upon pebble lithology.

In 1875 STERNBERG (p. 486—487) obtained a formula, subsequently termed STERNBERG's law, which linked decrease in pebble size to distance of transport. It is

$$W = W_0 e^{-aL}$$

where W is weight of the largest observed pebble, W_0 the largest initial weight, a is a constant for a particular stream and L is the distance of transport from the source. Maximum pebble diameter, A, or intermediate diameter, B, can also be used as can median size or any other measure of central tendency. This formula embodies the idea that loss of weight is proportional to the weight of the particle and distance of travel. STERNBERG's law fits the observed pattern of size decrease in some rivers but not that in others (cf. PLUMLEY, 1948 p. 546; HACK, 1956, figs. 27, 28 and 33). Reduction of pebble size in alluviating streams may be in better agreement with STERNBERG's Law than is the case with gravels in streams actively eroding their channels. For example, the debris on alluvial fans shows a down-current decline in accordance with STERNBERG's rule (BLISSENBACH, 1954, fig. 5). In ancient deposits, it is necessary to assume an initial weight, W_0, or diameter in order to calculate L. Although reasonable estimates for W_0 can sometimes be made, uncertainty about W_0 limits the accuracy of estimates of L in ancient deposits.

In 1897 LOKHTIN (cited in FLAMANT, 1900, p. 289—292 and LELIAVSKY, 1955, p. 8—10) proposed the relationship

$$C_F = d/S$$

where C_F is a constant for a particular stream and is called the *coefficient of fixation*, d is the diameter of the pebbles and S the gradient of the stream. This formula embodies the idea that it is essentially the gradient that determines size of a pebble at a particular point (SCHEIDEGGER, 1961, p. 171). Utilizing SHULITS' (1941, p. 622) negative exponential expression for downstream change in gradient, SCHEIDEGGER (1961, p. 172) expressed LOKHTIN's formula as

$$W = k e^{-3aL}$$

where W, a, and L are defined as above and k is a constant.

SCHEIDEGGER (1961, p. 172—173) applied a formula, developed by SUNDBORG (1958, p. 250) for sedimentation of suspended material, to pebble size particles to obtain an equation relating size distribution to distance of transport (cf. SAURAMO, 1923, p. 93—96). In this formula, differential transport according to size is the principal factor.

The foregoing studies indicate beyond question that the size of pebbles decreases in the down-current direction, although the relative importance of the factors that produce this decrease are as yet not fully understood. LELIAVSKY (1955, p. 5-10) and SCHEIDEGGER (1961, p. 168—175) review the history and development of efforts to relate decrease of pebble size to distance of travel and to isolate the most significant causal factors, as does PETTIJOHN (1957, p. 533—542).

In ancient sediments areal variation of the gravel size has been specified by the median or modal size in unconsolidated gravels or by some measure of maximum size in consolidated conglomerates or pebbly sandstones. Rigorously, use of the maximum pebble size requires that sample size be constant at each sample point of the area of study. Hence a definite sampling procedure at each sample point is desirable. Some investigators have averaged the diameters of the ten largest pebbles observed at each outcrop.

Areal variation of size of pebbles in ancient sediments has been mapped by FORCHE (1935, fig. 8), PICKEL (1937, fig. 2), FÜCHTBAUER (1954, fig. 2), POTTER

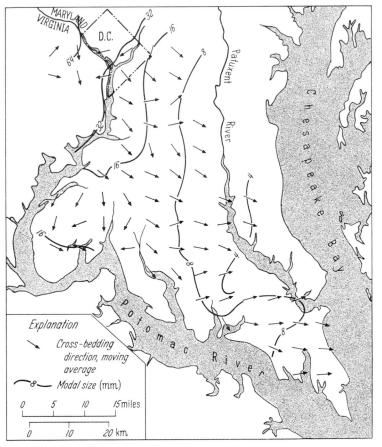

Fig. 8-9. Moving average of cross-bedding orientation and modal grain size in Brandywine gravel. Note that cross-bedding is generally perpendicular to gradient (modified from SCHLEE, 1957, fig. 11)

(1955, fig. 12), BLISSENBACH (1957, figs. 4 and 5), McDOWELL (1957, fig. 12), SCHLEE (1957, figs. 9 and 12), PELLETIER (1958, fig. 14), YEAKEL (1959, p. 159—167) and others in deposits mainly of fluvial or deltaic origin. Far-traveled, rather than locally-derived pebbles, were mapped. The studies of FORCHE, POTTER, SCHLEE, PELLETIER and YEAKEL convincingly show the decline in size of pebbles in the down-current direction inferred from cross-bedding. Figure 8-9 shows these relations in the Brandywine gravel of southern Maryland as does figure 4-10 for the Buntsandstein of northern France and Germany. Size was plotted against distance by BLISSENBACH (1957, fig. 9), SCHLEE (1957, fig. 14), PELLETIER (1958, fig. 15), YEAKEL (1959, fig. 65) and found to decline according to STERNBERG's Law with varying degrees of approximation. By assuming an initial size, W_0, McDOWELL (1957, p. 27—28), SCHLEE (1957, p. 1386—1387), PELLETIER (1958, p. 1056) and YEAKEL (1959, p. 165—166) estimated distance to the margin of the basin.

The foregoing studies suggest that areal variation of size of pebbles is most useful, if the site of deposition has a sharply defined fixed, up-slope, boundary. This situation prevails in those basins separated from their principal source areas by sharply defined mountain fronts. It also seems reasonable to expect a regular size decline of granules and the fine gravel down the slope on sub-marine fans of turbidite origin, although none has as yet been thoroughly demon-strated. Air-sorted ash deposits derived from a single volcano vent also show good size-distance relations (see fig. 8-13). On the other hand, in basal trans-gressive deposits above unconformities, as occur on many stable cratons, pebble size variations may be of little value. Also, if the pebbles are restricted to ancient channel systems, opportunities to sample may be limited and hence the pattern of down-current decline may not be easily observed.

There appear to be no studies of size variation in ancient marine gravels. However, KRUMBEIN (1937, fig. 2) found gravel on a modern lake beach to decline exponentially in size in direction of longshore transport.

Roundness variation of pebbles in conglomerates is of little value in identifying current systems, mainly because only 10 to 20 miles (16 to 32 km), of transport is required to round most pebbles. Thus, unless the edge of the basin has been preserved, marked variations in roundness are not to be expected. PLUMLEY (1948, p. 565—570) and others developed differential equations relating roundness to distance of transport.

Thickness and Lithofacies Maps

Areal variations of thickness and facies were noted early in the study of sedi-mentary rocks. In 1836 facies distribution in the Carboniferous Yoredale Series was represented by a diagram by PHILLIPS (pl. 24). However, apparently HULL (1862, pl. 7) was the first geologist to plot and contour thickness based on field measure-ments on a map. His map, truly a landmark in the study of sedimentary basins, shows the thickness of clastic sediments and limestones in the Carboniferous System in Great Britain (fig. 8-10). The regional scope of this map, the well-displayed complimentary relations between thickness of limestone and the clastic sediments, and the use of contours are particularly striking. HULL's map could easily be the product of research in the latter half of the 20th century rather than that of 100 years ago. Summaries of modern methods and principles of interpre-tation of facies maps are given by DAPPLES et al. (1948), KRUMBEIN (1952), HAUN and LE ROY (1958), BISHOP (1960), LEVORSEN (1960), WELLER (1960), and others.

Maps may show only the variation of thickness of a given stratigraphic unit or lithology, or they may show ratios of different lithologies. Thickness and litho-logic ratios may be shown on the same map. Mathematical functions of litho-logic components such as entropy or distance functions may also be mapped (Chapter 10/I). Whatever is plotted, however, it is ultimately based on thickness measurements, expressed in feet or meters, at each sample point.

Thickness Variation. Paleocurrent inferences that can be drawn from thickness maps vary markedly. A single clastic bed such as a varve, a clastic unit such as

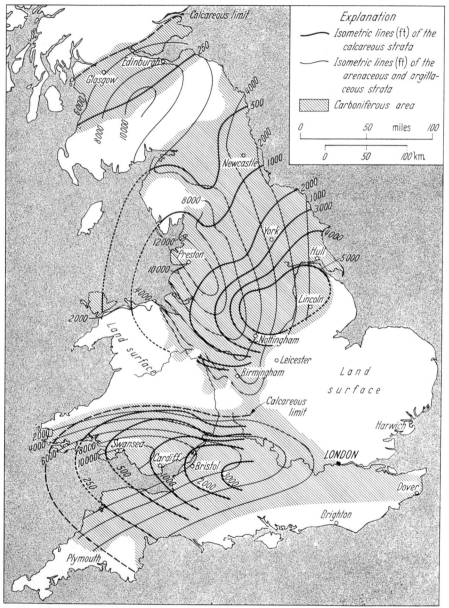

Fig. 8-10. Thickness of clastics and limestones (isometric lines) in Carboniferous system of Great Britain (HULL, 1862, pl. 7)

an ash fall, a loess sheet, and perhaps a graded turbidite bed, or in other words, any clastic deposit that records a single geologic event can supply much information about current systems. In contrast, the relations between total thickness of a clastic sequence in a basin and current direction are usually much more difficult to interpret.

A subaqueous turbidity flow, a volcanic ash deposit, a loess sheet or a glacial varve are examples of single clastic units, usually with appreciable lateral continuity, whose thickness is a function of distance from the source.

Figure 8-11 shows the theoretical relations between thickness and distance of transport in a glacial varve. Actual observation (fig. 8-12) confirms the rapid

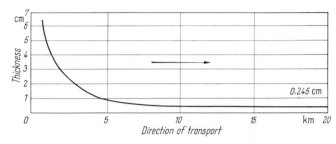

Fig. 8-11. Theoretical thickness of a varve, in direction of transport away from ice front, according to Sauramo (1923, fig. 9). Note rapid nonlinear decrease from source and compare with fig. 8-12

nonlinear decrease in glacial varve thickness away from the source of supply. In general, the thicker a bed, the coarser its grain, and the more rapid its deposition. Positive correlation between grain size and bed thickness was observed by KURK (1941) in glacial outwash and by SCHWARZACHER (1953, fig. 4) in cross-

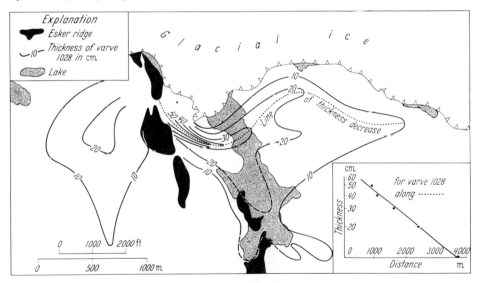

Fig. 8-12. Areal variation of thickness of varve 1028 (modified from DEGEER, 1940, pl. 57). Note approach to exponential decay and compare with fig. 8-11

bedded sands. Thickness and median grain size of volcanic ash deposited downwind from the volcanic explosion in Iceland also demonstrate these relations for both median size and thickness (fig. 8-13); both approximate straight lines, if their logarithms are plotted against distance; both lines display a similar slope. Rapid nonlinear decrease in thickness away from the crater is characteristic of most ash falls (WILCOX, 1959, figs. 68, 71, and 72). Thickness of wind-blown

loess also appears to decline exponentially away from the source area (KRUM-
BEIN, 1937, fig. 4; SMITH, 1942, fig. 4-10; SIMONSON and HUTTON, 1954). Thickness
of individual turbidite beds can also be expected to decrease down-current and
WOOD and SMITH (1959, p. 8 and figs. 1 and 8) cite some evidence that this is
true. Unfortunately, few systematic measurements of change in thickness of
turbidite layers, in the direction of transport, appear to have been made. However, it may be that decrease in thickness of turbidite beds is as rapid as that of
varves and hence changes in bed thickness will not be apparent in most flysch basins, unless the original edge of the basin and proximal part of the bed has escaped erosion. Preservation of the original edge of a flysch basin is probably unusual, especially in turbidites of pre-Mesozoic age.

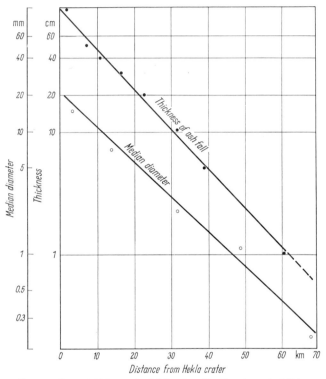

Fig. 8-13. Plots of thickness of ash fall and median diameter against distance from Hekla eruption in Iceland (data from THORARINSSON, 1954, Table 4 and pl. 2). Note correlation between grain size and thickness

Thus, under the rather specialized conditions of a geographically fixed source of supply and of uniform subsidence of the basin of deposition, a clastic sequence consisting of many beds, each of which represents a single pulse of deposition, will become both thicker and coarser toward its source. Under these conditions, maps of grain size, bed thickness and total thickness of section will delineate the current systems. Deposits of varves, turbidites, loess and volcanic ash best show these characteristics.

Maps of single pulses or cycles of sand deposition into a basin usually represent an integrated response to a current system. Fluvial and deltaic sands usually provide the most information. Figure 8-14 shows a map of the Pennsylvanian Booch sands in Oklahoma. Because there is a deep basin to the south, the dendritic pattern shown by this sand is interpreted as a delta distributary system rather than a tributary system. Figure 8-15 illustrates another example of sand pattern in a basin. The complex meandering pattern of sand thicker than 40 feet suggests fluvial deposition. Cross-bedding along the south edge of the basin indicates that flow was to the south. Usually some knowledge of either regional stratigraphic relations or directional structures is necessary to distinguish between fluvial tributary and deltaic distributary sand patterns. "Bundles" of turbidite sands (SULLWOLD, 1961, p.67—74) also appear to be elongated parallel to direction of current flow. On the other hand, barrier beach deposits and both elongate and nonelongate quartz or carbonate sand bodies of many ancient shallow marine seas usually are much more difficult to relate to paleocurrents.

The thickness of clastic deposition within a basin is commonly affected by differential subsidence and hence usually does not lend itself to unambigious

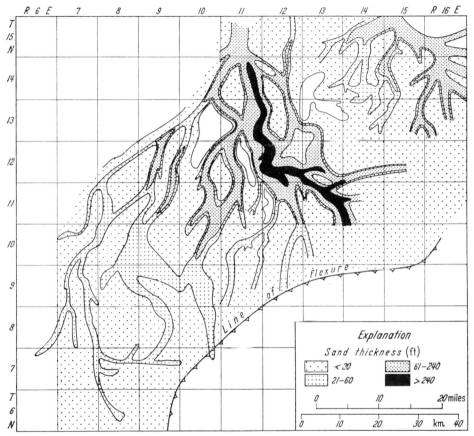

Fig. 8-14. Thickness of sandstone in Booch delta of Pennsylvanian age. Line of flexure marks rapidly subsiding basin to south (modified from BUSCH, 1959, fig. 12)

paleocurrent interpretation. Figure 8-16 shows the possible relations between thickness and paleocurrents in a thinning clastic wedge such as commonly occurs along basin margins. Figure 8-17 shows diagrammatically some relations of current system to isopachs in an intracratonic basin. Obviously, the regional isopach pattern of clastics within a basin is not, by itself, a reliable guide to paleocurrents.

Lithofacies Maps. Many types of facies maps can be made by systematic measurement of stratigraphic sections. Areal variations in thickness or relative proportions of sand, shale and limestone, or variations in thickness or relative abundance of a particular lithology, such as limestone, are most commonly mapped. Usually an isopach map is helpful for the interpretation of a lithofacies

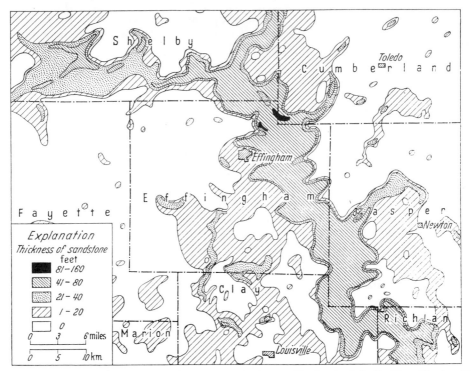

Fig. 8-15. Thickness of Pennsylvanian sandstone between the Summun (No. 4) coal and the Harrisburg (No. 5) coal in east central Illinois, U.S.A. (modified from POTTER, 1962 b, pl. 2). Note meandering pattern of sandstone thicker than 40 feet. Transport direction to south and east

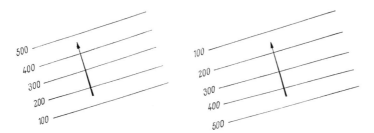

Fig. 8-16. Transport direction (arrow) and isopach along basin margin. Interval thickens basinward, commonly from distal, low yield source (left). Interval thins basinward, commonly from well defined high yield source at basin margin (right)

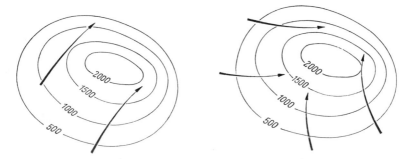

Fig. 8-17. Two subcircular basins with contrasting transport patterns: uniform (left) and laterally opposing (right)

map. Lithofacies maps may be made of either a time or rock unit. Misleading inferences may be obtained, however, if the stratigraphic limits of the mapped unit are not everywhere uniform, e.g., if sections truncated by erosion, as along a basin margin, are included.

Facies maps that show the areal distribution and character of clastic sediments usually contribute most to paleocurrent analysis. Per cent of clastics, clastic ratio, and sand-shale ratio are usually easiest to interpret. However, unless at least some independent supporting evidence from directional structures, or,

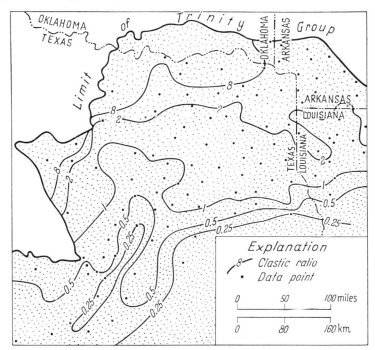

Fig. 8-18. Clastic ratio map of Trinity group (Cretaceous) in portions of Arkansas, Oklahoma, Texas and Louisiana, U.S.A (modified from FORGOTSON, 1960, fig. 1). Note higher concentrations of clastics proximal to northern limit. Mapped unit thickens to south and east and attains a thickness of approximately 2,500 or more feet where clastic ratio is 0.25

say maximum pebble size variation is available, inferences about paleocurrents from facies maps usually are not definitive.

Clastic ratio is defined as ratio of thicknesses of clastic deposits to thickness of nonclastics:

$$\frac{\text{conglomerate} + \text{sandstone} + \text{shale}}{\text{limestone} + \text{dolomite} + \text{evaporite} + \text{coal}}$$

and sand-shale ratio as ratio of thickness of sandstone to thickness of shale, or more generally:

$$\frac{\text{conglomerate} + \text{sandstone}}{\text{shale}}$$

Although not invariably true, higher clastic ratio along the margins of an ancient basin has been taken to indicate proximity to the strand line. In fluvial sediments, and in some marine sediments, higher sand-shale ratio is similarly interpreted. This interpretation usually is supported by other evidence in ancient sediments even though it is at variance with data gathered from many modern marine shelves. Both the irregular topography and erratic distribution of grain size of sediments on many modern shelves have been attributed to the effects of Pleistocene glaciation and related sea level change (SHEPARD, 1959, p. 103—107).

Figure 8-18 illustrates higher clastic ratio proximal to the present northern limit of deposition of the Cretaceous Trinity Group. A map of percentage of coarse

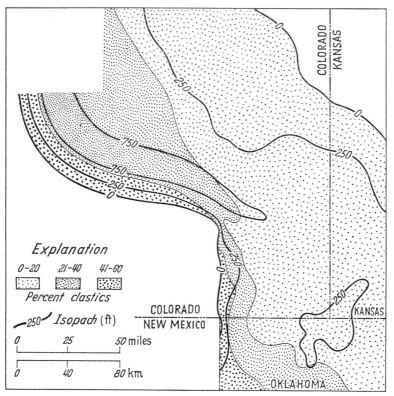

Fig. 8-19. Per cent coarse clastics and isopach of Pennsylvanian sediments of Atoka age (modified from MAHER, 1953, fig. 13). Source presumed to be from the west

clastics in Atoka sediments of Pennsylvanian age illustrates a similar situation (fig. 8-19). Where directional structures have been mapped in outcrop and related to maps of clastic distribution, current system has been found to be at right angles to clastic distribution (AMSDEN, 1955, fig. 1; YEAKEL, 1959, fig. 68; PRYOR, 1960, fig. 18). However, many more integrated basin studies are needed before these relations can be presumed to be generally true.

Figure 8-20 illustrates a "clastic ratio" map of the Mississippian Salem limestone in Indiana. As this formation thickens to the southwest, the ratio of carbonate sand (calcarenite) to carbonate mud (calcilutite) decreases in an irregular manner. In reality, figure 8-20 is a species of sand-shale ratio map. PINSAK

(1957, p. 18) related the changing ratio to water depth. Knowledge of cross-bedding orientation or other directional structures in the Salem limestone could contribute to the interpretation of this map. Systematic study of both directional structures and distribution of carbonate facies is needed for better understanding of relations between paleocurrent systems and carbonate deposition.

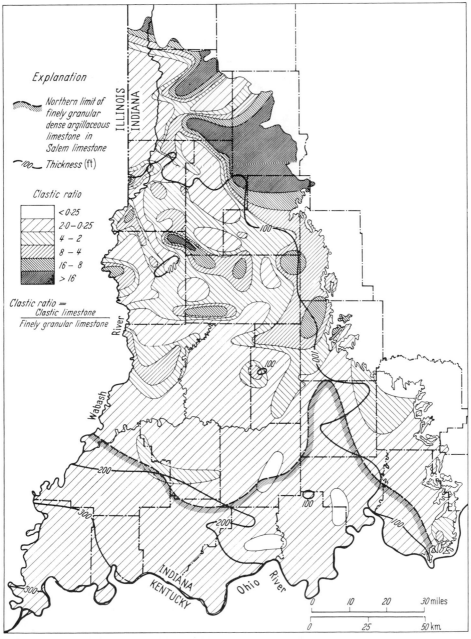

Fig. 8-20. Ratio of coarse to fine-grained limestone and isopach of Salem (Mississippian) limestone in southwestern Indiana, U.S.A. (modified from Pinsak, 1957, pl. 3)

Miscellaneous Examples

The number of discrete sandstone bodies or "beds," as measured by electric logs, has been systematically mapped (KRUMBEIN and NAGEL, 1953, fig. 7; POTTER, 1962a, fig. 11) and may be of some supplementary use in delineating

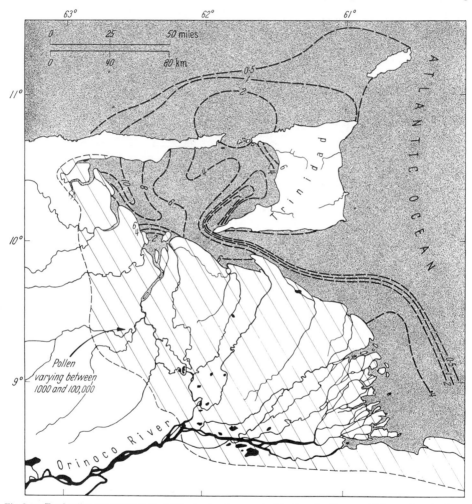

Fig. 8-21. Total pollen concentration (in thousands of grains per gram) in Recent sediments of Orinoco delta (modified from MULLER, 1959, fig. 3). Note rapid nonlinear decrease seaward

paleocurrent systems. For example, KRUMBEIN and NAGEL found the number of sand bodies to be a maximum in a zone parallel to and east of the source of Upper Cretaceous sediments in the Rocky Mountain and Great Plains regions of the United States. They interpreted the maximum to be the mean position of the maximum marine oscillation (p. 957). The number of sand units, however, may be closely related to thickness of section. The mean thickness of sand beds, cross-bedded or not, has been found to increase in the direction of supply (SCHWARZACHER, 1953, p. 329; PELLETIER, 1958, fig. 16).

The concentration (number of particles per gram or unit volume of sediment) of water and air-bourne pollen or microspores is reported to show a rapid non-linear decrease from the strand line and hence be useful in basin analysis (HOFF-MEISTER, 1954 and 1960, and MULLER, 1959). MULLER studied the Recent sediments of the Orinoco delta and found that total pollen content decreased seaward (fig. 8-21) and showed some size decrease away from the delta, which was the principal source. He believed (p. 28), however, that most of the pollen was transported by marine rather than air currents. MULLER suggested that palynological provinces in marine sediments could be defined and used just as mineral provinces have been used to reconstruct clastic dispersal patterns in sands.

Another possible means of delineating a dispersal pattern by the areal variation of a scalar property is by determining the "average age" of sands based on minerals such as zircon or say potash feldspar. Such study might, for example, distinguish between sands derived from an ancient massif versus those from a Mesozoic orogeny. Future study based on geochemical methods will doubtless reveal many other types of dispersal systems.

Evidence from Ancient Buried Landscapes

Important information about paleocurrents and paleoslopes can be obtained by systematic mapping of the ancient buried landscapes or unconformities.

The regional truncation displayed on a paleogeologic map of an unconformity surface is related to regional tilting and hence to the paleoslope during the erosional interval. In general, the more negative the area, the smaller the stratigraphic hiatus across an unconformity. Thus direction of progressive increase in stratigraphic hiatus of an unconformity can serve, as a first approximation, as indicating the direction of a positive area and hence the probable source of clastics in the immediately overlying sediments.

The paleogeologic map at the base of the Pennsylvanian System in the Illinois basin illustrates the principle stated above (fig. 8-22). The clastic sediments of both the underlying Mississippian and the basal overlying Pennsylvanian strata were deposited by currents which flowed southwest down the paleoslope. Only in the northwest corner of the basin was transport to the southeast (SIEVER and POTTER, 1956, fig. 3). During the development of the unconformity, erosion was greater at the northeast, north and northwest portions of the basin than at the south end. In the southern part of the basin southwest-trending pre-Pennsylvanian erosional channels cut into the Mississippian sediments. The effect of channelling is well shown by the distribution of the Kinkaid limestone. The pattern of truncation shown by paleogeologic maps is very useful in reconstructing ancient dispersal systems and paleoslopes and even for appraising possible distribution of sand bodies (cf. BUSCH, 1959, p. 2832—2835). LEVORSEN (1931) early recognized the significance that regional mapping of unconformities have for location of source areas.

Detailed mapping of an unconformity may also reveal an entrenched channel system which, if present, would indicate the direction of paleoslope with precision during at least the last part of the erosional interval. Figure 8-23, shows such a channel system in a portion of the Illinois basin. These channels extend as much as 200 feet (65 m) into underlying sediments and exhibit dendritic outlines. Such a channel system exerts a strong influence on the direction of transport of the immediately overlying sands, especially, if these sands are fluvial. Channel

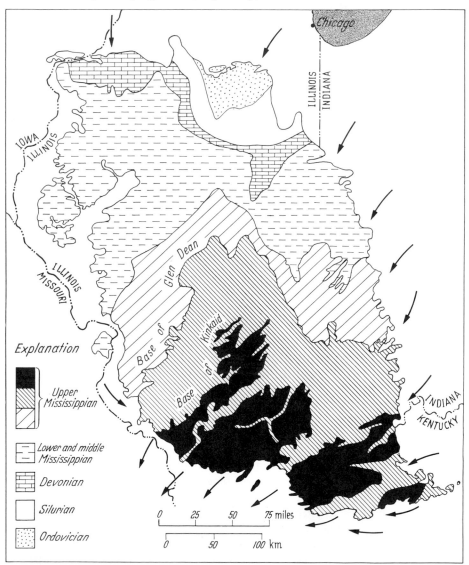

Fig. 8-22. Paleogeologic map at base of Pennsylvanian system in Illinois basin (modified from WANLESS, 1955, fig. 2) and direction of basal Pennsylvanian currents (arrows) inferred from cross-bedding and quartz pebble distribution (POTTER and SIEVER, 1956, fig. 7). Note greater truncation of pre-Pennsylvanian sediments to the northwest, north and northeast of more negative southern part of basin

systems at unconformities may be delineated by paleogeographic maps, by sand thickness maps, or by isopach maps (cf. ANDRESEN, 1961). With sparse subsurface control, one should take care to distinguish integrated channel systems from a buried karst topography.

The direction in which unconformities merge and become compound indicates the direction of a positive area (fig. 8-24). This direction is a probable source of clastics and may well be the up-current direction.

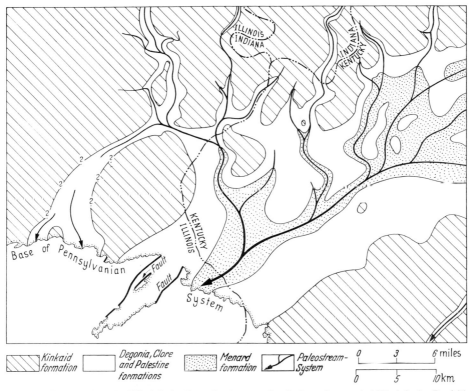

Fig. 8-23. Erosional channels of Mississippian-Pennsylvanian unconformity in southern part of Illinois basin. Dendritic patterns indicate paleoslope to southwest

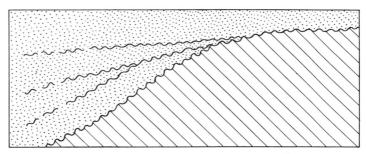

Fig. 8-24. Development of compound unconformities along a basin margin

In both tectonically passive (cratons) and tectonically active (mobile belts) regions, different unconformities, widely separated in time, may show similar patterns of truncation and have similar paleodrainage systems and thus indicate that the distribution of major tectonic elements changed but little. If such is the case, paleocurrent systems recorded in the intervening sediments may be expected to have a corresponding stability and to be related to paleoslopes inferred from the unconformities. Paleocurrent systems of similar orientation can persist for long periods of time so that large segments of the crust can record a uniform transport direction (POTTER and PRYOR, 1961).

Detailed mapping of unconformities can contribute to knowledge of paleocurrents, especially if paleogeologic mapping is combined with critical geomorphic and tectonic analysis.

Summary

The current systems of air, ice or water that deposited clastic particles can be reconstructed by systematic mapping of clastic dispersal patterns. Knowledge of dispersal pattern is often the key to successful exploration of oil and gas and of some types of ore deposits, especially in glaciated areas.

A clastic dispersal pattern may be defined by the areal variation of pebble or mineral composition, grain size, grain roundness, concentration of microspores, bed thickness, lithologic composition and other attribute and scalar properties of sediments. Evidence from ancient buried landscapes can also contribute information about ancient current systems. Both dispersal patterns and unconformities can be especially useful in subsurface studies, where evidence from directional structures has usually been meager.

The study of clastic dispersal patterns began in the 18th Century, in glaciated regions, with the tracing of glacial erratics. Subsequently, boulders and pebble counts have been widely used to reconstruct flow pattern of glacial ice, identify source regions of major glacial lobes, as aids to stratigraphic correlation of till sheets, and as guides to prospecting for ore deposits. Some use has also been made of light and heavy minerals to delineate glacial dispersion patterns.

Dispersion of sands, especially heavy minerals, was noted early in the development of sedimentary petrology. Different methods of analysis, of both the heavy and light minerals, were tried and evaluated so that now comparatively standardized methods permit rapid and trustworthy study of mineral dispersal. Systematic mapping of provinces defined by mineral or pebble content suggests that there are only a few common types of dispersal patterns. These patterns have been used primarily to locate the sources of modern and ancient sands. Distribution patterns in a basin usually gives good information about paleocurrents and, if combined with data from directional structures, a better understanding of basin filling can nearly always be achieved.

Some textural properties have proved useful in delineating ancient current systems. Maximum pebble size commonly shows a measurable decline in direction of transport. On the other hand, pebble roundness has not been very useful as its limiting value is rapidly acquired and hence, in most ancient deposits, there is comparatively little variation of this property. Nor has size variation of water-deposited sands been very rewarding chiefly because such variations are difficult to sample or because they are non-systematic. Sand-sized particles abrade slowly and size-sorting is not a major factor in many environments. Size-sorting is important in air-dispersed ash falls, however, which show good size-distance relations. Roundness of sand grains has been used to delineate mineral provinces, but roundness differences observed are related to source rocks rather than to wear in transit. Down-current decrease in bedding thickness has been noted but not widely used.

Subsurface mapping of sand patterns of fluvial, deltaic and possibly the turbidite origin have proved useful. Formation thickness, on the other hand, is difficult to interpret because one cannot always distinguish between variations due to differential supply from those produced by differential subsidence. Among lithofacies maps, clastic and sand-shale ratio maps are probably the easiest to relate to current system. However, many more integrated basin studies of directional structures and lithofacies are needed before current systems can be inferred with confidence from lithofacies maps alone.

Studies of major unconformities contribute to paleocurrent analyses. One can usually outline positive and negative areas and thus suggest a source of supply and orientation of paleocurrents from mapped unconformities. Detailed mapping may also reveal buried drainage patterns that precisely determine paleoslope. Both the slope of an old erosion surface as well as its local relief may have exercised strong control on current systems which deposited the overlying clastic sediments.

Knowledge of dispersal patterns and of ancient buried landscapes, either alone or integrated with that of directional structures, makes possible a better understanding of paleocurrents and basin sedimentation.

References

AMSDEN, T. W., 1955: Lithofacies map of Lower Silurian deposits in central and eastern United States and Canada. Bull. Am. Assoc. Petrol. Geologists 39, 60—74.
ANDEL, TJ. H. VAN, 1950: Provenance, transport and deposition of Rhine sediments. Wageningen: Veeman & Sons. 129 p.
— 1959: Reflections on the interpretation of heavy mineral analyses. J. Sediment. Petrol. 2 , 153—162.
— 1960: Sources and dispersion of Holocene sediments, northern Gulf of Mexico in Recent sediments, northwest Gulf of Mexico. Tulsa: Am. Assoc. Petrol. Geologists, p.34—55.
ANDRESEN, M. J., 1961: Geology and petrology of the Trivoli sandstone of the Illinois basin. Illinois Geol. Survey, Cir. 316, 31 p.
ANDERSON, R. C., 1955: Pebble lithology of the Marseilles till sheet in northeastern Illinois. J. Geol. 63, 228—243.
— 1957: Pebble and sand lithology of the major Wisconsin glacial lobes of the Central Lowland. Bull. Geol. Soc. Am. 68, 1415—1450.
BAUSCH VAN BERTSBERG, J. W., 1940: Richtungen der Sedimentation in der rheinischen Geosynkline. Geol. Rundschau 31, 328—364.
BLISSENBACH, E., 1954: Geology of alluvial fans in semi-arid regions. Bull. Geol. Soc. Am. 65, 175—190.
— 1957: Die jungtertiäre Grobschotterschüttung im Osten des bayerischen Molassetroges. Geol. Jahrb., Beih. 26, 9—48.
BISHOP, M. S., 1960: Subsurface mapping. New York: John Wiley and Sons. 198 p.
BUSCH, D. A., 1959: Prospecting for stratigraphic traps. Bull. Am. Assoc. Petrol. Geologists 43, 2829—2843.
CAROZZI, ALBERT, 1957: Tracing turbidity current deposits down the slope of an alpine basin. J. Sediment. Petrol. 27, 271—282.
CHARLESWORTH, J. K., 1957: The Quaternary Era, with special reference to its glaciation, vol. 1. London: Edward Arnold Ltd. 591 p.
CURRAY, J. R., 1961: Tracing sediment masses by grain size modes in 21st Internat. Geol. Congr., Session Norden, pt. 23, p.119—130.

DAPPLES, E. C., W. C. KRUMBEIN and L. L. SLOSS, 1948: Tectonic control of litho-facies associations. Bull. Am. Assoc. Petrol. Geologists **32**, 1924—1947.

DAUBRÉE, A., 1879: Études synthétiques de géologie expérimentale, vol. **1**. Paris: Dunod. 828 p.

DAVIS, S. N., 1951: Studies of Pleistocene gravel lithologies in northeastern Kansas. Kansas Geol. Survey, Bull. **90**, 173—192.

DE GEER, G., 1940: Geochronologia suecica, principles, atlas. Kungl. Svenska Vet. Akad. Handl. [3] **18**, 360 p.

DREIMANIS, A., 1956: Steep Rock iron ore boulder train. Proc. Geol. Assoc. Can. **8**, p. 27—70.

ERDMANN, E., 1879: Bidrag till kännedomen om rullstenars bildande: Ett geologiskt experiment. Geol. Fören i Förhandl. **4**, 407—417.

FORGOTSON jr., JAMES M., 1960: Review and classification of quantitative mapping techniques. Bull. Am. Assoc. Petrol. Geologists **44**, 83—100.

FLAMANT, A., 1900: Hydraulique, 2nd ed. Paris: Encyclopédie des travaux publics, Béranger. 689 p.

FLINT, R. F., 1957: Glacial and Pleistocene geology. New York: John Wiley and Sons. 553 p.

FORCHE, F., 1935: Stratigraphie und Paläogeographie des Buntsandsteins im Umkreis der Vogesen. Mitt. geol. Staatsinst. Hamburg **15**, 15—55.

FÜCHTBAUER, H., 1954: Transport und Sedimentation der westlichen Alpenvorlands-molasse. Heidelberger Beitr. Mineral. u. Petrog. **4**, 26—53.

— 1958: Die Schüttungen im Chatt und Aquitan der deutschen Alpenvorlands-molasse. Eclogae Geol. Helv. **51**, 928—941.

GOODELL, H. G., and D. S. GORSLINE, 1961: A sedimentologic study of Tampa Bay, Florida. 21st Internat. Geol. Congr., Session Norden, pt. 23, p. 75—88.

GRIP, ERLAND, 1953: Tracing of glacial boulders as an aid to ore prospecting in Sweden. Econ Geol. **48**, 715—725.

GROSSGEYM, V. A., 1959: Some petrographic and paleogeographic features of sedi-ments in geosynclinal formations (as demonstrated in the Caucasian Fold Province). Izvest. Acad. Sci. U.S.S.R., Geol. Ser., p. 51—62 [English translation by Am. Geol. Inst.].

HACK, J. T., 1956: Studies of longitudinal stream profiles in Virginia and Maryland. U.S. Geol. Survey, Prof. Paper No. 294-B, 45—97.

HARRISON, W., 1960: Original bedrock composition of Wisconsin till in central In-diana. J. Sediment. Petrol. **30**, 432—446.

HAUN, J. D., and L. W. LEROY, 1958: Subsurface geology in petroleum exploration. Golden: Colorado School of Mines, 887 p.

HAWKES, L., and J. A. SMYTHE, 1931: Garnet-bearing sands of the Northumberland coast. Geol. Mag. **69**, 345—361.

HITCHCOCK, E., 1843: The phenomena of drift or glacio-aqueous actions in North America, between the Tertiary and Alluvial Periods. Rept. Assoc. Am. Geol. and Naturalists, Proc. and Trans., p. 169—221.

HOFFMEISTER, W. S., 1954: Microfossil prospecting for petroleum. U.S. Patent Office, 2,686,108.

— 1960: Palynology has important role in oil exploration. World Oil **150**, 101—104.

HOLMES, C. D., 1952: Drift dispersion in west-central New York. Bull. Geol. Soc. Am. **63**, 993—1010.

HORBERG, LELAND and P. E. POTTER, 1955: Stratigraphic and sedimentologic aspects of the Lemont drift of northeastern Illinois. Illinois Geol. Survey, Rept. Inv. **185**, 23 p.

HSU, K. J., 1960: Texture and mineralogy of the Recent sands of the Gulf Coast. J. Sediment. Petrol. **30**, 380—403.

HULL, EDWARD, 1862: On iso-diametric lines, as means of representing the distribution of sedimentary clay and sandy strata, as distinguished from calcareous strata, with special reference to the Carboniferous rocks of Britain. Quart. J. Geol. Soc. London **18**, 127—146.

HYYPPÄ, ESA, 1948: Tracing the source of the pyrite stones from Vihanti on the basis of glacial geology. Bull. comm. géol. Finlande, No. 142, 97—122.

KAGAMI, HIDEO, 1961: Sources of shelf sediments off Sakata, Yamagata, Japan. Japan. J. Geol. and Geography 32, 411—420.

KAURANNE, L. K., ERIC LINDBERG and ERKKU LYYTIKÄINEN, 1961: Heavy metal analysis of humus in prospecting. Bull. comm. géol. Finlande No. 196, 456—472.

KNECHTEL, M. M., 1942: Snake Butte boulder train and related glacial phenomena, north-central Montana. Bull. Geol. Soc. Am. 53, 917—936.

KRUMBEIN, W. C., 1937: Sediments and exponential curves. J. Geol. 45, 577—601.

— 1941: The effects of abrasion on the size, shape, and roundness of rock fragments. J. Geol. 49, 482—520.

— 1952: Principles of facies map interpretation. J. Sediment. Petrol. 22, 200—211.

—, and F. C. NAGEL, 1953: Regional stratigraphic analysis of "Upper Cretaceous" rocks of Rocky Mountain region. Bull. Am. Assoc. Petrol. Geologists 37, 940—960.

KUENEN, PH. H., 1956: Experimental abrasion of pebbles 2. Rolling by current. J. Geol. 64, 336—368.

KURK, E. H., 1941: The problem of sampling heterogeneous sediments. Unpublished M.S. thesis, University of Chicago, 37 p.

LELIAVSKY, SERGE, 1955: An introduction to fluvial hydraulics. London: Constable & Co. 257 p.

LEVORSEN, A. I., 1931: Pennsylvanian overlap in United States: Bull. Am. Assoc. Petrol. Geologists 15, 113—148.

— 1960: Paleogeologic maps. San Francisco: W. H. Freeman & Co. 174 p.

LIPPITT, L., 1959: Statistical analysis of regional facies change in Ordovician Coburg limestone in northwestern New York and southern Ontario. Bull. Am. Assoc. Petrol. Geologists 43, 807—816.

LUNDQVIST, G., 1935: Blockundersökningar, historic och metodik. Sveriges Geol. Undersökn. [3] No. 390, 45 p.

MAHER, J. C., 1953: Permian and Pennsylvanian rocks of southwestern Colorado. Bull. Am. Assoc. Petrol. Geologists 37, 913—939.

MARSHALL, P. E., 1929: Beach gravels and sands. Trans. Proc. New Zealand Inst. 60, 324—365.

McDOWELL, J. P., 1957: The sedimentary petrology of the Mississagi Quartzite in the Blind River area. Ontario Dept. Mines, Geol. Cir. No. 6, 31 p.

MILTHERS, KELD, 1942: Indicator boulders and morphology of the landscape in Denmark. Danmarks Geol. Unders. [2] No. 69, 122—128.

MILTHERS, V., 1909: Scandinavian indicator-boulders in the Quaternary deposits; extension and distribution. Danmarks Geol. Undersøgelse [2] No. 23, 153 p.

MÖLDER, KARL, 1948: Die Verbreitung der Dacitblöcke in der Moräne in der Umgebung des Lees Lappajärvi. Bull. comm. géol. Finlande No. 142, 45—52.

MULLER, JAN, 1959: Palynology of Recent Orinoco delta and shelf sediments. Report of the Orinoco Shelf Expedition; Micropaleontology 5, 1—32.

PELLETIER, B. R., 1958: Pocono paleocurrents in Pennsylvania and Maryland. Bull. Geol. Soc. Am. 69, 1033—1064.

PETTIJOHN, F. J., 1941: Persistence of heavy minerals and geologic age. J. Geol. 49, 610—625.

— 1957: Sedimentary rocks, 2nd ed. New York: Harper & Brothers. 718 p.

— 1960: Some contributions of sedimentology to tectonic analysis. 21st Internat. Geol. Congr., Session Norden, pt. 18, p.446—454.

PHILLIPS, J., 1836: Illustrations of the geology of Yorkshire, pt. II, The Mountain Limestone District. London: John Murray. 253 p.

PICKEL, W. 1937: Stratigraphie und Sedimentanalyse des Kulms an der Edertalsperre. Z. deut. geol. Ges. 89, Abh. A., 233—280.

PINSAK, A. P., 1957: Subsurface stratigraphy of the Salem limestone and associated formations in Indiana. Indiana Geol. Survey, Bull. 11, 62 p.

PLUMLEY, W. J., 1948: Black Hills terrace gravels: a study in sediment transport. J. Geol. 56, 526—577.

POTTER, P. E., 1955: The petrology and origin of the Lafayette gravel, pt. I, Mineralogy and Petrology. J. Geol. **63**, 1—38.
— 1962a: Regional distribution patterns of Pennsylvanian sandstones in Illinois basin. Bull. Am. Assoc. Petrol. Geologists **46**, 1890—1911.
— 1962b: Sand body shape and map patterns of Pennsylvanian sandstones in Illinois. Illinois Geol. Survey, Cir. 339, 36 p.
—, and W. A. PRYOR, 1961: Dispersal centers of Paleozoic and later clastics of the Upper Mississippi valley and adjacent areas. Bull. Geol. Soc. Am. **72**, 1195—1250.
—, and RAYMOND SIEVER, 1956: Sources of basal Pennsylvanian sediments in the Eastern Interior Basin: 1. Cross-bedding. J. Geol. **64**, 225—244.
PRYOR, W. A., 1960: Cretaceous sedimentation in upper Mississippi Embayment. Bull. Am. Assoc. Petrol. Geologists **44**, 1473—1504.
— 1961: Petrography of Mesaverde sandstones in Wyoming. Wyoming Geol. Assoc., 16th Ann. Field Conf., p. 34—46.
RITTENHOUSE, GORDON, 1943: Transportation and deposition of heavy minerals. Bull. Geol. Soc. Am. **54**, 1725—1780.
RUSSELL, R. DANA, 1937: Mineral composition of Mississippi River sands. Bull. Geol. Soc. Am. **48**, 1307—1348.
SAKSELA, M., 1950: Das pyroklastische Gestein von Lappajärvi und seine Verbreitung als Geschiebe. Bull. comm. géol. Finlande No. 144, 17—30.
SAURAMO, M., 1923: Studies on the Quaternary varve sediments in southern Finland. Bull. comm. géol. Finlande No. 60, 164 p.
— 1924: Tracing of glacial boulders and its application in prospecting. Bull. comm. géol. Finlande No. 67, 37 p.
SCHEIDEGGER, A. E., 1961: Theoretical geomorphology. Berlin-Göttingen-Heidelberg: Springer. 333 p.
SCHLEE, JOHN, 1957: Upland gravels of southern Maryland. Bull. Geol. Soc. Am. **68**, 1371—1410.
SCHOKLITSCH, A., 1930: Der Wasserbau, vol. 1. Vienna: Springer. 484 p.
SCHWARZACHER, W., 1953: Cross-bedding and grain size in the Lower Cretaceous sands of East Anglia. Geol. Mag. **90**, 322—330.
SEDERHOLM, J. J., 1910: Sur la géologie Quaternaire et la géomorphologie de la Fennoscandia. Bull. comm. géol. Finlande No. 30, 66 p.
SHALER, N. S., 1893: The conditions of erosion beneath deep glaciers, based upon a study of the boulder train from Iron Hill, Cumberland County, Rhode Island. Harvard Coll. Mus. Comp. Zool. Bull. **16**, 185—225.
SHEPARD, F. P., 1959: The earth beneath the sea. Baltimore: The Johns Hopkins University Press. 275 p.
SHULITS, SAMUEL, 1941: Rational equation of river-bed profile. Trans. Am. Geophys. Union **22**, 622—631.
SIEVER, R., and P. E. POTTER, 1956: Sources of basal Pennsylvanian sediments in the Eastern Interior Basin. 2. Sedimentary Petrology. J. Geol. **64**, 317—328.
SIMONSON, R. W., and C. E. HUTTON, 1954: Distribution curves for loess. Am. J. Sci. **252**, 99—105.
SLAWSON, C. B., 1933: The jasper conglomerate, an index of drift dispersion. J. Geol. **41**, 546—552.
SMITH, G. D., 1942: Illinois loess — variations in its properties and distribution: A pedologic interpretation. Illinois Agr. Exp. Sta. Bull. **490**, 139—183.
STERNBERG, H., 1875: Untersuchungen über Lang- und Querprofil geschiebeführender Flüsse. Z. Bauwesen **25**, 483—506.
SULLWOLD jr., W. H., 1960: Tarzana fan, deep submarine fan of late Miocene age, Los Angeles County, California. Bull. Am. Assoc. Petrol. Geologists **44**, 433—457.
— 1961: Turbidites in oil exploratation *in* Geometry of sandstone bodies — a symposium. Tulsa, Okla.: Amer. Assoc. Petrol. Geologists 63—51.
SUNDBORG, A., 1958: A method for estimating the sedimentation of suspended material. Int. Assoc. Sci. Hydrology, Pub. 43, 249—259.

SWINEFORD, ADA, and JOHN C. FRYE, 1951: Petrography of the Peoria loess in Kansas. J. Geol. **59**, 306—322.

THORARINSSON, SIGURDUR, 1954: The eruption of Hekla 1947—1948, II, 3, The tephra-fall from Hekla on March 29, 1947. Soc. Sci. Islandica, Reykjavik, 68 p.

TILAS, D., 1740: Tankar om Malmletande i anleding af löse gratstenar. Handlingar Kongl. Swenska Vetenskapsakad, p. 190—193.

U.S. Dept. Commerce, Weather Bureau, 1955, Meteorology and Atomic Energy. Washington: U.S. Government Printing Office. 169 p.

UNRUG, R., 1957: Recent transport and sedimentation of gravels in the Dunajec valley (western Carpathians). Acta Geol. Polonica **7**, 217—257.

WANLESS, H. R., 1955: Pennsylvanian rocks of Eastern Interior Basin. Bull. Am. Assoc. Petrol. Geologists **39**, 1753—1820.

WELLER, J. M., 1960: Stratigraphic principles and practice. New York: Harper & Bros. 725 p.

WILCOX, R. E., 1959: Some effects of Recent volcanic ash falls with special reference to Alaska. U.S. Geol. Survey, Bull. **1028**-N, 476 p.

WOOD, A., and A. J. SMITH, 1959: The sedimentation and sedimentary history of the Aberystwyth Grits (Upper Llandoverian). Quart. J. Geol. Soc. London **64**, 163—195.

WRIGHT, W. B., 1914: The Quaternary Ice Age. London: Macmillan & Co. 464 p.

YEAKEL, L. S., 1959: Tuscarora, Juniata and Bald Eagle paleocurrents and paleo-geography in the central Appalachians. Unpublished Ph.D. thesis, The Johns Hopkins University, 454 p.

Dispersal and Current Systems (1963—1976)

A wide diversity of techniques are used in the study of sediment dispersal. Probably the greatest advance since 1963 has been the study of sediment dispersal in the modern oceans.

Studies of dispersal patterns, that is the areal variation of scalar sedimentary properties resulting from sediment transport, are a major part of sedimentology. The literature on this subject is exceptionally widely scattered; it is found in papers on meteorology, pollution and the environment, marine geology, prospecting for metallic minerals, clay mineralogy, glacial geology, paleogeomorphology, and engineering geology, to name but a few. The subject receives the continuing attention of sedimentologists in their study of the provenance of both modern and ancient sands and sandstones. And, if we are willing to include dispersion by ancient man — his trading of valuable products — we should also examine some of the archeological literature!

Annotated References

Rather than write a summary of this vast subject, we have instead chosen to simply give a selected, annotated bibliography organized into what we believe are useful subheadings. We believe perusal of these annotated references will provide an effective initial insight to post-1963 work on sediment dispersal; the statistical techniques of analysis are deferred to the end of the supplement.

Airborne Dust, Ash Falls and Loess

Thickness variations of ash falls and loess have received some attention by sedimentologists, although probably much less than they deserve. For example, how many sedimentology courses discuss these topics? Knowledge of thickness and/or abundance of inorganic airborne materials could be meaningfully mapped in a widespread shale, if one had a fine scale stratigraphy, and thus possibly related to paleowinds and paleogeography — two fundamental aspects of sedimentology.

EATON, G. P., 1964: Windborne volcanic ash. A possible index to polar wandering. J. Geol. 72, 1—35.
Probably still one of the best reviews of thickness patterns of volcanic ash and their controls.
FEHRENBACKER, J. B., J. L. WHITE, H. P. ULRICH and R. T. ODELL, 1965: Loess distribution in southeastern Illinois and southwestern Indiana. Proc. Soil Sci. Soc. Am. 29, 566—579.
Thickness maps plus regression plots of thickness versus distance (for different stratigraphic units) from the Wabash River, a glacial outwash stream.

Frazee, C. C., J. B. Fehrenbacker and W. C. Krumbein, 1970: Loess distribution from a source. Proc. Soil Sci. Am. **34**, 296—301.
Relative merits of logarithmic, exponential and hyperbolic equations to describe loess dispersal.

Glass, H. D., J. C. Frye and H. B. Willman, 1968: Clay mineral composition, a source indicator of Midwest loess *in* R. E. Bergstron, ed., The Quaternary of Illinois. Urbana, University of Illinois, College Agriculture, Sp. Pub. **14**, 35—40.
No maps, but tables and qualitative discussion give the essential ideas. The only study we know of like this — but surely there are (should be) more.

Rose, W. I., Jr., S. Bonis, R. E. Stoiber, M. Keller and T. Bickford, 1973: Studies of volcanic ash from two recent Central American eruptions. Bull. Volcanol. **37**, 338—364.
Maps of thickness and a helpful table of volumes of ash blankets (p. 348); authors use interesting double log plots. Good literature source.

Rozycki, S. Z., 1968: The directions of winds carrying loess dust as shown by analysis of accumulative loess forms in Bulgaria *in* C. B. Schultz and J. C. Frye, eds., Loess and related eolian deposits of the world, **12** Int. Assoc. Quaternary Res., Proc. 7th Congr. Lincoln, University of Nebraska, 233—245.
Maps on various scales show, by thinning and linearities, strong trends and thus paleowind directions. Figure 13-5 is a loess-based Pleistocene paleowind map for eastern Europe.

Ruhe, R. V., 1969: Quaternary landscapes in Iowa. Ames, Iowa State University Press, 255 p.
This monograph contains two sections, p. 29—37 and 114—127, on mathematical and geological analysis of thickness and grain size decline in loess away from its floodplain source. Good as an example and as basic background reading.

— 1973: Background of model for loess-derived soils in the Upper Mississippi River Basin. Soil Sci. **115**, 250—253.
About 100 references on loess in the Upper Mississippi Valley.

Scheidegger, A. E., and P. E. Potter, 1968: Textural studies of grading: volcanic ash falls. Sedimentology **11**, 163—170.
An analytical approach expressing bed thickness as a function of time and grain size (fall velocity).

Turner, D. B., 1970: Workbook of atmospheric dispersion estimates. U.S. Environmental Protection Agency, Research Triangle Park, N.C., AP-26, 7th Printing, 84 p.
Six chapters include the basic equations and needed tables plus problems and their solutions. Reference material for those studying loess patterns as well as ash falls.

Windom, H. L., 1975: Eolian contributions to marine sediments. J. Sediment. Petrol. **45**, 520—529.
Summary and review showing that illite, quartz, and kaolinite are the chief eolian contributors to marine sediments. Key ideas: zonal winds and eolian dust in marine muds.

Arenites and Rudites

A very, very large literature exists. Size decrease of pebbles and boulders in water-transported rudites is an excellent direct guide to source area (Bradley *et al.*, 1972; Füchtbauer, 1967; Hoffman, 1969; Meckel, 1967) and even roundness variation has been used (Laming, 1967). Systematic study of the mineralogy and roundness of sands provides a somewhat more general guide (Davies and Moore, 1970; Dietz, 1973; Gazzi *et al.*, 1973; Ross, 1970). Absolute age dating of individual mineral species (Allen, 1972) offers promise, but studies are still rare. Regional variation of grain size in sands and sandstones continues to be the least studied.

ALLEN, P., 1972: Wealden detrital tourmaline: implications for northwestern Europe. J. Geol. Soc. London 128, 273—294.
A most unusual paper — one solely devoted to provenance and dispersion based on tourmaline types and their $^{40}Ar/^{39}Ar$ dating by Fitch and Miller. Table 2 correlates major orogenic events in much of western Europe with later detritus. Model for the future?

BLATT, H., 1967: Provenance determinations and recycling of sediments. J. Sediment. Petrol. 37, 1031—1044.
A thoughtful philosophical paper about some of the fundamental problems of provenance, especially for sands.

BRADLEY, W. C., R. K. FAHENSTOCK and E. T. ROWEKAMP, 1972: Coarse sediment transport by flood flows on Knik River, Alaska. Bull. Geol. Soc. Am. 83, 1261—1284.
Major changes in coarse gravel occur in the first 16 miles of a glacial river's valley train — plots of size, sorting, and roundness for various lithologies. Much tabulated data.

CARDIGAN, R. A., 1967: Petrology of the Morrison Formation in the Colorado Plateau Region. U.S. Geol. Survey Prof. Paper 556, 113 p.
Paleocurrents and trend surface studies of detrital, light and heavy minerals effectively outline converging flow into a large sedimentary basin. Model study.

CONREY, B. L., 1967: Early Pliocene sedimentary history of the Los Angeles Basin. California Div. Mines Geol., Sp. Rept. 93, 63 p.
Isopach and lithofacies maps, grain size, roundness, sedimentary structures and some petrology of Pliocene turbidites in a tectonically active basin. Notable for emphasis on dispersion as a guide to turbidite fans. Cf. PAYNE (1972).

DAVIES, D. K., and W. R. MOORE, 1970: Dispersal of Mississippi sediment in the Gulf of Mexico. J. Sediment. Petrol. 40, 339—353.
An integration of much existing data as well as many new analyses of the dispersal shadow of a large river into a small ocean.

DIETZ, V., 1973: Experiments on the influence of transport on shape and roundness of heavy minerals. Contr. Sedimentol. 1, 69—102.
One aspect of dispersion that has received little attention in the past 30 years is the experimental study of abrasion, especially abrasion of heavy minerals. This paper compares experimental with natural transport in different environments for short and long distances. Much tabulated data and virtually all the relevant references.

EHRLICH, R., J. J. ORZECK and B. WEINBERG, 1974: Detrital quartz as a natural tracer — Fourier grain shape analysis. J. Sediment. Petrol. 44, 145—150.
Closed-form Fourier amplitude spectra on quartz grain shape plus statistical analysis are used in a detailed provenance study of a river-beach and cliff system in southern California. Innovative, but not yet widely used.

FAUPL, P., W. GRÜN, G. LAUER, R. MAURER, A. PAPP, W. SCHNABEL and M. STURM, 1970: Zur Typisierung der Sieveringer Schichten im Flysch des Wienerwaldes. Austr. Geol. Bundesanst. Wien Jahrb., 113, 73—158.
Unusually well-integrated study of paleocurrents, heavy and light minerals, nannofossils, grain size and pebble morphology. Detailed pebble petrography. Good example of what can be achieved in provenance studies by integration of different disciplines. Figure 4, a provenance-paleocurrent diagram, is unusual.

FÜCHTBAUER, H., 1967: Die Sandsteine in der Molasse nördlich der Alpen. Geol. Rundschau 56, 266—300.
Summary of the author's classic work on sand dispersal related to the Tertiary orogenesis of the Alps. Also shows downcurrent decline of pebble size. Very well referenced.

GAZZI, S. P., G. G. ZUFFA, G. GANDOLFI and L. PAGANELLI, 1973: Provenienza e dispersione litorannea delle sabbie delle spiagge adriatiche fra le foci dell'Isonzo e del Foglia: iquadramento regionale. Mem. Soc. Ital. 12, 1—37.

Very comprehensive study of light and heavy mineral dispersion along the north-west part of the Adriatic coast plus composition of inputting rivers. Very complete tabulated data. Excellent model.

HAHN, C., 1969: Mineralogisch-sedimentpetrographische Untersuchungen an den Fluß-bettsanden im Einzugsbereich des Alpenrheins. Eclogae Geol. Helv. **62**, 227—278.
Exhaustive mineralogical-size study of 226 samples of the Rhine river above Lake Constance. In this rather small region seven heavy mineral provinces exist. Good as a model, if you are planning a super-detailed study!

HOFFMAN, P., 1969: Proterozoic paleocurrents and depositional history of the East Arm fold belt Great Slave Lake, Northwest Territories. Canadian J. Earth Sci. **6**, 441—462.
Exponential decrease in size of pebbles in a Precambrian fanglomerate away from a boundary fault of an aulacogen.

HUBERT, J. F., and W. J. NEAL, 1967: Mineral composition and dispersal patterns of deep-sea sands in the western North Atlantic petrologic province. Bull. Geol. Soc. Am. **78**, 749—752.
Heavy and light minerals of 204 samples from the modern sands of the western North Atlantic, mostly from the deep sea. Three petrographic provinces (includes compositional data from some rivers onshore U.S. and Canada) with Figure 12 being an excellent summary.

INGLE, J. C., JR., 1966: The movement of beach sand in Developments in sedimentology 5. Amsterdam: Elsevier Publ. Co., 221 p.
A general source for a subject widely studied by engineers.

KIRCHNER, H. J., 1974: Die Sedimentverteilung des strandnahen Seebereiches vor Westerland/Sylt. Meyniana **24**, 57—62.
Artificial nourishment of an 8-km block along the North Sea permits good defini-tions of sediment dispersion (defined by 946 sample points), which is both parallel and perpendicular to the shore. Informative illustrations.

LAMING, D. J. C., 1957: Imbrication, paleocurrents and other sedimentary features in the Lower New Red Sandstone, Devonshire, England. J. Sediment. Petrol. **36**, 940—959.
Interesting map of cross-bedding and roundness of limestone pebbles, the latter increasing down current. One of the few dispersal maps based on roundness of pebbles and cobbles. See also LUSTIG (1965).

LUSTIG, L. K., 1965: Clastic sedimentation in Deep Springs Valley, California. U.S. Geol. Survey Prof. Paper **352-F**, 131—192.
All the key variables mapped in detail on a small fan — maximum and mean pebble size, roundness, lithology, percent granules and silt-clay ratios are all combined with slope to estimate tractive force and paths of sediment transport. For different scales of dispersion. Compare Lustig's Figure 112 with Hubert's and Neal's Figure 12.

McMANUS, D. A., J. C. KELLEY and J. S. CREAGER, 1969: Continental shelf sedimen-tation in the Arctic environment. Bull. Geol. Soc. Am. **80**, 1961—1983.
Dispersion in an Arctic sea, one that is ice covered for nine to ten months each year, is mapped with the help of factor analysis.

MECKEL, L. D., 1967: Origin of Pottsville conglomerates (Pennsylvanian) in the central Appalachians. Geol. Soc. Am. Bull. **78**, 223—258.
Dispersion based on maximum pebble size and cross-bedding.

MILLIMAN, J. D., 1972: Atlantic continental shelf and slope of the United States — Petrology of the sand fraction of sediment, northern New Jersey to southern Florida. U.S. Geol. Survey Prof. Paper **529-J**, 39 p.
Dispersion of light minerals of sands on a continental shelf; most sands are residual except for some near shore. Compare with Ross (1970).

NAGAHAMA, H., and A. IIJIMA, 1965: The petrography and sources of the later Tertiary sandstones in northwest Kyushu. Japanese J. Geol. Geog. **36**, 61—75 and 89—134.
A three-part study of paleocurrents and dispersal with very detailed petrography (especially good to see use of varietal types of minerals).

—, O. Hirokawa and T. Enda, 1968: History of researches on paleocurrents in reference to sedimentary structures — with paleocurrent maps and photographs of sedimentary structures. Bull. Geol. Survey Japan **19**, 1—17. (Japanese with English subtitles.)
Figures 9, 10, 11 and 12 are maps that show maximum pebble diameter of different lithologies in a conglomerate. Also additional paleocurrent maps and beautiful photographs of sedimentary structures.

Payne, J. N., 1972: Hydrologic significance of lithofacies of the Cane River Formation or equivalents of Arkansas, Louisiana, Mississippi and Texas. U.S. Geol. Survey Prof. Paper **569-C**, 17 p.
Subsurface lithofacies mapping of a Tertiary unit shows that most of its water resources occur in deltaic and bar sands. Good example of sand dispersion and its relation to hydrology.

Pelletier, B. R., 1974: Sedimentary textures and relative entropy and their relationship to the hydrodynamic environment of the Bay of Fundy *in* B. R. Pelletier, ed., Offshore geology of eastern Canada. Geol. Survey Canada, Paper **74-30**, 77—95.
Textural parameters related to energy fields of a large bay.

Pilkey, O. H., 1963: Heavy minerals of the U.S. south Atlantic continental shelf and slope. Bull. Geol. Soc. Am. **74**, 641—648.
Mapped distribution and percentages of heavy minerals in shelf sands; no well-defined provinces other than epidote-rich and epidote-poor areas.

Ross, D. A., 1970: Atlantic continental shelf and slope of the United States. Heavy minerals of the continental margin from southern Nova Scotia to northern New Jersey. U.S. Geol. Survey Prof. Paper **529-G**, 40 p.
Systematic study of heavy minerals from 229 surface samples — recent and relict sands recognized. Dispersal of sediment is mainly offshore. See also Milliman (1972).

Stanley, K. O., and W. J. Wayne, 1972: Epeirogenic and climatic controls of early Pleistocene fluvial sediment dispersal in Nebraska. Geol. Soc. Am. **83**, 3675—3690
Climatic interpretation of widerspread gravel is based on change in clast size and composition. Activity of a regional arch is also expressed in gravel thickness (epeirogenesis).

Todd, T. W., 1966: Darton Ridge, Pennsylvanian feature of Wyoming Shelf. Am. Assoc. Petrol. Geologists Bull. **50**, 2519—2546.
One of the few areal studies of size distribution in an ancient sandstone. Also some maps of detrital-provenance indicators.

Muds and Mudstones

Long neglected, lateral variations in argillaceous sediments are gradually receiving more attention. However, there is as yet no clearly defined methodology of study comparable to that for arenites and rudites. One possible solution would be to collect systematically all the relevant literature and try to formulate a general methodology for defining their lateral variations — which surely would give significant clues to the origin of muds, modern and ancient.

Bjorlykke, K., 1974: Geochemical and mineralogical influence of Ordovician Island Arcs on epicontinental clastic sedimentation. A study of Lower Paleozoic sedimentation in the Oslo Region, Norway. Sedimentology **21**, 215—272.
Provenance study of a 1000-m thick sequence based on chemical and clay mineral analyses. Vogt's maturity index, defined as $(Al_2O_3 + K_2O)/(MgO + Na_2O)$, plus trace elements, are used to help determine provenance.

Chamley, H., and F. Picard, 1970: L'héritage détritique des fleuves provencaux en milieu marin. Tethys **2**, 211—226.

Clay mineralogy and heavy minerals of small streams that flow into the Mediterranean Sea. Cf. QUAKERNAAT (1968).

CONANT, L. C., and V. E. SWANSON, 1961: Chattanooga Shale and related rocks of central Tennessee and nearby areas. U.S. Geol. Survey Prof. Paper **357**, 91 p.
Mostly careful internal stratigraphy, but contains some maps of sandstone thickness within the shale and cross sections that reveal systematic internal facies variations.

DAVIS, J. C., 1970: Petrology of Cretaceous Mowry shale in Wyoming. Bull. Am. Assoc. Petrol. Geologists **54**, 487—502.
Mineral distribution patterns and other evidence suggests that this widespread shale is a transgressive deposit. One of the very few regional studies of shales. Many maps. A possible model study for this forgotten sediment?

GRIFFIN, J. J., H. WINDOM and E. D. GOLDBERG, 1968: The distribution of clay minerals in the world ocean. Deep Sea Res. **15**, 433—459.
One of the most definitive papers ever written on the origin of clay minerals. This paper is based on geology's oldest technique — systematic mapping — which shows that climate plays a significant role in clay mineral composition at low latitudes, but not at high latitudes. This and other evidence indicate that the vast majority of clays in the recent sediments of the world ocean are detrital.

HATHAWAY, JOHN C., 1972: Regional clay mineral facies in estuaries and continental margin of the United States East Coast *in* BRUCE W. NELSON, ed., Environmental framework of coastal plain estuaries. Geol. Soc. Amer. Mem. **133**, 293—316.
Eleven maps of diverse minerals in Holocene sediments plus two inferred paleo-current systems and one of bottom drift directions. More than 400 samples from Key West to the Gulf of Maine.

KABATA-PENDIAS, A. 1967: Geochemical characteristics of Triassic formations from the north-western areas of Poland. Kwartalnik Geol. **11**, 509—517.
Trace elements and their different ratios, clay minerals plus Eh and pH data form the basis of this study of 368 samples from the subsurface. The illite-chlorite ratio was considered most significant for paleoenvironments.

LISITZIN, A. P., 1972: Sedimentation in the World Ocean. Soc. Econ. Paleontol. Mineral., Sp. Pub. **17**, 218 p.
The best single source for dispersion, in air and water, of land-derived material in the world ocean, mostly but not entirely using Russian references.

NEIHEISEL, J., 1966: Significance of clay minerals in shoaling problems. Comm. Tidal Hydraulics, U.S. Army Corps of Engineers, Tech. Bull. **10**, 30 p.
Clay minerals studied in modern sediments to help define shoaling in harbors.

—, 1976: Techniques for use of organic and amorphous materials in source investigations of estuary sediments *in* BRUCE W. NELSON, ed., Environmental framework of coastal plain estuaries. Geol. Soc. Amer. Mem. **133**, 359—381.
An unusual provenance paper because it is concerned with the fine fraction and its organics as well as amorphous materials. Useful model and technique paper for environmental sedimentology.

PARHAM, W. E., 1966: Lateral variations of clay mineral assemblages in modern and ancient sediments *in* L. HELLER and A. WEISS, eds., Proc. Int. Clay Conf., 1966, **1**, 135—145.
An unusual and interesting paper that compiles much literature about a neglected subject. Author's illustrations effectively tell his story. Many more studies like this needed for mudstones and shales.

—, and G. S. AUSTIN, 1969: Clay mineralogy, fabric and industrial uses of the shale of the Decorah Formation, southeastern Minnesota. Minnesota Geol. Survey, Rept. Invs. **10**, 32 p.
Figures 4 and 5 show clay mineral facies in part of a very widespread shale.

PELZER, E. E., 1966: Mineralogy, geochemistry and stratigraphy of the Besa River Shale, British Columbia. Bull. Canadian Petrol. Geol. **14**, 273—321.
Systematic areal variation of Lower Mississippian mineral and chemical facies of a black shale that varies from 1,000 to 7,000 ft in thickness.

PEVEAR, D. R., 1972: Source of Recent nearshore marine clays *in* B. NELSON, ed., Environmental framework of coastal plain estuaries. Geol. Soc. Am. Mem. 133, 317—336.
Clay minerals of rivers of southeastern U.S. reflect Piedmont weathered soils and contain chiefly kaolinite plus some vermiculite and smectite. River clays nearly identical to those of Piedmont clays. Good example of systematic dispersion–provenance study along a coast line.

PILKEY, O. H., and D. NOBLE, 1966: Carbonate and clay mineralogy of the Persian Gulf. Deep Sea Research 13, 1—16.
Maps of clay minerals plus calcite, aragonite and dolomite all based on 45 samples. Mineralogical trends roughly parallel gulf axis.

QUAKERNAAT, J., 1968: X-ray analysis of clay minerals in some Recent fluvial sediments along the coasts of central Italy. University of Amsterdam, Phys. Geog. Lab., Pub. 12, 105 p.
Systematic mapping of clay minerals in many small rivers indicates mineralogical variation to depend primarily on source rocks.

RATEEV, M. A., Z. N. GORBUNOVA, A. P. LISITZYN and G. L. NOSOV, 1969: The distribution of clay minerals in the oceans. Sedimentology 13, 21—43.
Study of the less than 0.001 mm friction of the clays of the Indian and Pacific Oceans shows kaolinite, gibbsite, and smectite to be maximal in tropical-humid zones whereas chlorite and illite are most abundant in moderate to high latitudes. See also GRIFFIN *et al.* (1968).

Paleoflow from Maps of Unconformities

Paleogeologic mapping of unconformities can show beautifully, if you have enough subsurface data, ancient river systems, which are sensitive guides to source areas, paleoslopes and differential subsidence. It is surprising how hard it is to find examples of paleoflow inferred from paleogeologic maps.

ANDRESEN, MARVIN J., 1962: Paleodrainage patterns: their mapping from subsurface data and their paleogeographic value. Am. Assoc. Petrol Geologists Bull. 46, 398—405.
A short methodology paper with Table 1 describing and comparing and contrasting the five principal methods.

BEUF, S., B. BIJU-DUVAL, O. DE CHARPAL, P. ROGNON, O. GARIEL and A. BENNACEF, 1971: Les grès du Paléozoique inférieur au Sahara. Paris: Éditions Technip, 464 p.
A remarkable study of terrigenous sedimentation on a craton, where both Lower Paleozoic marine and fluvial sheet sandstones have downslope paleocurrents that are parallel to paleovalleys within the sequence and are parallel to the orientation of interbedded proglacial outwash related to a continental ice sheet.

BRISTOL, H. M., and R. H. HOWARD, 1971: Paleogeologic map of the sub-Pennsylvanian Chesterian (Upper Mississippian) surface in the Illinois Basin. Illinois Geol. Survey Cir. 458, 14 p.
Over 53,000 wells used to map paleochannels in Illinois Basin — certainly one of the most detailed subsurface studies of an unconformity ever made. Cross-bedding in overlying and underlying quartz arenites parallels trend of paleovalleys.

CHENOWETH, P. A., 1967: Unconformity analysis. Bull. Am. Assoc. Petrol. Geologists 51, 4—27.
A comprehensive paper, with many examples from North America, of how offlap, overlap, tilting and truncation may be interpreted from unconformities and thus define paleoslopes even though only regional, rather than detailed subsurface, mapping is available.

— 1968: Early Paleozoic (Arbuckle) overlap, southern Mid-continent, United States. Bull. Am. Assoc. Petrol. Geologists 52, 1670—1688.

Basement rocks, basement topography and Cambro-Ordovician overlap on a craton — classical aspects of dispersion studies that should not be forgotten.

CHRISTOPHER, J. E., 1974: The Upper Jurassic Vanguard and Lower Cretaceous Mannville Groups. Dept. Min. Resources, Saskatchewan Geol. Survey Rept. **151**, 349 p.
Contains numerous colored subcrop maps of a large area which are related to paleoslopes and orientation of intervening clinobeds. Also a good source to see how depositional environments are recognized from cores and wire line logs.

DEGRAW, H. M., 1975: The Pierre-Niobrara unconformity in western Nebraska *in* W. G. E. CALDWELL, ed., The Cretaceous System in the western interior of North America. Geol. Assoc. Canada, Sp. Paper **13**, 589—600.
Early Pierre paleotopography appears to have been structurally controlled. Multi-directional drainage suggests proximity to a major divide.

MARTIN, HENNO, 1975: Structural and palaeogeographical evidence for an Upper Paleozoic Sea between southern Africa and South America *in* K. S. W. CAMPBELL, ed., Gondwana Geology. Canberra, Australian Nat. University Press, 37—59.
Paleozoic valleys in the northwestern part of southwest Africa have relief up to 1,000 m and have been filled in by Permo-Carboniferous ice and are now being exhumed after further burial by Lower Cretaceous plateau basalts. This drainage system, oriented toward the present African coast, suggests the presence of a proto south Atlantic Ocean in late Paleozoic time. Ten years ago, how many of us would thought that mapping paleovalley systems could lead to such a conclusion ?

MARTIN, R., 1966: Paleogeomorphology and its application to exploration for oil and gas (with examples from western Canada). Bull. Am. Assoc. Petrol. Geologists **50**, 2277—2311.
Paleogeologic and other maps of buried erosion surfaces can outline buried topography on both a regional and local scale and thus effectively define regional paleoslope as well as helping to find local sandstone bodies.

MCMILLAN, N. J., 1973: Shelves of Labrador Sea and Baffin Bay, Canada *in* R. G. MCCROSSAN, ed., The future petroleum provinces of Canada. Canadian Soc. Petrol. Geologists, Mem. **1**, 473—517.
Figure 23 shows a reconstruction of Tertiary paleodrainage for much of Canada and its associated offshore basins. Certainly one of the most wide-scale drainage reconstructions ever attempted.

STAPP, R. W., 1967: Relationship of Lower Cretaceous depositional environment to oil accumulation, northeastern Powder River Basin, Wyoming. Bull. Am. Assoc. Petrol. Geologists **51**, 2044—2055
Uses a sandstone thickness map to infer a paleovalley system cut into shale.

Pleistocene Tills and Their Ancient Equivalents

There is much continued interest in the study of till dispersion for prospecting purposes as well as for general regional provenance. Here we primarily emphasize Pleistocene tills but pebble counts and petrology have also been used to study the provenance of ancient tills (diamictite) in Paleozoic and older deposits. Perhaps here more than anywhere else in sedimentology, the concept of an "indicator particle", be it a jasper boulder, a diamond or a rich ore, is meaningful — and most practical.

DREIMANIS, A., and W. J. VAGNERS, 1969: Lithologic relation of till to bedrock *in* H. E. WRIGHT JR., ed., Quaternary geology and climate. Washington Nat. Acad. Sci. 93—98.
Clear exposition of changes of till composition in direction of glacial flow. Bimodal frequency distributions near a source rock ledge are replaced by fine-grained "terminal mode" distributions down current.

GUNN, C. B., 1968: A descriptive catalog of the drift diamonds of the Great Lakes Region, North America. Gems and Gemology, 297—303 and 333—334.
Dispersion at its best or worst ? Forty catalogued occurrences and descriptions of diamonds in glacial drift in the Midwest and Ontario. Probably a lower, outer limit of dispersion from one or two point sources in Canada ? See also SKINNER (1972).

HÄKLI, T. A., and P. KEROLA, 1966: A computer program for boulder train analysis. C.R. Soc. géol. Finlande 38, 219—235.
Combining a "down current" negative exponential function with a transverse normal distribution, the authors model the distribution of boulders in a boulder train.

HYVÄRUNEN, L., K. KAURANNE and V. YLETYINEN, 1973: Modern boulder tracing in prospecting in M. J. JONES, ed., Prospecting in areas of glacial terrain. London, Inst. Mining and Metallurgy, 87—95.
An up-to-date review with Scandinavian examples and 28 references, some from 1912. Direct counts supplemented by use of soil maps and air photos. The 12 other papers of this symposium should also be consulted.

MUTANEN, T., 1971: An example of the use of boulder counting in lithologic mapping. Bull. Geol. Soc. Finland 43, 131—140.
357 boulder (> 20 cm) counts in an 8 by 9 km area were contoured. Each rock type was separately mapped. The axes of the boulder fans correlated well with striations and pedogeochemical anomalies. The method gives information about bedrock lithology in poorly exposed areas and clues to the location of concealed rock bodies and their associated ores.

RIDLER, R. H., and W. W. SHILTS, 1974: Mineral potential of the Rankin Inlet, Ennadai Belt. Canadian Mining J., July.
Application of drift prospecting techniques to a specific mineral producing area. Good bibliography.

SHILTS, W. W., 1971: Till studies and their application to regional drift prospecting. Canadian Mining J. 192, 45—50.
Dispersal of till primarily using trace metals. Case histories and methodology.

— 1973: Drift prospecting; geochemistry of eskers and till in permanently frozen terrain: District of Keewatin, Northwest Territories. Geol. Survey Canada Paper 72-45, 34 p.
Interesting and very practical application of glacial dispersion and ore prospecting.

— 1973: Glacial dispersal of rocks, minerals and trace elements in Wisconsinan till, southeastern Quebec, Canada in R. F. BLACK, ed., The Wisconsinan Stage. Geol. Soc. Am. Mem. 136, 189—228.
Very detailed and comprehensive study of a fairly small area. Dispersal assessed by petrology of pebbles, sand and silts plus trace elements in soil, all of which are related to bedrock. Outstanding.

SKINNER, R. G., 1972: Prospecting for diamonds in northern Ontario — a suggestion. Geol. Survey Canada, Rept. Activities Pt. A, Paper 73-1, 218—219.
Brief discussion of the Moose River Basin as the source of the Great Lakes Diamond fan. (See GUNN, 1968).

Dispersal of Spores and Pollen

Gradients of airborne spores and pollen in marine basins have been used to help delimit shorelines (and thus source areas) and to help map paleocurrent systems. But in all truth this use is unfortunately still all too rare. And again, how often is this aspect of palynology a part of a sedimentology course? Both the type and preservation of spore and pollen also can be helpful in determining the depositional environment. Hence sedimentologists and palynologists should routinely become better acquainted with each other.

BECKER, G., M. J. M. BLESS, M. STREEL and J. THOREZ, 1974: Palynology and ostracode distribution in the Upper Devonian and basal Dinantian of Belgium and their dependence on sedimentary facies. Meded. Rijks Geol. Dienst (New Ser.) **25**, 9—99.

Sedimentologists and palynologists work together to produce a masterly analysis of a depositional environment incidental to which there is discussion of the lateral distribution of spores and acritarchs (p. 21—24) and ostracodes (p. 35—36).

BIRKS, H. J. B., and M. SAARNISTO, 1975: Isopollen maps and principal components analysis of Finnish pollen data for 4,000, 6,000, and 8,000 years ago. Boreas **4**, 77—96.

Principal component analysis used to help simplify maps of pollen spectra from 8,000 to 4,000 years.

CROSS, A. T., G. G. THOMPSON and J. B. ZAITZEFF, 1966: Source and distribution of palynomorphs in bottom sediments, southern part of Gulf of California. Mar. Geol. **4**, 467—524.

Based on about 150 cores and 100 grab samples, this paper discusses in detail the factors that control palynomorph distribution — wind, water currents, terrigeneous input (dilution), and size sorting. Numerous maps. Depending on type, some palynomorphs diminish from shore whereas others increase.

DAVEY, R. J., 1970: Palynology and palaeo-environmental studies, with special reference to the continental shelf sediments of South Africa *in* A. FARINACCI, ed., Proc. 2nd Planktonic Conf., Rome: Edizioni Tecnoscienza **1**, 331—347.

Distribution of spores and pollens is related to the type of sediment and distance of deposition from the landmass. Informative discussion, and figures and 35 references.

JEKHOWSKY, B. DE, 1963: Répartition quantitative des grands groupes de "micro-organontes" (spores, hystrichospheres, etc.) dans les sédiments marins du plateau continental. C.R. Soc. Biogeog. France **349**, 29—47.

Off the Orinoco delta the 4,000 sporomorph per gram contour parallels closely the coastline; in addition there is a seaward decrease in concentration.

MANTEN, A. A., 1966: Some current trends in palynology. Earth Sci. Rev. **2**, 317—343.

Pages 324—326 contain some discussion of pollen dispersion in the marine realm.

NEEDHAM, H. D., D. HABIB and B. C. HEEZEN, 1969: Upper Carboniferous palynomorphs as a tracer of red sediment dispersal patterns in the northwest Atlantic. J. Geol. **77**, 113—120.

A fascinating short note that uses palynomorphs — pollen, spores, acritarchs, chitinozoans, etc. – derived from red, on shore Carboniferous sediments in the Maritimes to help identify reworked red sediment carried southward by contour currents along the eastern continental margin of North America — almost to Florida. Thus the reworked palynomorphs serve as unusual, source-area-specific, indicator particles. How many more similar problems invite a comparable solution?

SMITH, N. D., and R. S. SAUNDERS, 1970: Paleoenvironments and their control of acritarch distribution: Silurian of east central Pennsylvania. J. Sediment. Petrol. **40**, 324—333.

Acritarch distribution is in part controlled by prevailing currents; they also give some information about the depositional environment. Authors recommend more awareness of their potential by sedimentologists. Key Ideas: paleogeography, paleoenvironments and acritarchs.

WILLIAMS, D. B., and W. A. S. SARJEANT, 1967: Organic-walled microfossils as depth and shoreline indicators. Marine Geol. **5**, 389—412.

Evidence regarding depth and proximity to shorelines provided by Chitinozoa, acritarchs, spores, and pollen and dinoflagellate cysts is reviewed. New observations on the Niger delta region and the North Atlantic Ocean are presented. The concentration of these groups does show a potential for indicating trends of shorelines but they are of doubtful value as depth indices.

Basin Analysis and the Sedimentary Model up to 1963

Introduction

General Statement

Geologists have not been content just to observe and describe. They have concurrently attempted to collate their observations, to formulate general principles or laws, and to build more comprehensive theories. Students of sedimentary deposits are no exception. The first efforts at interpretation of sediments were directed toward reconstruction of the environment of deposition at a particular time and place. With ever-widening comprehension, the efforts at interpretation have been extended to embrace longer periods of time and to include, at last, the whole basin in which the sediments in question were deposited. *Consideration of the basin as a whole provides a truly unified approach to the study of sediments.*

Basin analysis is closely related to paleogeography. Paleogeography is the geography of the past, usually a particular epoch of the past for which one can depict, on a map, the distribution of land and sea, can delineate the basins of sedimentation, the position of the shoreline and the source lands that provided the sediments which fill the basins. To achieve these ends, data on the present and presumed past distributions of the sediments concerned are required. The character of the sediments needs to be known as well as do the contained faunas. From these the nature of the environment and the distribution of marine and nonmarine deposits can be determined. Although paleogeographic reconstruction has been the concern of geologists since the beginning of stratigraphy as a geological discipline, it has been pursued primarily by stratigraphers. Well known to American students are the paleogeographic maps of WILLIS (1909) and SCHUCHERT (1910). A more modern approach to the problems of paleogeography is given by RUKHIN (1958).

As the efforts and skill with which basins are delineated have been increased and more satisfactory paleogeographic reconstructions have become possible, so also has the nature of the fill within these basins come under closer scrutiny. The student of sediments has come to realize that the distribution of sediments in time and space is not random. As in igneous petrology, where the concept of "consanguinity" (IDDINGS, 1892, p. 128) has long been applied to those igneous rocks which are members of a group associated in time and space and with a community of characters or family likeness, so also has the notion of a "consanguineous association" been extended to sediments. A consanguineous association is a natural group of sediments related to one another by origin. Such cogenetic associations have also been termed "facies", although this term is commonly applied in a very different sense, namely to the areally segregated homogeneous parts of a specific stratigraphic unit.

Terms like "flysch" and "molasse," when used outside of the type area, denote consanguineous associations. They are kindred groups of sediments, of somewhat restricted petrologic character, arranged in a well-defined manner, and believed to be characteristic of certain types of basins or of certain stages in the filling of a specific basin. Flysch is a cogenetic assemblage of rhythmically interbedded marine turbidite sands and pelagic shales. Molasse is likewise an interbedded sequence of sands and shales but differently structured. Both illustrate the concept of a consanguineous association.

The notion of a consanguineous association is an empirically established concept. Like the concept of "race" in anthropology or of "personality" in psychology, difficulties are encountered in defining the term and in applying it. Such difficulties arise from hybrid and atypical associations as well as inadequate knowledge. However, the theories about these associations, such as their tectonic significance or their presumed relation to kinds of basins or their temporal relation or place in a hypothetical geosynclinal cycle, are independent of the existence of such associations. The interpretative aspects are still incomplete, vague in some particulars, and not universally agreed upon.

Another approach to the study of the sedimentary fill within a basin is to study its spatial arrangement. The vast amount of subsurface information now available makes possible what has come to be called "lithofacies analysis." Lithofacies analysis consists in plotting, on a map, the thickness or appropriate thickness ratios of various lithologies within a given stratigraphic interval throughout a basin. Such maps display patterns of greater or lesser regularity. The interpretation of these facies patterns is based on certain assumptions or postulates and on empirically observed relations between the facies distribution and other phenomena. One may, for example (DICKEY and ROHN, 1955, p. 2308), note the relation between the occurrence of oil pools and the sand-shale ratio. If an empirically established relation is found to exist, the facies distribution can then be used to predict as yet untested favorable areas for exploration. The real situation in Nature is no doubt complex, but clearly lithofacies analysis is an attempt to analyze the arrangement of the basin fill and to utilize the pattern for predictive purposes.

As our knowledge of the sedimentary basins of the past has increased, efforts have been made to define and classify the several types of basins, especially the geosynclinal basins. This development is rooted in the recognition of the geosynclinal theory — a unifying concept of great importance in geology (GLAESSNER and TEICHERT, 1947; KNOPF, 1948; AUBOUIN, 1959) — despite the difficulty of formulating and applying a definition of geosyncline. The geosynclinal fill was recognized as an accumulation in a basin with special characteristics. The problems of basin architecture and classification of geosynclines has received special attention in recent years. The interested reader is referred to more extended reviews of the problem by BUCHER (1933), KAY (1944, 1947) and WEEKS (1952)

The study of basins and the fill contained therein has led to the conclusion that there is no one universal principle that can be applied indiscriminantly to all sedimentary basins. Most of the earlier efforts at synthesis are attempts to generalize, to synthesize, and to relate many phenomena. Lacking specificity they provide minimal predictivity. Consequently in more recent years attention has been focussed on the "model concept" (NANZ, 1957).

The Model Concept in Sedimentology

What is a "sedimentary model"? It is an intellectual construct which, as in much of geology, is based on a prototype. The concept of a geosyncline is based on the Appalachian model; the concept of nappes is based on the Alpine originals. Such models have proved useful, stimulating but also limiting and misleading.

The sedimentary model differs from the earlier attempts at synthesis or basin analysis in that the paleocurrent system provides the integrating framework. It determines the distribution, orientation, and make-up of the sedimentary bodies which constitute the basin fill. A sedimentary model in essence describes a recurring pattern of sedimentation. With drastic alteration of the paleocurrent system and the establishment of a new pattern, a different model comes into being. The distribution of facies and sand bodies, the sedimentary fabric and directional structures are all a response to a current system which is closely correlated with basin geometry. Basin geometry or architecture is, in turn, the product of tectonism. Knowledge of the regional tectonic patterns is essential for full understanding of the problem.

The size of the basin is not considered an essential parameter of the sedimentary model. If it were, few if any basins would be alike and the possibility of generalization would not exist.

In short, the model concept embodies the idea that the fill of sedimentary basins is an organized response to a relatively few major dispersal patterns which can be defined and identified by systematic study. With the proper choice of model one should be able to make more successful predictions about those portions of a basin which are concealed and unexplored.

Paleocurrents and the Sedimentary Model

It is our task here to show the relations between the study of the directional properties of sediment, "paleocurrents," and the model concept.

The value of paleocurrent studies and paleogeographic interpretation was clearly seen by SORBY. In one of his first papers (1859, p. 137) he wrote: "It is now several years since I first became convinced that a diligent study of the various structures produced by the action of the currents present during the formation of stratified rock would lead to the knowledge of many very valuable and remarkable facts in connexion with... ancient physical geography." He further states that (p. 145) "a comparison of what may be seen in progress in modern currents... with the structure of deposits formed at earlier epochs, is sufficient to convince anyone that the mere direction of the current can be readily determined in those cases in which its velocity was sufficient to have any decided effect... and that such observations would enable us to learn... the quarter from whence their materials were drifted." SORBY was clearly looking at the sedimentary basin as a whole and saw the now obvious relations between the paleocurrent system and the paleogeography of the basin. This was indeed the prelude to basin analysis.

BRINKMANN (1933) likewise saw the intimate connection between paleocurrents and paleogeography. He made a study of the cross-bedding of the Triassic Buntsandstein and was especially interested in the relations between basinal configuration and the transport direction. But BRINKMANN also saw that other attributes such as grain size, fossil orientation, thickness and peculiarities of the bedding might also, if measured adequately, increase the reliability of paleogeographic reconstruction, a view which he reaffirmed more recently (1955, p.562). BRINKMANN's thesis was soon confirmed by FORCHE (1935), for the Buntsandstein.

In more recent years the problem of basin analysis has been more formally restated and presented in terms of the model concept. NANZ (1957), for example, presented a philosophy of study which involved the formulation of models from which extrapolations or predictions can be made. As noted by NANZ, "when related to tectonic framework and depositional cycle, one sees that the number of sedimentologic 'models' needed is not infinite."

The model concept was further developed at an informal conference (POTTER, 1959), and presented in a more explicit manner still later by PRYOR (1960, p.1501; 1961), PETTIJOHN (1962, p.1469), POTTER (1962a, p.1909) and SLOSS (1962). PRYOR (1960, Table 8) was the first to explicitly develop a model for a specific basin. He listed the different elements of the model and emphasized their interrelationships. POTTER's study of the regional distribution patterns of the Pennsylvanian sandstones of the Illinois basin (POTTER, 1962a, Table 2 extended PRYOR's analysis and led to the formulation of a model of sedimentation integrating the regional patterns of deposition with other aspects of sedimentation. The model includes basin geometry, the lithic fill, the arrangement or distribution of the fill within the basin and the tectonic setting. Earlier PRYOR (1961) noted the similarity between the late Cretaceous sediments of the Mississippi embayment and the Chester sediments (Mississippian) of the Illinois basin. The relations between the major sand bodies in these two depositional basins and the paleoslopes, depositional strikes and basin axes are very similar. Based on the parallelism of these features, a depositional model was developed for this type of intracratonic basin. Common to both the conceptual models of POTTER and PRYOR is the notion that the sand pattern is related to the sediment transport direction which in turn is closely controlled by the paleoslope.

SLOSS (1962) has further developed the model concept. Not only did SLOSS outline the philosophical background for model formulation, but he outlined the salient features of several basin models including those marginal to the craton, the clastic wedge and the turbidite basin models.

It is in respect to paleocurrents that the sedimentation model concept presented in this book differs from the earlier attempts at basin analysis. The studies of the architecture of sedimentary basins, of the distribution of lithofacies within the basin, of consanguineous associations and cycles of basin filling are but partial, incomplete and rather static approaches to the problem of model formulation. Knowledge that one is dealing with a particular sedimentary association, for example, usually does not convey information about the organization or arrangement of the sedimentary elements of the association. Knowledge of such arrangement is vital for successful exploration. The integrating factor heretofore missing is the paleocurrent system — reconstructed by mapping of the directional properties of sediments. Knowledge of the current system is essential to understanding and prediction of the various model elements. The fundamental assumption of the model concept is that *there is a close relationship between the arrangement of major sedimentation elements in a basin and directional structures inasmuch as both are a product of a common dispersal pattern*. It is our intent to develop this thesis in this chapter.

The Continental Glacial Model

Illustrative of the model concept is the glacial model best displayed by the deposits of the Late Glacial epoch in North America and Europe. This model was perhaps the first to be fully worked out — the first in which the movement plan or paleocurrent system was fully understood and mapped. Perhaps this is so because the deposits are so well exposed, because of the freshness of the record, and because of the interest generated by the controversy on the origin of the Drift. The deposits, the topographic forms and the bedrock markings form a related whole — an ordered pattern in terms of distribution, composition, and orientation relative to the movement plan. A resumé of the salient features of the glacial model is most instructive.

Glacial deposits form a consanguineous association — a cogenetic group of materials. Included here are the ice-deposited till, fluvial and eskerine outwash sands and gravels, glaciolacustrine varved silts and clays, and peat and perhaps also the aeolian silt-loess.

These deposits, petrologically distinct entities, are arranged spatially and temporally in an orderly manner (fig. 9-1). In part the arrangement is expressed by topographic form. This arrangement, however, is best understood in terms of the relations to the movement of the ice. The ice front is normal to the lines of flow; hence the terminal and recessional morainal belts are transverse to the line of movement as are the annual moraines. The drumlins, on the other hand, are longitudinal deposits parallel to the flow direction. The outwash aprons flanking the transverse moraines are also transverse accumulations of sand and gravel. The sand and gravel of the eskers is a longitudinal accumulation — a "shoestring sand" parallel to movement. There is also a normal glacial cycle and hence a standard stratigraphic order. This order is characterized by scored and fluted bedrock overlain by till, which in turn is overlain by outwash and/or aeolian silt or loess. Superimposed on these materials is the profile of weathering. Multiple glaciations lead to superimposition of similar cycles.

The composition of the drift reflects the movement pattern. The material derived from the bedrock has a distribution and concentration gradient closely related to the direction and distance of ice flow (figs. 8-2, 8-3, and 8-4). Boulder trains lie down-current and diverge from the source ledges. The concentration of a particular component falls off exponentially (KRUMBEIN, 1937, p. 590—594) in the same direction.

The paleocurrent system is best discovered by examination of the scorings and markings of the bedrock over which the ice moved. The grooves and striations, gouge and chatter marks and knob-and-trail all record the direction of ice flow.

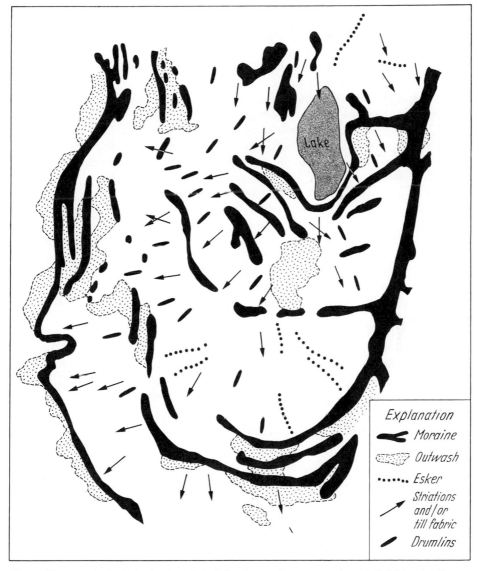

Fig. 9-1. Diagrammatic sketch showing theoretical relations between ice-movement plan and glacial deposits. Compare with figure 5-16

Mapping of such features reveals the regional movement pattern. Although locally parallel, the striations show a regional divergence reflecting the advance of ice lobes, or, in a larger view, a radial dispersion from one or several ice centers. The till fabric is closely correlated with the flow pattern — the long axes of till stones being generally parallel to the ice flow (fig. 3-6). Fabric analyses in the absence of bedrock markings are sufficient to establish the flow pattern. Cross-bedding in the outwash deposits likewise probably shows a transport direction in harmony with the general transport scheme. In summary, the movement plan is the integrating framework in which the distribution of facies — till, outwash, etc., orientation of the sand bodies — eskers and outwash aprons, longitudinal and transverse, respectively, and topographic forms such as drumlins, morainal belts and others, are set. All of the features associated with an ice sheet or ice lobe are meaningless except when viewed in light of the flow pattern or paleocurrent system.

Working out of the glacial model involves measurement and mapping. Included are (1) mapping of lineations on bedrock such as striations, grooves and others, (2) mapping of till fabric, (3) boulder and pebble counts and mapping of boulder trains, (4) mapping of the deposits and delineation of the land forms closely correlated with those deposits such as drumlins, annual and other moraines, eskers, and outwash aprons, (5) determination of drift thickness.

A full understanding of the sedimentary framework and the movement plan enables one to make predictions — predictions in some cases of economic value. In the latter category are the varied examples of location of the source ledges of drift boulders illustrated by the spectacular discovery of ore deposits in Finland, Canada and other places. The several "sand bodies" — the eskers and outwash deposits — are of economic worth as sources of sand and gravel and as aquifers.

Is there an analogue of the glacial model in ancient sedimentary basins? Is the distribution of facies in these basins likewise related to the movement plan? Are the sand bodies longitudinal or transverse or both? Does the dispersion of material show the same distribution pattern and decline of concentration as in the glacial model? And do the deposits display internal features, fabric, sole markings and cross-bedding reflecting the transport pattern? It is our contention that these questions can all be answered in the affirmative.

Analysis of Sedimentary Basins

The model concept and the problems of basin analysis are best illustrated by examples. Hence we present, in capsule form, the salient facts about several well-worked out basins which serve as prototypes for several models of sedimentation.

We first present the Paleozoic molasse basins of the central Appalachians, then the Mississippi Embayment type basins, and finally the salient features and problems of turbidite basins. As is common practice in geology, we have named the several models after particular basins which serve as prototypes and which have been well described and documented.

Central Appalachian Molasse Basins

The filling and deformation of the Appalachian geosyncline is complex. At two times in the development of the central Appalachians were flysch-like sequences succeeded by a molasse accumulation (fig. 9-2). The first of these molasse deposits was formed at the close of the Ordovician and during the early Silurian

(YEAKEL, 1959). The deposits consists of the red mudstones and sandstones of the Juniata formation and the conglomeratic quartzites and protoquartzites of the Tuscarora formation. The second molasse consists of the Catskill red clastics and the overlying conglomerates, proto-quartzites and shales of the Pocono of Devonian and Mississippian ages respectively (PELLETIER, 1958).

These molasse deposits followed a long period of marine sedimentation which began with marine carbonates and dark shales which pass upward into a turbidite flysch. Clastic sedimentation culminated in the deposition of the coarse nonmarine molasse — a dominantly alluvial accumulation, a product of wastage and erosion of the adjacent orogenic uplift.

These molasse accumulations resemble each other very closely and display both similar petrology and sedimentary structures (Table 9-1). The red clastics are indistinguisable from one another; both exhibit marked cross-bedding; intraformational shale-pebble conglomerates, and other structures. The conglomeratic quartzites of both accumulations

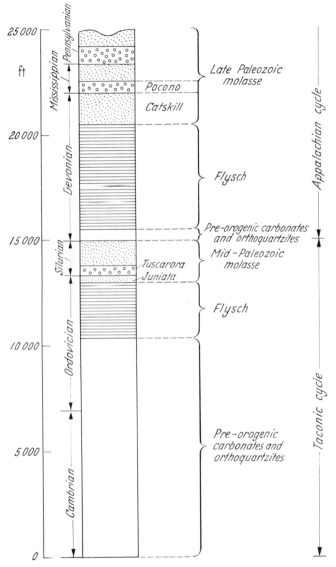

Fig. 9-2. Simplified stratigraphic column, central Appalachians

form the two most prominent resistant ridge-forming units in the central Appalachians. Both likewise are cross-bedded, contain shale fragments, and exhibit *Arthrophycus* or similar structures. Cross-bedding is the most common directional structure. It is of the same scale and type in both the Tuscarora and the Pocono. The conglomerates in each contain mainly vein quartz and quartzite pebbles. The Pocono, unlike the Tuscarora, is coal-bearing. It seems probable that

Table 9-1. *Central Appalachian Basin Model*

Unit	Basin Geometry	Directional Structures	Lithic Fill	Arrangement	Tectonic Setting
A. Midpaleozoic (Taconic) Molasse					
Juniata (Ordovician) Tuscarora (Silurian)	Basin elongated parallel to tectonic strike; asymmetrical transverse cross-section.	Chiefly thick cross-beds which reflect paleoslope transverse to axis of basin; paleocurrents flow north-west (280°—290°).	Juniata: Red impure sandstones, conglomerate and mudstone. Tuscarora: Protoquartzites, subgraywacke, conglomerate and shale.	Total section thins down paleoslope; sand-shale ratio diminishes down slope; fluvial becomes marine westward.	Moderate to strong subsidence; strong uplift of source area on proximal margin of basin.
B. Late Paleozoic (Appalachian) Molasse					
Catskill (Devonian) Pocono (Mississippian)	Basin elongated parallel to tectonic strike; asymmetrical transverse cross-section.	Chiefly thick cross-beds which reflect paleoslope transverse to axis of basin; paleocurrents flow north-west (290°—305°).	Catskill: Coarse, red, impure sandstones and mudstones. Pocono: Protoquartzites and subgraywacke, conglomerate, shale and a little coal.	Total section thins down slope; sand-shale ratio diminishes down slope; fluvial becomes marine westward.	Moderate to strong subsidence; strong uplift of source area on proximal margin of basin.

these deposits are largely nonmarine over most of their outcrop area. If traced far enough westward, however, both contain marine fossils.

The molasse sands are ill-sorted, generally angular or poorly rounded. Rock particles, mainly low-rank metamorphic materials, phyllites and slates, and sedimentary debris — shale and silt stone — are most abundant. The Pocono is a distinct "salt-and-pepper" sandstone. On the whole, the Tuscarora is a cleaner sand, in places well-sorted and rounded, but in its eastern most exposures it more closely resembles the Pocono.

Not only are the two sequences alike in their gross and microscopic aspects, but they also display similar patterns of sedimentation (figs. 9-3 and 9-4). The Pocono is thickest at its eastern or southeasternmost limits where it attains a thickness near 2,000 feet (650 m). North and westward it thins (in part due to erosional truncation) until some two hundred miles (320 km) down-dip it is reduced to less than two hundred feet (65 m). The Tuscarora likewise attains its maximum thickness at its southeastern limit. Its maximum thickness is doubtful owing to stratigraphic uncertainties but it is probably in excess of 1,000 feet (300 m).

Both the Silurian and Mississippian formations show a facies change westward. The proportion of shale increases; that of sand decreases. The sand-shale ratio of 1.0 is attained near the place where the marine fauna appears. The facies strike is northeast-southwest parallel to both the tectonic strike and to the lines of equal thickness. Sand greatly exceeds shale in the exposed parts of the two molasse accumulations. Other facies attributes show a similar parallelism. The maximum pebble size in the conglomerates of both deposits is greatest where these deposits are thickest. The pebble size diminishes exponentially northwestward. The line of equal pebble size — which defines the sedimentary strike — is parallel to the facies strike which in turn follows the present trend of the Appalachians.

The size decrease in the gravels is a down-current phenomenon as is readily verified by mapping the cross-bedding azimuths. In both the Juniata-Tuscarora and the Catskill-Pocono the cross-beds show a transport pattern from southeast to northwest. Both the changing sand-shale ratio and the decline in the maximum pebble size are a response to these paleocurrent systems.

The similarities in the patterns of sedimentation in these two Paleozoic formations is further shown by the distribution of the oil and gas fields. The so-called "Clinton" gas fields of Silurian age occur in a well-defined linear, northeast-trending belt parallel to and somewhat west of the 1:1 sand-shale ratio line. Similarly the late Devonian-early Mississippian oil and gas pools occur in a belt likewise parallel to the sedimentary strike and west of the 1:1 sand-shale ratio line. In both cases also the individual pools are stratigraphic traps — elongate sand lenses in a shale matrix. The elongation of the individual pools parallel to the present Appalachians is not structural. It is controlled by a primary sand accumulation pattern. The fields are presumed to be sand bars in a marine environment deposited parallel to the shoreline.

The paleogeographic reconstructions (figs. 9-3 and 9-4) best illustrate the basic similarities in the pattern of sedimentation in these two epochs of molasse deposition. The position of the shoreline — approximated from fossil evidence — can be drawn parallel to the sedimentary strike. The margin of the molasse basin — the line separating the area of sedimentation from the area of erosion — can also be drawn parallel to the sedimentary strike, if its position can be estimated. The latter can be done by extrapolation of the pebble-size change in an up-current direction, if one assumes a reasonable maximum pebble size and that the size change is exponential. These boundaries, the shoreline and the "fall-line," then divide the area into three major parts: 1) an area of supply or erosion, 2) an area of continental alluviation, and 3) an area of shallow marine sedimentation.

Clearly the sedimentary model appropriate for the late Paleozoic molasse basin is the appropriate model also for the mid-Paleozoic molasse basin. It is the model for an elongate miogeosynclinal basin in which the transport direction is transverse to the axis of the basin, in which the sediments thicken toward the source, thin seaward, and in which the linear sand bodies in the marine part of the section are parallel to the sedimentary strike.

Scott W. Starratt
Dept. of Paleontology
U. C. Berkeley
Berkeley, Ca. 94720

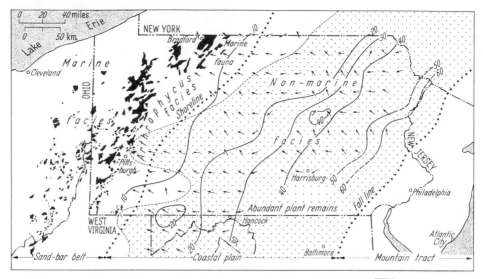

Fig. 9-3. Sedimentary framework, Pocono formation (Mississippian) in central Appalachians (modified from PELLETIER, 1958, fig. 16)

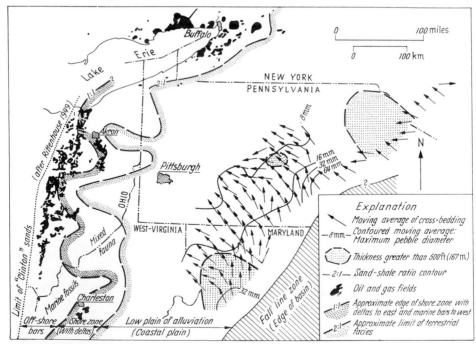

Fig. 9-4. Sedimentary framework, Tuscarora quartzite (Silurian) in central Appalachians (modified from Yeakel, 1959, fig. 68)

Our knowledge of the sand bodies within the dominant nonmarine portion of the basin is, at present, nil. Very probably, in this alluvial phase, the linear sands are related to fluvial channels and are, therefore, longitudinal; i.e., parallel to the transport direction. Individual sand bodies, however, are difficult to recognize in a formation dominantly or wholly sand.

The model outlined above is probably a model of considerable generality and may apply to some molasse basins elsewhere. On the other hand, it is not applicable to all areas of alluviation. It is not the appropriate model for the sedimentation in the Gulf Coast geosyncline, for example, where the fill thickens in the down-current seaward direction and where the alluvial materials are derived from areas far removed from the site of deposition.

Chesterian (Illinois Basin) and Gulfian (Mississippi Embayment) Model

The Cretaceous sediments (Gulfian Series) of the Mississippi Embayment and Chesterian Mississippian sediments of the Illinois basin constitute another model that is quite unlike the central Appalachian molasse (PRYOR, 1960, 1961; POTTER, 1962b). In terms of basin geometry, directional structures, lithic fill, arrangement and tectonic setting these deposits exhibit striking similarities. Table 9-2 summarizes these features. The main differences between them are both nonessential ones; size and structural position in the continental framework. The upper Mississippi Embayment, whose present boundaries are essentially those of the original limits of sedimentation, is much smaller than the area of Chester sedimentation of which the Illinois basin is only a small part. Secondly, the upper Mississippi Embayment is still connected with the present continental margin, whereas the Illinois basin has long been separated from it and is now structurally an intracratonic basin. Both were "open ended" basins.

The Gulfian series was deposited in an oblong basin that widened and deepened down its plunge to the south. Transverse cross-section is essentially symmetrical. The Chesterian sediments were also deposited in a trough like basin with a weakly defined negative axis plunging to the southwest; its transverse cross-section is likewise essentially symmetrical. Both the Gulfian and Chesterian series thicken down the plunge. The Gulfian series varies from 200 feet (65 m) in the north to 1,200 feet (400 m) or more, 150—175 miles (240—280 km) to their known limits to the south. Chesterian sediments, truncated at their top by major unconformity that cuts progressively lower to the north, have maximum thickness of approximately 1,400 feet along their southern edge. In general, sediments are thicker along the axial portions of both basins.

Cross-bedding and ripple marks predominate in the sands. Cross-bedding indicates that average transport direction closely coincides with orientation of basin axes and that longitudinal transport down a paleoslope to the southwest prevailed. Chesterian sediments show a tendency for centripetal transport along present basin margins.

Although there are some differences, the character of the fill is much the same in both basins. The sediments are predominantly of shelf and deltaic origin. Gulfian sediments consist of approximately 30 per cent protoquartzitic sand, 50 per cent clay, and 20 per cent impure calcilutites and chalks. Lignites are present. Chesterian sediments consist of approximately 25 per cent protoquartzitic sands, 50 per cent shale, largely marine, and 25 per cent pure to impure carbonates. Some thin coal beds are present.

Table 9-2. *Gulfian and Chesterian Model*

Unit	Basin Geometry	Directional Structures	Lithic Fill	Arrangement	Tectonic Setting
A. Mississippi Embayment					
Eocene	Oblong basin widening and deepening down plunge to south; symmetrical transverse cross-section.	Chiefly thick cross-beds which reflect a paleoslope parallel or subparallel to basin axis.	Protoquartzitic sands, largely nonmarine, 75%. Clay and shale 25%. Lignite and coal.	Total section expands down paleoslope. Longitudinal clastic filling predominates. Fluvial and delta pattern at north becomes more marine southward. Some cyclic deposition.	Mild basin subsidence. Mild to moderate uplift of distal source area.
Cretaceous	Oblong basin widening and deepening down plunge to south. Symmetrical transverse cross-section.	Chiefly thick cross-beds which reflect a paleoslope parallel or subparallel to basin axis.	Protoquartzitic sands principally nonmarine, 30%. Clay and shale, 50%. Impure calcilutites and chalks, 20%. Minor lignites and coals.	Total section expands down paleoslope. Longitudinal clastic filling. Delta pattern at up dip end; carbonates and marine clays down dip. Cyclic deposition not pronounced.	Mild to moderate subsidence in basin. Mild uplift of distal source area.
B. Illinois Basin					
Pennsylvanian	Broad trough-like basin originally widening and deepening down plunge to gentle plunge to southwest. Symmetrical and asymmetrical transverse cross-section.	Chiefly thick cross-beds which reflect paleoslope. Slope generally parallels basin axis but has centripetal components along east and west margins.	Protoquartzitic and subgraywacke sands, principally nonmarine, 30%. Marine and nonmarine shales 60%. Impure non-reef carbonates 4%. Many mineable coals, 1%.	Integrated clastic dispersal by streams and deltas as total section expands down paleoslope and toward basin axis. Longitudinal plus some lateral supply, especially from east. Carbonates have weak relationship	Intracratonic basin of mild to moderate subsidence. Mild to moderate uplift of distal source region.

(row above, cut off)		to paleoslope. Wide persistence of limestone and coals. Marked cyclic deposition.	Integrated clastic dispersal by streams and deltas. Weak negative axis localizes sand input. Limestones have highly lateral continuity and thicken down-slope as does total section. Cyclic deposition.	Intracratonic basin of mild subsidence. Mild uplift of distal source area.
Chesterian	Broad troughlike basin originally widening and deepening down gentle plunge to southwest. Symmetrical, transverse cross-section.	Chiefly thick cross-beds which reflect paleoslope. Slope generally parallels weak basin axis but has some centripetal components on east and west margins.	Protoquartzitic sands, marine and nonmarine, 25%. Pure and impure carbonates 25%. Shale, principally marine 50%. Some nonmineable coals.	

With the exception of contrasts in cyclic sedimentation, the arrangement of patterns of sedimentation in the two basins are very similar. Longitudinal transport of clastic sediments down a paleoslope to the south and southwest is the dominant control. Figure 9-5 shows the facies distribution and cross-bedding of the McNairy formation in the Mississippi Embayment. The sand-clay ratio is greater than 1:1 in the north and decreases to less than 1:2 in the south. The finer clastics become increasingly calcareous southward and reflect the transition from largely nonmarine sedimentation in the north to wholly marine in the south, as the total section expands down paleoslope. Figure 9-6 shows cross-bedding directions in the Chester series and the thickness of interbedded limestone (below the Menard limestone). Carbonates thicken downslope in the direction of transport. The clastic ratio decreases southward so that it is less than 2:1 along the southern edge of the basin. The facies distribution is clearly related to the south-west flowing paleocurrents in both the Gulfian embayment and the Illinois basin.

Because of extensive drilling, subsurface sand distribution in the Chester basin is well known (POTTER, 1962b). Elongate lenticular sand bodies of several types predominate. The vast majority trend down the paleoslope to the southwest parallel to cross-bedding direction. Figure 9-7 shows the relations between a birdfoot delta and crossbedding in a Chesterian sand, the Palestine, and the thickness of an underlying limestone. Limestone thickens to the south and defines depositional strike whereas cross-bedding and delta pattern point down the paleoslope. Patterns such as these are typical of many elongate Chesterian sands and indicate that carbonate distribution in the Chesterian series are closely correlated. Presumably drilling would reveal a similar relationship in the Gulfian series where four major longitudinal sand trends are already known.

The tectonic setting of both basins was very similar. Mild subsidence in the basin was coupled with mild uplift. The source region was remote from the basin of deposition.

In summary, the similarities between the character, the arrangement, and the paleocurrent system and its relation to the architecture in these two basins are impressive. Both basins are oblong with southward plunging axes, both are

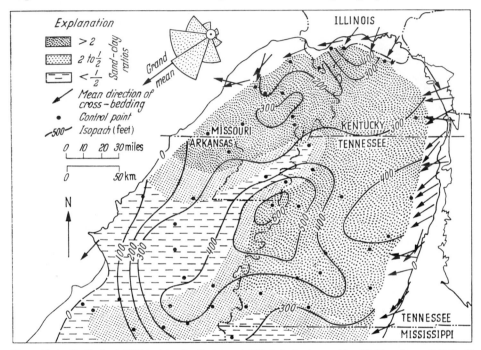

Fig. 9-5. Facies distribution in McNairy formation (Cretaceous) of Mississippi embayment (modified from PRYOR, 1960, fig.18)

open-ended to the south, and both received sediments longitudinally into the closed northeastern end. Particularly striking are the similarities between facies distribution, paleoslope, transport direction, depositional strike and spatial distribution of marine and nonmarine sediments. The linear sand trends are directly related to this model. The sand bodies are regarded as deltaic-fluviatile in origin.

The Chesterian sediments of the Illinois basin and the Gulfian series of the Mississippi embayment, therefore, are two examples of a model — an oblong basin plunging toward the open end, filled with sediments of a shallow, marine shelf — deltaic origin (fig. 9-8).

The model described is also the appropriate one for the early Tertiary fill of the Mississippi embayment and for the Pennsylvanian sediments of the Illinois basin (POTTER, 1962a, Table 2). PRYOR (1961, p.130) suggested that the present Mississippi River delta complex, as it developed in the Tertiary, is the descendent of the stream that supplied clastics to the head of the Mississippi embayment in Cretaceous time. Like the Tertiary fill of the Embayment, the Pennsylvanian

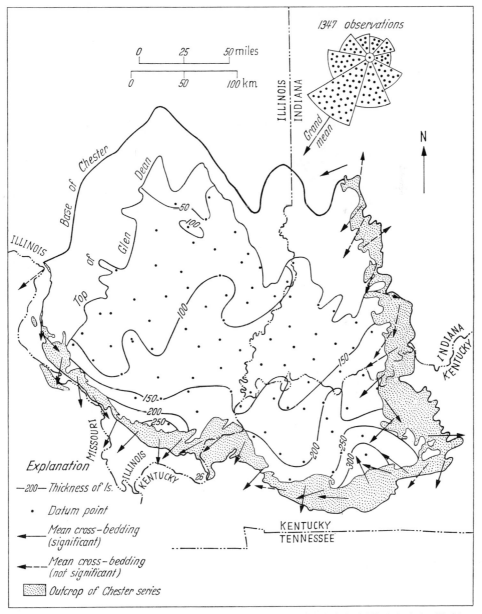

Fig. 9-6. Cross-bedding directions and limestone thickness in Chester series (Mississippian), Illinois basin (modified from Potter *et al*, 1958, fig. 15)

sediments of the Illinois basin have less carbonate than in the Chester series, but the basin elements of the model — longitudinal shallow water fill down a weak trough-like basin — are still applicable.

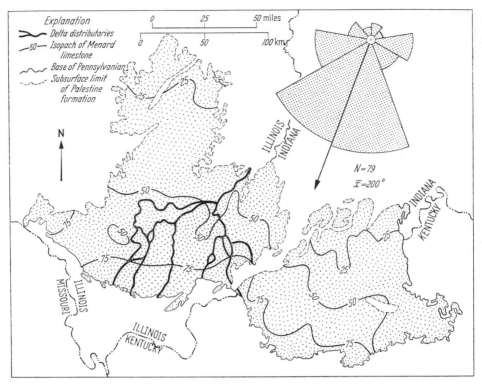

Fig. 9-7. Relations between birdfoot delta and cross-bedding in Palestine sandstone, Chester series (Mississippian), Illinois basin (Illinois Geological Survey)

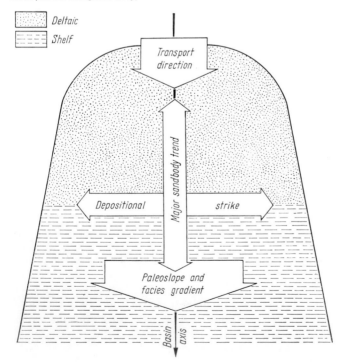

Fig. 9-8. Diagrammatic sketch of Mississippi embayment model (PRYOR, 1961, fig. 9)

Turbidite Basins

The basins previously described are those in which the current system is mainly fluvial, — basins in which the sediment is distributed by water flowing down a subaerial slope. It is of interest to present, as a contrast, a basin in which subaqueous turbidity currents are the principal agents of sediment distribution.

The concept of turbidity current action is relatively new and as yet no turbidite basin has been studied in its entirety nor have all elements of the turbidite model been investigated. Hence our knowledge is incomplete although various partial studies and some theoretical considerations enable us to sketch in outline form the nature of the model (Table 9-3). We cannot even be sure, however, that there is but a single model for all turbidite basins.

Nearly all generally accepted turbidites were deposited in flysch basins associated with orogenic belts or in basins of the California type such as the Ventura

Table 9-3. *Turbidite Basin Model*

Basin Geometry	Directional Structures	Lithic Fill	Arrangement	Tectonic Setting
Basin elongated parallel to tectonic strike; deep; small to moderate size.	Chiefly sole markings: flutes and grooves; crossbedding on a small scale. Paleocurrents flow both transverse and longitudinal to basin axis.	Graded turbidite graywackes and lithic sandstones; rhythmically interbedded shales; olistostromes rare; calcareous facies rare.	Sand-shale ratio increasing upwards; marginal facies rarely recognized.	Strongly negative; usually has a proximal source area.

and Los Angeles basins. It is not yet clear whether the patterns of sedimentation in a California-type basin and a flysch basin are basically alike or significantly different.

The geometry of a flysch basin is very difficult to determine. As KSIAZKIEWICZ (1958, p. 424) has pointed out, most flysch sediments were sheared off at their base and therefore, they are nowhere in contact with their former coasts. Hence next to nothing is known about the near-shore sediments of the flysch basin. It is generally agreed, however, that the basins were deep, not only in a structural sense, but also in actuality and that the sedimentation took place mainly in depths of the order of 3,000 feet (1,000 m) or more. The arguments for a deep-water origin of the flysch sediments have been presented in detail many times (e.g. KUENEN, 1959, p.1011) and are not reviewed here. The total sediment accumulation may be very great. The basins are generally elongate parallel to the tectonic strike of the region.

The fill is the characteristic flysch assemblage — a thick sequence of marine geosynclinal sediments, consisting of evenly stratified pelagic shales and turbidite sandstones (graywackes and similar rocks), which generally exhibit marked graded bedding. Transition to calcareous types also occur (KUENEN, 1958, p.329). In addition to repetitive grading, many of the sandstones display such accessory characters as flute casts, groove casts and other sole markings, convolute bedding, small-scale ripple cross-lamination and the like. Normally the sands are unfossiliferous or at most contain redeposited shell fragments or foraminiferal remains. The shales are likewise non-fossiliferous, or in the younger systems, contain only scanty pelagic and deep-water benthonic remains.

Locally the fill is conglomeratic — the so-called "wildflysch" (pebbly mudstones) of some sections — with small- and large-scale slump structures. Such a facies is perhaps a proximal facies; i.e., one deposited near the margin of the basin closest to the source of the sediment.

Very little is known about the arrangement of the fill. Owing to complex deformation of most flysch deposits and to the absence of any marker beds in

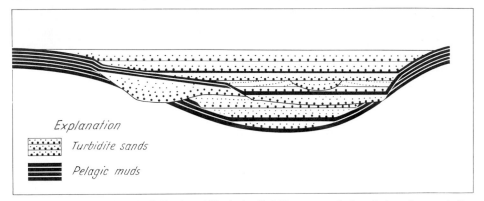

Fig. 9-9. Hypothetical section of a California turbidite basin. Turbidites: coarse stipple; pelagic muds: gray shading (SULLWOLD, 1961, fig. 2)

this very thick, monotonous sand-shale succession, study of lithofacies is difficult. General considerations and some data on a few basins suggest that the proximal or marginal facies is marked by thicker and coarser sands, by slumped beds, larger and more prominent flute casts, and perhaps a higher per cent of total sand. In the distal facies, the sands would be thinner, finer-grained, more generally characterized by groove rather than flute casts and with few or no slump structures. Studies of modern sediments in the offshore basins of southern California (GORSLINE and EMERY, 1959, p. 283) suggest that the sand facies in some cases may be thickest in the deepest part of the basin and isolated from the shallow-water sands of the margins by pelagic muds (fig. 9-9). This distribution may reflect the tendency of the turbidity flows to seek the lowest areas where they stagnate and deposit for, as previously noted, marginal, centripetally supplied facies are least likely to be preserved. In other cases, the sands may form subaqueous fan-like bodies or tongues near points of entry into the basin (SULLWOLD, 1960, fig. 6).

A great deal more is known about the paleocurrent systems in flysch basins than about the other elements of the model excepting perhaps only the petrology of the fill. Systematic mapping of the directional features in the flysch basins has been summarized in Table 5-2.

As noted by KUENEN (1958, p. 331), we might expect to find centripetal supply in round basins and lateral supply in oblong ones. However, in many flysch basins the transport was mainly parallel to the long axis of the basin and hence generally parallel to the tectonic strike of the beds. A very good example is shown by TEN HAAF's study (1959) of the Apennines (fig. 9-10). Most generally also the flow pattern in the Polish Carpathians is longitudinal (DZULYNSKI et al., 1959). The longitudinal pattern could be due to introduction of the fill by a large river entering the trough at one end — a view advocated by KUENEN (1957; 1959; p. 331), but it might also be attributed to lateral introduction without deposition followed by longitudinal reorientation of the current accompanied

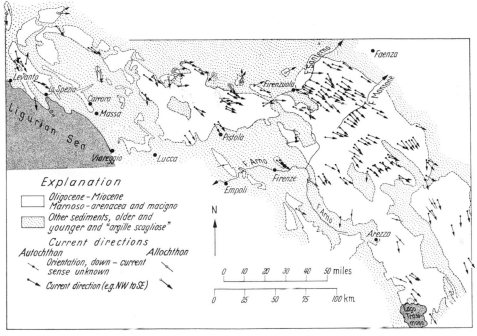

Fig. 9-10. Paleocurrent system in Apennine flysch (Miocene) (TEN HAAF, 1959)

by deposition. Commonly the direction of slumping is transverse and the current flow longitudinal (cf. p. 160). Though longitudinal flow is the rule, both lateral and oblique flow are also known.

The concept of sedimentary strike is somewhat difficult to apply to some turbidite basins. The relations between current flow and direction of slope are difficult to interpret. Mention has been made in Chapter 5 of the problem arising from diverse flow directions on the same sole and that arising from varied flow directions on successive beds.

The source land problem has been much discussed. There is some evidence of medial geanticlinal uplifts in some flysch basins and these are said to have been major sources of sediments (HSU, 1960; DZULYNSKI et al., 1959). On the other hand cogent objections to such intrabasinal source lands have been raised (KUENEN, 1958, p. 337).

In summary, the turbidite basin is a deep, rather poorly ventilated, though not anaerobic, basin. Shallow-water sediments flank its margins; turbidity currents generated by slumps or other causes move rapidly down the marginal slopes, and deposit fan-like accumulations at the foot of the slopes, or seek the lowest parts of the basin where they deposit a multitude or "bundle" of turbidite sands (fig. 9-9). Longitudinal flow is the rule and hence longitudinal sand bodies may be expected in the axial portion of the basin. The distal parts of the basin will receive finer-grained graded silts. Continued deposition will presumably lead to a very flat basin floor across which turbidity currents may flow at will — moving in diverse directions depending on their place of entry into the basin. The basin sands may grade up-dip into the marginal shallow-water sediment or intertongue with them in the later stages of basin filling; in the earlier stages, however, the basin sands may be surrounded by mud-covered slopes and hence be wholly separated from the basin-margin sands.

Other Clastic Models

Two of the best known sedimentary basins are the Alpine and the Appalachian. Both have been termed "geosynclines." Both have many similarities; a thick sedimentary fill, folded and overthrust sheets, and metamorphism and granitic magmatism in the core region. The sedimentary fill has many similarities and the terms flysch and molasse can aptly be applied to both. Neither are simple; both are complex basins. The Alpine and Appalachian models have many counterparts elsewhere.

As sedimentologists, we are interested in the filling of basins and are inclined to take the ideas and observations generated by the study of one basin and apply these to other basins. Thus we look for the equivalents of the Alpine molasse and flysch in other sedimentary basins. We are successful in some cases but not in others. There is no flysch or molasse in the Triassic Newark basins. The later are not Alpine basins. Instead, the Newark is geometrically, tectonically, and sediment-wise another type for which the Newark basins constitute a model (REINEMUND, 1955). The sedimentary fill, though resembling the molasse in most respects, is not a molasse.

The Newark basin is a half graben, with simple tilt or step-faulted, and is not involved in Alpine-type orogeny. It is characterized by diabase dikes, sills, and flows. Moreover, the sedimentation was active during the development of the border fault. Locally ponding of drainage produced lakes or swamps in which marls or coals formed. The Gondwana coal basins in India are Newark-type basins, half grabens with terrestrial clastics, coal beds, also cut by diabase dikes and sills (KRISHNAN, 1956, p. 302, 319—320). The Keweenawan clastics of the Lake Superior region belong to the Newark model: they too filled a basin with border fault, were never involved in Alpine-type folding, contain lava flows, dikes and sills, and some lake beds (Nonesuch formation).

Similarly the most thoroughly drilled and explored geosyncline — the Gulf Coast geosyncline, if indeed this is a geosyncline, is another example (MURRAY, 1961, p. 79). It is no doubt a prototype, one from which a model can be formulated, which is useful in interpreting the strata and structure of the Gulf Coastal Plain. Certainly this is true of its Cretaceous and Tertiary history. The appropriate model is perhaps similar to that described for the Mississippi Embayment except that the basin geometry is different.

How many models are there? Although many more basins require integrated study before this question can be answered, we believe only a small number of additional clastic models will be defined.

Non-Clastic Models

It might be presumed that the nonclastic sediments would lack directional structures and that their deposition would be largely independent of current systems. Lacking systematic study of the directional properties of these rocks, if they indeed have any, it is impossible to prove or disprove this presumption. Insofar as the carbonates, for example, are particulate materials they will respond to currents in exactly the same way as noncarbonate particles and will exhibit the same primary directional fabrics and structures. Indeed ripple-marked and cross-bedded limestones are common and as noted elsewhere in this volume (Chapter 4) these current structures display well-defined regional patterns. The carbonate sand bodies of some areas, such as those in the Bahamas, display a dune morphology and spatial arrangement that is clearly a response to currents. No doubt these accumulations would exhibit both preferred internal fabrics and structures closely correlated with their external forms as do their subaerial counterparts. Inasmuch as most carbonate deposits are a product of shallow waters, especially on stable shelves or platforms, there should be a carbonate or shelf model of sedimentation. Such models, however, have yet to be worked out.

Biogenic sedimentation, moreover, may also be current controlled or at least show some current response. The position, orientation, and shape of reefs may be in part a response to wind-driven currents. In fact, the shape of "fossil" reefs has been used to infer wind direction (LOWENSTAM, 1950, p.466, for example). Even the distribution of evaporites in an evaporite basin may be systematically related to the circulation pattern within the basin. BRIGGS (1957 and 1958), for example, formulated such a model and applied it to the Silurian salt beds in the Michigan basin.

At the present time, however, our knowledge of nonclastic models of sedimentation, especially those which relate the arrangement of the deposits within the basin to paleocurrents, is too incomplete to permit any significant generalization.

Implications of the Model Concept

The idea that the vast majority of the earth's sediments can be represented by a comparatively few recurring sedimentary models carries with it a number of implications.

If it is true that the model, based on readily identifiable and measureable properties, provides a large scale framework, then the areal distribution of innumerable other properties must also be closely related to this plan of organization. The composition and distribution of clay minerals, for example, should also be closely correlated with the model plan, to the extent that they behave as other detrital materials and have not undergone too much post-depositional change (cf. WEAVER, 1960).

If, also, the model is closely correlated with basin geometry, then any alteration of that geometry will drastically alter the model and, conversely, inasmuch as the geometry is tectonically controlled, a significant change in the model is evidence of tectonic movement of some kind. The relations between growing structures, penecontemporaneous faulting and paleocurrents remain to be further explored.

In stable areas the paleocurrent systems should persist through very long periods of geologic time. Such indeed seems to be the case in the areas peripheral to the Precambrian shields (*cf.* POTTER and PRYOR, 1961). On the other hand, in tectonically active mobile belts with numerous small flysch basins, models will be short lived.

Finally, the model concept should be useful in the exploration and development of sedimentary basins. Correct matching of an unexplored basin with a known prototype should provide for a much more perceptive delineation of paleoslope,

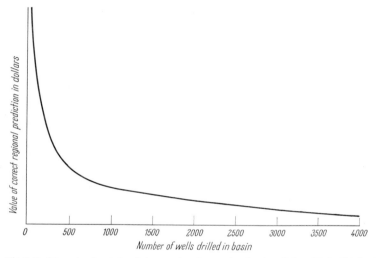

Fig. 9-11. Diagrammatic relations between value of a correct regional prediction of facies distribution in a basin and number of wells drilled in the basin

facies distribution and sedimentary trends, including contouring of sand bodies. The actual value, measured in dollars, of correct regional prediction varies greatly in the different stages of basin development. For instance, in a relatively unknown basin with little or no subsurface data, a successful regional prediction of facies distribution and paleoslope based on outcrop observation alone is many times more valuable than a similar prediction in that same basin after 1,000 or 4,000 wells have been drilled. Figure 9-11 shows the value of a correct regional prediction as number of wells increase. The probable position and trend of the productive belt of oil and gas· pools of late Devonian and early Mississippian in the Appalachian basin could presumably have been predicted if the Catskill-Pocono model, worked out largely from outcrop studies, were understood as it now is.

Much more meaningful data can be collected if one knows the depositional strike because lithology and thickness are more likely to change more drastically down-dip than they do parallel to the sedimentary strike. Information gathered from outcrops along the basin margin will determine the depositional strike and provide the proper orientation for further geological and geophysical studies. Once the proper model is known, one should be able to better predict both probable areas of favorable stratigraphic traps, such as sand bodies enclosed in a shale matrix and their probable orientation. The local anomalies disclosed by empirical methods, such as trend analyses, may be more readily interpreted, if the framework of the model is known.

Beyond Basin Analysis

Some aspects of paleocurrent analysis transcend any particular basin and involve larger segments of the earth's crust, even the continent as a whole.

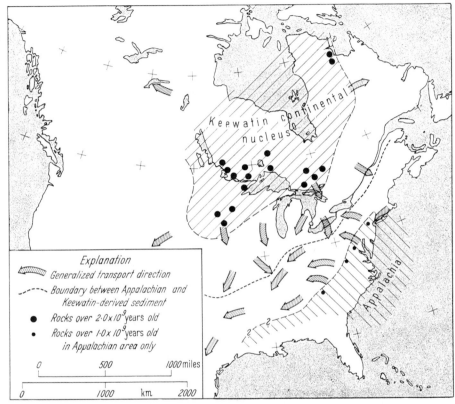

Fig. 9-12. Precambrian and Paleozoic paleocurrents in eastern North America

Paleocurrent data can be useful, for example, in studying the evolution of the larger structures in the earth's crust. The elevation of an uplift, such as the Black Hills of South Dakota, should be reflected in the paleocurrent patterns of related sediments. Paleocurrent data can provide a means for dating this and similar uplifts. As noted by KUENEN (1959b), paleocurrent studies can also be of value in solving the problem of the foundering of large segments of continental crust (cf. PETTIJOHN, 1960, p. 452). Coastal outcrops of Tertiary flysch in the Maritime Alps of France and Italy, for example, show directional structures indicating flow from south to north (BOUMA, 1959). The presumed source of these sediments was south of the present shoreline from an area now covered by waters of considerable depth. Thus, the reality and importance of borderlands, such as Appalachia, can be better assessed, if paleocurrent data are available (fig. 9-12).

Compilation of cross-bedding or other paleocurrent data, such as that made by POTTER and PRYOR (1961), for large regions and for a long span of time, will disclose patterns of sediment transport which are closely related to the evolution and growth of the continent. The evolution of the continent can be better understood, if these paleocurrent data are combined with absolute age determination. The important role played by the Canadian shield, for example, as a source of sediment from late Precambrian to the present is well shown by such an approach (fig. 9-12).

Summary

A sedimentary model has as its *elements:* 1) basin geometry, 2) type of lithic fill (sedimentary association), 3) arrangement of lithic fill (lithofacies), 4) current system, and 5) tectonic setting. The model concept relates all five elements to one another to provide an *integrated description of a recurring pattern of sedimen-*

Table 9-4. *Kinds and Sources of Data for Basin Analysis*

Source of data	Elevation	Thickness	Lithology	Scalar properties	Directional properties
Outcrops	Yes	Yes	Yes	Yes	Yes
Cores	Yes	Yes	Yes	Yes	No, unless oriented
Cuttings	Yes	Yes	Yes	Yes	No
Geophysical well surveys	Yes	Yes	By interpretation	No	No*

* Use of remote sensing device in well may be possible.

tation. In general, a particular model may prevail during the complete filling of a sedimentary basin but usually it does not. Size of the basin is not an essential element in the model.

Like the concept of a sedimentation unit which supposes essentially constant physical conditions during the deposition of the unit (bed), the model concept as here used supposes the maintenance of a particular set of conditions which produces a particular basin-wide pattern of sedimentation during a significant interval of geologic time. With a change in these conditions, a new pattern is established and a new model prevails. Hence, within a particular basin, we may have a deep-water turbidite regime — the flysch model, for example — which is transformed into a shallow water or nonmarine regime — the molasse model.

Four of the five elements of the model, basin geometry, type of lithic fill, arrangement of the fill and current system, are completely descriptive. In contrast, tectonic setting is less tangible and may be highly interpretative, especially in pre-Cretaceous time. However, it is a desirable feature of the model because it relates the descriptive features of the model to the tectonic cycle.

It is presumed that the several models or patterns of sedimentation succeed one another in an orderly predictable way. If true, such a concept would lead to the formation of a supermodel. It is not our task here to formulate such a model but rather to consider only those fundamental patterns of sedimentation created by a current system that persisted for a significant interval of time during the filling of a basin.

"Basin analysis" as here presented has been but an extension of the kinds of analyses familiar to many geologists. These analyses involve the collection of data from various parts of the basin and presentation of such data on appropriate maps. Data may be of several kinds so that several types of maps can be constructed (Table 9-4). The new aspect of basin studies is in the recognition of the interrelation between the paleocurrent system, the arrangement of the fill and the basin geometry. Paleocurrents add a new dimension; they are the aspect missing from the older studies.

Data are derived from various sources. As is clear from inspection of Table 9-4, paleocurrent data are most readily derived by study of outcrops although directional properties can be potentially obtained from oriented cores and conclusions about paleocurrents can be obtained from the areal variation of attributes and scalars such as mineral provinces and maximum pebble size. Limitation of paleocurrent data to outcrops along the margin of the basin makes the problem of down-dip prediction difficult. If, however, the basin belongs to a well-known type, the geologist can, with the proper model, extrapolate from his outcrop data and with knowledge of the transport plan, as shown by paleocurrent measurements, make a rational interpretation of the concealed and as yet unexplored parts of the basin. The likelihood of correct prediction may be further enhanced by adapting the empirical methods of trend surface analysis (Chapter 10) to the different types of known sedimentary models.

References

AUBOUIN, J., 1959: A props d'un centenaire: Les aventures de la notion de géosynclinal. Rev. Geogr. Phys. et Géol. Dynam. [2] **11**, 135—188.

BOUMA, A. H., 1959: Flysch Oligocène de Peira-Cava (Alpes-Maritimes, France). Eclogae Geol. Helv. **51**, 893—900.

BRIGGS, L. I., 1957: Quantitative aspects of evaporite deposition. Mich. Acad. Sci. Arts and Letters **42**, 115—123.

— 1958: Evaporite facies. J. Sediment. Petrol. **28**, 46—56.

BRINKMANN, R., 1933: Über Kreuzschichtung im deutschen Buntsandsteinbecken. Nachr. Ges. Wiss. Göttingen, Math.-physik. Kl., Fachgruppe IV, No. 32, 1—12.

— 1955: Gerichtete Gefüge in klastischen Sedimenten. Geol. Rundschau **43**, 562—568.

BUCHER, W. S., 1933: The deformation of the earth's crust. Princeton: Princeton Univ. Press. 518 p.

DICKEY, P. A., and R. E. ROHN, 1955: Facies control of oil occurrence: Bull. Am. Assoc. Petrol. Geologists **39**, 2306—2320.

DZULYNSKI, ST., M. KSIAZKIEWICZ and PH. H. KUENEN, 1959: Turbidites in flysch of the Polish Carpathian Mountains. Bull. Geol. Soc. Am. **70**, 1089—1118.

FORCHE, FRITZ, 1935: Stratigraphie und Paläogeographie des Buntsandsteins im Umkreis der Vogesen. Mitt. Geol. Staatsinst. Hamburg **15**, 15—55.

GLAESSNER, M. F., and C. TEICHERT, 1947: Geosynclines: A fundamental concept in geology. Am. J. Sci. **245**, 571—591.

GORSLINE, D. S., and K. O. EMERY, 1959: Turbidity-current deposits in San Pedro and Santa Monica Basins off southern California. Bull. Geol. Soc. Am. **70**, 279—290.

HAAF, E., TEN, 1959: Graded beds of the northern Apennines. Ph.D. thesis, Rijks University of Groningen, 102 p.

HSU, K. J., 1960: Paleocurrent structures and paleogeography of the Ultrahelvetic flysch basins, Switzerland. Bull. Geol. Soc. Am. **71**, 577—610.

IDDINGS, J. P., 1892: The origin of igneous rocks. Phil. Soc. Wash. Bull. **12**, 89—213.

KAY, G. M., 1944: Geosynclines in continental development. Science **99**, 461—462.

— 1947: North American geosynclines. Geol. Soc. Am. Mem. **48**, 143 p.

KNOPF, ADOLPH, 1948: The geosynclinal theory. Bull. Geol. Soc. Am. **57**, 649—570.

KRISHNAN, M. S., 1956: Geology of India and Burma, 3rd ed. Madras: Higgenbothams. 555 p.

KRUMBEIN, W. C., 1937: Sediments and exponential curves. J. Geol. **45**, 577—601.

KSIAZKIEWICZ, M., 1958: Sedimentation in the Carpathian Flysch sea. Geol. Rundschau **47**, 418—424.

KUENEN, PH. H., 1957: Longitudinal filling of oblong sedimentary basins. K. Nederl. Geol. Mijnb. Gen. Verh., Geol. Ser. **18**, 189—195.

— 1958: Problems concerning source and transport of flysch sediments. Geol. en Mijnbouw, n.s. **20**, 329—339.

— 1959a: Turbidity currents a major factor in flysch deposition. Eclogae Geol. Helv. **51**, 1009—1021.

— 1959b: La topographie et la géologie des profondeurs océanique. Colloq. internat. centre nat. recherche sci. (Paris) **83**, 157—163.

LOWENSTAM, H. A., 1950: Niagaran reefs of the Great Lakes area. J. Geol. **58**, 430—487.

MURRAY, G. E. **1961**: Geology of the Atlantic and Gulf Coastal Province of North America. New York: Harper & Bros. 692 p.

NANZ, R. H., 1957: Philosophy and technique for the study of sandstones (abstract). Inst. on Lake Superior Geology, Program, May, 1957.

PELLETIER, B. R., 1958: Pocono paleocurrents in Pennsylvania and Maryland. Bull. Geol. Soc. Am. **69**, 1033—1064.

PETTIJOHN, J. F., 1960: Some contributions of sedimentology to tectonic analysis. 21st Intern. Geol. Congr., Norden, pt. 18, p. 446—454.

— 1962: Paleocurrents and paleogeography. Bull. Am. Assoc. Petrol. Geologists **46**, 1468—1493.

POTTER, P. E., 1959: Facies model conference. Science **129**, 1292—1294.

— 1962a: Regional distribution pattern of Pennsylvanian sandstones. Bull. Am. Assoc. Petrol. Geologists **46**, 1890—1911.

— 1962b: Late Mississippian sandstones of Illinois. Illinois Geol. Survey Cir. 340, 36 p.

— EDWARD NOSOW, N. M. SMITH, D. H. SWANN and F. H. WALKER, 1958: Chester cross-bedding and sandstone trends in Illinois basin. Bull. Am. Assoc. Petrol. Geologists **42**, 1013—1046.

— and W. A. PRYOR, 1961: Dispersal centers of Paleozoic and later clastics of the upper Mississippi Valley and adjacent areas. Bull. Geol. Soc. Am. **72**, 1195—1250.

PRYOR, W. A., 1960: Cretaceous sedimentation in upper Mississippi Embayment. Bull. Am. Assoc. Petrol. Geologists **44**, 1473—1504.

— 1961: Sand trends and paleoslope in Illinois basin and Mississippi Embayment *in* Geometry of Sandstone Bodies. Tulsa: Am. Assoc. Petrol. Geologists, p. 119—133.

REINEMUND, J. A., 1955: Geology of the Deep River coal field, North Carolina. U.S. Geol. Survey Prof. Paper No. 246, 150 p.

RUKHIN, L. B., 1959: Principles of general paleogeography. Leningrad, 557 p. [Russian].

SCHUCHERT, CHARLES, 1910: Paleogeography of North America. Bull. Geol. Soc. Am. **20**, 427—606.

SLOSS, L. L., 1962: Stratigraphic models in exploration. Bull. Am. Assoc. Petrol. Geologists **46**, 1050—1057.

SORBY, H. C., 1859: On the structures produced by the current present during the deposition of stratified rocks. Geologist **2**, 137—147.

STANLEY, D. J., 1961: Études sédimentologiques des grès d'Annot et leurs équivalents latéraux. Inst. Français du Pétrole, Ref. 6821, 158 p.

SULLWOLD jr., H. H., 1960: Tarzana fan, deep submarine fan of late Miocene age, Los Angeles County, Calif. Bull. Am. Assoc. Petrol. Geologists **44**, 433—457.
— 1961: Turbidites in oil exploration *in* Geometry of Sandstone Bodies. Tulsa: Am. Assoc. Petrol. Geologists, p. 63—81.
WEAVER, C. E., 1960: Possible uses of clay minerals in search for oil. Bull. Am. Assoc. Petrol. Geologists **44**, 1505—1518.
WEEKS, L. G., 1952: Factors of sedimentary basin development that control oil occurrence. Bull. Am. Assoc. Petrol. Geologists **36**, 2071—2124.
WILLIS, BAILEY, 1909: Paleogeographic maps. J. Geol. **17**, 203—208, 253—256, 342—343, 403—409, 503—508, 600—602.
YEAKEL, L. S., 1959: Tuscarora, Juniata and Bald Eagle paleocurrents and paleogeography in the central Appalachians. Unpublished Ph.D. thesis, The Johns Hopkins University, 454 p.

Basin Analysis and the Sedimentary Model (1963—1976)

Plate tectonic theory and new techniques of data acquisition have revolutionized the study of sedimentary basins solving some problems as well as creating at least one new, very fundamental one — the development of criteria to identify today's plate tectonic regimes in the ancient, sedimentary record, which now goes back almost four billion years.

Basin analysis in the past thirteen years has greatly expanded and has been transformed so that today there is a vast, diversified literature to which geophysicists and students of tectonics, especially plate tectonics, as well as sedimentologists and stratigraphers contribute. During these thirteen years developments have been so great that, if one had been a Rip Van Winkle asleep in a library, awakening today and having to catch up would be a staggering task. Seismic surveys yielding a seismic stratigraphy, magnetic and gravity surveys, more detailed bathymetric studies, and deep-sea drilling have greatly enhanced our understanding of the modern ocean basins and their relation to the continents so that today we know a great deal about continental margins and very much more about ocean basins.

Most significant has been the dramatic change in thinking about sedimentary basins, both modern and ancient, engendered by the arrival and general acceptance of the plate tectonic paradigm. The old classification of basins and the familiar nomenclature of Marshall Kay and others have been replaced by a new one. The geosynclinal concept and all it implies has to be re-examined and reconciled with the new knowledge. Geosynclinal theory was based rather largely on type examples. From several well-described basins a general evolutionary model was formulated that was believed applicable to all similar basins. But where deviations from the supposed norm occurred, the theory was unable to offer clear insights. Plate tectonics, on the other hand, treats the problem of basin origin, fill and evolution in terms of a universal theory so that now individual basins are but one of several end-results of world-wide plate interactions. As noted by DICKINSON (1974, p. 10), given the overall framework of varied plate motions, deductive reasoning from the theory has the potential to shed some insight on quite unfamiliar evolutionary trends, as well as to explain, in a fairly coherent fashion, a range of basin events that might issue in different circumstances from any particular stage in the evolution of moving plates. By considering basin evolution in terms of plate motions and interactions, one enlarges the scope of theoretically controlled analysis and reduces the need for wholly intuitive suggestions.

Before we look at basin analysis in light of plate tectonics, it is desirable to summarize briefly — independent of any theory — some general principles about basins. We hope these suggested principles will stimulate more discussion of basin characteristics and the factors responsible for their origin.

1. The evidence collected during the last decade has shown rather clearly that the continents are permanent, but that the ocean basins are ephemeral so that

the earth's longer history is to be found in the continental rocks, even though its post-Paleozoic history is well recorded in the deep oceans.

2. Because sedimentation is dependent upon tectonic uplift and/or subsidence, every basin and its fill is a consequence of tectonic movement and thus should be related to a particular tectonic model or setting.

3. Ratio of supply to basin subsidence largely determines water depth within the basin and consequently, the kinds and proportions of the major depositional environments of its fill.

4. Nearly all major sedimentary accumulations or prisms, when seen in largest perspective, are wedge-shaped, be they clastic or carbonate.

5. These prisms accumulate in asymmetric basins — an asymmetry described in terms of the *basin polarity;* that is the asymmetry of facies patterns and structural deformation within most major basins.

6. Internal facies and paleocurrents inferred from directional structures and dispersion patterns have a rational relation to external wedge geometry and basin polarity.

7. Most basins have a fill that is organized into one or more major temporal sequences.

8. Basins can be either long- or short-lived with either persistent or variable paleoslopes with the former much more common than the latter.

As one approaches the study of a particular basin, certain questions almost always arise. What kind of basin was it? What size and extent? What kinds and proportions of sediments filled it? Where were the source lands that supplied its clastic sediments? What was the nature of the source land? What was the history or evolution of the basin and its filling? How was this history related to later deformation? If we are studying a basin on a continent, how does it relate to continent-wide paleogeography? If, on the other hand, the basin were along a continental margin, what can we learn, if anything, about its open ocean margin? And what can the sediments of such a marginal basin tell us about the size and current system of this former ocean? What relation, if any, does the basin have to a plate boundary? And finally, what type of crust underlies the basin, continental or oceanic?

These and other questions require a *systematic inventory of facts*. We need to know:

1. the facies distribution and thickness within the basin at various stages of its filling,

2. the paleocurrent pattern of each facies,

3. the relation of facies and paleocurrents to both contemperaneous and later deformation, and

4. the petrology of its fill.

SPENCER (1974, p. 788—802) provides an example of a comprehensive inventory, which although directed at the tectonics and deformation of Mesozoic and Cenozoic mobile belts, should be a useful model for sedimentologists inventorying sedimentary basins.

From this inventory of facts we can infer paleocurrent patterns, ascertain the polarity of the basin, and as noted above, study basin evolution through time.

We show in schematic form (fig. 9-13) the kinds of things we observe, the inferences to be made from these observations, and the relation of these inferences to the basin itself. For the basin itself, it does not matter whether the basin is called a geosyncline or given the name of one or another of the basins marginal to the continent (such as a back-arc basin).

The principal advances during the past thirteen years have been in the techniques of data acquisition which have become more diverse and sophisticated, especially the collection of data by geophysical methods, the study of larger areas, the use of computers to help digest the large volume of resulting data, and above all, plate tectonic theory.

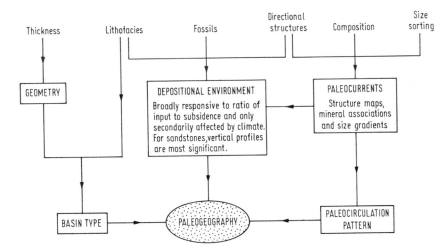

Fig. 9-13. Simplified flow diagram for basin analysis, the practical exploration "payoff" usually being a paleogeography for as small a time interval as possible. Thickness, lithofacies, fossils, directional structures, composition and size-sorting are the "facts"; geometry, depositional environment and paleocurrents represent direct inferences; and basin type, paleocirculation system and paleogeography are highest levels of abstraction

As noted, the plate tectonic concept has given us a new terminology, offering an alternative to the older nomenclature of Kay and others (Table 9-5). We now have not only the deep ocean basins themselves, but basins related to divergent plate margins, basins related to convergent plate boundaries, and intraplate basins (figs. 9-14 and 9-15). As with many other aspects of geology, the plate tectonic paradigm has caused renewed interest in basin classification. Some of these recent papers, like Dickinson's, are based entirely on plate tectonic concepts, whereas others are only partially so. BALLY's (1975) classification of basins, a classification with three major types and twelve subtypes, is an example of one in which plate tectonic theory plays a major role. On the other hand, the classifications by PORTER and McCROSSAN (1975) and of KLEMME (1975) tend to be more descriptive — and thus likely to longer survive the test of time. Klemme recognized seven major basin types. The classifications used by Bally, McCrossan and Porter, and Klemme are all oriented toward estimating the petroleum potential of sedimentary basins; i.e., to what extent is the petroleum potential of a basin related to its type?

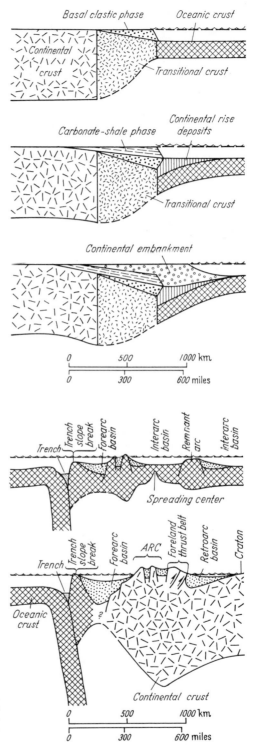

Fig. 9-14. Schematic diagrams show successive depositional phases in evolution of rift-margin prism at continent-ocean boundary with ten-fold exaggeration. (Redrawn from DICKINSON, 1974, fig. 7, Soc. Econ. Paleontol. Mineral. Sp. Pub. 22). Basal clastics unit of miogeocline contains rift valley red beds, proto-oceanic evaporites, rift lavas, etc.

Fig. 9-15. Schematic diagrams with ten-fold exaggeration show tectonic settings of sedimentary basins *(stippled)* associated with two different arc-tranch systems. (Redrawn from DICKINSON, 1974, fig. 10, Soc. Econ. Paleontol. Mineral. Sp. Pub. 22)

Currently we probably have more classifications of basin types than we probably need so that usage and testing — throughout the geologic column — will finally decide which, if any, are the most useful. One is tempted to compare the state of basin classification today with that of sandstone classification ten to twenty years ago, when there was but little agreement. Today, fortunately, there is now, after considerable testing and usage, considerable agreement about the petrographic classification of sandstones.

The classifications of basins which grew out of the plate tectonic concept are largely based on the present-day earth so it is fairly easy to identify present-day basins noting their relation to the continental margin and to island arcs. Not so easy, however, is recognizing their counterparts in the geologic past. We are, in

Table 9-5. *Plate Tectonic Classification of Sedimentary Basins* (after DICKINSON, 1974)

Type	Subtype	Tectonic setting	Fill
Oceanic basins	Rise crest Rise flank Deep basin	Mid-ocean rift with trench Thermally elevated tract	Volcanic turbidites, pelagic carbonates, and siliceous oozes, probably in that order
Basins along rifted continental margins (fig. 9-14)	Pre-rift arch	Thermal rise along incipient rift	Small fans on flanks of uplift
	Rift valley	Grabens and half grabens	Continental red clastics with some ephemeral lake deposits and volcanics
	Proto-oceanic gulf	Expanded rift zone with intermittent marine invasions	Marginal clastics intercalated with central evaporites
	Narrow ocean	Flooded by marine waters; may be divided by mid-ocean rise	Marine sediments; transitional to open ocean
	Open ocean	Miogeoclinal embankment on margins; well-developed mid-ocean rise and rift	Miogeocline consists of: a) basal clastics; b) carbonate-shale sequence; and c) continental embankment or paralic phase Note: all sediments derived from continent with transport oceanward
	Aulacogen	A failed or aborted rift valley or proto-oceanic gulf	Very thick section; early volcanic phase (often rhyolitic); later carbonate-shale sequence; final phase clastic fill

Table 9-5 (Cont.)

Type	Subtype		Tectonic setting	Fill
Arc-trench systems (convergent margins) (fig. 9-15)	Trench		Subduction zone; upthrust mélange on inner trench wall	Mixture of "scraped-off" pelagic sediments and land-derived turbidites plus ophiolite scraps
	Fore-arc basin		Between trench and magmatic arc	Variable; shelf, fluvial to deltaic to turbidite; up to 5 to 12 km fill
	Intra-arc basin		Between either two volcanic island arcs or between continent and arc	Volcanic sediments but including continent-derived sediments in those basins between arc and continent
	Back arc areas	Inter-arc basin	Basin on oceanic crust to rear of magmatic arc	Volcaniclastic turbidites plus minor pelagic materials
		Retro-arc basin	Basin on continental crust to rear of magmatic arc	Fluvial, deltaic, and marine up to 5 km thick; typical molasse or exogeosynclinal; derived from fold-thrust belt
	Suture belt		Complexly deformed join between crustal blocks; *peripheral* basins may form at juncture	Fluvial and deltaic strata in peripheral basins; similar to deposits in retro-arc basins
Intra-continental basins	Difficult to relate to plate tectonic theory. Bounded on all sides by anorogenic terraine. Possibly related to deep, aborted rifts. Essentially cratonic basins			

fact, as yet far from having a workable set of criteria for recognizing the several kinds of basins identified in plate tectonic theory.

Sandstone composition and, to some degree, clay mineralogy will disclose whether or not volcanic debris was supplied to the basins and thus tell us something about the nature of the source land. As noted by DICKINSON and RICH (1972), the composition of the sandstones in the Great Valley sequence in the Sacramento Valley of California shows a progressive change with time, a change attributed to the unroofing of the Sierra Nevada magmatic complex.

Earlier, READING (1972) deductively explored, in a pioneering paper, some of the relations between plate tectonics and flysch and molasse successions. He suggested that the petrographic composition of the flysch could be related to the type of continental margin, the associated igneous rocks, and nature of the pre-flysch facies (Table 9-6). Although much more documentation is needed, this type of analysis should go far to help identify plate tectonic sequences in the pre-Mesozoic history of the earth. Hence, for the terrigenous fill of basins, sandstone petrology and clay mineralogy are essential for a complete interpretation.

Table 9-6. *Classification of Flysch Basins* (modified from READING, 1972, Tables 1 and 2)

Plate junction and margin	Type of continental margin, if present		Type of crust	Flysch type	Pre-flysch	Compositionally mature turbidites	Volcanic turbidites	Associated igneous rocks and molasse
None	Passive	Atlantic	Continental and oceanic	Atlantic	Ophiolites, oceanic sediments or shelf and continental rise sediments	Abundant	Absent	Rare, except ophiolites underlying flysch; no molasse
		Japan Sea	Intermediate, or possibly continental	Japan Sea	Possibly ophiolites and oceanic sediments or an earlier island arc	Abundant to rare	Rare to abundant	Tuffs common; no molasse
Convergent junction with over-riding and destructive[a] margin	Active	—	Oceanic or intermediate	Island Arc	Ophiolites or an earlier island arc	Rare	Abundant	Abundant extrusive and intrusive; no molasse
		Andean	Continental and oceanic	Andean	Ophiolites or earlier shelf and continental rise sediments	Rare to common	Common to abundant	Abundant extrusive and intrusive, especially granites; molasse
		Himalayan	Continental or possibly oceanic	Mediterranean	Varied	Common to abundant	Absent to common	Rare to common; molasse
Strike-slip junction and conservative[b] margin	Active		Continental, intermediate or oceanic	Californian	Varied	Common to abundant	Absent	Rare; no molasse
Divergent junction and constructive[c] margin	Active		Oceanic	None	—	—	—	—

[a] Destruction of crust by subduction into and consumption by the earth's mantle.
[b] No losses or additions to crust.
[c] Formation of new crustal materials at rift or at line of plate separation.

In addition, environmental inferences based on the carbonate fill can yield useful insights about water depths and paleocirculation patterns.

DICKINSON (1974, p. 12) has attempted to deduce the sequences to be expected in a basin formed at a divergent plate margin. What kinds of sequences characterize the various basins associated with convergent plates? How do these deductive sequences compare with reality? Certainly these questions deserve much more attention by sedimentologists. In addition to continued deductive thinking, it seems clear that we still need to describe and interpret carefully many more sedimentary basins.

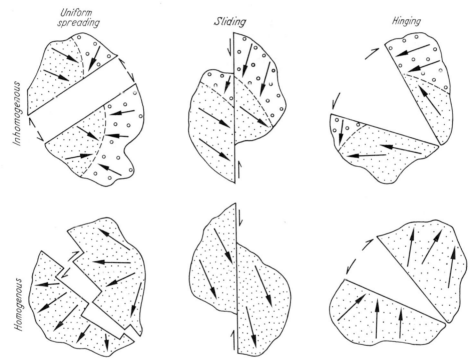

Fig. 9-16. Diagramatic representation of ways in which continents may separate and how paleocurrents and mineral associations can be used to match separated parts. (POTTER, 1974, fig. 4)

Along with facies, structural trends, age dating, and continental morphology, paleocurrent patterns are proving a useful tool for matching continents which have drifted apart (fig. 9-16). Of particular interest are the large-scale paleo-glacial and paleowind patterns. These have proved helpful in the Gondwanaland reconstruction (fig. 9-17). One can use paleocurrents and other sedimentologic criteria for matching particular basins such as the coastal eastern margin of Canada with Greenland (Fig. 9-18).

Closure of the two sides of an ocean leading to the formation of a suture zone, may not be synchronous throughout its length. Progressive closure may lead to shifting depocenters and migration of flysch and other facies in an orderly manner as postulated by GRAHAM et al. (1975) as shown in figure 9-19.

An aspect of paleocurrents, still far from being fully exploited, is compilation of more region-wide paleocurrent and facies maps — maps of regions exceeding a thousand or more kilometers in length and width. To supplement maps based on ancient glacial deposits and ancient eolianites, one can, for example, conceive

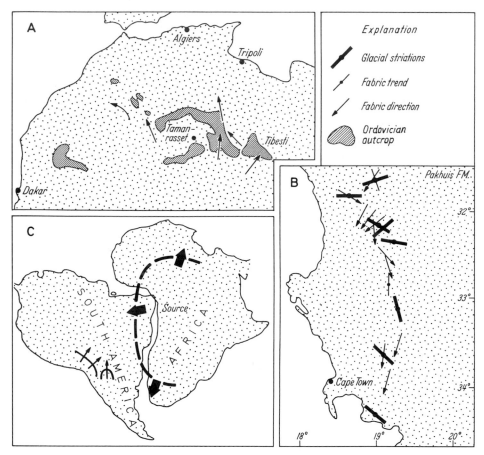

Fig. 9-17. Ordovician and/or Silurian paleoglacial flow in South America and Africa. (Redrawn from BIGARELLA, 1973, fig. 7.) Paleoflow in eastern Brazil and Africa suggest a common source — a large continental ice sheet

of maps of all the Mesozoic and Cenozoic molasse of the Great Plains of Canada and the United States, or of the Cambro-Ordovician of northern Africa, or of the cross-bedded oolitic limestones of Jurassic age in Europe. A map fully in this spirit is that of the Devonian of the Brooks Range in northern Alaska (fig. 4-28).

An entirely different aspect of paleocurrent analysis on the basin scale, one that was really only beginning in 1963, are *paleoceanic circulation patterns* based

Fig. 9-18. Successive stages of inferred separation of Greenland from the main North American continent (Redrawn from BEH, 1975, Figs. 10, 11, 12, and 13). Paleocene basalt indicates inferred position of possible mantle plume while shading shows region of inferred continental crust. Careful matching of marginal basins using faunal evidence, lithologic similarity, and tectonic styles would be helpful in confirming this sequential picture. Additionally, how many types of paleocurrent systems in these coastal basins are consistent with a prerift connection as well as with the inferred rifting?

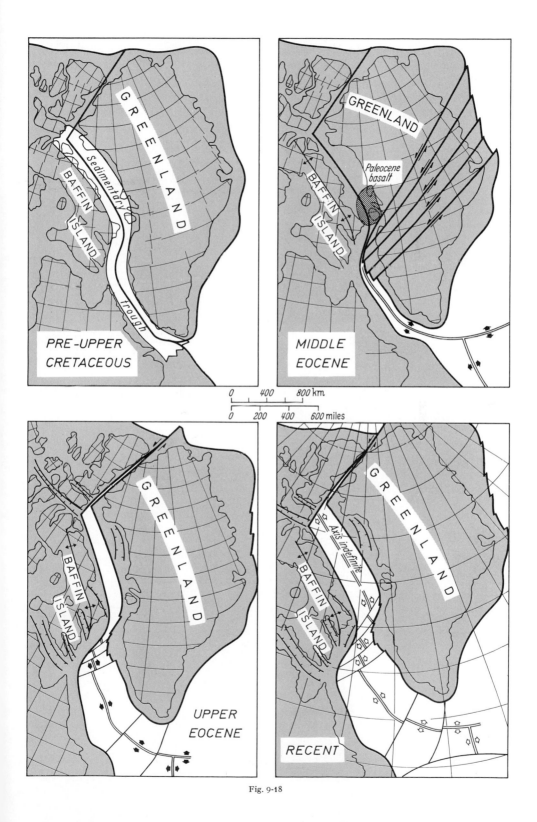

Fig. 9-18

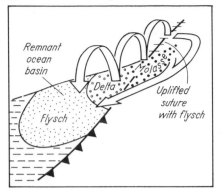

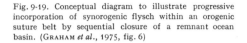

Fig. 9-19. Conceptual diagram to illustrate progressive incorporation of synorogenic flysch within an orogenic suture belt by sequential closure of a remnant ocean basin. (GRAHAM *et al.*, 1975, fig. 6)

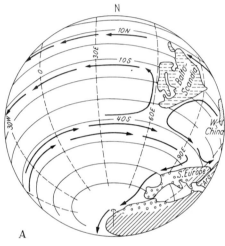

A

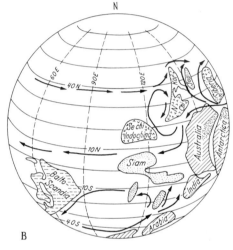

B

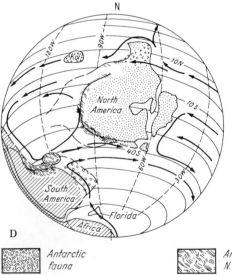

D

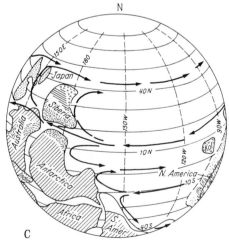

C

Antarctic fauna		Argentina N.America fauna
Iberia-Bolivia fauna		Alaska-China fauna

Fig. 9-20

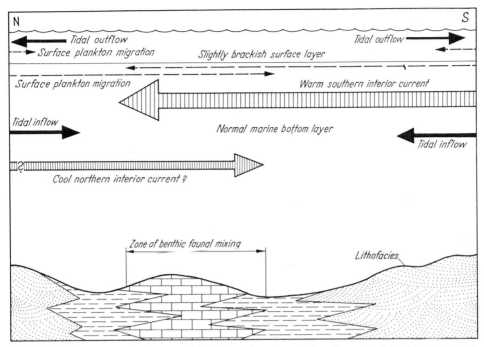

Fig. 9-21. A stratified system of paleocurrents inferred from paleoecology and depositional facies in the Cretaceous seaway of the western interior of North America. (Redrawn from KAUFFMAN, 1975, fig. 3)

on systematic, world-wide studies of fossil communities, especially those of Paleozoic age (fig. 9-20). Such maps of global scope far surpass in scale — except perhaps in only a sedimentologist's most fanciful dream — even those based on ancient eolian and glacial deposits. To supplement observational studies such as those by ROSS (1975), some experiments on world-wide paleoceanic circulation systems, with different positioning of the continents, have also been made (LUYENDYK et al., 1972, fig. 5). In years to come we look forward to paleontologists contibuting even more to paleocurrent studies — from fossil orientation in single outcrops to studies of single onshore basins and beyond. Another example of the new dimensions of paleocurrent study embodying paleontological data is that by BRISKIN and BERGGREN (1975, p. 191—192), who studied the paleoecology of foraminifera in deep sea cores, and inferred different circulation patterns in the North Atlantic Ocean during the Pleistocene. Earlier RUDDIMAN and McINTYRE (1973) studied the rate of lateral migration of polar waters in the North Atlantic during the last deglaciation — and thus indirectly documented a changing paleocurrent system.

Another aspect of paleocirculation patterns that has developed since 1963 is recognition of the fact that in marine basins and large lakes the current system could vary with depth — a fact long known to oceanographers, but rarely recognized by most paleocurrent mappers. One example of this is in a Cretaceous

Fig. 9-20. Hypothetical positions of continental plates in the Cambrian with the south equatorial current deflected into three separate counter-clockwise gyres in A, B, and D. Patterns indicate distribution of different trilobite associations. (Redrawn from Ross, 1975, fig. 6)

epiric sea, where KAUFFMANN (1975) inferred different current systems from paleontological data (fig. 9-21). The general problem presented by his diagram certainly needs much more attention by sedimentologists. However, the internal consistency of most paleocurrent systems and facies distributions in basins suggests that the current system that deposited the preserved sediment must have been the dominant one — if not the only one.

We believe that paleocurrent analysis — be it systematic mapping of directional structures, facies mapping and matching, or studies of global paleocirculation systems in either air or water — will play a progressively greater role in virtually all aspects of basin analysis in the years to come, especially for the role of plate tectonics in pre-Mesozoic time.

Annotated References

We found a rich, varied, and vast literature on megasedimentology — the study of sedimentation in basins and larger areas — and here include only a very, very small sample of it. Anyone studying this literature will soon find need for some guides to organizing it; the following subheadings reflect our pragmatic approach to this difficult problem.

Lithologic Atlases, Source Books and Maps plus Bibliographies

The references in this section fall into two groups: one with tectonic evolution in light of the plate tectonic paradigm; the other with specific basins summarizing what is known about their sedimentary fill. There are only a few papers in which both aspects, tectonic and sedimentologic, are jointly considered.

BRGM, ELF-RE, ESSO-REP and SNPA, 1974: Atlas du bassin d'Aquitaine. Bur. Rech. Geol. Mineral., Boîte Postale 6009, 45018 Orleans, Cedex, France. (French and English).
    Twenty-six beautiful colored plates of southwestern France (Aquitaine is called the "Texas" of France) most of which contain more than one geologic map and cross section and all of which contain a complete English as well as French text. The Aquitaine Basin is principally a Mesozoic basin rich in carbonates and flysch followed by Tertiary molasse. An outstanding example of the cooperation of many different organizations and people and a rich source for classroom basin studies.
BURK, C. A., and C. L. DRAKE, eds., 1974: The geology of continental margins. New York: Springer, 1009 p.
    Thirteen parts with 71 papers, including many descriptions of selected continental margins and the geology of selected small ocean basins plus 9 interesting papers on ancient continental margins. Excellent example of contribution of marine geophysics to basin sedimentology.
DICKINSON, W. R., ed., 1974: Tectonics and sedimentation. Soc. Econ. Paleontol. Mineral., Sp. Pub. 22, 204 p.
    Good collection of review articles and description of specific types of sedimentation on one of sedimentology's age-old, but major, questions. Full emphasis on plate tectonics. Cf. DOTT and SHAVER (1974).
DOTT, R. H., JR., and R. H. SHAVER, eds., 1974: Modern and ancient geosynclinal sedimentation. Soc. Econ. Paleontol. Mineral., Sp. Pub. 19, 380 p.

Twenty-nine articles in honor of Marshall Kay, one of North America's great students of geosynclines. Six major sub-headings include much material on this topic. Excellent source book. See also DICKINSON (1974).

EXPLORATION DEPT. SHELL OIL CO., 1975: Stratigraphic atlas North and Central America. Houston: Shell Oil Company.
A collection of 226 plates (31 × 40 cm) of black and white cross-sections and maps of the major subdivisions of all the geologic systems in North and Central America arranged in order of geologic age. Included are paleogeologic maps, outcrop, isopach, and lithofacies maps and maps showing the occurrences of oil and gas, radiometric ages, age of zero edge, stratigraphic and structural cross sections, and additional data of special interest in one or another geologic system. References for data taken from published papers are given at end of sections for each geologic system. These extend to 1970, with a few to 1971 and 1972.

HAMILTON, W., 1974: Map of sedimentary basins of the Indonesian region. U.S. Geol. Survey, Misc. Invs. Ser. Map I-875-B.
Isopachs of sedimentary fill plus distribution of Neogene volcanics, severely deformed Cenozoic sediments, basin complexes and location of major faults, including subduction zones, in a vast area. Brief discussion of geologic history of individual basins.

HOLLAND, C. H., ed., 1971 and 1974: Cambrian of the new world, 1 and Cambrian of the British Isles, Norden, and Spitsbergen, 2. London: Wiley Interscience, 456 and 300 p.
These two volumes, the first of a series on the lower Paleozoic rocks of the world, contain a vast wealth of material; Volume 1 has five chapters and Volume 2, seven chapters. Little about paleocurrents, but numerous semi-schematic maps, abundant cross sections and much emphasis on paleogeography and stratigraphic paleontology. Good source materials for a seminar on megasedimentology.

KASBEER, T., 1973: Bibliography of continental drift and plate tectonics. Geol. Soc. Am., Sp. Paper 142, 96 p.
Although oriented toward plate tectonics, you will find many papers on megasedimentology in this volume.

LINDEN, W. J. M. VAN DER, and J. A. WADE, eds., 1975: Offshore geology eastern Canada, vol. 2 Regional geology. Geol. Survey Canada, Paper 74-30, 258 p.
Useful as a source for basin development and geology along part of a trailing continental margin. Especially notable are the studies by Jansa and Wade of a basin analysis of the continental margin and the large collection of maps and seismic profiles. An outstanding source of data for teaching.

McCROSSAN, R. G., ed., 1973: The future petroleum provinces of Canada — their geology and potential. Canadian Soc. Petrol. Geologists, Mem. 1, 720 p.
Twenty-seven authors effectively describe the Phanerozoic sedimentary on- and offshore basins of Canada. Many informative cross sections and regional maps, some of the latter in color. Superb collection of case histories.

—, and R. P. GLAISTER, eds., 1964: Geologic history of western Canada. Calgary, Alberta Soc. Petrol. Geologists, 232 p.
Sixteen chapters including the Precambrian, all the geologic systems and the Quaternary plus a chapter on formation fluids and a concluding historical summary. Area studied includes about 15 percent of North America. Many illustrations (colored maps and cross sections) plus many references. Excellent source for basin studies.

NAIRN, A. E. M., and F. G. STEHLI, eds., 1973 and 1974: The ocean's basins and margins, vol. 1, The South Atlantic, and vol. 2, The North Atlantic. New York: Plenum Press Inc., 583 and 598 p.
These two volumes are part of a series planned for all the world's oceans and their margins. Many papers relevant to megasedimentology, continental drift and geophysics.

PRICE, R. A., and R. J. W. DOUGLAS, eds., 1972: Variations in tectonic styles in Canada Geol. Assoc. Canada, Sp. Paper No. 11, 688 p.

A very good source book for those interested in the relations between tectonics and sedimentation, all in 11 chapters. If you are in a hurry, be sure and read p. 639—665, a summary of sedimentary basins and geotectonic evolution. Many helpful summary tables.

ROCKY MOUNTAIN ASSOC. GEOLOGISTS, 1972: Geologic atlas of the Rocky Mountain region. Denver: A. B. Hirschfeld Press, 331 p.
The complete geologic history of a large part of the United States with many maps and cross sections by many different local experts. Key ideas: paleogeography, tectonics, basin descriptions, paleoenvironments, and economic resources.

SCHOTT, W., ed., 1969: Paläogeographischer Atlas der Unterkreide von Nordwest-deutschland. Bundesanst. Bodenforsch. Hannover, 315 p.
Three parts: an "Explanation" of 315 pages and two large paleogeographic atlases, one of 289 pages and the other a large number of folded maps. Over 1,100 references that are cross-indexed by regional geology, stratigraphy, paleogeography, paleontology, economic geology, etc. An in-depth study that contains all the elements of a basin analysis.

SPENCER, A. M., ed., 1974: Mesozoic-Cenozoic orogenic belts. Edinburgh and London: Scottish Acad. Press and Geol. Soc., Sp. Pub. 4, 809 p.
Something of a world atlas with 42 articles by local experts plus a general analysis by Spencer himself. Emphases is primarily on tectonic deformation, but virtually every article has a standardized stratigraphic table giving thicknesses, lithology and age of fill. Also an appendix consisting of a very interesting questionnaire for making inventories of orogenic belts. This questionnaire could be easily adapted to a more detailed inquiry into basin filling. Rich source book for megasedimentology.

VARIOUS AUTHORS, 1967: Paleotectonic investigations of the Permian system in the United States. U.S. Geol. Survey Prof. Paper 315, Parts A to K, 271 p.
Eleven parts (for different geographic regions) provide some source materials for students of basin analyses.

VINOGRADOV, A. P., ed., 1967, 1968 and 1969: Atlas-litologo-paleogeograficheskikh kart USSR, 1, 2, 3 and 4. Moscow, Glav, Upravlenie Geodezii i Kartografii (Russian and English).
An unparalled compilation of many diverse maps including the Precambrian to Quaternary. All maps are colored and show dominant lithologics and thicknesses and distribution of units. Also some maps show intensity of tectonic movements, plus distribution of oil and gas and other mineral resources. A vast wealth of material for classroom studies of basins by those willing to take the time to digest it all.

YOUNG, G. M., ed., 1973: Huronian stratigraphy and sedimentation. Geol. Assoc. Canada, Sp. Paper 12, 271 p.
Twelve papers on diverse sedimentological aspects of the Huronian (Precambrian) of Canada with some emphasis on paleocurrents.

ZIEGLER, A. M., B. R. RICKARDO and W. S. MCKERROW, 1974: Correlation of the Silurian rocks of the British Isles. Geol. Soc. Am., Sp. Paper 154, 154 p.
An unusual source book for a basin analysis. Very brief discussion of correlation, distribution, lithofacies, sediment sources, marine red beds, fossil communities and unconformities in 18 pages, including 7 figures followed by correlation of 450 Silurian stratigraphic units and more than 800 references.

Basins, Geosynclines and Plate Tectonics

Most papers in this group deal with the tectonic evolution of specific orogenic belts marginal to the continent. Several present general principles, two deal with present-day oceans or embryonic oceans (Red Sea), and one with how the separation of South America from Africa affected the former's coastal basins.

Asmus, H. E., 1975: Controle estrutual da deposicão Mesozocia nas bacias da margem continental Brasileira. Rev. Brasileira Geociencias 5, 160—175.
Relates fill of the Brasilian continental margin to structural and depositional consequence of separation of South America from Africa beginning in the Upper Jurassic. Good example of the "new breed" of basin studies. Key ideas: plate tectonics, continental drift, sedimentary sequences, and basin models.

Atwater, T., 1970: Implications of plate tectonics for the Cenozoic tectonic evolution of western North America. Bull. Geol. Soc. Am. 81, 3513—3536.
Primarily a study of the tectonic patterns and their relation to motion at boundary of North American and Pacific plates, both present and past.

Berggren, W. A., and C. D. Hollister, 1974: Paleogeography, paleobiogeography and the history of circulation in the Atlantic Ocean in W. W. Hay, ed., Studies in paleoceanography. Soc. Econ. Paleontol. Mineral., Sp. Pub. 20, 126—186.
An interpretation of the tectonic evolution of this classic "geosyncline" in terms of the plate tectonic paradigm, with special emphasis on the northern Appalachians.

Dewey, J. F., W. C. Pitman III, W. B. F. Ryan and J. Bonnin, 1973: Plate tectonics and the evolution of the Alpine System. Bull. Geol. Soc. Am. 84, 3137—3180.
Detailed analysis of tectonic history of zone of Alpine deformation related to interaction of European and African plates; a very complex history

Dickinson, W. R., 1972: Evidence for plate tectonic regimes in the rock record. Am. J. Sci. 272, 551—576.
Useful schematic diagrams nicely display the possible variation of continental separation and types of basins associated with a "leading" edge and plate boundaries. The author is a sedimentary petrologist, one who early leaped aboard the plate tectonics train, perhaps because he lives in California.

Editorial Committee, 1975: Problems on geosynclines in Japan. Tokyo, Association for the Geological Collaboration in Japan, Organizing Committee for the Symposium, 1974, Monograph/19, 262 p. (Japanese).
Twenty-nine articles ranging from sedimentology to structure to metamorphic and igneous petrology. Useful reference to see what basin analysis can be like in complexly intruded and folded geosynclinal belt. English abstracts and figure captions.

Falvey, D. A., 1974: The development of continental margins in plate tectonic theory. J. Australian Petrol. Explor. Assoc. 14, 95—106.
Attempts to apply plate tectonic theory to the so-called Atlantic-type continental margins and to formulate a general model. The new element in this model is the attribution of the initial subsidence in the rift-valley zone to metamorphism in the deep crust which leads to denser rock and hence subsidence. Heretofore the depression was attributed to thinning related to crustal stretching. From this model theoretical time-stratigraphic and structural cross sections are constructed for each of the stages in the rifting and continental separation.

Fischer, A. G., and S. Judson, eds. 1975: Petroleum and global tectonics. Princeton Uni. Press, 322 p.
Nine papers with many directly related to basin analysis.

Gussow, W. C., 1976: Sequence concepts in petroleum engineering. Geotimes, September, 16—17.
Brief summary of an exciting new development — the worldwide recognition of eustatic cycles from seismic sections. See Sloss (1974) as well.

Hallam, A., 1971: Mesozoic geology and the opening of the North Atlantic. J. Geol. 79, 129—157.
Includes generalized facies and environmental interpretations for western Europe and eastern North America during Mesozoic in light of plate tectonics concepts.

Hollister, C. D., D. A. Johnson and P. F. Lonsdale, 1974: Current-controlled abyssal sedimentation: Samoan Passage, equatorial west Pacific. J. Geol. 82, 275—300.
Figure 1 shows the inferred probable path of strongest Antarctic bottom water flow in the deep ocean to extend over nearly 2,400 miles. Integrated geophysical and geologic study indicate intense flow in narrow passage — as inferred from

sediment types — beginning in Eocene. A good example of a new breed of deeper ocean basin analysis studies.

Hsu, K. J., 1973: Mesozoic evolution of the California Coast Ranges *in* K. A. De Jong and R. Scholten, eds. Gravity and tectonics: New York: John Wiley and Sons, p. 379—396.
Excellent summary of a fascinating area with emphasis on the types and sequences of its flysch fill as well as on largest scale of all deformational structures, melanges. Plate tectonic interpretation stressed.

Kahle, C. F., ed., 1976: Plate tectonics, assessments and reassessments. Am. Assoc. Petrol. Geologists Mem. **23**, 514 p.
Twenty-four papers including some on sedimentology such as Global Tectonics and the Sediments of Modern and Ancient Trenches: Some Different Interpretations, Early Evidence of Continental Drift: Pro and Con, and Marine Sedimentary Environments and Their Faunas in Gondwana Area.

Kinsman, D. J. J., 1975: Rift valley basins and sedimentary history of trailing continental margins *in* A. G. Fischer and Sheldon Judson, eds., Petroleum and global tectonics. Princeton, Princeton Univ. Press, 83—126.
Strong emphasis on geophysics but includes discussion of sediment source areas and supply as well as the fill of rifted continental margins, including the author's favorite sediment-evaporites.

Klemme, H., 1971: Giants, supergiants and their relation to basin types. Oil and Gas J., March 1, 8 and 15.
Relation of oil production to plate tectonics and classification of eight types of basins; observes that 85 percent of world's oil comes from 5 percent of the world's oilfields.

Le Pichon, X., J. Francheteau and J. Bonnin, 1973: Plate tectonics (Developments in Tectonics, **6**). Amsterdam: Elsevier Publ. Co., 302 p.
Parts of Chapters 6 and 7, processes at accreting and consuming plate boundaries, have direct relevance to megasedimentology.

Lowell, J. D., B. Cenik and J. Gerard, 1972: Sea-floor spreading and structural evolution of southern Red Sea. Bull. Am. Assoc. Petrol. Geologists **56**, 247—259.
A good example of modern rifting and geothermal gradients so high that oil exploration was locally abandoned. Because no major river enters the Red Sea, instead of a thick clastic section — as in the head of the Gulf of California — its fill consists mostly of evaporites, carbonates, and only minor clastics plus lavas.

Meyerhoff, A. A., and H. A. Meyerhoff, 1973: Tests of plate tectonics *in* C. F. Kahle, ed., Plate Tectonics — assessments and reassessments. Am. Assoc. Petrol. Geologists Mem. **23**, 43—145.
Summary of a large amount of data by a creative antidrift thinker, one who should certainly not be overlooked. But drift or antidrift, paleocurrents have a role to play.

Mitchel, A. H., and H. G. Reading, 1969: Continental margins, geosynclines and ocean floor spreading: J. Geol. **77**, 629—646.
Contintntal margins divided with three types: Atlantic (no trench), Andean (trench) and island arcs plus "other", the unaccounted for "error".

Naylor, D., and S. N. Mounteney plus Pergrum, R. M., G. Rees and D. Naylor, 1975: Geology of the northwest European continental shelf **1** and **2**. London, Graham Trotman Dudley Ltd., 162 and 224 p.
Volume 1 has two parts—structural setting and geology of the west British shelf (subdivided into geographic areas) while Volume 2 is devoted to the North Sea and describes its basin evolution as well as its history of oil and gas exploration. Well illustrated, not excessively technical and very well done.

Sestini, G., ed., 1970: Development of the northern Apennines geosyncline. Sediment. Geol. **4**, 203—642.
Ten articles on a fascinating geosyncline — its eugeosynclinal, miogeosynclinal, and postgeosynclinal aspects, plus topics such as olistostromes, flysch, and significance for continental drift. Best single English-language description of northern Apennines. See also Ricci Lucchi (1975).

SLOSS, L. L., 1972: Synchrony of Phanerozoic sedimentary-tectonic events of the North American craton and Russian platform. Int. Geol. Congr. 24th Sess. Sec. **6**, 24—32.
After comparing the geologic histories of two widely separated cratons and finding much evidence for broadly synchronous periods of basin subsidence, it is suggested that even laterally moving plates have possibly had similar epeirogenic histories.

WOODLAND, A. W., ed., 1975: Petroleum and the continental shelf of north-west Europe. **1**, Geology. New York, John Wiley and Sons, 501 p.
This volume, an outstanding source book for basin analysis, concludes 38 papers that touch on almost all aspects of megasedimentology—tectonics, sedimentary fill, geophysics, thermal history, diagenesis, and petroleum potential. Article by Ziegler (p. 131—149) gives a short overview relating the North Sea's basin history to the tectonic framework of north-western Europe. Highly recommended.

YORATH, C. J., E. R. PARKER, and D. J. GLASS, 1975: Canada's continental margins and offshore petroleum exploration. Canadian Soc. Petrol. Geologists, Mem. **4**, 898 p.
Many case histories and much more, organized in Atlantic, Baffin Bay, Arctic and Pacific facing margins plus sections on engineering studies and general topics, the the latter containing two articles of special interest: Some Remarks on Regression and Transgression in Deltaic Sediments plus North Atlantic Old Red Sandstone—Some Implications for Devonian Paleography. Excellent source book.

ZIEGLER, W. H., 1975: Outline of the geological history of the North Sea *in* A. W. WOODLAND, ed., Petroleum and the continental shelf of north-west Europe. London, Applied Science Publishers Ltd. and the Institute of Petroleum **1**, 165—187.
The North Sea lies on cratonic crust and has been affected by four diastrophic periods. To explain how these affected the North Sea basin, the author considers regional tectonics and geology as far away as Africa and North Africa and brings it all together with liberal use of plate tectonics. Sixteen informative facies and tectonic maps plas an excellent summary of geologic events. See also his article North Sea Basin History in the Tectonic Framework of Northwestern Europe of the same volume.

Paleocurrents and Continental Reconstruction

Planetary winds and continental glaciers are on large enough scale to impose a paleocurrent pattern of planetary proportions. To these can be added reconstructions based on world-wide studies of faunal provinces and the presumed current systems to which they responded. Hence, rifting and separation of continents should disrupt and displace these ancient paleocurrent systems. Mapping them should provide confirmation of such continental fragmentation; and when an individual cratonic basin lies astride a zone of rifting, paleocurrent analysis may assist in identifying the separated parts of such a basin. See also Table 5-4 for some additional references.

BIGARELLA, J. J., 1970: Continental drift and paleocurrent analysis *in* 2nd Gondwana Symp., Proc. and Papers. Int. Union Geol. Sci., Comm. on Stratigraphy, Subcommission on Gondwana Stratigraphy and Paleontology (Geol. Soc. South Africa, Marshalltown, Transvaal, South Africa), p. 73—97.
One of the best — and one of the few — efforts using paleocurrent data to see if continents fit.

—, 1973: Paleocurrents and the problem of continental drift. Geol. Rundschau **62**, 447—477.
Paleocurrent analysis on a super-regional, grand scale is used to investigate the former proximity between South America and Africa, by the leading student of

cross-bedding in the southern hemishpere. In a general way paleocurrents indicate radial transport from a common source. Over 100 references.

BOUCOT, A. J., 1974: Early Paleozoic evidence of continental drift: Pro and con *in* C. F. KAHLE, ed., Plate tectonics — assessments and reassessments. Am. Assoc. Petrol. Geologists Mem. **23**, 273—294.
Early Paleozoic lithofacies, structural, biogeographic, and animal community data do not indicate intercontinental connection, nor do they disprove it. Very interesting example of contribution of paleontology and other evidence to problems of continental drift.

WHITTINGTON, H. B., and C. P. HUGHES, 1972: Ordovician geography and faunal provinces deduced from trilobite distribution. Roy. Soc. London Philos. Trans., Biol. Sci. **263**, 235—278.
Statistical analysis of fauna at both the generic and family levels plus world maps of the reconstructed continents.

Turbidite Basins

Paleocurrent studies seem to have been particularly helpful in unravelling the evolution of turbidite basins. Only Graham et al. and Morris, however, attempt to relate the results to plate tectonics. The organization of the turbidite fill in these basins has become much better known as a result of the work of the Italian school, E. Mutti and F. Ricci Lucchi in particular, and of Walker and his students in Canada.

BRIGGS, G., and L. M. CLINE, 1967: Paleocurrents and source areas of late Paleozoic sediments of the Ouachita Mountains, southeastern Oklahoma. J. Sediment. Petrol. **37**, 985—1000.
Facies relations, sedimentary petrology and paleocurrents all indicate that the Ouachita geosyncline derived its sediments from marginal sources. Most was derived from the south but transported longitudinally by turbidity currents. See also MORRIS (1974).

CONTESCU, L. R., 1974: Geologic history and paleogeograhpy of eastern Carpathians, example of Alpine geosyncline evolution. Bull. Am. Assoc. Petrol. Geologists **58**, 2436—2476.
Best English-language overview of the complexly deformed eastern Carpathian geosyncline. Paleocurrents are very complex, because they respond to local uplifts within and marginal to basin as well as general longitudinal slope of principal basin. Author emphasizes the order within the geosynclinal fill.

GRAHAM, S. A., W. R. DICKINSON and R. V. INGERSOLL, 1975: Himalayan-Bengal model in flysch dispersal in the Appalachian-Ouachita system. Bull. Geol. Soc. Am. **86**, 273—286.
This paper is a good example of deductive thinking and synthesis typical of the impact of plate tectonics on our concepts of basins.

LAJOIE, J., 1970: Flysch sedimentology in North America. Geol. Assoc. Canada Sp. Paper **7**, 272 p.
Fourteen papers ranging from the Recent to the early Paleozoic plus some experimental results. Paleocurrent data abundant.

McIVER, N. L., 1970: Appalachian turbidites *in* G. W. FISHER, F. J. PETTIJOHN, J. C. REED JR., and K. N. WEAVER, eds., Studies of Appalachian geology — central and southern. New York: Interscience, 69—81.
Primarily a review of paleocurrent patterns in Martinsburg (Ordovician) and Upper Devonian turbidites in Appalachian Basin.

MORRIS, R. C., 1974: Sedimentary and tectonic history of the Ouachita Mountains *in* W. R. DICKINSON, ed., Tectonics and sedimentation. Soc. Econ. Paleontol. Mineral., Sp. Publ. **22**, 120—142.

Outstanding integration of paleocurrents, stratigraphy, and petrology of Paleozoic Ouachita turbidite basin combined with explanation in terms of plate tectonics.

OJAKANGAS, R. W., 1968: Cretaceous sedimentation, Sacramento Valley, California. Bull. Geol. Soc. Am. **79**, 973—1008.
Analysis of 35,000-foot (10,668 m) fill of Cretaceous trough by use of sandstone mineralogy and paleocurrent structures. Determination of source areas and paleocurrent flow patterns.

PAREA, G. C., 1965: Evoluzione della parte settentrionale della Geosinclinale Appenninica dall'Albiano all'Eocene superiore. Acc. Naz. Sci. Lett. Art. Ser. **VI, 7**, 97 p.
One of the earliest basin analysis of turbidites and their related sediments. Particularly noteworthy are the criteria for distribution of the different facies within the basin (p. 57—66), where different ages of sub sea fans are delineated, and the regional paleogeographic reconstruction. Early big thinking, very well done.

SCOTT, K. M., 1966: Sedimentology and dispersal pattern of a Cretaceous flysch sequence, Patagonian Andes, southern Chile. Bull. Am. Assoc. Petrol. Geologists **50**, 72—107.
A well-integrated study of clast size (in conglomerates), sandstone provenance, grain fabric, sole markings and slump folds in a turbidite basin.

TRETTIN, H. P., 1971: Geology of Lower Paleozoic formations, Hazen Plateau and southern Grant Land Mountains, Ellesmere Island, Arctic Archipelago. Geol. Surv. Canada, Bull. **203**, 134 p.
Conceptual model with lateral and longitudinal fill. Details of how structural attitude affects reorientation of bedding.

WALKER, R. G., and E. MUTTI, 1973: Turbidite facies and facies associations *in* G. V. MIDDLETON and A. H. BOUMA, eds., Turbidites and deep-water sedimentation, Los Angeles. Pacific Section Soc. Econ. Paleontol. Mineral. 119—157.
A good resumé of slope-fan-basin floor system. The turbidite model is described in terms of various turbidite facies, their areal distribution and vertical profile or sequence.

WALKER, R. G., 1975: Generalized facies models for resedimented conglomerates of turbidite association. Bull. Geol. Soc. Am. **86**, 737—748.
Proposes three conglomerate models and relates these to a general submarine turbidite fan model.

YERKES, R. F., T. H. McCULLOH, J. E. SCHOELLHAMER and J. G. VEDDER, 1965: Geology of the Los Angeles Basin California — an introduction. U.S. Geol. Survey, Prof. Paper **420-A**, 57 p.
Evolution of a prolific oil-producing turbidite basin and much of its geology. Cf. CONREY (1967) in Chapter 8.

Molasse Basins

The sedimentary, including paleocurrent, history of molasse basins is the best-documented and understood of all types of basins. The papers included in this section effectively demonstrate the benefits of an integrated study of stratigraphy, facies, and paleocurrents.

EISBACHER, G. H., 1974a: Sedimentary history and tectonic evolution of the Sustut and Sifton Basin, north-central British Columbia. Geol. Survey Canada Paper **73-31**, 57 p.
Very well-documented, integrated study of clastic fill of successor basins (north-central British Columbia). Cross-bedding, clast size, provenance, and paleohydraulics.

—, 1974b: Evolution of successor basins in the Canadian Cordillera of British Columbia *in* R. H. DOTT JR., and R. H. SHAVER, eds., Modern and ancient geosynclinal sedimentation. Soc. Econ. Paleontol. Mineral., Sp. Publ. **19**, 274—291.

A well-documented effort to relate sedimentology to basin geometry and evolution in light of plate tectonics.

Füchtbauer, H., 1967: Die Sandsteine in der Molasse nördlich der Alpen. Geol. Rundschau 56, 266—300.
The last of a series of papers on the molasse basin of southern Germany. Mainly petrographic but including data on size decline of pebbles, percent of conglomerate, and paleocurrent flow based on mapping heavy mineral provinces.

Meckel, L. D., 1967: Origin of Pottsville conglomerates (Pennsylvanian) in the central Appalachians. Bull. Am. Geol. Soc. 78, 223—258.
An analysis of that part of the Appalachians lying in Pennsylvania and adjacent areas utilizing facies, petrography, and paleocurrents. Reconstruction of basin geometry, source areas, and depositional pattern.

—, 1970: Paleozoic alluvial deposition in the central Appalachians: A summary in G. W. Fisher, F. J. Pettijohn, J. C. Reed Jr., and K. N. Weaver, eds., Studies of Appalachian geology — central and southern. New York: Interscience, 49—67.
A summary of the current and dispersal patterns in three nonmarine Paleozoic alluvial wedges in central Appalachians. Presents depositional model and fining-upward fluvial cycles.

Royse, C. F., Jr., 1970: A sedimentologic analysis of the Tongue River-Sentinel Butte interval (Paleocene) of the Williston, Basin, western North Dakota. Sediment. Geol. 4, 19—80.
Integrated facies study — paleocurrents, sedimentary structures and grain size plus short section on basin analysis.

Schoeffler, J., 1973: Étude structurale des formations molassiques du piedmont nord des Pyrénées. Rev. Inst. Français du Pétrole 28, 515—665.
Combines sedimentology with the structure to provide an unusually complete, in-depth study of molasse along the north side of the Pyrenees. Excellent.

Veit, Erwin, 1963: Der Bau der südlichen Molasse Oberbayerns auf Grund der Deutung seismischer Profile. Bull. Ver. Schweiz. Petrol. Geol. Ing. 30, 15—52.
Essentially a paper on the geometry of the Tertiary molasse basin, which is 4,000 to 5,000 m thick, in southern Bavaria.

Shelf and Shelf-to-Basin Transition

The shelf environment is closely related to the cratonic basins in terms of tectonic stability. Sediments in both may be subject to strong tidal influences. Both may also contain carbonates, some of which may be imprinted with paleocurrent patterns. The paper by Galloway and Brown is notable for the presentation of a conceptual model for the shelf to basin transition zone.

Adams, R. W., 1970: Loyalhanna limestone — cross-bedding and provenance in G. W. Fisher, F. J. Pettijohn, J. C. Reed Jr., and K. N. Weaver, eds., Studies of Appalachian geology — central and southern. New York: Interscience, 83—100.
Depositional framework of Loyalhanna Limestone (Pennsylvanian) of southwestern Pennsylvania and adjacent areas based on facies, petrography and paleocurrents. A tide-dominated, shallow marine trough with multiple sources.

Asquith, D. O., 1970: Depositional topography and major marine environments, late Cretaceous, Wyoming. Bull. Am. Assoc. Petrol. Geologists 54, 1184—1224.
Very informative maps and cross sections illustrate specific slope deposits—how to recognize, map and interpret them. A very significant article for every student of sedimentary basins. Includes a discussion of how to decompact a slope sequence to obtain a better idea of its original configuration.

Galloway, W. E., and L. F. Brown Jr., 1973: Depositional systems and shelf-slope relations on cratonic basin margin, uppermost Pennsylvanian of north-central Texas. Bull. Am. Assoc. Petrol. Geologists 57, 1185—1218.

A good analysis of a three-component depositional system: (1) fluvial-deltaic, (2) shelf-edge, and (3) slope-fan (turbidite). Mainly based on subsurface data, geometry and character of sand bodies. A well-documented and well-integrated model for sediment dispersal of a basin margin province. Good discussion of intrabasinal tectonics versus extrabasinal eustatic controls on depositional facies.

THOMSON, A., M. R. THOMASSON, 1969: Shallow to deep water facies development in the Dimple Limestone (Lower Pennsylvanian), Marathon Region, Texas in G. M. FRIEDMAN, ed., Depositional environments in carbonate rocks. Soc. Econ. Paleontol. Mineral. Sp. Pub. **14**, 57—77.
An interesting example of shelf to basin transition of a carbonate sediment, the deep-water facies being a carbonate turbidite.

Cratonic and Other Basins

There are extensive, little-deformed cratonic basins ranging from Precambrian to Pleistocene in age and, although there have been many studies of individual carbonate or sandstone formations deposited in these intracratonic basins, there are few integrated studies of such basins as a whole. The studies of Beuf and coworkers on the Lower Paleozoic sandstones of the Sahara come closest to such an all-inclusive study. Cratonic basins have as yet to be interpreted in terms of plate tectonics.

BARRETT, P. J., 1970: Paleocurrent analysis of the mainly fluviatile Permian and Triassic Beacon rocks, Beardmore Glacier area, Antarctica. J. Sediment. Petrol. **40**, 395—411.
Based mainly on parting lineation and cross-bedding in Beacon sandstones, Antarctica. Over 3000 measurements.

BEUF, S., B. BIJU DUVAL, O. DE CHARPAL, P. ROGNON, O. GARIEL and A. BENNACEF, 1971: Les grès du Paléozoique inférieur au Sahara. Paris, Éditions Technip.,464p.
Description of Lower Paleozoic sandstones of the Sahara — a vast tectonically stable craton in sharp contrast to continental margins. Perhaps the standard for all cratonic areas. Very well illustrated and well written.

JORDAN, R. R., 1964: Columbia (Pleistocene) sediments of Delaware. Delaware Geol. Survey Bull. **12**, 59 p.
Grain size, petrology, and sedimentary structures in a Pleistocene coastal plain with diverse environments. Very rich in data and many maps. Good model for basin analysis of comparable deposits.

McKEE, E. D., and E. J., CROSBY, Coordinators, 1975: Paleotectonic investigations of the Pennsylvanian System in the United States. Part I, Introduction and Regional Analysis of the Pennsylvanian System; Part II, Interpretive summary and special features of the Pennsylvanian System; Part III, Plates. U. S. Geol. Survey, Prof. Paper **853**, 349 and 192 p.
A vast amount of stratigraphic information organized by regions as well as articles on special features and interpretation plus many, many maps. Mostly, but not completely, detrital sedimentation.

MIALL, A. D., 1975: Post Paleozoic geology of Banks, Prince Patrick, and Eglinton Islands, Arctic Canada in C. J. YORATH, E. R. PARKER, and D. J. GLASS eds., Canada's Continental Margins and Offshore Petroleum Exploration. Canadian Soc. Petrol. Geologists Mem **4**, 557—587.
Combines paleocurrent mapping with facies and thickness maps plus cross sections. Article nicely illustrates the utility of paleocurrent mapping *early* in a basin analysis.

MULDER, C. J., P. LEHNER, and D. C. K. ALLEN, 1974: Structural evolution of the Neogene salt basins in the eastern Mediterranean and the Red. Sea. Geol. Mijnbouw **54**, 208—221.

Notable for its seismic sections illustrating olistostromes (Fig. 5) in the eastern Mediterranean and the evolution of the evaporite fill of the Red Sea rift whose main subsidence began in the Oligocene. What will the future study of paleocurrent systems in evaporite basins contribute to our knowledge of rift sedimentation.

SCHWAB, F. L., 1970: Origin of the Antietam Formation (Late Precambrian (?) - Lower Cambrian) central Virginia. J. Sediment. Petrol. **40**, 354—366.
Provenance and paleocurrent (cross-bedding) study showing cratonic source of Antietam (Cambrian) orthoquartzite in central Appalachians.

STEWART, J. H., F. G. POOLE and R. F. WILSON, 1972: Stratigraphy and origin of the Chinle Formation and related upper Triassic strata in the Colorado Plateau region. U.S. Geol. Survey Prof. Paper **690**, 336 p.
A basin analysis of detrital Triassic conglomerates, sandstones, and mudstones (with many red beds) over large parts of Arizona and nearby states. The analysis integrating paleocurrents, isopach maps, and petrology. Figures 31 and 33 especially informative. Many carefully described stratigraphic sections make this an excellent general reference study. See also U.S. Geol. Survey Prof. Papers **691** and **692**.

WEIMER, R. J., 1970: Rates of deltaic sedimentation and intrabasin deformation, Upper Cretaceous of Rocky Mountain region *in* J. P. MORGAN, ed., Deltaic sedimentation, modern and ancient. Soc. Econ. Paleontol. Mineral. Sp. Publ. **15**, 270—292.
Some principles of basin analysis developed by regional study of Cretaceous sedimentation in 14 states of the United States. Stresses importance of penecontemporaneous growing structures as traps for petroleum.

References

BALLY, A. W., 1975: A geodynamic scenario for hydrocarbon occurrences *in* Proc. 9th World Petroleum Congr., Tokyo. Barking, England, Geol. and Applied Sci. Publ. Ltd., Sec. **2**, 33—44.

BEH, R. L., 1975: Evolution and geology of western Baffin Bay and Davis Strait, Canada *in* C. J. YORATH, E. R. PARKER, and D. J. GLASS, eds., Canada's Continental Margins and Offshore Petroleum Exploration. Canadian Soc. Petrol. Geologists Mem. **4**, 453—476.

BIGARELLA, J. J., 1973: Paleocurrents and the problem of continental drift. Geologische Rundschau **62**, 447—477.

BRISKIN, M., and W. S. BERGGREN, 1975: Pleistocene stratigraphy and quantitative paleoceanography of tropical North Atlantic core V16-205 *in* T. SAITO and L. H. BURCKLE, eds., Symposium on Late Neogene Epoch Boundaries, 24th Int. Geol. Congr. Montreal (1972), Am. Mus. Natural History, New York, 167—198.

DICKINSON, W. R., 1974: Plate tectonics and sedimentation *in* W. R. DICKINSON, ed., Tectonics and sedimentation. Soc. Econ. Paleontol. Mineral, Sp. Pub. **22**, 204 p.

—, and E. J. RICH, 1972: Petrologic intervals and petrofacies in the Great Valley sequence, Sacramento Valley, California. Bull Geol. Soc. Am. **83**, 3007—3024.

GRAHAM, S. A., W. R. DICKINSON and R. V. INGERSOLL, 1975: Himalayan-Bengal model for flysch dispersal in the Appalachian-Ouachita System. Bull. Geol. Soc. Amer. **86**, 273—286.

KAUFFMAN, E. G., 1975: Dispersal and biostratigraphic potential of Cretaceous benthonic bivalvia in the Western Interior *in* W. G. E. CALDWELL, ed., The Cretaceous System in the Western Interior of North America. Geol. Assoc. Canada, Sp. Paper **13**, 163—194.

KLEMME, H. D., 1975: Giant oil fields related to their geologic setting; a possible guide to exploration. Bull. Canadian Petrol. Geol. **23**, 30—66.

LUYENDYK, BRUCE, P., D. FORSYTH and J. D. PHILLIPS, 1972: Experimental approach to the paleocirculation of the oceanic surface waters. Bull. Geol. Soc. Am. **83**, 2649—2664.

PORTER, J. W., and R. G. McCROSSAN, 1975: Basin consanquinity in petroleum resource estimation *in* J. D. HAUN, ed., Methods of estimating the volume of undiscovered oil and gas resources. Am. Assoc. Petrol. Geologists, Studies in Geology 1, 50—75.

POTTER, P. E., 1974: Sedimentology: Past, present and future. Die Naturwissenschaften **61**, 461—512.

READING, H. G., 1972: Global tectonics and the genesis of flysch successions. 24th Int. Geol. Congr. (Montreal), **6**, 59—66.

ROSS, R. J., JR., 1975: Early Paleozoic trilobites, sedimentary facies, lithospheric plates and ocean currents. Fossils and Strata **4**, 307—329.

RUDDIMAN, W. F., and A. McINTYRE, 1973: Time-transgressive deglacial retreat of polar waters from the North Atlantic. Quaternary Res. **3**, 117—130.

SPENCER, A. M., ed., 1974: Mesozoic-Cenozoic orogenic belts. Edinburgh and London: Scottish Academic Press and Geol. Soc. Sp. Pub. **4**, 809 p.

Methods of Study up to 1963

Introduction

The objectives of paleocurrent research are the identification, description and interpretation of the current patterns of the past.

In previous chapters, methods of measurement at the outcrop have been described for each of the major directional structures. The techniques common to the study of all types of directional structures and properties are presented in this chapter. Some of the techniques are peculiar to geology, as for example the rotation of structurally deformed beds. Others, such as statistical methods are not. Statistical methods, so useful in many observational sciences, have proved helpful also in sampling and describing paleocurrent patterns. The techniques used range from elementary to moderately advanced. Most general statistical texts (cf. Dixon and Massey, 1957) provide ample background as do Miller and Kahn (1962).

Collecting the Data

Collecting paleocurrent data over a large area requires sampling. The objective of sampling is to determine whether or not an anisotropic regional pattern exists and to describe it if it does. In connection with the sampling problem, it is useful to distinguish between population parameters, say the mean azimuth of *all* flute marks in a formation, and *estimates* of such parameters based on a sample, generally a small one. Usually in paleocurrent studies, population parameters can never be evaluated but can only be estimated by sampling. Population parameters are denoted by capital letters and their sample estimates by lower case ones.

In general, we may assume a close identity between the orientation of directional structures in outcrop and the orientation of those not available for measurement. ·For example, the orientation of a directional structure has little or no control on the presence or absence of an outcrop in which the structure occurs.

Most desirable is some appreciable "two-dimensional" coverage; i.e., exposure over an appreciable area rather than exposure in a straight, narrow outcrop belt.

Two dimensional coverage is obtained in well-dissected areas of flat-lying sediments or in areas of outcrops repeated by folding as in the Appalachian or Jura Mountains, or in areas of basins with marginal exposure along several sides. If the current system diverges, converges, or is otherwise irregular, a narrow linear outcrop belt is usually inadequate to delineate such irregularities.

For either poorly-exposed formations or those restricted to small areas, it may be possible and necessary to visit every known outcrop. Commonly, however, many more outcrops are available than it is either possible or necessary to visit. The investigator, therefore, usually studies only a selected fraction of all outcrops.

Difficult accessibility and time (cost) may restrict sampling to principal roads, streams or shorelines; some effort, however, should always be made to cover the whole outcrop belt. Coverage is facilitated by use of a sampling grid. A grid is useful because: 1), it assures distribution of sampling effort across an outcrop belt, so that every part of the formation receives comparable attention, and (2), it can be useful in analysis and presentation of the collected data.

Although some workers have designed a grid *after* collecting the data, it is usually advantageous to do so *before* field work is begun. Grids may be designed by

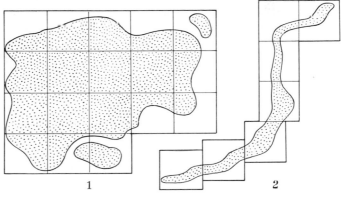

Fig. 10-1. Two outcrop patterns and grid designs

the investigator (PRYOR, 1960, fig. 13) or be based, as in parts of the United States and Canada, on the township and range system of land subdivision (POTTER and OLSON, 1954, fig. 3), or on latitude and longitude (FAHRIG, 1961, fig. 2), or on map sheets. Figure 10-1 shows examples of grids suitable for different outcrop patterns. Sampling points within a grid interval may be randomly selected in one of several ways (*cf.* KRUMBEIN, 1960, p. 359—361). However, such random selection is usually impractical, for no outcrops may exist at the predetermined sample points or the selected points may be difficult of access. In practice, outcrops are commonly sampled as encountered, some effort being made to distribute the sample points evenly within a given area.

The number of outcrops required depends upon the variability of the paleo-current pattern, on the objectives of study and on the accuracy desired.

Practically every clastic unit that has directional structures will show a preferred rather than random transport direction. Variability of the paleocurrent pattern usually can be assessed by a preliminary sample of a few outcrops. If these outcrops show closely similar current directions, an open spacing of outcrops will suffice. If, on the other hand, current direction is more variable, more outcrops will be required. Comparison of the preliminary results with previously published paleocurrent studies provides a guide for density of outcrop sampling. In any case, variability is measured quantitatively by *variance*, which is estimated by the *sample variance:*

$$s_x^2 = \sum_{i=1}^{n} (x_i - \bar{x})^2 / (n-1)$$

where the x_i are individual azimuths, $\bar{x}$ is the vector mean of all the observations and n the number of observations. The vector mean provides an approximate minimum sum of squares, if the data are approximately symmetrical about $\bar{x}$. The more variable the pattern, the greater variance and the more outcrops required to maintain the same level of accuracy. Accuracy of an estimate of current direction is measured by the standard deviation, the square root of the variance, or some other confidence limit based on the variance (see p. 256).

The number of samples needed varies markedly with the objectives of the study. If only a direction of transport is desired, studies of cross-bedding show that a regional mean can be estimated, to at least the correct quadrant and usually much less, from measurements in as few as a dozen or fewer widely scattered outcrops. An even smaller number of measurements would be required, if transport direction were to be inferred from directional structures such as glacial striations, which commonly have a small regional variance.

If the field area is easily accessible, it is usually good practice to sample in two stages, initially obtaining a fairly open distribution of sample points. If this sampling reveals a simple uniform pattern, additional sampling is commonly not required. If, on the other hand, the pattern appears to be more complex, additional sample points will be required. If the outcrop area is not easily accessible, all the data that may be needed should be obtained in a single sampling.

Two to four outcrops per six to twelve mile (10 to 20 km) interval have usually proved more than adequate to show regional paleocurrent patterns even though many patterns have been delineated with less dense control. Although more sample points could be obtained, it is usually better to study other aspects of the problem, such as mineral provinces or lithofacies, rather than simply add more sample points. Information from mineral provinces or lithofacies maps is especially helpful in the interpretation of complex paleocurrent patterns.

"Nested" or hierarchical sampling, or subsampling (COCHRAN, 1953, p. 215—267) as it is sometimes called, makes it possible to estimate the variability associated with different levels of sampling each of which contributes to the total variance. These different levels of subsampling form a hierarchy. For example, in studies of cross-bedding, variability

1. within beds,
2. between beds within outcrops,
3. between outcrops within specified areas, such as sections or townships,
4. between areas,

contributes to total variability. It may be desirable to analyze variance in geologic units which form a *natural* hierarchy. For example, in a study of cross-bedding in a sand-shale sequence with clearly defined, mappable sand bodies, total variability could be considered as arising from

1. between beds within outcrops,
2. between outcrops within sand bodies,

3. between sand bodies,

4. remaining unassigned sources.

Components of variance are additive so that in the sand body example the total variance, s_x^2, around the vector mean equals

$$s_x^2 = s_{x_1}^2 + s_{x_2}^2 + s_{x_3}^2 + e$$

where $s_{x_1}^2$, $s_{x_2}^2$, and $s_{x_3}^2$ are the estimates contributed by the different levels and e is the unassigned remainder, usually called "error," which, in this case, also includes the variability within beds.

POTTER and OLSON (1954, Table 3) estimated the components of variance of cross-bedding orientation in the sandstones of the Pennsylvanian Mansfield formation in the eastern part of the Illinois basin. They found (Table 3) that cross-bedding was most variable between outcrops within townships and least variable within cross-bedded units. More accurate estimates of mean direction are obtained by allocating most effort at those sampling levels that have the largest components of variance. Thus, if variability of cross-bedding is generally comparable to that of the sandstones of the Mansfield formation, the best procedure would be to sample a relatively large number of outcrops, each with relatively few measurements, and to make only one measurement for each cross-bedded unit. Hence early knowledge, either qualitative or quantitative, of variance components is useful in formulating a sampling plan. Because components of variance appear to be fairly stable for different directional structures, they need not be routinely computed in paleocurrent studies.

Quantitative estimates of the variance components may also contribute to our understanding of the processes of sedimentation. For example, total variance and its components may be different in marine and fluvial cross-bedding; significant differences may also appear, if components of variance from structures formed by turbidity and ordinary currents are compared.

Detailed maps of small areas of exceptional exposure usually provide considerable insight to local variability and knowledge of sedimentation. Such maps should be made, if possible.

The number of measurements required to estimate mean current direction at a given outcrop depends on the outcrop variance of the particular directional structure. As in regional studies, the more variable the structure, the more measurements that are required to obtain a reliable mean. The standard deviation is a conventional measure of dispersion. The sample standard deviation $s_{\bar{x}_n}$, has a probability of containing the mean azimuth of a particular structure of approximately two-thirds. The value of the standard deviation decreases with increasing number of observations as

$$s_{\bar{x}_n} = \frac{s_x}{\sqrt{n}}.$$

In figure 10-2, for example, one sees how the *sample* standard deviation of cross-bedding decreases for a standard deviation of 24°, a fairly typical value for an outcrop, as the number of measurements increase. This figure also shows the same relations for a standard deviation of 57°, an unusually high value (*cf.* fig. 7-2). Four observations reduce the sample standard deviation to one-half, nine to one-third and so on. Even with a sample standard deviation as high as 57°, more than 10 observations are rarely necessary. Hence in most regional studies, five to ten measurements of cross-bedding per outcrop are ample. If other structures with lower variability are studied, even fewer individual measurements are required or an average trend is estimated by eye for a particular bed (sole marks) or an entire outcrop (glacial striations). If, however, the outcrop has more than ordinary significance, as for example, if it is an outlier or if it is to be used for predicting a sand trend, more measurements should be obtained. In no case, however, should an outcrop ever be omitted because only one or two measurements can be made.

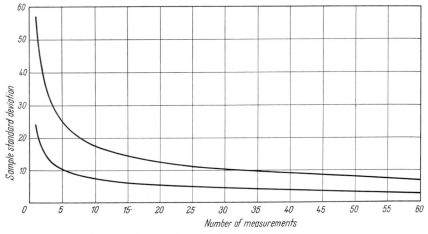

Fig. 10-2. Sample standard deviation $s_{\bar{x}_n}$, and number of measurements

More measurements are required, for a given level of accuracy, to estimate a variance than a mean. Thus, if one plans to contour the variability at each outcrop in order to relate variability to paleoslope (*cf.* TANNER, 1959, fig. 6), more observations should be taken than if only a mean transport direction were to be estimated.

A standardized notebook form greatly facilitates systematic collection and processing of the data. Usually only a brief reconnaissance of the unit to be sampled is necessary to design a form or to evaluate the usefulness of one. In addition to measurements of directional structures, features that have proved useful include:

1. kinds and abundance of sedimentary structures,
2. grain size: either maximum pebble size or average sand size,
3. thickness of sedimentation units,
4. thickness of outcrop,
5. abundance, kinds, and orientation of fossils, if any,
6. dip and strike of bedding, if structurally deformed,
7. stratigraphic position and location.

Figures 10-3 and 10-4 show examples of field notebook forms that have been used, each designed for a specific problem. The notebook form of figure 10-3 stresses the different directional structures and their relative abundance, whereas figure 10-4 emphasizes cross-bedding and maximum pebble size, and was designed for use in an area of folded strata which necessitate tilt correction. Use of a field form that is similar to a data processing card, or perhaps using such a card itself, will facilitate handling the data with a computer (*cf.* KRUMBEIN and SLOSS, (1958).

When angles are to be treated arithmetically, a compass reading from 0 to 360° is most convenient. If the angles are to be treated vectorially, however, a compass graduated in 90° intervals is more useful.

Formation   *Grain size*   *Thickness*

Sedimentary structures

Relative Abundance

_____cross-bedding   _____rib and furrow
_____current lineation   _____ripple mark
_____load casts   _____ripple scour
_____plant casts   _____shale pebble congl.

Orientation

cross-bedding   ripple scour   current parting
1   1   1
2   2   2
3   3   3

ripple mark   plant casts   micro-cross-bedding
1   1   1
2   2   2
3   3   3

Bedding

General

Operator	Outcrop
Quadrangle	Section
County	Township
State	Range

Fig. 10-3. Field notebook form emphasizing kinds and abundance of different directional structures

Samples for petrographic or clay mineral study should be collected, if regional mineral provinces are to be studied. Number of samples required depend on the number of provinces, their size, and their petrographic contrast. For example, four provinces with moderate contrast will require sampling of more localities than two provinces of greater contrast. Sampling density also depends on the objectives of the study. If a formation in a very large basin or if the sands of a modern continental shelf are to be studied, samples may be widely spaced, say one every 12 miles (20 km) or more. On the other hand, the use of pebble counts in

Station: Location:

Special features:

No	Bedding		Cross-bed		Tilt corrected		Thick-ness	No	Bedding		Cross-bed		Tilt corrected		Thick-ness
	Strike	Dip	Strike	Dip	Strike	Dip			Strike	Dip	Strike	Dip	Strike	Dip	
	Total								Total						
	Average								Average						

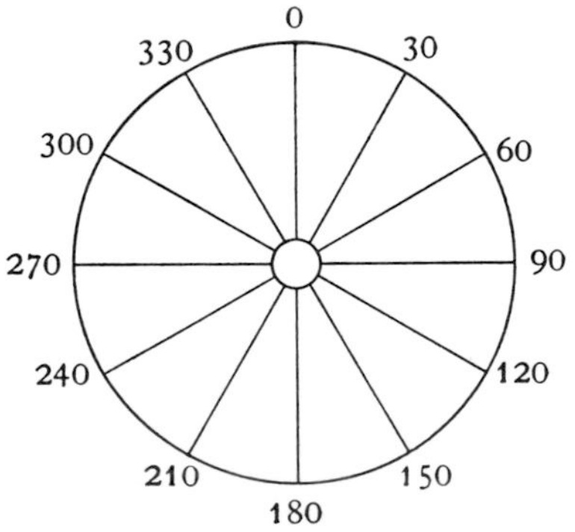

Maximum pebble size:

0
330   30
300   60
270   90
240   120
210   150
180

Modal class: _____

Average:

Average inclination
of cross-bed:

Bearing of mean azimuth:

Average thickness of bed:

Fig. 10-4. Field notebook form for tilted beds emphasizing cross-bedding and maximum pebble size (modified from YEAKEL, 1959, fig. 3)

prospecting for ore bodies may require that practically all outcrops be examined. Usually more closely spaced samples are required to accurately delineate zones of mixing.

Correction of Data for Tectonic Tilt

In regions of tilted or folded strata, directional sedimentary structures no longer retain their original orientation. It is necessary, therefore, to correct their present attitude for tectonic deformation and to "restore" them to their original position before any paleocurrent analysis can be made.

A directional sedimentary structure may have been rotated about one or more of three mutually perpendicular axes. Rotation about a vertical axis is possible but normally it is impossible to detect, although, exceptionally, TEN HAAF (1959, p. 78) was able to demonstrate such rotation in most of the allochtononous blocks in the Apennine flysch in Italy. Proof of this was possible only because the sole markings of these beds were so generally uniform in the undisturbed strata that they constituted a reference direction, departure from which signified rotation.

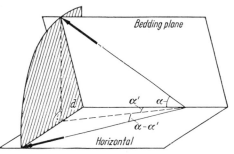

Rotation about a horizontal axis is self-evident inasmuch as the bedding planes, which were essentially horizontal at time of deposition are now inclined. Structures observed in tilted strata can by restored to their original position by rotation about the strike of the strata in which they are found through an angle equal to the dip of these beds. The rotation of a *line*, as in figure 10-5, may represent a fabric direction, a ripple crest or trend of current parting.

Fig. 10-5. Relation between an inclined direction and its horizontal projection (modified from TEN HAAF, 1959, fig. 52). The structural dip is d, α is the angle between the strike of the bed and current direction and α' is the angle between the projection of the current direction on the horizontal and the strike of the bed

Rotation can be achieved graphically, by trigonometric calculation, or by use of a rotation apparatus or device. The graphic solution may be accomplished by the conventional methods of descriptive geometry or by the use of the stereographic projection.

The problem of rotating a *plane*, involved in correcting the strike and dip of cross-bedding for tilt, for example, is essentially the same in treatment as the "two-tilt problem" or the problem of finding the strike and dip of a set of beds below an angular unconformity before tilt of the overlying strata. Solutions of this problem have been given by HARKER (1884), FISHER (1938), SPIEKER (1938), BUCHER (1944) and more recently by DONN and SHIMER (1958) and HIGGS and TUNELL (1959). The solution is achieved by rotating the pole of the plane representing the cross-bed around the strike of the true bedding through the angle of dip of that bedding (fig. 10-6). Where cross-bedded strata in the same outcrop have the same strike and dip, as many as six cross-beds can be rotated at the same time. Trigonometric solution of the problem is more laborious unless programmed for computer solution. EINSELE (1960, p. 554—555) shows, however, a simple trigonometric solution, provided the azimuth of the true maximum dip direction, projected onto the tilted bedding plane, is measured. Unfortunately this azimuth usually cannot be measured directly in tilted beds in the field. Several devices have been designed which make possible mechanical restoration of the original

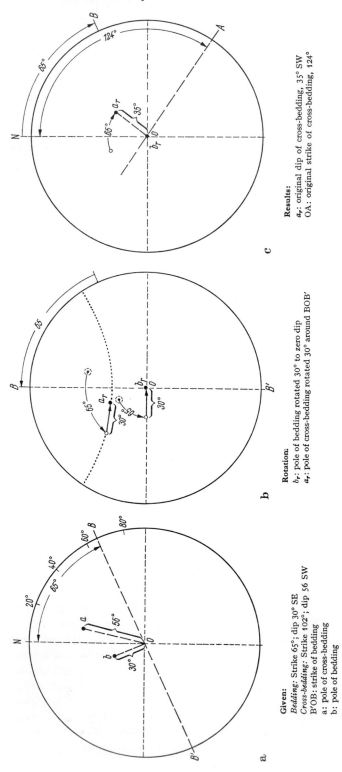

Given:
Bedding: Strike 65°; dip 30° SE
Cross-bedding: Strike 102°; dip 56 SW
B'OB: strike of bedding
a: pole of cross-bedding
b: pole of bedding

Rotation:
b_r: pole of bedding rotated 30° to zero dip
a_r: pole of cross-bedding rotated 30° around BOB'

Results:
a_r: original dip of cross-bedding, 35° SW
OA: original strike of cross-bedding, 124°

Fig. 10-6. Problem of tilt correction of cross-bedding by use of bedding normals

azimuth and inclination of a primary sedimentary structure (CLOOS, 1938; TEN HAAF, 1959, fig. 51).

If one considers the errors of measurement and the irregularities of the structure being measured, it seems likely that no azimuth correction need be made for linear structures such as strike of ripple mark in low dipping strata. TEN HAAF (1959, p. 73) and RAMSAY (1961, fig. 2) have shown, for example, that the azimuth of a bedding plane lineation is altered less than 3 degrees in beds tilted up to 25 degrees (fig. 10-7). For a planar structure, however, such as cross-bedding, these limits are different. If the foreset beds dip in essentially the same direction as the structural dip, rotation will change the azimuth of maximum foreset dip direction but little. On the other hand, if foreset dip direction is approximately at right angles to structural dip, restored azimuth of the foreset bedding can change appreciably, even if the structural dip is as low as 10°.

Tilt correction as described above presupposes simple rotation about a horizontal axis; but if the fold axis is not horizontal, there has been rotation about two horizontal axes. Correction for tilt in this case involves first rotation of the structure to correct for plunge and a second correction for the residual tilt as outlined above. The double rotation may be achieved by use of the stereographic projection as demonstrated by TEN HAAF (1959, p. 75) and RAMSAY (1961, p. 85). If the fold plunge is slight its effects may be neglected. As shown in figure 10-8, the angular

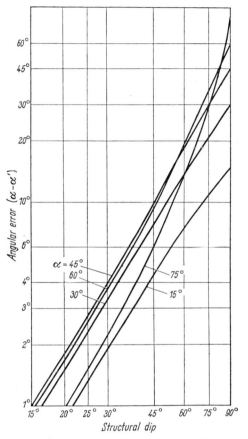

Fig. 10-7. Angular error, $\alpha - \alpha'$, resulting from neglect of dip of a linear measurement (TEN HAAF, 1959, fig. 53). Relation of angular error to structural dip and horizontal is shown in fig. 10-5

error, ε, due to neglect of fold plunge, δ, is under 10 degrees for beds dipping up to 45 degrees (cf. MIKKOLA, 1955, p. 27; RAMSAY, 1961, fig. 5; PLESSMAN, 1961, fig. 38).

The preceding discussion assumes that the deformation was uncomplicated tilt or simple flexure folding with or without plunge. In more severely deformed and faulted areas, shear folding and marked compression also distort the sedimentary structures and the correction for such distortion is more difficult and subject to many uncertainties. A rather complete analysis of these problems is given by RAMSAY (1961) who also presents charts for making appropriate corrections. Another consequence of close folding and marked comparison is telescoping and distortion of sedimentary gradients (Chapter 8).

In summary, rotation of azimuth about a vertical axis is negligible except in rare cases. Rotation of azimuth of a linear structure about a horizontal axis can be neglected, if structural dips are under 25 degrees. If over 25 degrees, an azimuth correction, which can be made in the field, is required; the rotation of cross-beds can best be done graphically on the stereo net. If angle of inclination of cross-bedding is to be studied, rotation should be made for any measurable structural dip. Tectonic deformation has been found to effect "restored" angle of inclination (Chapter 4). Lineation can be similarly treated, although tilt-correction of these features can easily be made in the field at the time of measurement as outlined in Chapter 5. Correction for plunge need not be made, if the plunge is under 10 degrees or if the beds dip less than 45 degrees.

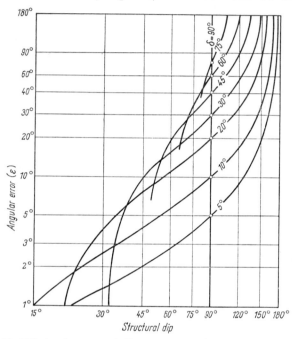

Fig. 10-8. Angular error, ε, of a linear measurement due to ignored axial pitch (δ) (modified from TEN HAAF, 1959, fig. 55)

Statistical Summary and Analysis

Review of published studies indicates a wide range in treatment of the data collected in the field and laboratory. Some workers have done little whereas others have made extended analyses. Processing the data statistically involves estimating means and variances and hypothesis testing.

Directional Structures

The orientation of grain axes and of cross-bedding, can be specified by two angles, azimuth and inclination. The grain axes or cross-bedding poles constitute three-dimensional distributions. On the other hand, structures such as sole markings are specified by but one angle, azimuth, and form only a two-dimensional distribution. In most studies of cross-bedding and grain fabric, it is advantageous to separate the analysis of azimuth from that of inclination. Hence two-dimensional azimuth distributions are most widely used.

Measurements of grain azimuths in a thin section or of directional structures in an outcrop or over an entire basin, yield a series of angles

$$x_1, \ x_2, \ x_3 \ldots, \ x_n$$

which can be summarized by a frequency distribution. For data which only indicate a *line of movement*, such as grain orientation, symmetrical, transverse ripple mark, current parting, and some fossil fabrics such as plant fragments, the frequency distribution ranges only through 180°, from 0° to 180° or 90° to 270°. It generally is subdivided into class intervals of 10 to 20°. For those data which indicate *a direction of movement* of which flute marks, cross-bedding and some fossil debris such as orthoceracone cephlapods and *Turritella* are examples, the frequency distribution ranges from 0 to 360°. In such distributions, 30°, 40° and 45° class intervals have been commonly used. The size of the class interval depends on variability and on the number of observations. For a given number of observations, the smaller the variability the smaller the class interval. Too small a class interval can lead to class intervals with irregular frequencies.

The observed data are grouped into appropriate class intervals from which a histogram, or more commonly a "current rose," is constructed (fig. 10-9).

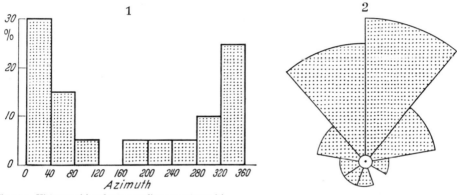

Fig. 10-9. Histogram (1) and corresponding current rose (2)

A current rose is a histogram converted to a circular distribution. Although the actual number of observations for each class can be plotted, a better procedure is to plot the percentage of observations in each class and then indicate the total number of observations used in construction of the diagram. Use of percentage data facilitates comparison. In the geological literature, the current rose conventionally indicates the direction toward *which current moved* unlike the wind rose of the meteorologists, which indicates the direction from which the winds came.

The simplest measure of prevailing current direction, one that is attractive because it involves no computation, is an *interval estimate*, the *modal class*, defined as that class interval which contains the most observations. In figure 10-9 the modal class lies between 0 to 40 degrees. This figure illustrates the advantage of the current rose over a histogram for visual, graphic interpretation. In the conventional histogram, the choice of origin may separate the observed data into two groups, when in fact the distribution is unimodal. The midpoint of the modal class can serve as a *point estimate*, a specific direction rather than an interval estimate, of average current direction. The midpoint of the modal class is primarily useful for making regional and subregional compilations for it is usually inconvenient to plot a current rose for the many individual outcrops on a regional map, even if enough data were collected at each outcrop to make a current rose.

TANNER (1959) has stressed the value of modes when two or more preferred directions of transport occur in a sample. Usually, however, a substantial number of observations, say 25 to 50 or even more, are necessary to obtain a current rose which can be reasonably duplicated by subsequent samplings.

The vector mean was used by REICHE (1938, p.912—916) for cross-bedding data. The resultant or mean vector may be computed graphically (REICHE, 1938, fig.3; RAUP and MIESCH, 1957, fig.1) by assigning a unit length to each vector, but it may be computed more readily from either the individual observations or from grouped data (number or per cent). The calculations for grouped data are

$$V = \sum_{i=1}^{n} n_i \cos x_i$$

$$W = \sum_{i=1}^{n} n_i \sin x_i$$

$$\bar{x}_v = \arctan W/V$$

$$R = (V^2 + W^2)^{\frac{1}{2}}$$

$$L = (R/n)\,100$$

where x_i is the mid-point azimuth of the ith class interval, $\bar{x}_v$ is the azimuth of the resultant vector, n_i the number of observations in each class, n the total number of observations, R the magnitude or length of the resultant vector, and L is the magnitude of the resultant vector in terms of per cent. The magnitude of the vector is a measure of the concentration of the azimuths; the greater L the greater the concentration. The ratio R/n was called the consistency ratio by REICHE (1938, p.913). For ungrouped data the sines and cosines of each angle are used directly, in a manner similar to the calculation of latitudes and departures in plane surveying. PINCUS (1956, fig.7) shows a computation table for a short method of computing $\bar{x}$ and R.

Table 10-1 compares the arithmetic with the vector mean. For the data of group A computation of the arithmetic means is straightforward. The arithmetic mean is 226° and the vector mean 229°. For the data of group B, in which the measurements lie on both sides of 360°, a similar direct averaging yields an arithmethic mean, 188°, unrelated to the vector mean, 344°. A meaningful arithmetic average can be easily computed in this situation, however, by simply assigning a negative sign to observations in the third and fourth quadrants, measured counter-clockwise from the origin, and adding them to the observations of the first quadrant. Thus

$$(-120 - 70 - 50 - 29 - 10 + 10 + 12 + 29 + 90)/9 = -11 \text{ or } 349°,$$

which is close to the vector mean 346°. For the data of group C, however, the arithmetic mean, no matter how computed, fails because the data are so disperse.

Thus the arithmetic mean is a quick and easily computed average, especially for those outcrops where the data lie within 180°. However, the arithmetic mean should not be used for data scattered around the compass as they commonly are in regional and subregional compilations and in some outcrops. In such cases, it is a very poor measure of central tendency. *The best procedure always is to compute the vector mean.*

The distribution shown in Table 10-1 raises the important question of the reliability of measures of central tendency, when the data are somewhat dispersed as they nearly always are. Several questions arise. For example, could the observed distribution of azimuths have been reasonably obtained from a formation in which there were equal numbers of azimuths in each class interval or, in other words, a formation wholly lacking in any preferred transport direction? Equal numbers of azimuths in each class interval define a uniform statistical distribution. Obviously, no geologic significance should be attached to a mean calculated from a formation whose paleocurrents had non-preferred orientation. One might also wish to know whether or not an observed mean value could have been obtained from a distribution with a specified azimuth determined by other geologic evidence. For example, the mean azimuth of elongate pebbles in a modern stream could be tested against the direction of stream flow.

Ordinary statistical tests which would answer these questions must be modified before they can be applied to the circular distribution in which the limits of angular deviation, 0° and 360°, are the same direction. One modification was

Table 10-1. *Azimuths and Measures of Central Tendency and Dispersion*

Group	Azimuths in Degrees									$\bar{x}_a$	$\bar{x}_v$	R	L
A . .	130	210	230	220	300	270	260	240	170	226	229	6.21	69.0
B . .	231	310	260	290	350	20	29	90	12	349	338	4.23	47.0
C . .	320	170	220	10	100	40	300	70	254	165	350	1.17	13.0

x_a = arithmetic mean; x_v = vector mean; R = vector magnitude; L = consistency ratio in per cent.

proposed by TUKEY (1954, cited in HARRISON, 1957, Table 1 and RUSNAK, 1957, p. 53—54). TUKEY's solution combines the chi square test for grouped data (DIXON and MASSEY, 1957, p. 226—228) with calculation of the vector mean to compare the observed distribution with the uniform distribution of equal number of azimuths in each class interval. If the chi square test indicates that the observed distribution does not depart significantly from the uniform distribution, the vector mean is not considered significant.

Another procedure is to compute the sample variance around the vector mean. The ratio of the observed sample variance, s_o^2, to the variance of the uniform distribution, s_u^2, which is 10,800 (GRIFFITHS and ROSENFELD, 1953, p. 210—214), is called the F ratio and provides a test of the statistical hypotheses, H_o versus H_a

$$H_o: S_o^2 = S_u^2$$
$$H_a: S_o^2 < S_u^2.$$

The actual ratio for the data of group C, Table 10-1, is

$$F = 10,800/10,614 = 1.0175.$$

In this example, both numerator and denominator have eight degrees of freedom. Because the observed value of F is less than the critical value of $F_{0.05}(8, 8) = 3.44$, the observed distribution does not differ significantly from a uniform one at the 0.05 level of significance and H_o is accepted. If the observed variance does not differ significantly from the variance of a uniform distribution, obviously no

geologic significance should be attached to the vector mean. KRUMBEIN and MIL-
LER (1953, p. 513—514), DIXON and MASSEY (1957, p. 147—183), MILLER and
KAHN (1962, p.108) and many others explain the use of the F ratio in hypothesis
testing. Tables of F are widely available in most elementary statistics books.
Significance values of 0.05, 0.01, and 0.001 are most commonly used in
paleocurrent studies. These values indicate the probability of rejecting H_o when
it is, in fact, correct. The F test can be used for either grouped or ungrouped
data regardless of sample size. Tests of significance employing the F ratio assume
the data to be normally distributed. Commonly directional data appear to
warrant this assumption. However, as used here, the F ratio is calculated from
the vector rather than the arithmetic mean. Since the data are approximately
symmetrical, the vector and arithmetic means should tend to coincide.

If the observed variance differs significantly from that of the uniform distribution, one may wish to calculate a *confidence interval* for the estimated mean.
This interval may be the sample standard deviation, s_x, or a broader confidence
interval may be desired, say a 90 per cent interval which has a probability of 90 per
cent of containing the true mean in repeated sampling. Such confidence limits
are obtained from the sample standard deviation by the formula

$$\bar{x} \pm (t_{\alpha/2} \, s_x)/n$$

where t_α is the value of the t distribution (DIXON and MASSEY, 1957, p. 127—128)
for a given level of significance a, say 0.05.

The magnitude of the consistency ratio in per cent, L, negatively correlates
with variance. This correlation is nonlinear and has been investigated by CURRAY
(1956, figs. 2 and 3) and FRANKS *et al.* (1959, fig. 2), who give curves to convert,
one to the other over certain ranges.

A third possible procedure is a test devised by RAYLEIGH (1894, p. 41) which
uses R, the magnitude of the vector, to calculate the quantity:

$$1 - e^{-R^2/n}$$

RAYLEIGH's procedure, for large samples, tests the null hypothesis H_o against
the alternative H_a where

H_o: the x_i are uniformly distributed in the interval 0 to 360°

versus

H_a: the x_i are not uniformly distributed in the interval 0 to 360°.

If the above quantity is greater than the desired level of significance, say 0.05,
H^0 is rejected. R can be tested graphically using L (CURRAY, 1956, fig. 3) or,
as shown by DURAND and GREENWOOD (1958, p. 230), the simplest procedure
is to compare $2R^{2/n}$ with the chi square table using 2 degrees of freedom. DURAND
and GREENWOOD (1958, Appendix 2) also show that the term $C^2 + S^2$ of TUKEY's
test is equal to $2R^{2/n}$. Hence most of the large sample techniques, that test the
hypothesis that the vector mean could have arisen from a uniform distribution,
are identical.

A uniformly most powerful test against a particular orientation is available,
if one can geologically specify a preferred orientation independent of the measurements gathered. Orientation of long axes of pebbles in glacial till provides an

example. Knowledge of mean orientation of glacial striations in the vicinity of the till fabric sample provides an average direction, y, against which the hypothesis of preferred pebble orientation can be tested. Thus

H_o: The azimuths of long axes of pebbles x_i are uniformly distributed

versus

H_a: The x_i are obtained from a nonuniform distribution with preferred direction, y, the average direction of ice movement as indicated by glacial striations.

The test statistic is:

$$V' = V \cos y + W \sin y$$

The hypothesis of a uniform distribution is accepted if

$$V' \, (2/n)^{1/2} \leq z$$

and rejected if

$$V' \, (2/n)^{1/2} > z$$

where z is the probability obtained from cumulative normal table for a given level of significance.

DURAND and GREENWOOD (1958, fig. 1) provide modifications to both the Rayleigh test and the V' distribution so that these tests can be used for n as small as 5 and 6. Values of R are given in Table 10-1. Using the modifications for small samples, calculations show that there is less than 1 chance in 20 of obtaining data so nonuniform as the azimuths of group A and B of Table 10-1 from a uniform distribution. In contrast, at the same level of significance, the azimuths of group C represent a uniform distribution.

STUDENT's t distribution (DIXON and MASSEY, 1957, p. 123—124) can be used to test whether or not two populations have the same vector mean, as, for example, whether the mean orientation of flute marks in two adjacent turbidite formations is the same or not. Use of STUDENT's t test assumes that the directional data are normally distributed. This assumption commonly appears to be justified.

To summarize, the best procedure is to calculate the vector average and use a Rayleigh or modified Rayleigh test of significance. For large samples with no clearly specified alternative, the chi square table can be used to evaluate R and for small samples the modifications of DURAND and GREENWOOD (1958, fig. 1) are appropriate. If an observed distribution is to be tested against an alternative distribution with a specified mean, the V'-distribution of DURAND and GREENWOOD (1958, p. 231—233) can be used for samples with as few as five measurements.

The two-dimensional circular normal distribution (GUMBEL et al., 1953) has been used by some and has the form

$$\frac{1}{2 \pi I_0 (k)} \; e^{k \cos (x-y)}$$

where $I_0(k)$ is a Bessel function of the first kind of pure imaginary argument, y denotes the preferred orientation, k is a measure of concentration about y, and x is a random angle. The mean of this distribution is the vector mean. The length of the vector mean, R, uniquely determines k. The relation of k to vector strength is tabulated by GUMBEL *et al.* (1953, Table 2) and is reproduced in Table 10-2. WATSON and WILLIAMS (1956) present confidence limits for the vector mean, tests for a common k, and other tests.

Table 10-2. *Relation of k to Vector Strength R.* (Modified from GUMBEL *et al.*, 1953, Table 2)

R	k	R	k	R	k
.00	0.00000	.30	0.62922	.60	1.51574
.01	.02000	.31	.65242	.61	.55738
.02	.04001	.32	.67587	.62	.60044
.03	.06003	.33	.69958	.63	.64506
.04	.08006	.34	.72356	.64	.69134
.05	.10013	.35	.74783	.65	1.73945
.06	.12022	.36	.77241	.66	.78953
.07	.14034	.37	.79730	.67	.84177
.08	.16051	.38	.82253	.68	.89637
.09	.18073	.39	.84812	.69	.95357
.10	0.20101	.40	0.87408	.70	2.01363
.11	.22134	.41	.90043	.71	.07685
.12	.24175	.42	.92720	.72	.14359
.13	.26223	.43	.95440	.73	.21425
.14	.28279	.44	.98207	.74	.28930
.15	.30344	.45	1.01022	.75	2.36980
.16	.32419	.46	.03889	.76	.45490
.17	.34503	.47	.06810	.77	.54686
.18	.36599	.48	.09788	.78	.64613
.19	.38707	.49	.12828	.79	.75382
.20	0.40828	.50	1.15932	.80	.87129
.21	.42962	.51	.19105	.81	3,00020
.22	.45110	.52	.22350	.82	.14262
.23	.47278	.53	.25672	.83	.30114
.24	.49453	.54	.29077	.84	.47901
.25	.51649	.55	1.32570	.85	.68041
.26	.53863	.56	.36156	.86	.91072
.27	.56097	.57	.39842	.87	4.17703
.28	.58350	.58	.43635		
.29	.60625	.59	.47543		

The circular distribution has been extended to three dimensions to include inclination as well as orientation and has been widely used in studies of paleomagnetic data (COX and DOELL, 1960, p. 668—673). Although a few workers have compiled all regional cross-bedding data on a Schmidt net, it is usually more effective to separate the analysis of cross-bedding azimuth from dip angle of foresets (*cf.*, STEINMETZ, 1962). The three-dimensional circular distribution could be effectively used, however, to evaluate three-dimensional studies of grain fabric and geophysical properties such as dielectric susceptibility.

Line of movement data of 0 to 180 range (θ system), must be transformed by doubling the angles (2θ system), before the above tests can be used (KRUMBEIN, 1939, p. 688). Doubling the angles transforms a distribution with two modes 180° apart into a single unimodal distribution. In practice one transforms the data of the θ system into the 2θ system, performs the computations and tests as before, and then converts the mean and variance of the 2θ system into that of θ system. In the θ system, the standard deviation of a uniform distribution is only 51.96° or 52° (GRIFFITHS and ROSENFELD, 1953, p. 210—214).

OLSON and POTTER (1954) and DIXON and MASSEY (1957, p. 127—128) explain in detail procedures to estimate variance components.

Mineral Provinces

Although one mineral species or one pebble type will define a province, most provinces are based on two or more components and are *multicomponent* or *multivariate* systems.

If either heavy or light minerals or pebble types are studied, the results of a count of a particular sample are of the form

$$p_1 + p_2 + p_3 + \cdots + p_n = 100\%$$

where the p_i, $i = 1$ to n, is the percentage of each species counted.

A province may be defined by selecting some combination of the i's, say the stable heavy mineral suite i_1 (zircon), i_2 (tourmaline), and i_3 (rutile) such that

$$p_1 + p_2 + p_3 \geq a$$

define province A and

$$p_1 + p_2 + p_3 < a$$

define province B where p_1, p_2, and p_3 and a, their total percentage, are arbitrarily chosen by inspection. This procedure can be extended to several different combinations of the i's. In essence, several variables are combined into a single one thus permitting delineation of different dispersal patterns. Usually along the boundaries of the zones the values will be close to a. Such values have been considered to represent zones of mixing.

Ratios of components have also been used to define mineral provinces. For heavy minerals such ratios should always be based on minerals with comparable specific gravity in order to minimize effects of hydraulic sorting. Ratios of varietal types of a particular heavy mineral have been often used.

Multicomponent provinces can be delineated by inspection by preparing maps of each variable or of combinations of them. PRYOR (1960, figs. 3, 4, 9, 13 and 18) illustrates both approaches. Depending on the degree of statistical correlation between the variables examined, these maps will show varying degrees of agreement. Examples of classification by inspection, based on comparatively fine distinctions, are provided by SINDOWSKI (1958, figs. 4 and 5) and FÜCHTBAUER (1958, fig. 6).

It is also possible to use several types of mapping functions to advantage. Entropy-like functions (PELTO, 1954, p. 505—508; FORGOTSON, 1960, p. 91—94; MILLER and KAHN, 1962, p. 427—432)), functions which express the degree of mixing of three or more components that together form 100 per cent, are useful. *Relative entropy*, H_r, is defined by PELTO (1954, p. 507) as

$$H_r = -100 \left(\sum_{i=1}^{n} p_i \ln p_i \right) / H_m$$

where p_i is the proportion of the i-th component, n the number of components and H_m, the maximum entropy is

$$H_m = -\ln \frac{1}{n} = \ln n.$$

If only three components are studied, relative entropy values can be obtained from a triangular diagram with an overlay (FORGOTSON, 1960, p. 91) but, if there are four or more components, H_r must be calculated. When one component of the mixture is dominant, relative entropy is low; relative entropy is at a maximum, 100 per cent, when all components are present in equal amounts. Relative entropy is a measure of the extent to which the proportions are equal.

A departure map (KRUMBEIN, 1955) may also be useful. Such a map is based on a distance function that can be calculated between a given composition and any other. Here the proportions p_i are thought of as the components of a vector

$$\mathbf{X} = [p_1, p_2, \ldots, p_n].$$

For example, suppose directional structures indicate a particular point source as providing part of the fill of a basin. The average mineral and textural composition of this source defines a vector $\mathbf{Y}$ with, say 6 components p_1', p_2', p_3' and so on. Then the distance between $\mathbf{Y}$ and any other compositional and textural vector $\mathbf{X}$ is

$$(\mathbf{X}^T\mathbf{Y})^{1/2} = \sqrt{(p_1-p_1')^2 + (p_2 - p_2')^2 + \cdots + (p_6 - p_6')^2}.$$

This scalar quantity can then be calculated for each sample and the map contoured. If only three components are studied, the distance function can be obtained graphically (KRUMBEIN, 1955, fig. 2).

A classifying function described by PELTO (1954, p. 503—505) also may have some value for study of mineral provinces.

The above described procedures can also be applied to study of gross lithology or facies.

Another approach is to formulate a geologic *hypothesis* that divides the area into two or more parts and then use mineralogical or other data to *test statistically* this hypothesis. Contrasts in current systems based on directional structures or on contrasting subsurface patterns of sand distribution are relevant examples of classifying criteria that can be used. Examples of the former approach are provided by SIEVER and POTTER (1956, p. 320—328).

If provinces are based on a single variable, a variety of univariate tests may be used. These include STUDENT's t test and the analysis of variance (DIXON and MASSEY, 1957, p. 115—123), tests of all contrasts (SCHEFFÉ, 1959, p. 55—82) and some nonparametric tests such as the Kolmogorov-Smirnov two-sample test (SIEGEL, 1956, p. 127—135). SCHEFFÉ's tests of all contrasts are particularly suitable, if three or more regions are to be tested.

Usually, however, a mineral province is based on two or more variables and multivariate tests should be used. Such tests should only be used, however, if the variables do not sum to 100 per cent. In statistical terms, this requires that there be no linear constraint on the variables. In geologic literature variables that generate number systems without a linear constraint have been called *open* (KRUMBEIN, 1962, p. 2229) and those that have a linear contraint *closed* (CHAYES, 1962, p. 441). Thus *all* the components of a heavy mineral or thin section analysis should not constitute the variables of a multivariate test but a combination of say zircon percentage, per cent round zircon, feldspar percentage and angularity of quartz can. KRUMBEIN (1962) also discusses the effect of linear constraints on facies map interpretation. Usually, however, restrictions imposed by linear constraints on the data can be avoided by early selection of a relatively few geologically significant variables that do not sum to one hundred percent. Such selection also minimizes unnecessary study.

Multivariate tests that are appropriate include the analysis of dispersion (RAO, 1952, p. 258—264; MILLER and KAHN, 1962, p. 249) and HOTELLING's multivariate T test (ANDERSON, 1958, p. 101—121). The analysis of dispersion is the multivariate analogue of the analysis of variance and HOTELLING's T test is the multivariate analogue of STUDENT's T test. MILLER (1954) and SIEVER and POTTER (1956, p.335) give examples of applications of the analysis of dispersion, the latter to mineral provinces.

It may be necessary to normalize the original petrographic data by transformation (BARTLETT, 1947; KRUMBEIN and MILLER, 1954) so that the above described parametric tests may be used.

Moving Averages and Trend Surfaces

Two-dimensional moving averages have been constructed for maps that show the areal variation of scalar properties such as grain size as well as orientation

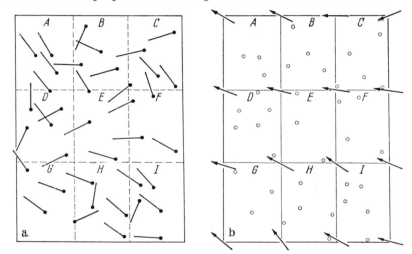

Fig. 10-10. Essential steps in construction of a two-dimensional moving average (modified from PELLETIER, 1958, fig. 2). Arrow at corner of A, B, D and E (right) is average of measurements from outcrop in areas A, B, D and E of original data (left). By successive overlapping, measurements in each area counted four times

of directional structures (POTTER, 1955, fig. 12; SCHLEE, 1957, fig. 13; PELLETIER, 1958, fig. 10; FAHRIG, 1961, fig. 2). Figure 10-10 shows the essential steps in their construction. Any arbitrary grid is a convenient one from which to construct the moving average. Moving averages smooth out local variations and thus provide a simple way to depict major sedimentary trends. The departure of the actual observed data from the smoothed map calls attention to local anomalies. A map of these anomalies may be constructed and is called a "residual" map.

Regional trends, as well as local anomalies, can be estimated analytically, especially, if a computer is available. In addition to grain size, maps of gravity, thickness, facies and mineral composition have been analyzed in this manner. OLDHAM and SUTHERLAND (1955), KRUMBEIN (1956), MILLER (1956), GRANT (1957) and MILLER and KAHN (1962, p. 390—406) describe the theory of trend surfaces and give examples of their application. Each point on the map is assigned geographic coordinates which, along with the value of variable, specify the trend surface. A least squares regression procedure is used to fit different surfaces, linear, quadratic and higher, to the observed data. The difference between the computed surface at a point and the actual value defines the residual at that point. Even if the data are irregularly spaced, as is commonly the case in outcrop and subsurface studies, linear and quadratic surfaces can be fitted to the observed data. A linear map surface has equally spaced contours of uniform strike but quadratic and higher order maps do not. KRUMBEIN (1959) discusses in detail trend surface analysis of contour maps with irregular spacing of control points. ALLEN and KRUMBEIN (1962) present a comprehensive trend surface study of a thin pebble bed in southeastern England. Doubtlessly, experience will show what different linear and quadratic map surfaces will he best suited to the different basin models (Chapter 9).

Presenting the Data

A complete presentation of a regional paleocurrent study should always include the following:

The unit studied should be briefly and concisely described. This stratigraphic description should include age, thickness, lithologic composition, and appropriate covering references.

The sampling procedure should be briefly summarized and mention made of how the data were processed.

A few easily accessible, representative outcrops should be cited in the text or in a short table.

All the measurements should be summarized by class interval in a frequency table the rows of which may be different formations or members of the unit studied or different geographic areas (cf. POTTER and PRYOR, 1961, Table 7). Appropriate average values should also be included in the table. Such a table provides, in compact form, a summary of all the observations from which inferences about paleocurrents were made.

The presentation of the data by map allows for considerable flexibility. Usually, however, it is desirable to present the actual data at several levels of factual generalization as well as a completely interpretative map, one that utilizes evidence from other sources than directional structures.

One of the factual maps should show the observed pattern of current direction at individual outcrops. At each outcrop an average value for a particular structure rather than individual measurements should be displayed. If the average value is not statistically significant, it should be indicated by a different symbol. It may be that more meaningful maps would be obtained, if current direction were weighted by bed thickness or grain size, two correlated scalar variables whose magnitude is related to current velocity. Such weighted maps would be true vector maps. Only in exceptional circumstances should current rose diagrams be shown for individual outcrops.

Various symbols for primary sedimentary structures, including directional structures, have been devised for use on maps and graphic columnar sections

(Bouma and Nota, 1960; Bouma, 1962, p. 5—18). As yet, however, none of these proposed symbols have become standardized.

Although there are many criteria for distinguishing top and bottom of vertical or overturned strata, only a few are common enough to warrant representation by a map symbol. A similar situation prevails with respect to directional properties. Use of a multiplicity of symbols tends to be confusing. There is need for only a very few. Figure 10-11 shows symbols that distinguish between directional structures that indicate a direction of movement and those that indicate a line of movement. Structures that show the direction of movement have an arrowhead, whereas those that indicate only a line of movement do not.

If the vector mean is calculated, the length of the symbol may be proportional to the vector magnitude in per cent. Directions which are not statistically signi-

Direction of Movement

 Current parallel to and in direction of arrow. Asymmetrical ripple mark, cross-bedding, flute casts, imbricated pebbles, *etc.*

Line of Movement

Current parallel to symbol. Parting lineation, groove casts, some two-dimensional fabrics, *etc.*

Current perpendicular to symbol. Symmetrical ripple marks and some two-dimensional fabrics

Fig. 10-11. Some suggested general types of map symbols for directional structures. Symbols show average orientations at an outcrop

ficant should be distinguished, perhaps, by broken lines. In each case, the symbol represents the current direction and the dot denotes the location of the outcrop. Measurements of doubtful validity should not be shown.

Unlike maps, it may be desirable to indicate in abbreviated or symbolic form the presence and nature of various sedimentary structures in stratigraphic tables or on graphic logs or stratigraphic columns. The various proposed symbols may serve this purpose.

In addition to an outcrop map, a map showing average values for grid segments or a moving average should be shown. This map should display appropriate current roses and the average of all the measurements in each grid unit. A map with grid segment or subregional averages is convenient to combine with isopach, facies, or mineral province maps. If the area is so large that only grid averages can be shown, several maps should be included to show the actual current pattern displayed at individual outcrops in local areas.

The final regional map should present the investigator's *interpretation* of paleocurrents, based on *all* the relevant evidence. Such a map permits a paleocurrent interpretation to be completely separated from the factual observation on which it is based.

Figure 10-12 shows, by hypothetical example, the above described types of maps.

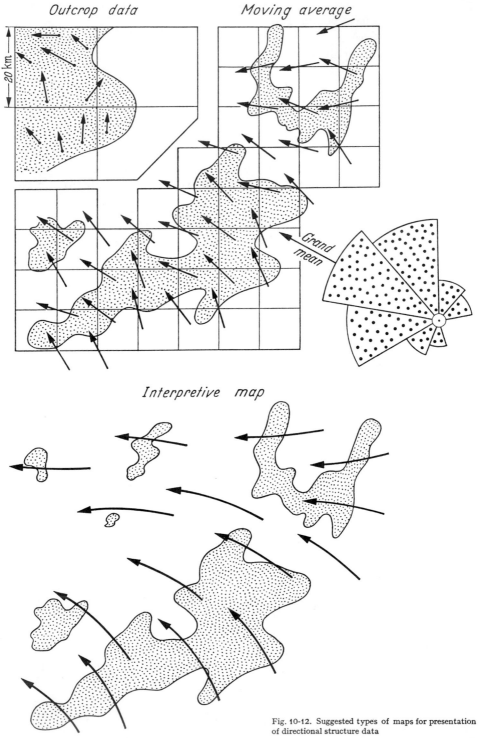

Fig. 10-12. Suggested types of maps for presentation of directional structure data

References

ALLEN, P., and W. C. KRUMBEIN, 1962: Secondary trend components in the Top Ashdown Pebble Bed: A case history. J. Geol. **70**, 501—538.

ANDERSON, T. W., 1958: An introduction to multivariate statistical analysis. New York: John Wiley and Son. 374 p.

BARTLETT, M. S., 1947: The use of transformations. Biometrics **3**, 39—52.

BOUMA, A. H., 1962: Sedimentology of some flysch deposits. Amsterdam: Elsevier Publishing Co. 168 p.

—, and D. J. G. NOTA, 1960: Detailed graphic logs of sedimentary formations. 21st Internat. Geol. Congr., Session Norden, pt. 23, p. 52—74.

BUCHER, W. H., 1944: The stereographic projection, a handy tool for the practical geologist. J. Geol. **52**, 191—112.

CHAYES, FELIX, 1962: Numerical correlation and petrographic variation. J. Geol. **70**, 440—452.

CLOOS, HANS, 1938: Primäre Richtungen in Sedimenten der rheinischen Geosynkline. Geol. Rundschau **29**, 357—367.

COCHRAN, W. G., 1953: Sampling techniques. New York: John Wiley and Son, 330 p.

COX, ALLAN, and R. R. DOELL, 1960: Review of paleomagnetism. Bull. Geol. Soc. Am. **71**, 645—768.

CURRAY, J. R., 1956: The analysis of two-dimensional orientation data. J. Geol. **64**, 117—131.

DIXON, W. J., and F. J. MASSEY jr., 1957: Introduction to statistical analysis, 2nd ed. New York: McGraw-Hill Book Co. 488 p.

DONN, W. L., and J. A. SHIMER, 1958: Graphic methods in structural geology. New York: Appleton-Century-Crofts, Inc. 180 p.

DURAND, DAVID, and J. A. GREENWOOD, 1958: Modifications of the Rayleigh test for uniformity in analysis of two-dimensional orientation data. J. Geol. **66**, 3, 229—238.

EINSELE, GERHARD, 1960: Schrägschichtung im Raumbild und einfache Bestimmung der Schüttungsrichtung. Neues Jahrb. Geol. Paläont., Monatsh., p. 546—559.

FAHRIG, W. F., 1961: The geology of the Athabasca formation. Geol. Survey Canada, Bull. **68**, 41 p.

FISHER, D. J., 1938: Problem of two tilts and the stereographic projection. Bull. Am. Assoc. Petrol. Geologists **22**, 1261—1271.

FORGOTSON jr., J. M., 1960: Review and classification of quantitative mapping techniques. Bull. Am. Assoc. Petrol. Geologists **44**, 83—100.

FRANKS, P. C., G. L. COLEMAN, N. PLUMMER and W. K. HAMBLIN, 1959: Cross-stratification, Dakota sandstone (Cretaceous), Ottawa County, Kansas. Kansas Geol. Survey, Bull. **134**, pt. 6, p. 223—238.

FÜCHTBAUER, H., 1958: Die Schüttungen im Chatt und Aquitan der deutschen Alpenvorlandsmolasse. Eclogae Geol. Helv. **51**, 928—941.

GRANT, FRASER, 1957: A problem in the analysis of geophysical data. Geophysics **22**, 309—344.

GRIFFITHS, J. C., and M. A. ROSENFELD, 1953: A further test of dimensional orientation of quartz grains in Bradford sand. Am. J. Sci. **251**, 192—214.

GUMBEL, E. J., J. A. GREENWOOD and DAVID DURAND, 1953: The circular normal distribution: theory and tables. J. Am. Statistical Assoc. **48**, 131—152.

HAAF, ERNST TEN, 1959: Graded beds of the Northern Apennines. Ph.D. thesis, Rijks University of Groningen, 102 p.

HARKER, A., 1884: Graphical methods in field geology. Geol. Mag., n.s. **1**, 154—162.

HARRISON, P. W., 1957: New technique for three-dimensional fabric analysis of till and englacial debris containing particles from 3 to 40 mm in size. J. Geol. **65**, 98—105.

HIGGS, D. V., and GEORGE TUNELL, 1959: Angular relations of lines and planes. Dubuque: W. C. Brown Co. 43 p.

KRUMBEIN, W. C., 1939: Preferred orientation of pebbles in sedimentary deposits. J. Geol. **47**, 673—706.

KRUMBEIN, W. C., 1955: Composite end members in facies mapping. J. Sediment. Petrol. **25**, 115—122.
— 1956: Regional and local components in facies maps. Bull. Am. Assoc. Petrol. Geologists **40**, 2163—2194.
— 1959: Trend surface analysis of contour-type maps with irregular control-point spacing. J. Geophys. Research **64**, 823—834.
— 1960: The "geological population" as a framework for analyzing numerical data in geology. Liverpool Manchester Geol. J. **2**, 341—368.
— 1962: Open and closed number systems in stratigraphic mapping. Bull. Am. Assoc. Petrol. Geologist **46**, 2229—2245.
—, and R. L. MILLER, 1953: Design of experiments for statistical analysis of geological data. J. Geol. **61**, 510—532.
— — 1954: A note on transformations of data for analysis of variance. J. Geol. **62**, 192—193.
—, and L. L. SLOSS, 1958: High-speed digital computers in stratigraphic and facies analysis. Bull. Am. Assoc. Petrol. Geologists **42**, 2650—2669.
MIKKOLA, T., 1955: Sedimentary transportation in Karelian quartzites. Bull. comm. géol. Finlande No. 168, 27—29.
MILLER, R. L., 1954: A model for the analysis of environments of sedimentation. J. Geol. **62**, 108—113.
— 1956: Trend surfaces: their application to analysis and description of environments of sedimentation. J. Geol. **64**, 425—446.
—, and J. S. KAHN, 1962: Statistical analysis in the geological sciences. New York: John Wiley and Sons. 483 p.
OLDHAM, C. H. G., and D. B. SUTHERLAND, 1955: Orthogonal polynomials: their use in estimating the regional effect. Geophysics **20**, 295—306.
OLSON, J. S., and P. E. POTTER, 1954: Variance components of cross-bedding direction in some basal Pennsylvanian sandstones of the Eastern Interior Basin: Statistical methods. J. Geol. **62**, 26—49.
PELLETIER, B. C., 1958: Pocono paleocurrents in Pennsylvania and Maryland. Bull. Geol. Soc. Am. **69**, 1033—1064.
PELTO, C. R., 1954: Mapping of multicomponent systems. J. Geol. **62**, 501—511.
PINCUS, H. J., 1956: Some vector and arithmetic operations on two-dimensional orientation variates, with application to geological data. J. Geol. **64**, 533—557.
PLESSMANN, WERNER, 1961: Strömungsmarken in klastischen Sedimenten und ihre geologische Auswertung. Geol. Jahrb. **78**, 503—566.
POTTER, P. E., 1955: The petrology and origin of the Lafayette gravel: Part 1, Mineralogy and petrology. J. Geol. **63**, 1—38.
—, and J. S. OLSON, 1954: Variance components of cross-bedding direction in some basal Pennsylvanian sandstones of the Eastern Interior Basin: Geological application. J. Geol. **62**, 50—73.
—, and W. A. PRYOR, 1961: Dispersal centers of Paleozoic and later clastics of the upper Mississippi Valley and adjacent areas. Bull. Geol. Soc. Am. **72**, 1195—1250.
PRYOR, W. A., 1960: Cretaceous sedimentation in upper Mississippi Embayment. Bull. Am. Assoc. Petrol. Geologists **44**, 1473—1504.
RAMSEY, J. G., 1961: The effects of folding upon the orientation of sedimentation structures. J. Geol. **69**, 84—100.
RAO, C. RADHAKRISHNA, 1952: Advanced statistical methods in biometric research. New York: John Wiley and Son. 390 p.
RAUP, O. B., and A. T. MIESCH, 1957: A new method for obtaining significant average directional measurements in cross-stratification studies. J. Sediment. Petrol. **27**, 313—321.
RAYLEIGH, LORD, 1884: The theory of sound, 2nd ed., vol. 1. New York: MacMillan & Co. 480 p.
REICHE, PARRY, 1938: An analysis of cross-lamination of the Coconino sandstone. J. Geol. **44**, 905—932.

RUSNAK, G. A., 1957: A fabric and petrologic study of the Pleasantview sandstone. J. Sediment. Petrol. **27**, 41—55.

SCHEFFÉ, HENRY, 1959: The analysis of variance. New York: John Wiley and Sons. 477 p.

SCHLEE, JOHN, 1957: Upland gravels of southern Maryland. Bull. Geol. Soc. Am. **68**, 1371—1410.

SIEGEL, SIDNEY, 1956: Nonparametric statistics. New York: McGraw-Hill Book Co. 312 p.

SIEVER, RAYMOND, and P. E. POTTER, 1956: Sources of basal Pennsylvanian sediments in the Eastern Interior Basin: Pt. 2, Sedimentary Petrology. J. Geol. **64**, 317—335.

SINDOWSKI, KARL-HEINZ, 1958: Schüttungsrichtungen und Mineral-Provinzen im westdeutschen Buntsandstein. Geol. Jahrb. **73**, 277—294.

SPIEKER, E. M., 1938: Problem of two tilts — HARKER's Solution corrected. Bull. Am. Assoc. Petrol. Geologists **22**, 1255—1260.

STEINMETZ, RICHARD, 1962: Analysis of vectorial data. J. Sediment. Petrology **32**, 801—812.

TANNER, W. F., 1959: The importance of modes in cross-bedding data. J. Sediment. Petrol. **29**, 211—226.

WATSON, G. S., and E. J. WILLIAMS, 1956: On the construction of significance tests on the circle and the sphere. Biometrika **43**, 344—352.

YEAKEL, L. S., 1959: Tuscarora, Juniata and Bald Eagle paleocurrents in the central Appalachians. Unpublished Ph.D. thesis, The Johns Hopkins University, 454 p.

Methods of Study (1963—1976)

Computers and pocket calculators now remove most of the drudgery from the analysis of paleocurrent data, and now, and additionly, sedimentologists are much more familiar with a wide range of statistical techniques. The consequence ? Easy, quick, and sophisticated data analysis

Sampling

There has been rather little attention given to sampling, although RAO and SENGUPTA (1970) reviewed at length the early work of OLSON and POTTER* (1954), using, however, the vector rather than the arithmetic mean. Rao and Sengupta also discuss sample size and costs. In practice, most field geologists have not given much attention to the details of sampling and probably considerably oversample most formations. Although not intended for the student of paleo-currents, WHITTEN's (1966) discussion of sampling mechanically deformed sediments is useful, comparative reading. His Chapter 3, *Sampling and Size in Data Collection*, gives comparative insight into sampling of directional structures, and his Chapter 13, *Sedimentary Characteristics Preserved in Folded and Metamorphosed Rocks*, is especially relevant for the sedimentologist working in metamorphic terrains — an area of study that will probably expand in years to come and bring sedimentologists profitably into closer contact with metamorphic petrologists.

What suggestions can be offered about sampling directional structures after thirteen years of study in diverse basins around much of the world? We suggest the following:

1. Lateral homogeneity: for a new basin always sample, if you can, in *two stages* — an initial, basin-wide reconnaissance followed by a second stage of more detailed effort, especially at boundaries between possible flow patterns. Many studies, for example, show that most basins are over sampled in terms of estimating a regional mean. This could be avoided by an initial, open sampling, if such is practical. If the unit is not vertically homogeneous, each part must be regionally subsampled.

2. Vertical homogeneity: in a thick unit always take the time to make several vertical profiles in key areas.

3. If possible, always relate mean and variance of flow pattern to *depositional facies*, both of which are responses to depositional environment. The main idea here is that a paleocurrent system is not an abstraction — but is instead always relatable to a particular depositional facies.

4. Sample directional structures proportional to their variance (see fig. 4-23).

a) Cross-bedding: One measurement per set and three to five measurements of the most accessible, first perceived sets for unidirectional cross-bedding; for bipolar and bimodal cross-bedding up to twenty to twenty-five per outcrop to fully define the two bipolar modes (ebb and flow currents in tidal deposits?). In one sampling experiment the half distributions, even with very small sample sizes, were good reflectors of the initial distributions (fig. 10-13). If possible, always relate variance to type of cross-bedding. Finally, trough axes, should they be exposed, are better estimators of paleocurrents than random foresets.

b) Solemarks. Typically, they have a low variance so that one general trend per bed and a few beds per outcrop will suffice. However, weakly divergent directions related to different stages of the eroding turbidity current do occur

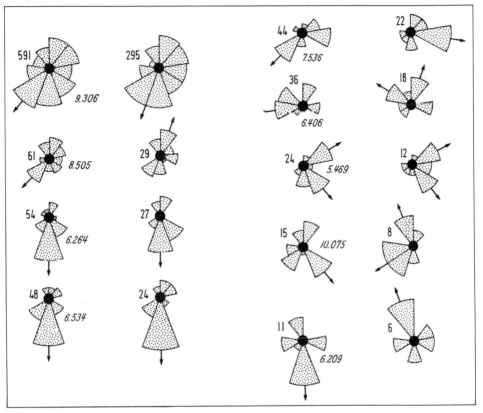

Fig. 10-13. Comparison of full current roses *(1st and 3rd columns)* with half distributions *(2nd and 4th columns)*. Observe similarities. Numbers of observations at upper left; and variance in lower right of each current rose. (HRABAR *et al.* 1971, fig. 11)

(fig. 5-12) as do some markedly diverse trends (much rarer) on adjacent beds related to the interfingering of two or more different subsea fans (fig. 5-13).

c) Ripple marks. Typically variable (fig. 4-15), although some studies indicate a remarkable preferred orientation (figs. 4-18 and 4-29). Again variability should be related to type of ripple mark and its depositional facies.

Exceptions to the above three sampling guides for directional structures occur when either costs of access are very high and/or an exposure has a special importance, such as an outlier.

Mathematical Techniques

For most two-dimensional data the vector mean $\bar{x}_v$ and its normalized length[1], L, have proven to be adequate even though various other parameters have been proposed (see annotated references). A very useful summary of the mathematics used in processing orientation data is that of BATSCHELET (1965), especially for paleoecologists; his summary starts with very basic, descriptive statistics and statistical inference and contains many diverse examples most of which are relevant to the earth sciences. For bipolar and polymodal distributions JONES and JAMES (1969) have developed a very helpful program that estimates the relative proportions of the two modes and calculates their respective vector means — a program that should be especially applicable to the study of tidally influenced, bipolar paleocurrents. Like most programs their procedure demands either an electronic desk calculator or a computer.

However, with the advent of electronic hand calculators, such as the Texas Instrument SR-51 or the Hewlitt Packard HP-125, both vector mean and vector strength, either weighted by some measure of magnitude or unweighted individual or grouped data, can be calculated easily in the field, if so desired. The following is a routine for the Texas Instruments SR-51 for finding the values of $\bar{x}_v$, R, and L with a minimum of effort (Table 10-3). The essential formulas are:

$$\bar{x}_v = \arctan W/V,$$

$$R = (V^2 + W^2)^{\frac{1}{2}},$$

$$L = (R/n)\ 100$$

where

$$V = \sum_{i=1}^{n} t_i^3 \cos \bar{x}_i,$$

$$W = \sum_{i=1}^{n} t_i^3 \sin \bar{x}_i.$$

t being the bed thickness measured in cm and n the total number of observations.

---

[1] See Chapter 10 for definitions of these terms, all the notation being the same.

Table 10–3 (Cont.)

b If the bearing or azimuth is to be entered in degrees-minutes-seconds, enter degrees (0 to 360), decimal point, minutes (two digits, 00 to 60), and seconds and decimal fractions of seconds (if applicable). For example: 279°-29′-35.7″ would be entered as 279.29357; 35°-07′ would be entered as 35.07

c If x_i is entered as a bearing (degrees east or west from north or south; e.g. N 30°W), convert the bearing to azimuth (after converting from degrees-minutes-seconds to decimal degrees, if required) as follows: If the bearing is

NE no action is required
NW press $+/-$
SE press $+/-$, $+$, 180, $=$
SW press $+$, 180, $=$

d If $\bar{X}_v$ is displayed as a negative number, convert to positive by pressing $+$, 360, $=$. The display represents decimal degrees; if conversion to degrees, minutes, and seconds is desired, press 2nd, FIX PT, 5, INV, 2nd, 17

Table 10-3. *Vector Program Key Sequence for Texas Instruments Calculator, SR-51*

Line	Enter	Press	Display shows	Remarks
0		2nd, CA, 2nd, FIX PT, 2	0.00	Clear calculator and set round-off for two decimal places
1 (a) or	t_i	y^x, 3, = (go to line 2)	t_i^3	See note a
1 (b)	1	(go to line 2)	1	
2	—	$x \leftrightarrows y$	ignore display	
3 (a) or	x_i (azimuth in degrees)	(go to line 4)	x_i in degrees azimuth	
3 (b)	x_i (bearing in degrees)	perform bearing conversion per note c, then go to line 4	x_i in degrees azimuth	
or				See notes b and c
3 (c) or	x_i (azimuth in degrees-minutes-seconds)	2nd, 17, = (go to line 4)	x_i in degrees azimuth	
3 (d)	x_i (bearing in degrees-minutes-seconds)	2nd, 17, = ; perform bearing conversion per note c; then go to line 4	x_i in degrees azimuth	
4	—	2nd, 18, SUM, 2	east component of t_i	
5	—	$x \leftrightarrows y$, SUM, 1	north component of t_i	
6	—	C		Return to line 1 (a) or line 1 (b) for additional values of t_i and x_i. When all values have been entered, go to line 7
7	—	RCL, 1	total of north components of all t_i's	
8	—	$x \leftrightarrows y$, RCL, 2	total of east components of all t_i's	
9	—	INV, 2nd, 18, STO, 2	$\bar{X}_v$	See note d
10	—	$x \leftrightarrows y$, STO, 1, ÷	R	
11	n (number of data points)	×, 100, =, STO, 3	L	The value of $\bar{X}_v$ is now in memory 2; R is in memory 1; and L is in memory 3

a If the weighted value of t_i^3 is to be used, use line 1 (a). If $t_i = 1$ is to be used, use line 1 (b)

The routine is applicable to either grouped or raw (ungrouped) data, because only the values of t_i and x_i are needed. For grouped data, grouping must be accomplished before using this routine; however, the calculator can be employed to reduce the amount of effort required to group the data. Alternate procedures are provided to accommodate directions in either degrees–minutes–seconds or a decimal number of degrees, and also directions in either azimuth readings (degrees clockwise from north) or bearings (degrees east or west of north or south). Although the SR-51 calculator performs all internal calculations to maximum accuracy, the calculated results can be rounded off to the accuracy desired by using the 2nd, FIX PT, n routine, where n is the desired number of digits to the right of the decimal point.[2] In the routine of Table 10-3 the calculator is set for two digits to the right of the decimal. Table 10-4 gives several additional worked examples.

Table 10-4. *Sample Calculations*

Example	Azimuths	$\bar{x}_v$	L
A	0, 280, 260, and 180	270	49.24
B	350, 190, 90, 340, 200, 310 and 230	270	22.32
C	45, 10, 25, 80, 65, 35, 55, 160 and 290	45	62.69

Completely programmable hand calculators are also available such as the Texas Instruments SR-52 and the Hewlitt-Packard HP-65. For these calculators a small magnetic card is inserted into the calculator and results are immediately forthcoming, obviating most of the key sequence of Table 10-3. Trend surface analysis for scalar data is now widely used both as polynomial and Fourier series and virtually always a computer is used. Although it has yet to receive wide use, a trend surface analysis program is also available for directional data (STURM, 1971). SCHUENEMEYER et al. (1972) also provide a computer program that not only tests for clusters on a sphere but calculates their means as well and hence is particularly useful for shape fabrics as well as for paleomagnetism.

Factor analysis is also finding greater use in the mapping of mineral associations, a classic expository paper being that of IMBRIE and VAN ANDEL (1964), whose account is still one of the most readable. More recent, and also very good, is KLOVAN's (1975) exposition.

Rotation

Basic background is given by RAMSAY (1967) in his Chapter 1. The effect of complicated folding on paleocurrent data has also been studied by DOTT (1974), who rotated a set of actual data in different ways depending upon the structural complexity. Depending upon the assumptions made in rotation, the resulting current roses differ somewhat. He recommends rotations in the field (Fig. 10-14). PARKS' (1970) program ROTATE saves a great deal of work; his 1974 program even prints out the current rose and its mean.

---

[2] This n is not to be confused with the total number of observations but is a symbol used in the SR-51's instruction manual.

Computer Programs

Many universities now have their computing programs of which we have listed only a very few. In addition, many can be purchased commercially in most cities, and virtually all of the needed computer programs are available from the Kansas Geological Survey (1966—1970) and in the BioMed package of DIXON (1973).

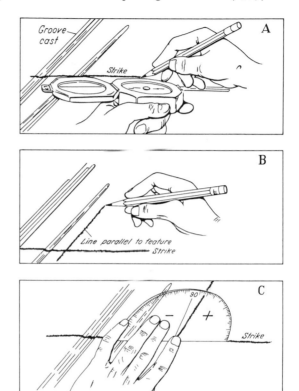

Fig. 10-14. Rotation of linear structures such as grooves and flutes in the field: (A) draw a strike line on lower surface of bed; (B) draw a line parallel to linear structure; and (C) measure the angle between the two and add or substract depending upon sign. (Redrawn from BRIGGS and CLINE, 1967, fig. 3, J. Sediment. Petrol., Soc. Econ. Paleontol. Mineral.)

Annotated References

Background

ELLIOT, I. L., and W. K. FLETCHER, eds., 1975: Geochemical exploration 1974 (Developments in economic geology, 1). Amsterdam: Elsevier Publ. Co., 720 p.
Seven statistical papers; considerable mathematical and statistical skill required.
HARBAUGH, J. W., and D. F. MERRIAM, 1968: Computer applications in stratigraphic analysis. New York, J. Wiley and Sons, Inc., 282 p.
Clear exposition of statistical methods used.
KING, L. J., 1969: Statistical analysis in geography. Englewood Cliffs, N. J., Prentice-Hall, Inc., 288 p.
Generally covers the subject, including factor and principal component analysis. Analysis of location patterns may be useful in some dispersion studies.
KRUMBEIN, W. C., and F. A. GRAYBILL, 1965: An introduction to statistical models in geology. New York, McGraw-Hill Book Co., 475 p.
One of the first books on this subject and a classic.

MARSAL, D., 1967: Statistische Methoden für Erdwissenschaftler. Stuttgart, E. Schweizerbart'sche Verlagsbuchhandlung, 152 p.
    Elementary, but with many good examples from geology. Easy German, too.
HAY, A. M., and M. A. A. RAHMAN, 1974: Use of chi-square for the identification of
peaks in orientation data. Comment: Bull. Geol. Soc. Am. **85**, 1963—1966.
    The latest of several recent papers on this subject; contains all the previous references.
JONES, T. A., 1968: Statistical analysis of orientation data. J. Sediment. Petrol. **38**,
61—67.
    Discussion of circular normal distribution with all the relevant references up to
1967. Paper discusses full circle (0—360°) as well as semi-circle (0—180°) data.
KELLING, G., 1969: The environmental significance of cross-stratification parameters
in an Upper Carboniferous fluvial basin. J. Sediment. Petrol. **39**, 857—875.
    One of the few uses of hierarchical sampling and analysis of variance to decipher a
paleocurrent system based on cross-bedding.
MARDIA, K. V., 1972: Statistics of directional data. London: Academic Press, 357 p.
    The only *book* that we know of on this subject.
REYMENT, R. A., 1971: Introduction to quantitative paleoecology. Amsterdam:
Elsevier Publ. Co., 226 p.
    Chapter 3, "Orientation Analysis" is clear, complete, and comprehensive as well
as having many worked, well-discussed examples. Primary discussion centers
about orientation in the plane of the bedding, using the vector mean and its
dispersion, but there are also a few examples involving the binomial distribution,
and chi-square, i.e., orientation without circles. Very strongly recommended.
SANDERSON, D. J., 1973: Some inference problems in paleocurrent studies. J. Sediment. Petrol. **43**, 1096—1100.
    Principally a review of the literature on this subject, most of the references being
included.
SAUNDERSON, HOUSTON C., 1975: A comparison of empirical and theoretical frequency
distributions for two dimensional palaeocurrent data from the Brampton esker
and associated sediments. Geografiska Annaler **57**, 189—200.
    A large number of climbing ripple observations are needed to estimate paleocurrents in eskers, if the one studied is at all representative.
SCHEIDEGGER, A. E., 1965: On the statistics of the orientation of bedding planes,
grain axes, and similar sedimentological data. U.S. Geol. Surv. Prof. Paper 525-C,
164—167.
    Computation of mean direction and inclination for three dimensional orientation
data using tensors. This is an extension to sedimentology from an earlier use of the
same method for finding fault plane solutions for earthquakes.
WATSON, G. S., 1966: The statistics of orientation data. J. Geol. **74**, 785—797.
    Basic paper, widely cited.

Sources for Computer Programs

CARLILE, R. E., and B. E. GEILLETT, 1974: Fortran and computer mathematics.
Tulsa, Petroleum Pub. Co., 400 p.
    Programming plus matrices, linear equations and least squares.
HARBAUGH, J. W., G. BONHAM-CARTER and W. M. MERRILL, 1971: Programs for
computer simulation in geology. Stanford Univ., Dept. Geol., Office Naval Res.,
Geog. Branch, Contract N 0014-57-A 0112-004, Task 389-154, various paging.
    Twenty programs and three appendices, mostly about things sedimentary; sums
up about five years work. Of interest, if you are considering simulating paleocurrent
patterns.
Laboratory for Computer Graphics and Spatial Analysis, 1971: Symap manual.
Cambridge, Harvard Univ., 157 p.
    Some selected Fortran programs, nine so far. Each has a brief expository account
of the technique and its application.

References

BATSCHELET, E., 1965: Statistical methods for the analysis of problems in animal orientation and certain biological rhythms. Am. Inst. Biol. Sci., 55 p.

BRIGGS, GARRETT, and L. M. CLINE, 1967: Paleocurrents and source areas of late Paleozoic sediments of the Ouachita Mountains, southeastern Oklahoma. J. Sediment Petrol. **37**, 985—1000.

DIXON, W. J., 1973: Biomedical computing programs. Los Angeles: Univ. Los Angeles Health Sci. Comp. Center, 773 p.

Dott, R. H., JR., 1974: Paleocurrent analysis of severely deformed flysch-type strata — a case study from South Georgia Island. J. Sediment. Petrol. **44**, 1166—1173.

HRABAR, S. V., E. R. CRESSMAN and PAUL EDWIN POTTER, 1971: Cross-bedding of the Tanglewood Limestone Member of the Lexington Limestone (Ordovician) of the Blue Grass region of Kentucky. Brigham Young Univ. Geol. Stud. **18**, 99—114.

IMBRIE, J., and TJ. H. VAN ANDEL, 1964: Vector analysis of heavy mineral data. Bull. Geol. Soc. Am. **75**, 1131—1156.

JONES, T. A., and W. R. JAMES, 1969: Analysis of bimodal orientation data. Math. Geol. **1**, 129—135.

KANSAS GEOLOGICAL SURVEY, 1966—1970: The computer contribution series. Lawrence, Kansas, 66044.

KLOVAN, J. E., 1975: R- and Q-mode factor analysis in R. B. McCAMMON, ed., Concepts in geostatistics. New York: Springer 21—69.

PARKS, J. M., 1970: Computerized trigonometric solution for rotation of structurally tilled sedimentary directional features. Bull. Geol. Soc. Am. **81**, 537—540.

—, 1974: Paleocurrent analysis of sedimentary crossbed data with graphic output using three integrated computer programs. Math. Geol. **6**, 353—372.

RAO, J. S., and S. SENGUPTA, 1970: An optimum hierarchical sampling procedure for cross-bedding data. J. Geol. **78**, 533—544.

RAMSAY, J. G., 1967: Folding and fracturing of rocks. New York: McGraw-Hill Book Co., 568 p.

SCHUENEMEYER, J. H., KOCH, G. S., Jr., and LINK, R. F., 1972: Computer program to analyse directional data based on the methods of Fisher and Watson. Math. Geol. **4**, 177—194.

STURM, E., 1971: High resolution paleocurrent analysis by moving vector averages. J. Geol. **79**, 222—233.

WHITTEN, E. H., 1966: Structural geology of folded rocks. Chicago: Rand McNally and Co., 663 p.

Author Index

Numbers in **bold face** denote complete citation in the References Cited and Annotated Bibliographies

A

Aario, R.   144, **151**, 230, **232**
Abbate, E.   226, 230, **232**
Abdel-Hady, M.   **81**
Adams, R.W.   143, 146, 147, **151**, 360
Agatson, R.S.   84, **126**
Albertson, M.L.   **133**
Allcock, J.   26, **28**
Allen, D.C.K.   **361**
Allen, J.R.L.   23, 24, 25, **28**, **126**, 136, 139, 140, 142, 150, **151**, 186, 187, 188, 189, 192, **193**
Allen, P.   85, 86, 112, **126**, 175, **181**, 303, **304**, 384, **387**
Almond, M.   **62**
Alvarez, W.   149, **153**
Amsbury, D.L.   147, **155**
Amsden, T.W.   290, **297**
Andel, T.H. van   14, **17**, 270, 271, 272, 278, **297**, 394, **397**
Andersen, S.A.   92, 93, **126**
Anderson, E.J.   **264**
Anderson, R.C.   276, **297**
Anderson, T.B.   189, **193**
Anderson, T.W.   383, **387**
Andreé, K.   92, 93, **126**
Andresen, M.J.   104, **126**, 294, **297**, **308**
Andrews, J.T.   72, **79**, 191, **193**
Angelucci, A.   **26**
Anketell, J.M.   226, 227, 228, 229, **231**, **232**
Antun, P.   203, 204, 205, **224**
Aoyama, H.   **234**
Aprodov   244
Arai, J.   219, **221**
Arbogast, J.L.   12, **17**, 50, 58, **61**, 238, **249**
Arkhanguelsky, A.D.   211, 213, **221**
Artyushkov, Ye.V.   227, 229, **232**
Asmus, H.E.   **355**
Asquith, D.O.   **360**
Asserto, R.   **263**
Atwater, T.   **355**
Aubouin, J.   313, **337**
Avenshine, A.T.   **151**

B

Baars, D.L.   89, **126**, 237, **249**, **263**
Bagnold, R.A.   79, 108, **126**
Bailey, L.T.   75, **79**
Bailey, R.J.   191, **193**
Baker, C.H., Jr.   **131**, 153, 154, **232**
Baker, D.W.   71, **79**
Baker, V.R.   144, **151**, 185, **193**
Ball, H.W.   85, **126**
Ball, M.M.   **263**
Bally, A.W.   342, **362**
Banks, N.L.   150, **151**, **263**
Barcelos, J.H.   **155**
Barrell, J.   **17**
Barrett, P.J.   26, **28**, **361**
Bartenstein, H.   112, 114, **126**
Bartish-Winkler, S.   139, **151**
Bartlett, M.S.   383, **387**
Bass, N.W.   8, **21**, 84, 85, **132**, 241, 245, **249**, **251**
Bassett, D.A.   10, **17**, 55, **61**, 84, **126**, 236, **249**
Bates, T.F.   55, **67**
Bathurst, R.G.C.   55, **61**
Batschelet, E.   392, **397**
Baumberger, E.   15, **17**
Bausch van Bertsbergh, J.W.B.   13, **17**, 86, 101, 107, 112, **126**, 280, **297**
Beadnell, H.J.H.   91, 124, **126**
Beasley, H.C.   166, **181**
Beaudoin, B.   27, **80**
Beaudoin, M.B.   75, 76
Becker, G.   **310**
Becker, G.F.   8, **17**, 244, **249**
Becker, R.D.   **151**
Beckman, D.D.   71, **82**
Beets, C.   209, **221**
Beh, R.L.   348, **362**
Bein, A.   143, **151**
Belderson, R.H.   136, **151**
Bell, D.   38, 40, **61**
Bell, W.A.   111, **126**
Belshe, J.C.   **267**
Benelli, F.   **263**
Bennacef, A.   **193**, **263**, **308**, **361**
Bennett, R.H.   36, **67**

Subject Index

The Plates

Plate 1 A: Imbrication of thin-bedded slabs of limestone in creek bed. Slabs dip into the current, which comes from the right. Ohio Coordinate System, South Zone, x: 1,513,400; y: 412,000, Union Township, Clermont County, Ohio, U.S.A.

Plate 1 B: Equal area net with 200 poles to slabs of limestone shown in plate 1 A. Note monoclinic symmetry

Plate 2: Cross-bedding in Navajo sandstone (Jurassic). Current from left to right. SW 1/4 SW 1/4 SW 1/4 sec. 20, T. 41 S., R. 9 W., Zion National Park, Kane County, Utah, U.S.A. (Photograph by W. K. HAMBLIN)

Plate 3 A: Micro-cross-lamination ("rib-and-furrow") in fine-grained flagstone of Devonian age. Current from upper right to lower left. Traces of foresets are concave down-current. Sidewalk at Latrobe Hall, The Johns Hopkins University, Baltimore, Maryland, U.S.A.

Plate 3 B: Micro-cross-lamination overlying parting lineation. Current from top to bottom, parallel to knife. Note parallelism of trough axes and trend of parting lineation. Freda sandstone (Precambrian). Northern Peninsula, Michigan, U.S.A. (Photograph by W. K. HAMBLIN)

Plate 4 A: Siltstone showing oversteepened and deformed cross-bedding. Current from right to left. Specimen 7.5 cm thick. Martinsburg shale (Ordovician). Near Middleton, Orange County, New York, U.S.A. (McBRIDE, fig. 9)[*]

Plate 4 B: Cross-bedding in Battery Rock sandstone (Pennsylvanian). Current from right to left. Note apparent tabular character of cross-bedded layers. SW 1/4 SW 1/4 NE 1/4 sec. 1, T. 11 S., R. 7 E., Hardin County, Illinois, U.S.A. (POTTER, 1962a, fig. 5)

Plate 5 A: Cross-bedding in middle Buntsandstein (Triassic). Current from left to right. Observer has left hand at base of large cross-bedded unit; his right hand is near the top of an underlying cross-bedded unit. At Beinhausen near Göttingen, Lower Saxony, Germany (Photograph by A. SCHNEIDER)

Plate 5 B: Cross-bedding in limestone of Paint Creek formation (Mississippian). Note change in apparent dip direction of foresets with different planes of section. SE 1/4 SE 1/4 SW 1/4 sec. 32, T. 13 S., R. 3 E., Massac County, Illinois, U.S.A. (Illinois Geological Survey)

Plate 6 A: Trough cross-bedding in sandstone of Spoon formation (Pennsylvanian). Current from lower left to upper right. Compare with figure 4-3-1. (POTTER and GLASS, 1958, pl. 1 A)

[*] Complete references to published pictures are given in appropriate chapters.

Plate 6B: Deformed, overturned, cross-bedding in Maxon sandstone (Cretaceous). Current from left to right. Pecos County, Texas, U.S.A. (Photograph by E. F. McBride)

Plate 7A: Large, transverse ripple marks in the Mississagi quartzite (Precambrian). Wave length is approximately 75 cm and amplitude is 15 cm. Striker Township. Ontario, Canada (McDowell, 1957, p. 11)

Plate 7B: Symmetrical, relatively uniform, transverse ripple marks. Chemung formation (Devonian). At Cumberland, Allegany County, Maryland, U.S.A. (Photograph by N. L. McIver)

Plate 8A: Linguoid or cuspate ripple marks, as seen from above, on thin-bedded sandstones of Hatch formation (Devonian). Current from upper left to lower right, parallel to pencil. Approximately 0.5 miles south of Hammondsport, Urbana Township, Steuben County, New York, U.S.A.

Plate 8B: Sand wave and internal cross-bedding at edge of point bar along Vermilion River. Current from right to left. Note asymmetrical ripple marks on top of sand wave. SW 1/4 SW 1/4 sec. 30, T. 18 N., R. 9 W., Vermillion County, Indiana, U.S.A.

Plate 9A: Large, symmetrical, transverse ripple marks in calcarenite, Cincinnatian series (Ordovician). Junction of Straight Creek and Ohio State Route 125, Ohio Coordinate System, South Zone; x: 1,614,698, y: 315,512, Brown County, Ohio, U.S.A.

Plate 9B: Equal area net with 196 poles to flanks of ripple marks shown in plate 9A

Plate 10A: Asymmetrical current ripples on point bar along Vermilion River. Current from upper left to lower right parallel to shovel handle. NW 1/4 SW 1/4 SW 1/4 sec. 30, T. 18 N., R. 9 W., Vermillion County, Indiana, U.S.A.

Plate 10B: Air photograph of underwater sand waves of carbonate sand. Crest distance between barchans in left center of pictures is 30 to 100 meters. Horns point to open sea. South end of Tongue of the Ocean, Bahama Banks, Bahama Islands (Shell Development Company)

Plate 11: Air photograph of large sand waves, transverse sand dunes and barchans of subaerial origin between Calexico, California and Yuma, Arizona, U.S.A. (Spence Air Photos)

Plate 12A: Flute casts on underside of sandstone of Martinsburg shale (Ordovician). Current from lower right to upper left. On State Route 254 four miles east of Staunton, Augusta County, Virginia, U.S.A. (McBride, 1962, fig. 14)

Plate 12B: Large irregular flute casts on base of graywacke of Martinsburg shale (Ordovician). Current from top to bottom. Some flutes have broadened into transverse scour marks (lower right). Pennsylvania Turnpike (Milepost 212), Cumberland County, Pennsylvania, U.S.A. (Photograph by E. F. McBride)

Plate 13A: Triangular to deltoid overlapping flute casts and prominent groove cast. Current flowed from upper right to lower left. In Hatch member of Naples group, Portage facies (Devonian), Grimis Glen, Naples, Ontario County, New York, U.S.A. (Photograph by N. L. McIver)

Plate 13B: Flute casts on base of sandstone of Caseyville formation (Pennsylvanian). Note that surface is not completely covered and that individual casts are broad and somewhat ill-defined. SW 1/4 SE 1/4 NE 1/4 sec. 28, T. 10 S., R. 2 W., Union County, Illinois, U.S.A. (Illinois Geological Survey)

Plate 14: Large bulbous flute casts. Current from left to right, Smithwick formation (Pennsylvanian). Three miles east of Marble Falls, Burnett County, Texas, U.S.A. (Shell Development Company)

Plate 15A: Current crescents, formed around pebbles on sandy beach along Chesapeake and Delaware Canal. Ebb current from left to right. At Bethal, Cecil County, Maryland, U.S.A.

Plate 15B: Current crescent casts on bottom of sandstone bed. Current crescents formed around shale pebbles which have weathered out leaving holes. Crescents are convex upcurrent. Current flowed from upper left to lower right. Juniata sandstone (Ordovician) exposed at Waggoners Gap on State Route 74, Cumberland County, Pennsylvania, U.S.A. (Photograph by N. L. McIver)

Plate 16: Drag groove cast with flute and squamiform load casts. Note rhythmic development of flute and load casts perpendicular to groove cast. Rhythmic pattern of flute and load casts called "dinosaur leather" (CHADWICK, 1948, p. 1315). Current from lower left to upper right. Normanskill shale (Ordovician). On U.S. highway 9 W just north of overpass above Interstate 87, Green County, New York, U.S.A. (Photograph by E. CHOWN)

Plate 17A: Groove and striation casts on base of siltstone bed. Note compound striated groove casts crossed by a second series of markings, principally poorly-developed flute casts. Current from lower right to upper left (Devonian) New York, U.S.A. (Photograph by N. L. McIver)

Plate 17B: Groove casts with *en echelon* drag rolls (feathered groove cast). Trimmers Rock sandstone (Devonian), west shore of Susquehanna River, near Sunbury, Pennsylvania, U.S.A. (Photograph by N. L. McIver)

Plate 18: Groove casts (a), brush casts (b), and ring cast (c). The ring casts constitute a set of skip casts. Each ring cast is slightly asymmetrical and shows current from right to left. Krosno beds (Oligocene) Rudawska Rumanowski, central Carpathians (Photograph from DZULYNSKI and SANDERS, 1962, pl. 14B)

Plate 19: Skip casts (a), brush casts (b) and problematic marking (c) thought to have been produced by fish. Current flowed from lower right to upper left. Krosno beds (Oligocene), central Carpathians, Katy, Poland (Photograph from DZULYNSKI and SLACZKA, 1958, pl. 29)

Plate 20A: Slide cast. Note mud lumps at down-current end of slide markings. Slide was from right to left. Krosno beds (Oligocene), Rudawska Rumanowski, central Carpathian Mountains, Poland (Photograph from ST. DZULYNSKI)

Plate 20B: Crescent cast around a piece of wood. Current from left to right. The piece of wood is 8 cm long. Flysch (Tertiary) Carpathian Mountains (Photograph from M. KSIAZKIEWICZ)

Plate 21 A: "Delicate flute casts" or furrow flute casts. Current from top to bottom. Hatch formation (Devonian) 0.5 miles south of Hammondsport, Steuben County. New York, U.S.A. (Photograph by W. HILLER)

Plate 21 B: Furrow casts in Trimmers Rock sandstone (Devonian). West shore of Susquehanna River near Sunbury, Pennsylvania, U.S.A. (Photograph by N. L. McIVER)

Plate 22 A: Rill mark on beach. Ebb current from to top to bottom. Along Chesapeake and Delaware Canal, at Bethal, Cecil County, Maryland, U.S.A.

Plate 22 B: Parting lineation in Devonian flagstones. Current parallel to lineation Sidewalk at Latrobe Hall, The Johns Hopkins University, Baltimore, Maryland, U.S.A.

Plate 23: Fine-textured, uniformly-oriented glacial striations on Livingstone limestone, Bond formation (Pennsylvanian), NW 1/4 sec. 21, T. 18 N., R. 13 W., Vermilion County, Illinois, U.S.A. (Illinois Geological Survey)

Plate 24 A: Load casts on underside of thin-bedded sandstone. Aux Vases sandstone (Mississippian). SW 1/4 NW 1/4 NW 1/4 sec. 3, T. 5 S., R. 9 W., Monroe County, Illinois, U.S.A. (Photograph by W. HILLER)

Plate 24 B: Flame structures at base of graded sand bed. Current from lower left to upper right. Note overturning of flame structure. Cross Lake group (Precambrian), Manitoba, Canada (Geological Survey of Canada)

Plate 25: Finely-textured, closely-packed load casts, some of which are elongated parallel to larger pencil, in sandstone. Note also faint alignment parallel to smaller pencil. Tar Springs sandstone (Mississippian). NE 1/4 SW 1/4 SW 1/4 sec. 25, T. 3 S., R. 1 W., Crawford County, Indiana, U.S.A. (Photograph by DALE FARRIS)

Plate 26: Large load casts, probably initiated as flutes with some overhanging margins. Current from right to left. Smithwick formation (Pennsylvanian), 3 miles east of Marble Falls, Burnett County, Texas, U.S.A. (Photograph by E. F. McBRIDE)

Plate 27 A: Asymmetrical deformed load casts. Note coarsest material is in load casts and that beds above and below are both graded and laminated. Current from left to right. Weldon Creek series, Ocoee group (Precambrian). On Tennessee State Route 73, near Kinzel Springs, Blount County, Tennessee, U.S.A.

Plate 27 B: Ball-and-pillow structure in clastic limestone of Cincinnatian series (Ordovician). Note conformity of bedding within pillow to external boundary of same; note also downward convexity of pillows and approach to reniform or hemispherical shape. Man-made stone wall rests on natural outcrop. Along Ohio State Route 763, 1.5 miles southwest of Decatur, Byrd Township, Brown County, Ohio, U.S.A.

Plate 28 A: Ball-and-pillow structure in laminated limestone. Note that the internal laminations conform to boundary of pillows. Laminations are concave up. Pillows sit in structureless fine-grained limestone. Cynthiana group (Ordovician), 2400 F.W.L., 1500 F.S.L., CC-65, Pendelton County, Kentucky, U.S.A.

Plate 28 B: Ball-and-pillow structure in sandstone. Note pillows are largest at base and become smaller and pass upwards into undisturbed beds. Chemung sandstone (Devonian) exposed along Juniata River on U.S. Highway 22-322 north of Amity Hall, Perry County, Pennsylvania, U.S.A.

Plate 29 A: Rudimentary ball-and-pillow structure in sandstone. Note that the pillows are confined to basal part of bed and are largely still attached to bed. The bed involved has, therefore, an undulatory pillow-form base and a flat top. Chemung sandstone (Devonian) along the Juniata River between Newport and Amity Hall, Perry County, Pennsylvania, U.S.A. (Photograph by N. L. McIver)

Plate 29 B: Underside of pillow-form sandstone. Note general resemblance to pillow lava and note also orientation of pillows. Largest pillow is approximately 2 feet (0.6 m) long. Chemung sandstone (Devonian). Along Juniata River between Newport and Amity Hall, Perry County, Pennsylvania, U.S.A. (Photograph by N. L. McIver)

Plate 30: Convolute lamination, as seen from above, in Upper Priabonien (Tertiary) flysch sedimentation. Near San Remo, Liguria, Italy (PLESSMANN, 1961, fig. 30)

The Plates

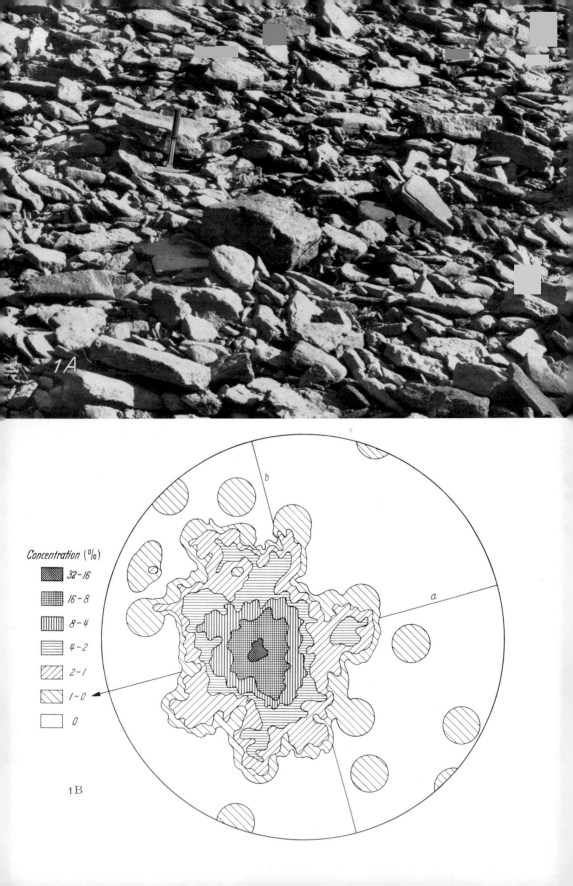

1A

Concentration (%)

	32-16
	16-8
	8-4
	4-2
	2-1
	1-0
	0

b

a

1B

2

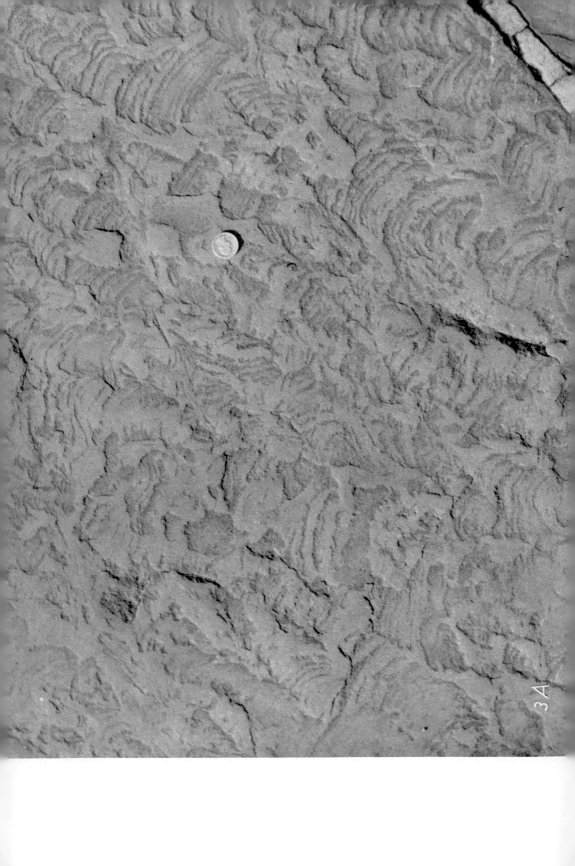

3A

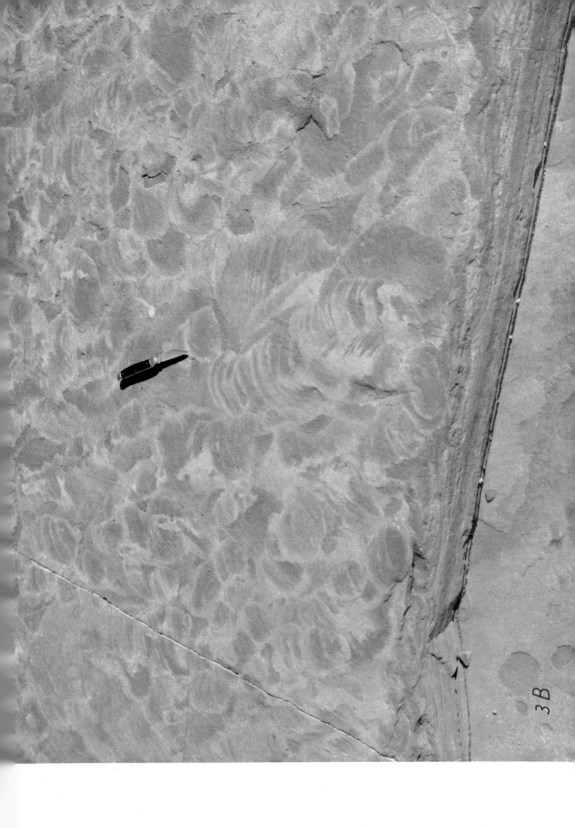

3B

4A

4B

5A

5B

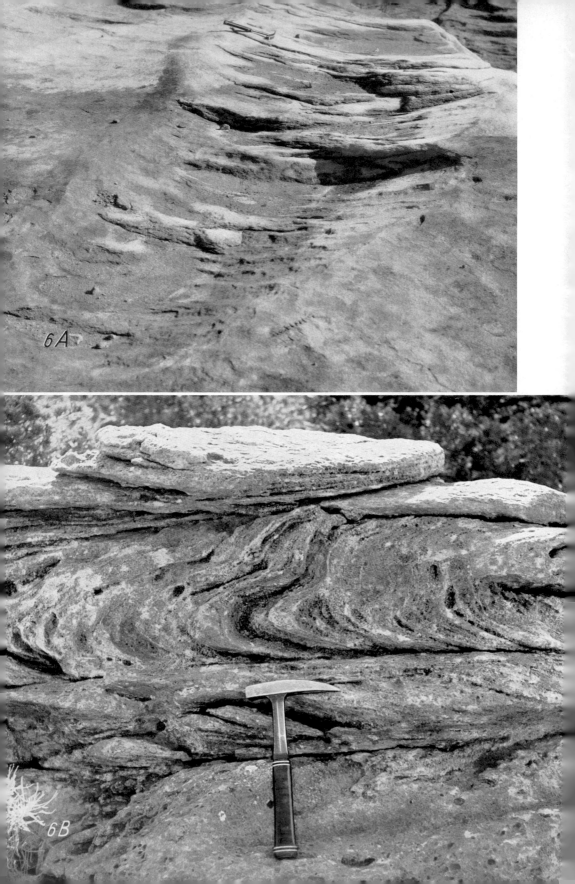

6A

6B

7A

7B

8 A

8 B

9A

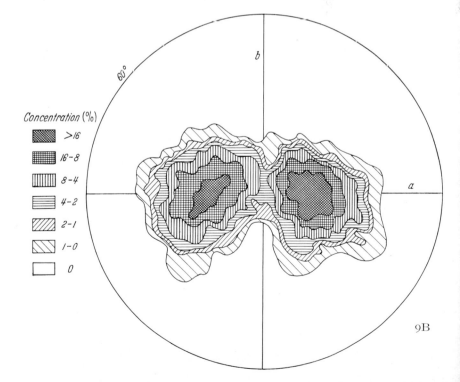

Concentration (%)

- ▨ >16
- ▦ 16-8
- ▥ 8-4
- ▤ 4-2
- ▨ 2-1
- ▨ 1-0
- ☐ 0

60°

b

a

9B

10 A

10 B

11

12 A

12 B

13 A

13 B

15 A

15 B

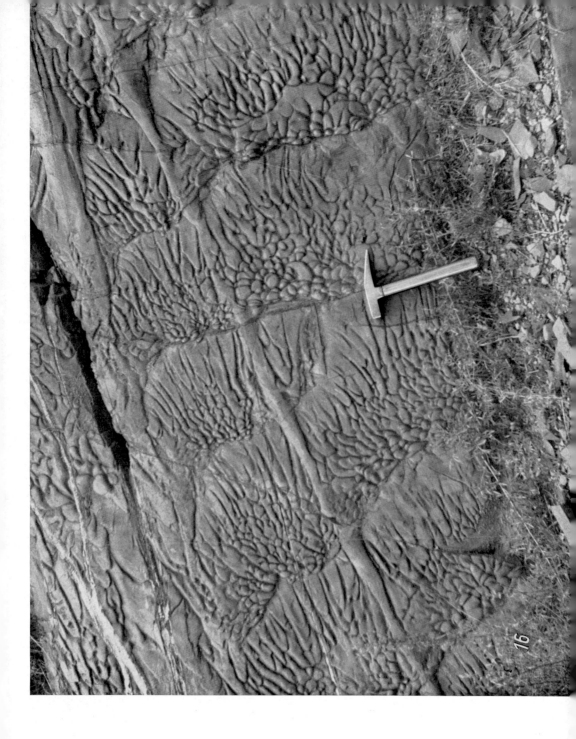

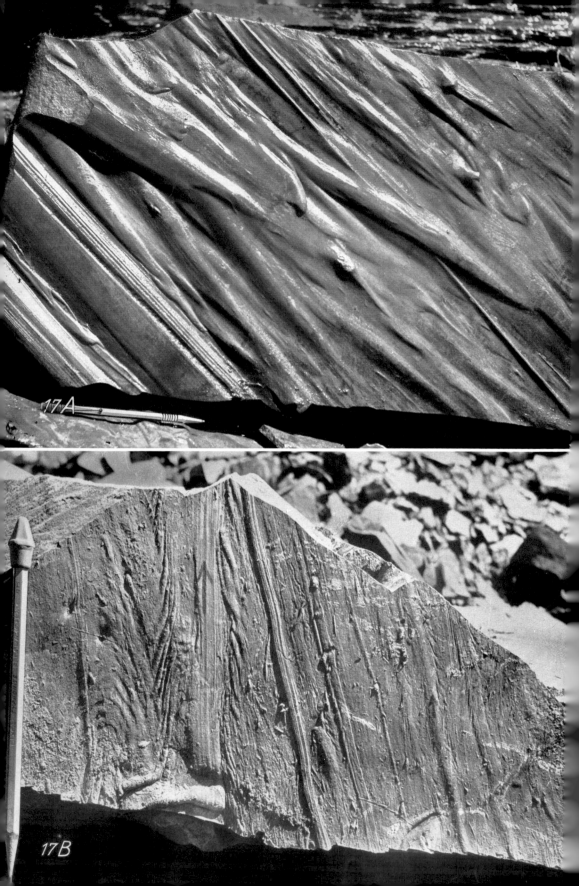

17A

17B

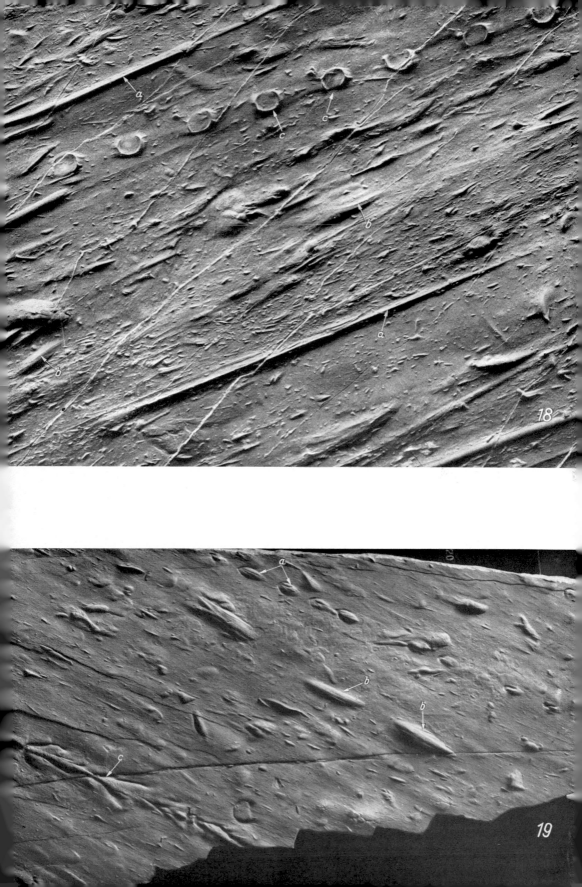

18

19

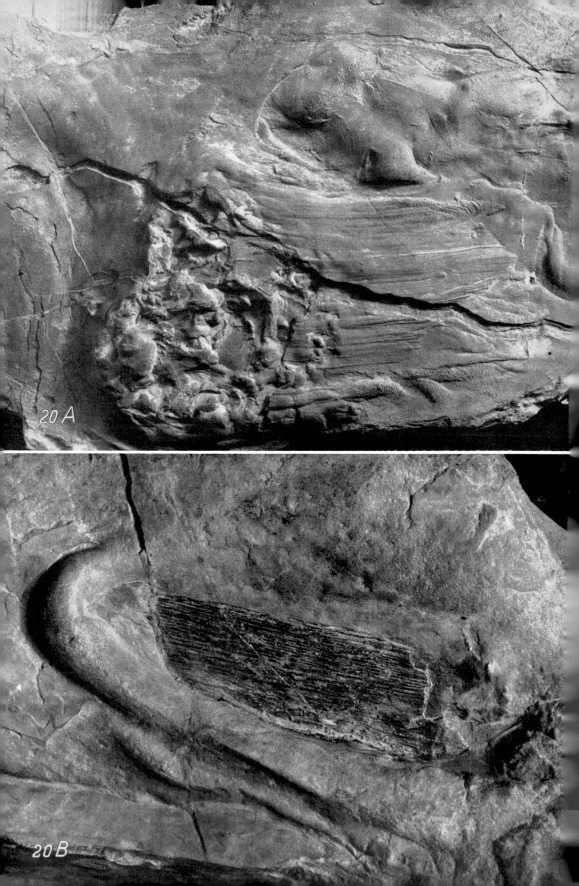

20 A

20 B

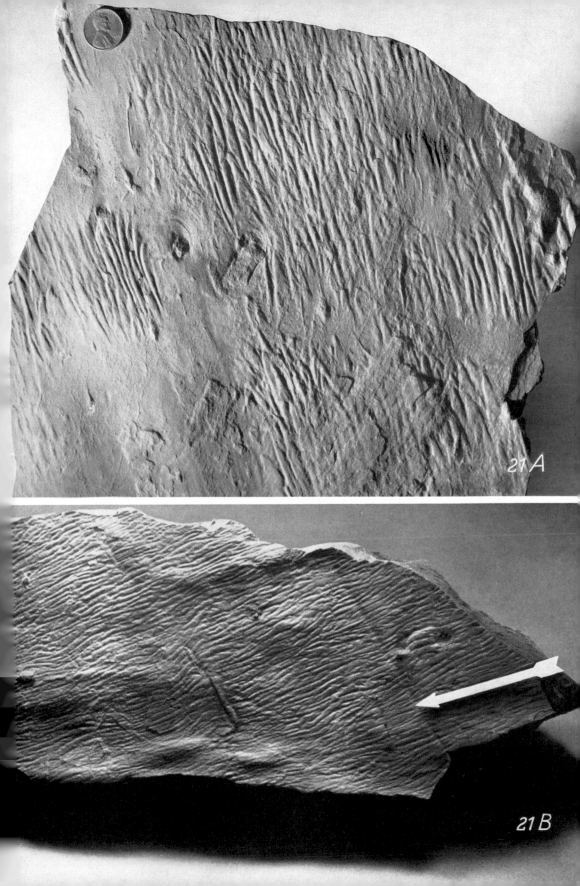

21 A

21 B

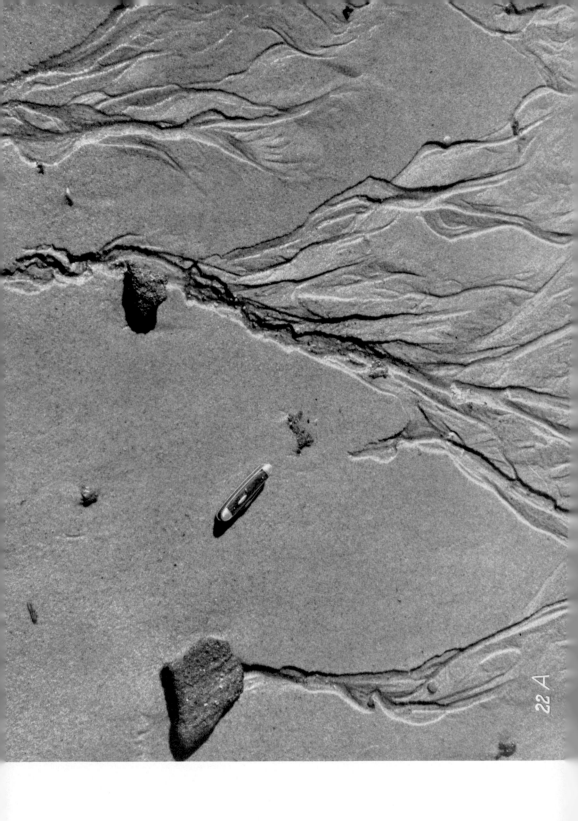

22 A

22 B

23

24 A

24 B

25

26

27 A

27 B

28 A

28 B

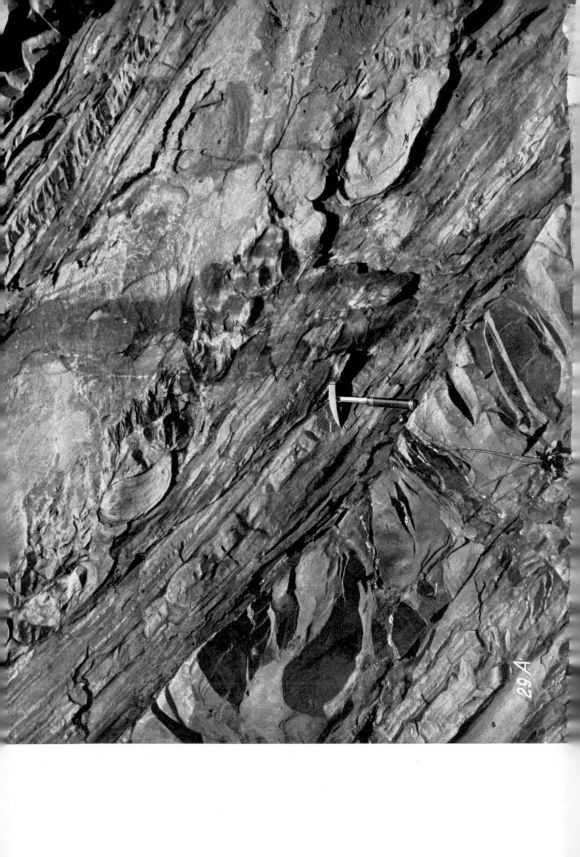

29 A

29 B